CONTINUOUS SIGNALS AND SYSTEMS WITH MATLAB™

Electrical Engineering Textbook Series

Richard C. Dorf, Series Editor
University of California, Davis

Forthcoming Titles

Electromagnetics
Edward J. Rothwell and Michael J. Cloud

Applied Vector Analysis
Matiur Rahman and Issac Mulolani

Optimal Control Systems
Subbaram Naidu

CONTINUOUS SIGNALS AND SYSTEMS WITH MATLAB™

Taan S. ElAli
Mohammad A. Karim

CRC Press
Boca Raton London New York Washington, D.C.

MATLAB is a registered trademark of The MathWorks, Inc.
For product information, please contact:
The MathWorks, Inc.
3 Apple Hill Drive
Natick, MA 01760-2098 USA
Tel: 508-647-7000
Fax: 508-647-7001
E-Mail: info@mathworks.com
Web site: www.mathworks.com

Library of Congress Cataloging-in-Publication Data

ElAli, Taan S.
Continuous signals and systems with MATLAB / Taan S. ElAli, Mohammad A. Karim.
p. cm. —(Electric engineering textbook series ; 1)
ISBN 0-8493-0321-4 (alk. paper)
1. Signal processing—Mathematics. 2. System analysis. 3. MATLAB. I. Karim, Mohammad A. II. Title. III. Series.

TK5102.9 .E34 2000
621.382′2—dc21 00-046802
CIP

Direct all inquiries to CRC Press LLC, 2000 N.W. Corporate Blvd., Boca Raton, Florida 33431, or visit our Web site at www.crcpress.com

International Standard Book Number 0-8493-0321-4
Library of Congress Card Number 00-046802
Printed in the United States of America 1 2 3 4 5 6 7 8 9 0
Printed on acid-free paper

Dedication

This book is dedicated first to the glory of the Almighty God. It is dedicated next to my beloved parents, my father, Saeed, and my mother, Shandokha—may God have mercy on their souls. It is dedicated then to my wife, Salam; my beloved children, Ali and Zayd; my brothers, Mohammad and Khaled; and my sisters, Sabha, Khulda, Miriam, and Fatma. I ask the Almighty God to have mercy on us and to bring peace, harmony, and justice to all.

Preface

All books on linear systems for undergraduates cover discrete and continuous systems material together in a single volume. Such books also include topics in discrete and continuous filter design, and discrete and continuous state-space representations. However, with this magnitude of coverage, the student typically gets a little of both discrete and continuous linear systems but not enough of either. Minimal coverage of discrete linear systems material is acceptable provided that there is ample coverage of continuous linear systems. On the other hand, minimal coverage of continuous linear systems does no justice to either of the two areas. Under the best of circumstances, a student needs a solid background in both these subjects. It is no wonder that these two areas are now being taught separately in so many institutions.

Continuous linear systems is a broad topic in itself and merits a single book devoted to that material. The objective of this book is to present the required material that an undergraduate student needs to master this subject matter and the use of MATLAB™ (The MathWorks Inc.) in solving problems in this area.

This book offers broad, detailed, focused comprehensive coverage of continuous linear systems, based on basic mathematical principles. It presents many solved problems from various engineering disciplines using analytical tools as well as MATLAB. This book is intended primarily for undergraduate junior and senior electrical, mechanical, aeronautical, and aerospace engineering students. Practicing engineers will also find this book useful.

This book is ideal for use in a one-semester course in continuous linear systems where the instructor can easily cover all of the chapters. Each chapter presents numerous examples that illustrate each concept. A distinguishing feature of this book is the wide range of engineering disciplines covered by the End-of-Chapter Examples, which demonstrate the theory presented. Most of the worked-out examples are first solved analytically and then solved using MATLAB in a clear and understandable fashion.

This book concentrates on explaining the subject matter with easy-to-follow mathematical development and numerous solved examples. The book covers traditional topics and includes chapters on system design, state-space representation, and linearization of nonlinear systems. The reader does not need to be fluent in MATLAB because the examples are presented in a self-explanatory way.

To the Instructor

In a semester-long course, Chapters 1 through 7 can be covered first. Chapters 8 and 9 can then be covered in any order. In one quarter and because of time constraints, Chapter 8 can be skipped, if desired.

To the Student

Familiarity with calculus, differential equations, and basic dynamics is desirable. If and where certain background material must be presented, that background material is presented right before the topic under consideration. This unique "just-in-time approach" helps the student stay focused on the topic. This book presents three forms of numerical solutions using MATLAB. The first form allows you to type any command at the MATLAB prompt and then press the Enter key to get the results. The second form is the MATLAB script, a set of MATLAB commands you type and save in a file. You can run this file by typing the file name at the MATLAB prompt and then pressing the Enter key. The third form is the MATLAB function form, where you create the function and run it in the same way you create and run the script file. The only difference is that the name of the MATLAB function should be the same as the file name.

To the Practicing Engineer

The practicing engineer will find topics in this book useful. In real life systems are nonlinear, and this book describes, step by step, the process of linearizing nonlinear systems. MATLAB, an invaluable tool for the practicing engineer, is used in solving most of the problems.

Acknowledgments

I would like to thank the CRC international team. Special thanks go to Nora Konopka, who greatly encouraged me when I discussed this project with her initially. She has reaffirmed my belief that this is a much-needed book. Thanks to Christine Andreasen, project editor, for her careful review of the text. Thanks also to Mr. R. Dlamini of Wilberforce University who helped edit the figures.

Finally, "nothing is perfect." Please forward any comments or concerns about this book to the author at telali@wilberforce.edu or to the publisher.

Authors

Taan S. ElAli, Ph.D., is an associate professor of engineering and computer science at Wilberforce University, Wilberforce, Ohio. He received his B.S. in electrical engineering from the Ohio State University, Columbus, his M.S. in electrical systems engineering from Wright State University, Dayton, Ohio, and his M.S. in applied mathematics and his Ph.D. in electrical engineering, with a specialization in systems, from The University of Dayton, Dayton, Ohio. He has more than 10 years teaching and research experience in the areas of continuous and discrete signals and systems. He was awarded the "Who's Who Among America's Teachers" for 1998 and 2000.

Dr. ElAli has contributed many journal articles and conference presentations in the area of systems. He has been extensively involved in the establishment of the electrical and computer engineering degree programs and curriculum development at Wilberforce University. He redesigned the freshman introduction to engineering course. He is the author of *Introduction to Engineering and Computing*. His next book, *Discrete Signals and Systems with MATLAB*®, will be published by CRC Press.

Mohammad A. Karim, Ph.D., is dean of engineering at City College of the City University of New York. He received his B.S. in physics from the University of Dacca, Bangladesh, in 1976, and his M.S. degrees in both physics and electrical engineering, and his Ph.D. in electrical engineering, from the University of Alabama in 1978, 1979, and 1981, respectively. He is active in research in the areas of information processing, pattern recognition, optical computing, displays, and EO systems.

Dr. Karim is the author of the books *Digital Design: A Pragmatic Approach, Electro-Optical Devices and Systems, Optical Computing: An Introduction,* and *Electro-Optical Displays,* eight book chapters, and more than 300 papers. He is the North American editor of *Optics & Laser Technology,* and an associate editor of *IEEE Transactions on Education*. He serves on the editorial board of *Microwave & Optical Technology Letters*. He has served as guest editor for nine journal special issues. He is a fellow of the Optical Society of America (OSA) and Society of Photo-Instrumentation Engineers (SPIE). He is a senior member of IEEE and a member of the American Society of Engineering Education (ASEE).

Contents

1

Signal Representation

CONTENTS

1.1 Examples of Continuous Signals

We experience signals of various types almost on a continual basis in our daily lives. The blowing of the wind is an example of a continuous wave. We can plot the strength of the wind wave, the velocity of the wave, and the distance it travels as functions of time. When we speak, continuous signals are generated. These spoken word signals travel from one place to another so that another person can hear them. These are our familiar sound waves.

When a radar system detects a certain object in the sky, an electromagnetic signal is sent. This signal leaves the radar system and travels the distance in the air until it hits the target object which then reflects back to the sending radar to be analyzed, where it is decided if the target is present or not. We understand that this electromagnetic signal, whether it is the one being sent or the one being received by the radar, is attenuated (its strength reduced) as it travels away from the radar station. Thus the attenuation of this electromagnetic signal can be plotted as a function of time. If you vertically attach a certain mass to a spring from one end while the other end is fixed and then pull the mass, oscillations are created such that the spring's length increases and decreases until the oscillations finally stop. The oscillations produced are a signal that also dies out with increasing time. This signal, for example, can represent the length of the spring as a function of time.

Signals also can appear as electric waves. Examples are voltages and currents on long transmission lines. Voltage value is reduced as the impressed voltage travels on transmission lines from one city to another. Therefore we can represent these voltages as signals as well and plot them in terms of time. When we discharge or charge a capacitor, the rate of charging and discharging depends on the time factor (among other factors). Charging and discharging the capacitor can be represented thus as voltage across the capacitor terminal as a function of time. These are a few examples of continuous signals that exist in nature that can be modeled mathematically as signals that are functions of various parameters.

1.2 The Continuous Signal

A signal, $v(t)$, is said to be continuous at time $t = t_1$ if and when $v(t_1^+) = v(t_1^-) = v(t_1)$. If, on the other hand, this is untrue, the signal $v(t)$ is referred to as a discontinuous signal. Figure 1.1 shows an example of a continuous signal.

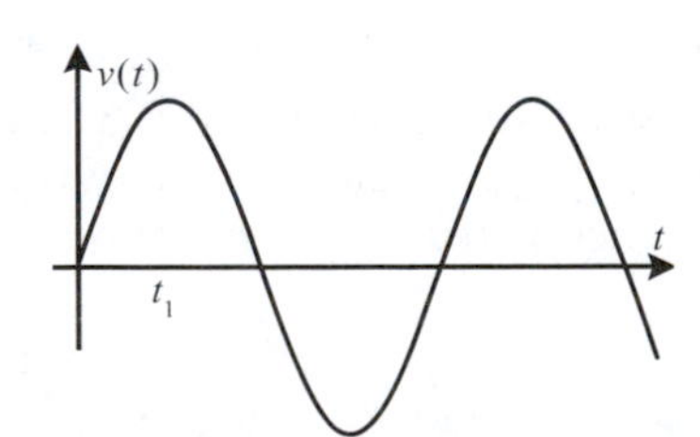

FIGURE 1.1
An example of a continuous signal.

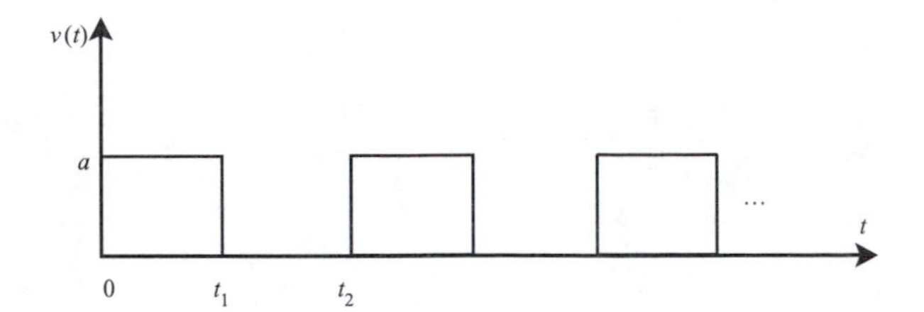

FIGURE 1.2
Signal for Example 1.1.

Example 1.1

Consider a signal of magnitude a as shown in Figure 1.2 and discuss its continuity.

Solution

As you can see in Figure 1.2, $v(t_1^+) \neq v(t_1^-)$ since $v(t_1^+) = 0$ and $v(t_1^-) = a$ where t_1^+ indicates time just after $t = t_1$ and t_1^- indicates time just before $t = t_1$. Also, $v(t_2^+) \neq v(t_2^-)$. Accordingly, in this example, the signal is piecewise continuous. It is continuous at all t except at $t = t_k$, where $k = 0, 1, 2, 3, \ldots$. It is customary to estimate a unique value for the signal at each of the discontinuities. For example, the value of $v(t)$ at $t = t_1$ is

$$v(t = t_1) = 1/2[v(t_1^+) + v(t_1^-)] = a/2$$

1.3 Periodic and Nonperiodic Signals

A signal $v(t)$ is periodic if $v(t) = v(t + kT)$, where k is an integer that has the values $\pm 1, \pm 2, \pm 3, \pm 4, \ldots$, and T is referred to as the fundamental period. In other words, if the signal $v(t)$ repeats itself every kT seconds or units of time it is called periodic with period T. If $v_1(t)$ is periodic with period T_1 and if $v_2(t)$ is periodic with period T_2, what can we say about the period of $v(t) = v_1(t) + v_2(t)$? Consider $v(t)$ to be periodic having period T. Accordingly, we can say that $v(t) = v(t + T)$ as per the definition of periodicity just given to us.

Since $v_1(t)$ and $v_2(t)$ are each periodic with period T_1 and T_2, respectively, then

$$v_1(t) = v_1(t + kT_1)$$

and

$$v_2(t) = v_2(t + mT_2)$$

where k and m are integers. Since

$$v(t) = v_1(t) + v_2(t)$$

we have

$$v(t+T) = v_1(t+kT_1) + v_2(t+mT_2)$$

But

$$v_1(t+T) + v_2(t+T) = v_1(t+kT_1) + v_2(t+mT_2)$$

For this last equation to be true, we must have

$$T = kT_1$$

and

$$T = mT_2$$

This implies

$$kT_1 = mT_2$$

which says that

$$T_1/T_2 = m/k$$

m/k is a rational number. Therefore, the combination signal

$$v(t) = v_1(t) + v_2(t)$$

is a periodic signal if $T_1/T_2 = m/k$ where m/k is a rational number, and the period of $v(t)$ is

$$T = kT_1 = mT_2$$

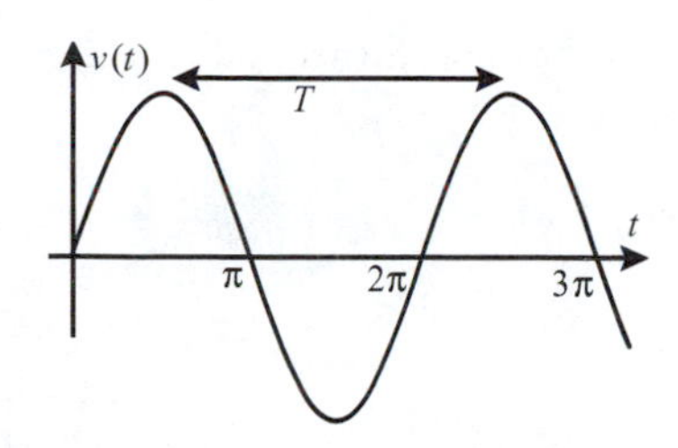

FIGURE 1.3
Sinusoidal signal.

1.4 General Form of Sinusoidal Signals

Consider the sinusoidal signal shown in Figure 1.3. The magnitude of the signal $v(t)$ at $\pi/2$ and $5\pi/2$ is the same. The time between these values is called the period T. Every T units of time the signal repeats with the same magnitude. The period of this signal is 2π units of time. A general form of a sinusoidal signal is

$$v(t) = A\sin(\omega t + \Phi) \tag{1.1}$$

where

- A is the magnitude of the signal $v(t)$
- ω is the angular frequency ($\omega = 2\pi f$)
- f is the frequency in hertz ($f = 1/T$)
- Φ is the phase angle in radians

Example 1.2

Let $v_1(t) = \sin(t)$ and $v_2(t) = \cos(t)$. Consider the sum of the two signals $v(t) = v_1(t) + v_2(t)$. Is $v(t)$ periodic? If so, what is the period of $v(t)$?

Solution

$v_1(t)$ has an angular frequency of 1 rad/sec and $v_2(t)$ has an angular frequency of 1 rad/sec, too. For $v_1(t)$ and with $w = 1$ we can write

$$1 = 2\pi/T_1 \Rightarrow T_1 = 2\pi$$

Also, for $v_2(t)$ and with $w = 1$ we can write

$$1 = 2\pi/T_2 \Rightarrow T_2 = 2\pi$$

Checking the ratio

$$T_1/T_2 = 2\pi/2\pi = 1$$

indicates that it is a rational number. Therefore, $v(t)$ is a periodic signal and has a period $T = kT_1 = mT_2$

$$T_1/T_2 = m/k = 1/1 \Rightarrow m = 1 \text{ and } k = 1$$

Thus,

$$T = kT_1 = 2\pi$$

or

$$T = mT_2 = 2\pi$$

Example 1.3

Let $v_1(t) = \sin(2\pi t)$ and $v_2(t) = \sin(t)$. Consider the sum $v(t) = v_1(t) + v_2(t)$. Is $v(t)$ periodic? If so, what is the period of $v(t)$?

Solution

For $v_1(t)$ we have

$$T_1 = 1$$

and for $v_2(t)$ we have

$$T_2 = 2\pi$$

The ratio

$$T_1/T_2 = 1/2\pi$$

is not a rational number. Therefore, $v(t) = v_1(t) + v_2(t)$ is not a periodic signal.

1.5 Energy and Power Signals

The signal energy over a time interval $2L$ is given by

$$E = \int_{-L}^{L} |v(t)|^2 \, dt \tag{1.2}$$

If the interval is infinite then

$$E = \int_{-\infty}^{\infty} |v(t)|^2 \, dt \tag{1.3}$$

and in terms of limits, it is given as

$$E = \underset{L \to \infty}{\text{limit}} \int_{-L}^{L} |v(t)|^2 \, dt \tag{1.4}$$

The average power is thus given by

$$P = \underset{L \to \infty}{\text{limit}} (1/2L) \int_{-L}^{L} |v(t)|^2 \, dt \tag{1.5}$$

Let us consider the following two cases.

1. If in Equation (1.4) we find that $0 < E < \infty$, then $v(t)$ is referred to as the energy signal and the average power is zero.
2. If in Equation (1.5) we find that $0 < P < \infty$, then $v(t)$ is referred to as a power signal and it produces infinite energy.

For periodic signals, if we set $2L$ in Equation (1.4) or (1.5) such that $2L = 2T = 3T = \ldots = mT$, where m is an integer then we have

$$P = \underset{m \to \infty}{\text{limit}} (m/mT) \int_{0}^{T} |v(t)|^2 \, dt$$

which simplifies to

$$P = 1/T \int_{0}^{T} |v(t)|^2 \, dt \tag{1.6}$$

Example 1.4

Let $v(t) = Ae^{-t}$ for $t \geq 0$. What kind of signal is given in Figure 1.4?

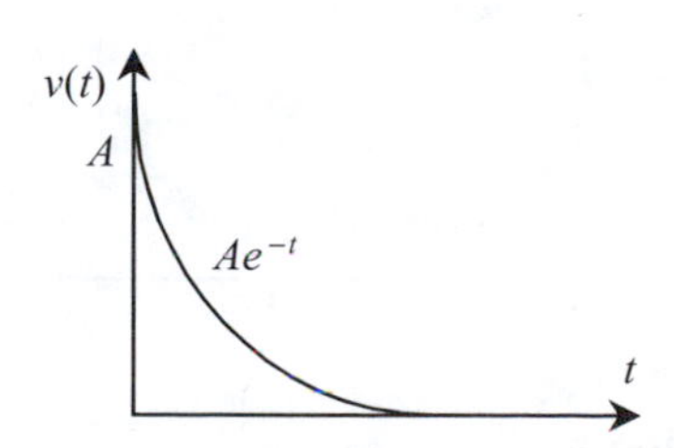

FIGURE 1.4
Signal for Example 1.4.

Solution

Let us calculate the energy in the signal first as

$$E = \int_0^\infty |Ae^{-t}|^2 \, dt$$

$$E = \int_0^\infty A^2 e^{-2t} \, dt$$

Carrying out the integration we get

$$E = (A^2/-2)e^{-2t}\Big|_0^\infty = (A^2/2).$$

Since $0 < E = A^2/2 < \infty$, $v(t)$ is an energy signal with zero power.

1.6 The Shifting Operation

Consider the signal $v(t)$ in general. The representation $v(t - t_0)$ implies that $v(t)$ is shifted to the right t_0 units provided $t_0 > 0$. On the other hand, when $t_0 < 0$, $v(t - t_0)$ implies that $v(t)$ is shifted to the left t_0 units.

Example 1.5

Consider the signal shown in Figure 1.5. Determine mathematically both $v(t - 2)$ and $v(t + 2)$.

Solution

Mathematically the signal in Figure 1.5 can be expressed as

$$v(t) = \begin{cases} t & t \geq 0 \\ 0 & t < 0 \end{cases}$$

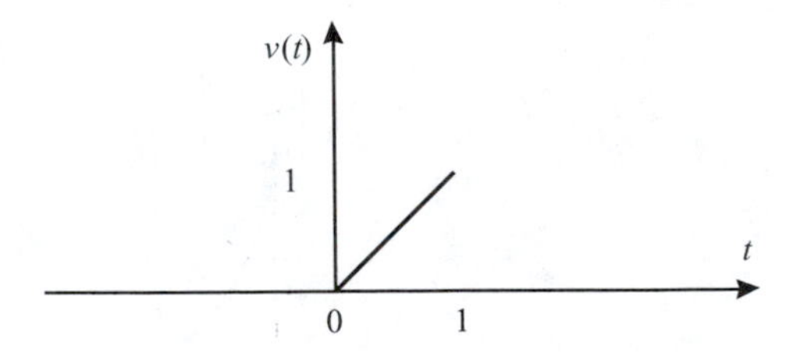

FIGURE 1.5
Signal for Example 1.5.

To plot $v(t-2)$ we slide the graph of $v(t)$ two units to the right. Likewise, to plot $v(t+2)$ we slide the graph of $v(t)$ two units to the left. Mathematically

$$v(t-2) = \begin{cases} t-2 & t-2 \geq 0 \\ 0 & t-2 < 0 \end{cases}$$

and finally

$$v(t-2) = \begin{cases} t-2 & t \geq 2 \\ 0 & t < 2 \end{cases}$$

Similarly,

$$v(t+2) = \begin{cases} t+2 & t+2 \geq 0 \\ 0 & t+2 < 0 \end{cases}$$

and finally

$$v(t+2) = \begin{cases} t+2 & t \geq -2 \\ 0 & t < -2 \end{cases}$$

The graphical and analytical solutions will readily agree.

1.7 The Reflection Operation

Given a signal $v(t)$, $v(-t)$ is referred to as its reflection. At times, it is also called an image of $v(t)$.

Example 1.6

For the signal shown in Figure 1.6, plot and find analytical expressions for $v(-t)$ and $v(1-t)$.

Solution

Mathematically the signal in Figure 1.6 can be expressed as

$$v(t) = \begin{cases} A & 0 \leq t \leq 1 \\ 0 & \text{otherwise} \end{cases}$$

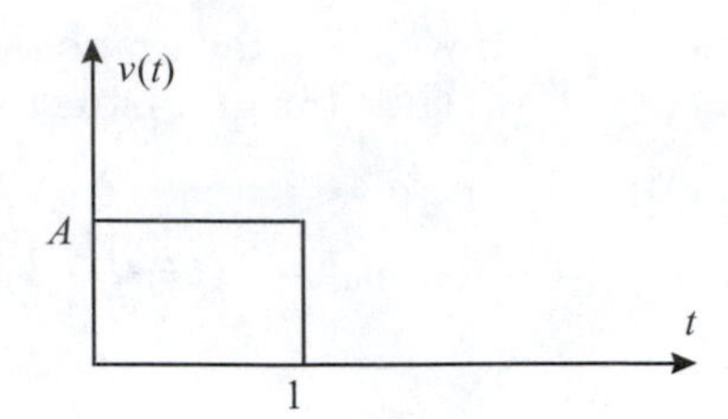

FIGURE 1.6
Signal for Example 1.6.

and

$$v(-t) = \begin{cases} A & 0 \leq -t \leq 1 \\ 0 & \text{otherwise} \end{cases}$$

By fixing the time range we get

$$v(-t) = \begin{cases} A & 0 \geq t \geq -1 \\ 0 & \text{otherwise} \end{cases}$$

or

$$v(-t) = \begin{cases} A & -1 \leq t \leq 0 \\ 0 & \text{otherwise} \end{cases}$$

Likewise,

$$v(1-t) = \begin{cases} A & 0 \leq 1-t \leq 1 \\ 0 & \text{otherwise} \end{cases}$$

By fixing the time interval we get

$$v(1-t) = \begin{cases} A & -1 \leq -t \leq 0 \\ 0 & \text{otherwise} \end{cases}$$

or

$$v(1-t) = \begin{cases} A & 1 \geq t \geq 0 \\ 0 & \text{otherwise} \end{cases}$$

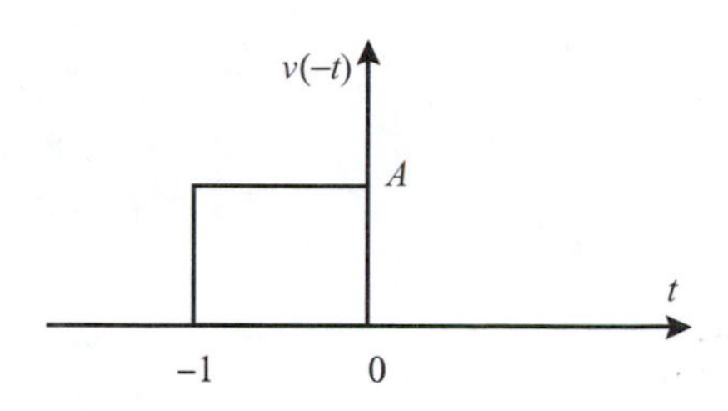

FIGURE 1.7
Signal for Example 1.6.

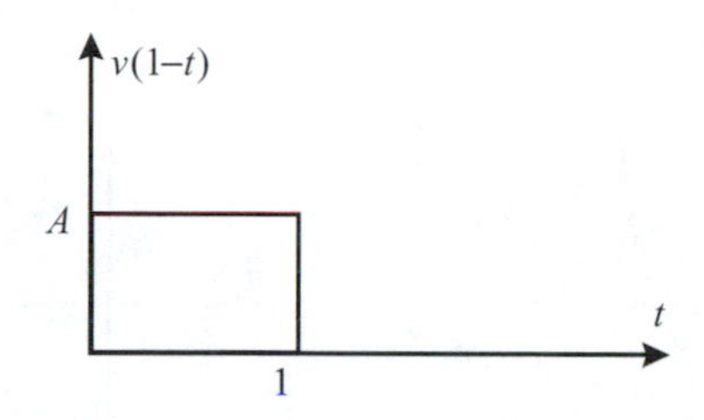

FIGURE 1.8
Signal for Example 1.6.

and finally

$$v(1-t) = \begin{cases} A & 0 \leq t \leq 1 \\ 0 & \text{otherwise} \end{cases}$$

The functions $v(-t)$ and $v(1-t)$ are shown plotted in Figures 1.7 and 1.8, respectively.

1.8 Even and Odd Functions

Consider the signal $v(t)$. It is said to be an even signal when $v(t) = v(-t)$. When $v(-t) = -v(t)$, then $v(t)$ is referred to as an odd signal. Any signal $v(t)$ can be represented as the sum of even and odd signals as

$$v(t) = v_{\text{even}}(t) + v_{\text{odd}}(t)$$

where

$$v_{\text{even}}(t) = 1/2[v(t) + v(-t)]$$

and

$$v_{\text{odd}}(t) = 1/2[v(t) - v(-t)]$$

Example 1.7

Consider the signals given in Figures 1.9 and 1.10. Are these signals odd, even, or neither?

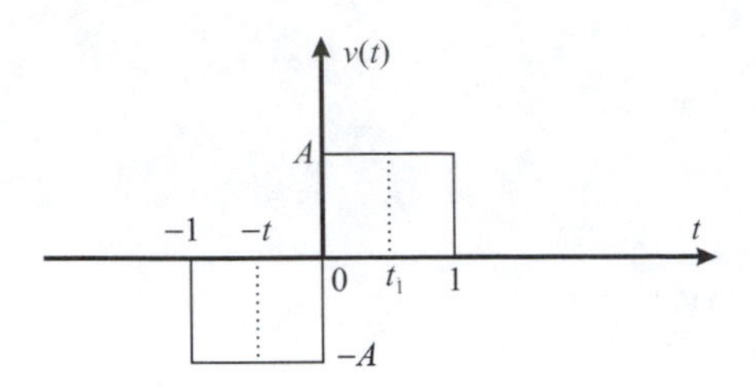

FIGURE 1.9
Signal for Example 1.7.

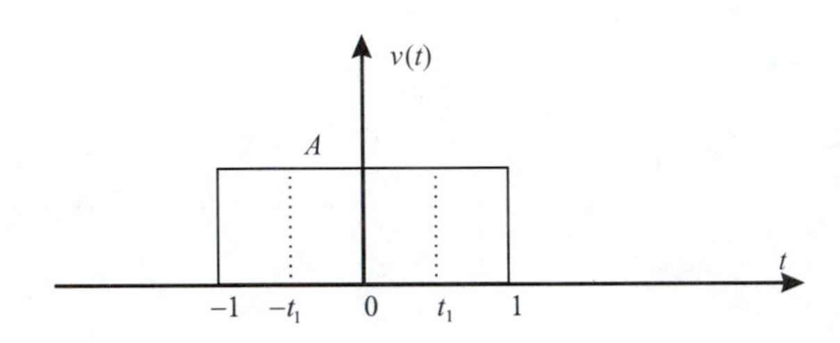

FIGURE 1.10
Signal for Example 1.7.

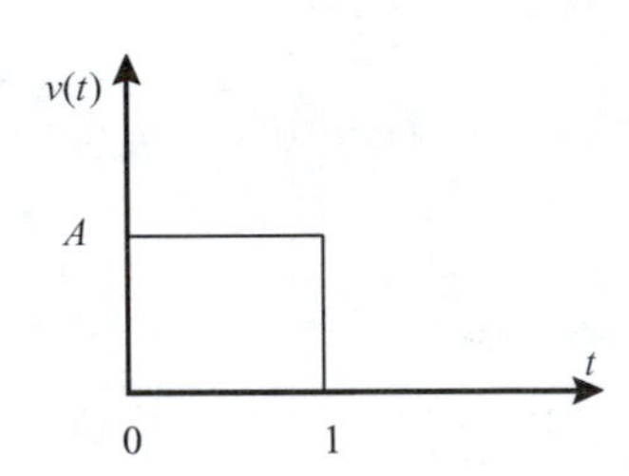

FIGURE 1.11
Signal for Example 1.8.

Solution

The signal shown in Figure 1.9 is odd since $v(t_1) = -v(-t_1)$. On the other hand, the signal shown in Figure 1.10 is even since $v(t_1) = v(-t_1)$.

Example 1.8

Express the signal shown in Figure 1.11 as the sum of odd and even signals.

Solution

As you can see, $v(t)$ is neither even nor odd. It can be written as the sum of an even and an odd signal where

$$v_{\text{even}}(t) = \tfrac{1}{2}[v(t) + v(-t)]$$

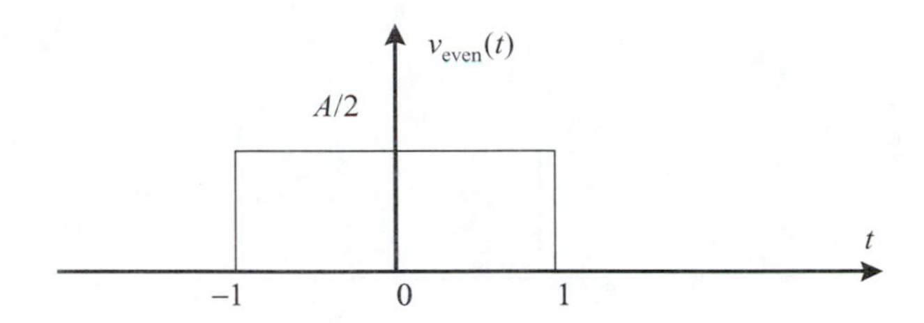

FIGURE 1.12
Signal for Example 1.8.

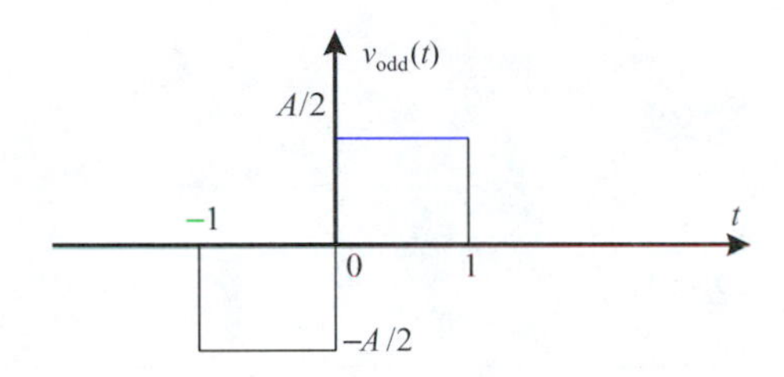

FIGURE 1.13
Signal for Example 1.8.

as shown in Figure 1.12, and

$$v_{odd}(t) = \tfrac{1}{2}[v(t) - v(-t)]$$

and is shown plotted in Figure 1.13. The sum of these two signals is seen to resemble the signal given.

1.9 Time Scaling

Given the signal $v(t)$, $v(nt)$ is a compressed signal of $v(t)$ if and when $|n| > 1$. On the other hand, $v(nt)$ becomes an expanded signal of $v(t)$ if and when $|n| < 1$.

Example 1.9

Given $v(t)$ shown in Figure 1.14a, find $v(t/2)$ and $v(2t)$.

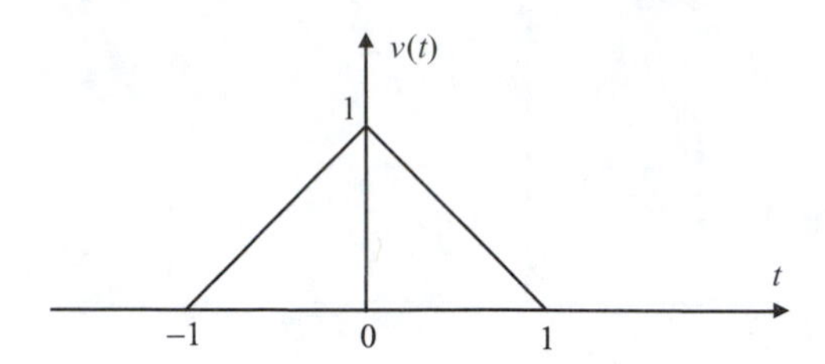

FIGURE 1.14a
Signal for Example 1.9.

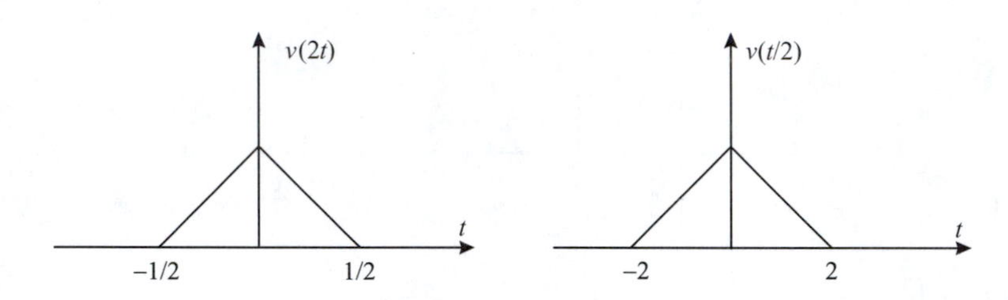

FIGURE 1.14b
Signal for Example 1.9.

Solution

The solution is presented graphically in Figure 1.14b.

Example 1.10

Consider the signal shown in Figure 1.15. Plot and analytically find $v(2t-1)$.

Solution

Mathematically, we can represent the signal shown in Figure 1.15 using the equation

$$v(t) = \begin{cases} t & 0 \le t \le 1 \\ 1 & 1 \le t \le 2 \\ 0 & t > 2 \end{cases}$$

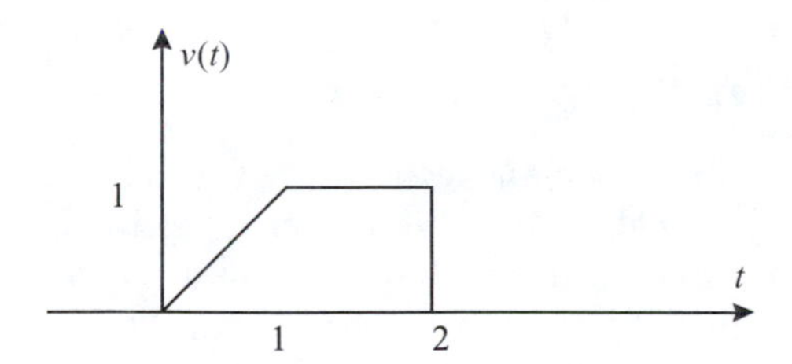

FIGURE 1.15
Signal for Example 1.10.

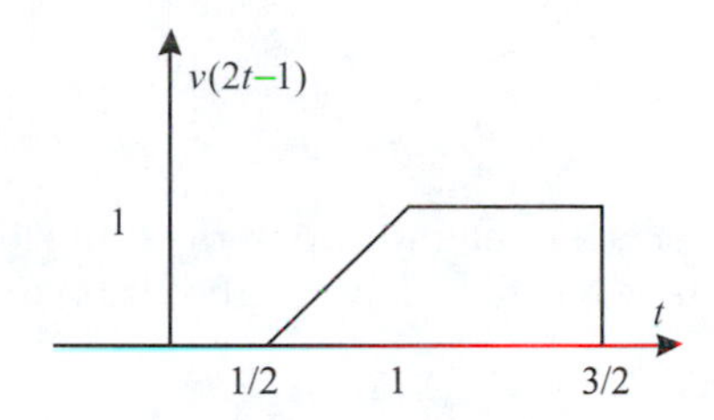

FIGURE 1.16
Signal for Example 1.10.

The signal shifted by 1 and scaled by 2 is

$$v(2t - t) = \begin{cases} 2t - 1 & 0 \leq 2t - 1 \leq 1 \\ 1 & 1 \leq 2t - 1 \leq 2 \\ 0 & 2t - 1 > 2 \end{cases}$$

By fixing the time scale, we get

$$v(2t - t) = \begin{cases} 2t - 1 & 1 \leq 2t \leq 2 \\ 1 & 2 \leq 2t \leq 3 \\ 0 & 2t > 3 \end{cases}$$

or finally

$$v(2t - 1) = \begin{cases} 2t - 1 & 1/2 \leq t \leq 1 \\ 1 & 1 \leq t \leq 3/2 \\ 0 & t > 3/2 \end{cases}$$

The signal $v(2t - 1)$ is thus obtained from $v(t)$ by compressing its slant segment by 1/2 and shifting the constant segment by $t = 1/2$. The resulting graph of $v(2t - 1)$ is shown plotted in Figure 1.16.

1.10 The Unit Step Signal

It is common to close and open an electrical switch to cause a sudden change in the signal of an electrical circuit. To effect changes in a fluid-based system you may actuate a valve. You may also release a spring which was compressed in a mechanical system. In these and similar situations the signal value changes quickly as compared to the response of the system. This is what we commonly refer to as the step signal. Mathematically, a unit step signal such as that shown in Figure 1.17 is represented as

$$Au(t) = \begin{cases} A & t > 0 \\ 0 & t > 0 \end{cases}$$

where A is the magnitude of the unit step signal. The signal is discontinuous at $t = 0$. $u(t)$ is like a switch that is on for $t > 0$.

Example 1.11

Express the rectangular pulse signal shown in Figure 1.18 as a sum of unit step signals.

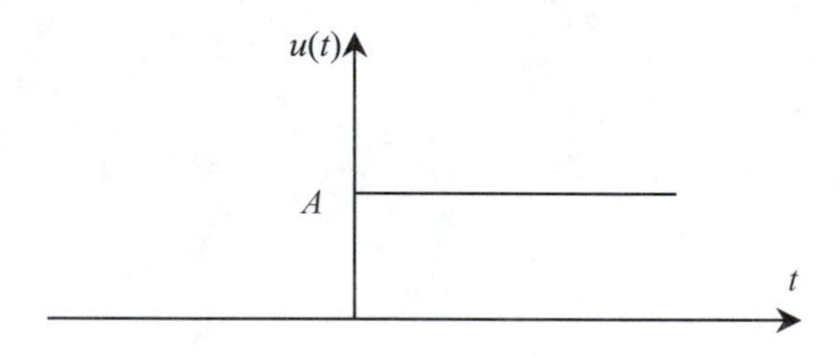

FIGURE 1.17
The step signal.

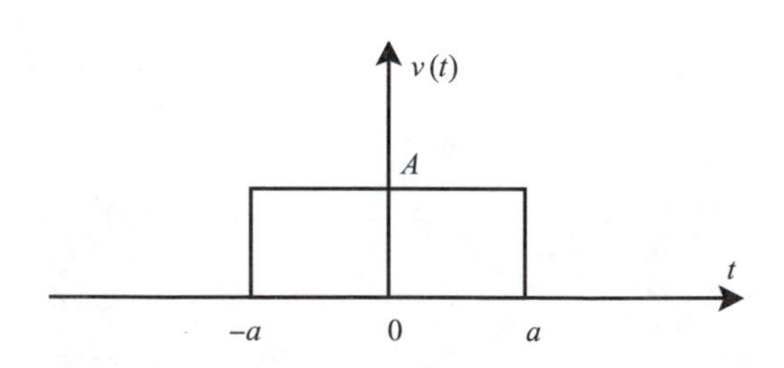

FIGURE 1.18
Signal for Example 1.11.

Solution

Let us look at the plots of $Au(t + a)$ and $Au(t - a)$ shown in Figures 1.19 and 1.20. By subtracting the signal of Figure 1.19 from that of Figure 1.20, we can write

$$Au(t + a) - Au(t - a) = A[u(t + a) - u(t - a)]$$

The graph of the difference shown in Figure 1.21 is identical to the rectangular pulse given in Figure 1.18.

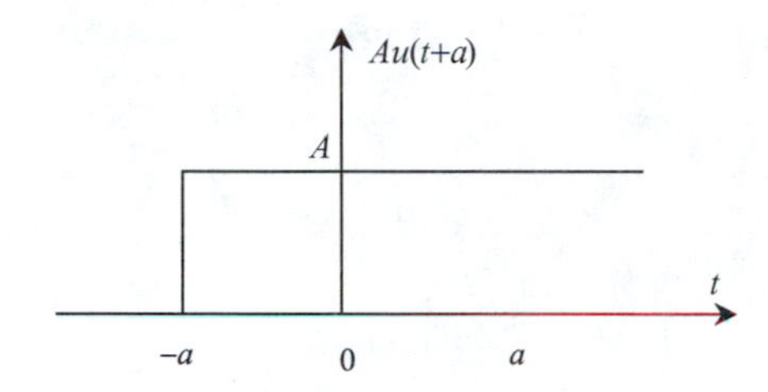

FIGURE 1.19
Signal for Example 1.11.

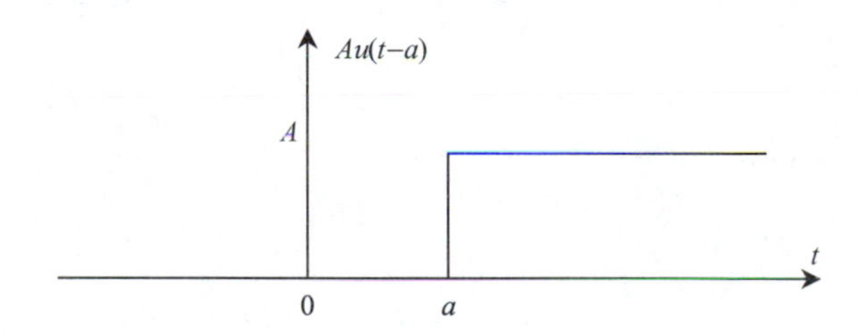

FIGURE 1.20
Signal for Example 1.11.

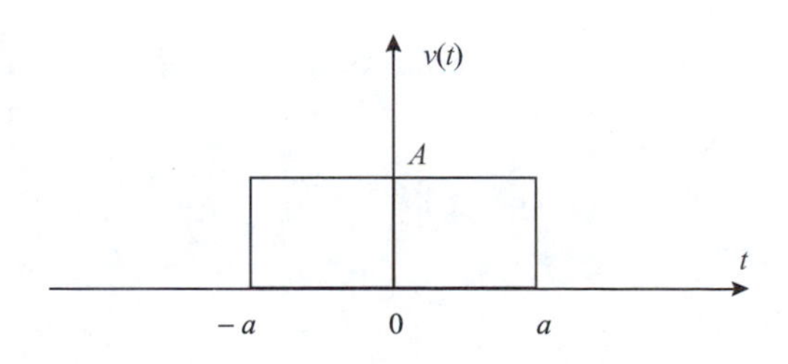

FIGURE 1.21
Signal for Example 1.11.

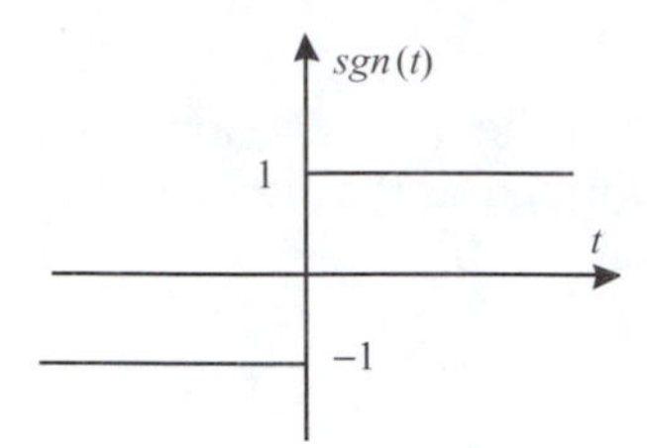

FIGURE 1.22
Signal for Example 1.11.

1.11 The Signum Signal

The signum signal is sketched in Figure 1.22. Mathematically, it can be expressed as

$$sgn(t) = \begin{cases} 1 & t > 0 \\ 0 & 1 = 0 \\ -1 & 1 < 0 \end{cases}$$

Is $sgn(t) = -1 + 2u(t)$? To check this, let us look at the following cases. If $t > 0$, $u(t) = 1$. In this case $-1 + 2(1) = 1$. When $t = 0$, $u(t) = 1/2$. Herein, $-1 + 2(1/2) = 0$. Finally, when $t < 0$, $u(t) = 0$. In this case $-1 + 2(0) = -1$. These cases do conform to the graph. Therefore, indeed, $sgn(t) = -1 + 2u(t)$.

1.12 The Ramp Signal

Ramp signals are also commonplace in many physical systems. A ramp signal can exist in a tracking-type situation, for example, when a target is moving at a constant velocity. A tracking antenna may need a ramp signal in order to track a plane traveling at a constant speed. In this case the ramp signal increases linearly as a function of time as shown in the plot in Figure 1.23.

Mathematically, the ramp signal can be written as

$$r(t) = \begin{cases} t & t \geq 0 \\ 0 & t < 0 \end{cases}$$

Notice that

$$\int_{-\infty}^{t} u(\tau)d\tau = \int_{0}^{t} 1 d\tau = \begin{cases} t & t \geq 0 \\ 0 & t < 0 \end{cases}$$

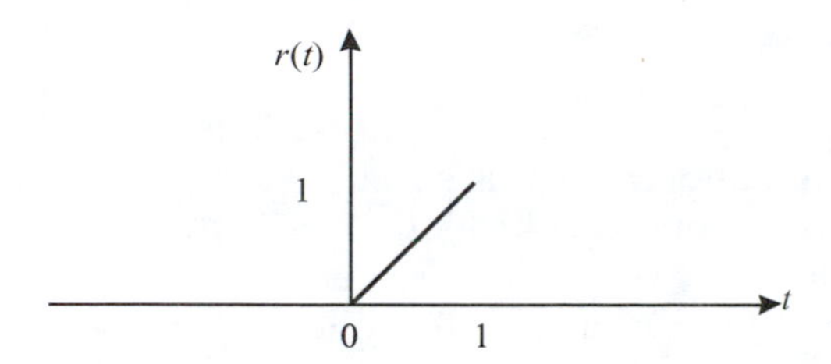

FIGURE 1.23
The ramp signal.

which is the same as the ramp signal described earlier. Accordingly, we can also write

$$\int_{-\infty}^{t} u(\tau)d\tau = r(t)$$

1.13 The Sampling Signal

The sampling signal shown in Figure 1.24 can be expressed mathematically as

$$S_a(x) = \frac{\sin(x)}{x}$$

The *sinc* signal is

$$sinc(x) = \frac{\sin(\pi x)}{\pi x}$$

The *sinc* signal is the sampling signal evaluated at $x = \pi x$ and written as

$$sinc(x) = S_a(\pi x)$$

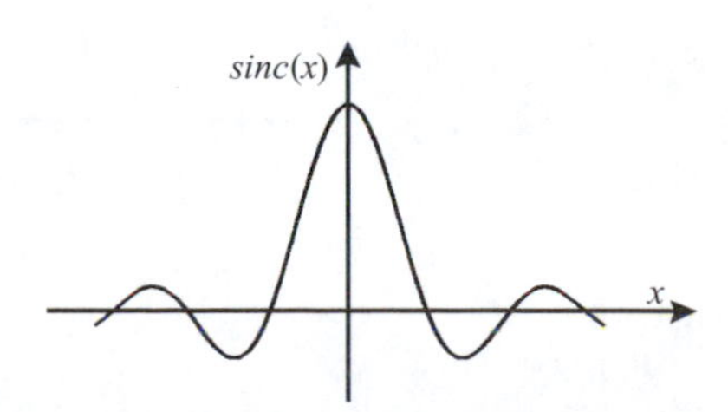

FIGURE 1.24
The sampling signal.

1.14 The Impulse Signal

The derivative of the step signal is defined as the impulse signal. Since the step signal is discontinuous at $t = 0$, the derivative of the step is of infinite magnitude at $t = 0$. This is shown in Figure 1.25 where A is the strength of the signal. The impulse signal, $\delta(t)$, has the following properties:

1. $\delta(t) = 0 \quad \text{for} \quad t \neq 0$
2. $\delta(0) \to \infty$
3. $\int_{-\infty}^{+\infty} \delta(t)dt = 1$
4. $\delta(t)$ is even; $\delta(t) = \delta(-t)$

Evaluating an integral where the integrand is multiplied by an impulse function is relatively easy if we use the sifting property of the impulse signal as

$$\int_{t_1}^{t_2} v(t)\delta(t - t_0)dt = \begin{cases} v(t_0) & t_1 < t_0 < t_2 \\ 0 & \text{otherwise} \\ (1/2)v(t_0) & t_0 = t_1, t_0 = t_2 \end{cases}$$

The scaling property of impulse signals is given by

$$\delta(at + b) = \frac{1}{|a|}\delta\left(t + \frac{b}{a}\right)$$

It can be used to evaluate an integral of $v(t)\delta(at + b)$ as

$$\int_{t_1}^{t_2} v(t)\delta(at + b)dt = \int_{t_1}^{t_2} \frac{1}{a} v(t)\delta\left(t + \frac{b}{a}\right)dt \text{ for } a > 0 \text{ and } t_1 < \frac{b}{a} < t_2$$

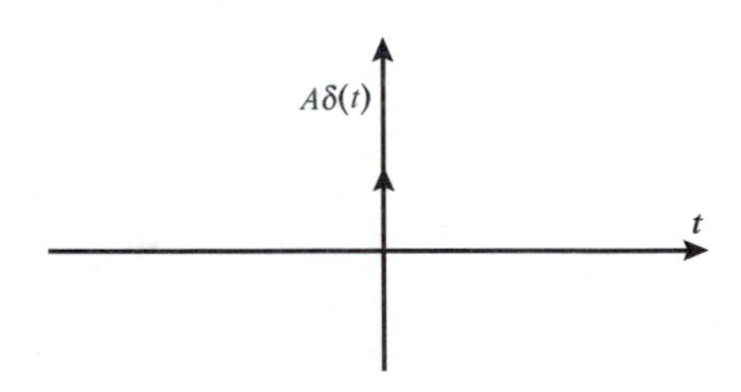

FIGURE 1.25
The impulse signal.

For $a < 0$, we have

$$\int_{t_1}^{t_2} v(t)\delta(at+b)dt = \int_{t_1}^{t_2} \frac{1}{|a|} v(t)\delta\left(t+\frac{b}{a}\right)dt$$

For $t_1 < \frac{b}{a} < t_2$, the integral simplifies to

$$\int_{t_1}^{t_2} \frac{1}{|a|} v(t)\delta\left(t+\frac{b}{a}\right)dt = \frac{1}{|a|} v\left(\frac{-b}{a}\right)$$

Example 1.12

Use the sifting property to evaluate the following integrals.

1. $\int_0^7 (t+1)\delta(t-1)dt$
2. $\int_0^\infty e^{-t}\delta(t-10)dt$
3. $\int_{-\infty}^{t} \delta(\tau)dt$

Solution

1. Since $0 < 1 < 7$ is true and because the impulse is applied at $t = 1$, the integral evaluates to

$$\int_0^7 (t+1)\delta(t-1)dt = 2$$

2. Since $0 < 10 < \infty$ is true and because the impulse is applied at $t = 10$, the integral reduces to

$$\int_0^\infty e^{-t}\delta(t-10)dt = e^{-10}$$

3. For $\int_{-\infty}^{t} \delta(\tau)dt$, we look at the following cases

   a) if $t < 0$, the point, $\tau = 0$, is not between $-\infty$ and t.
   b) if $t > 0$, the point, $\tau = 0$, is between $-\infty$ and t.

Therefore,

$$\int_{-\infty}^{t} \delta(\tau)d\tau = \begin{cases} 1 & t > 0 \\ 0 & t < 0 \end{cases}$$

What you see in the above equation is in fact the definition of the unit step signal, $u(t)$. This is to say that the integral of the impulse signal is the unit step signal.

1.15 Some Insights: Signals in the Real World

The signals that we have introduced in this chapter were all represented in mathematical form and plotted on graphs. This is how we represent signals for the purpose of analysis, synthesis, and design. For better understanding of these signals we will provide real-life situations and see the relation between these mathematical abstractions of signals and a feel of what they may represent in real life.

1.15.1 The Step Signal

The mathematical abstraction is repeated in Figure 1.26 where

$$Au(t) = \begin{cases} A & t > 0 \\ 0 & t < 0 \end{cases}$$

In real-life situations this signal can be viewed as a constant force of magnitude A newtons applied at time equals zero seconds to a certain object for a long time. In another situation, $Au(t)$ can be an applied voltage of constant magnitude to a load resistor R at the time $t = 0$.

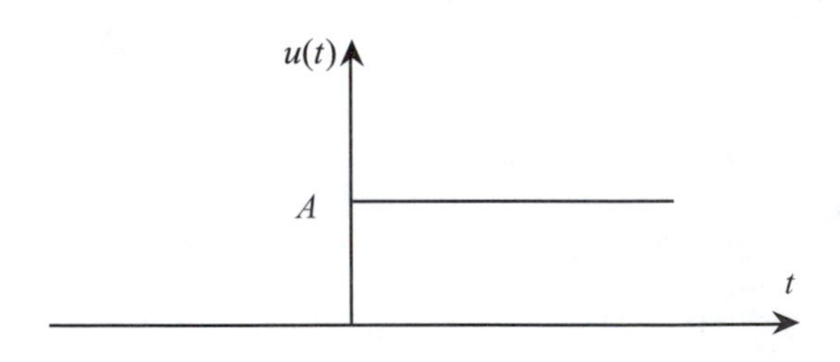

FIGURE 1.26
The step signal.

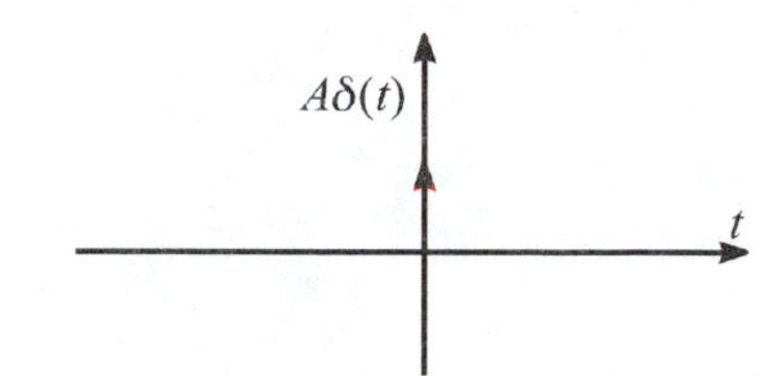

FIGURE 1.27
The impulse signal.

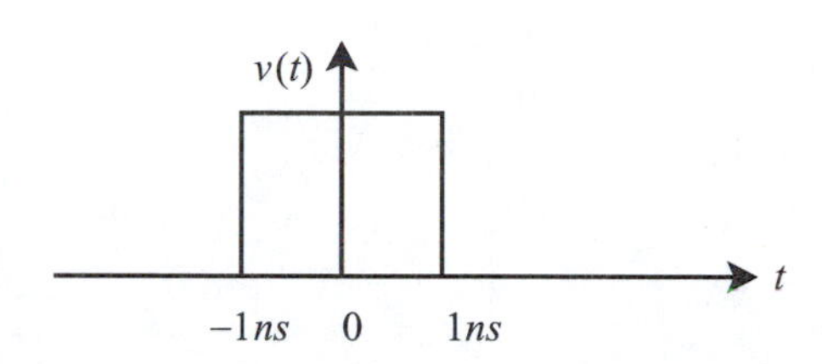

FIGURE 1.28
Modeling an impulse.

1.15.2 The Impulse Signal

The mathematical abstraction is repeated in Figure 1.27.

$$A\delta(t) = \begin{cases} 0 & t \neq 0 \\ A & t = 0 \end{cases}$$

Again, in real life, this signal can represent a situation where a person hits an object with a hammer with a force of A newtons for a very short period of time (pico seconds). We sometimes refer to this kind of signal as a shock.

In another real-life situation, the impulse signal can be as simple as closing and opening an electrical switch in a very short time. Another situation where a spring-carrying mass is hit upward can be seen as an impulsive force. A sudden oil spill like the one that occurred in the Gulf War can represent a sudden flow of oil. You may realize that it is impossible to generate a pure impulse signal for zero duration and infinite magnitude. To create an approximation to an impulse, we can generate a pulse signal of very short duration where the duration of the pulse signal is very short compared to the response of the system. Such a pulse is often approximated as shown in the plot of Figure 1.28.

1.15.3 The Sinusoidal Signal

The mathematical abstraction of a sinusoidal signal can be represented again as shown in Figure 1.29 and mathematically we can write

$$v(t) = A\sin(\omega t)$$

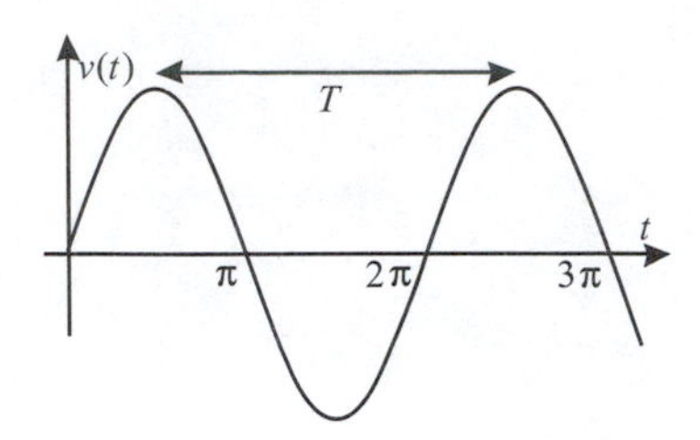

FIGURE 1.29
The sinusoidal signal.

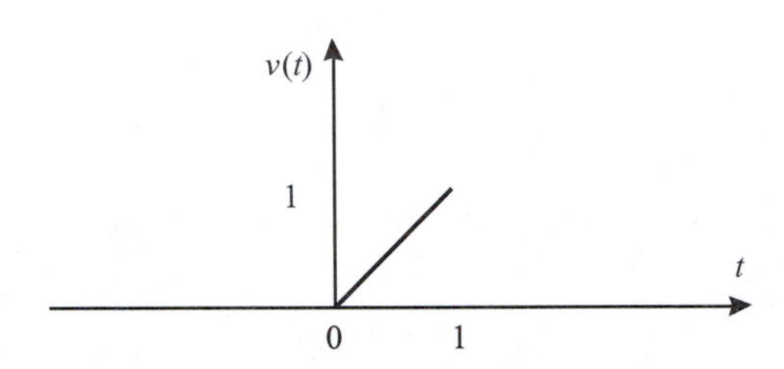

FIGURE 1.30
The ramp signal.

where A is the maximum magnitude of the signal $v(t)$, ω is the radian frequency ($\omega = 2\pi f$), and f is the frequency in hertz ($f = 1/T$). This signal can be thought of as a situation where a person is shaking an object regularly, where A is the magnitude of the force applied in newtons.

This is like pushing and pulling an object continuously with a period of T seconds. Thus, a push and pull forms a complete period of shaking. The distance the object covers during this shaking represents a sinusoidal signal. In the case of electrical signals, an AC voltage source is a sinusoidal signal.

1.15.4 The Ramp Signal

The ramp signal was represented earlier as a mathematical model and is repeated here as in Figure 1.30 where

$$r(t) = \begin{cases} t & t \geq 0 \\ 0 & t < 0 \end{cases}$$

In real-life situations this signal can be viewed as a signal that is increasing linearly with time. An example is where a person starts applying a force at time $t = 0$ to an object and keeps pressing the object with increasing force for a long time. The rate of the increase in the force applied is constant.

Consider another situation where a radar, an anti-aircraft gun, and an incoming jet are in one place. The radar antenna can provide an angular position input. In one case the jet motion forces this angle to change uniformly with time. This will force a ramp input signal to the anti-aircraft gun since it will have to track the jet.

1.15.5 Other Signals

A train of impulses can be thought of as hitting an object with a hammer continuously and uniformly. In terms of electricity, you may be closing and opening a switch continuously. A rectangular pulse is like applying a constant force to an object at a certain time and instantaneously removing that force. It is also like applying a constant voltage at a certain time and then instantaneously closing the switch of the voltage source. Other signals are the random signals where the magnitude changes randomly as time progresses. These signals can be thought of as shaking an object with variable force randomly, or as a gusting wind as time progresses.

1.16 End-of-Chapter Examples

EOCE 1.1

A triangular voltage signal of duration 2 ms is applied as an input to an RC circuit. The input signal is given by $x(t)$ as described in Figure 1.31.

1. Sketch $x(-t)$.
2. Sketch $x(-3 - t)$.
3. Sketch $x(t/3)$.
4. Find the even and the odd parts of $x(t)$ and sketch them.

Solution

1. Mathematically the signal $x(t)$ can be written as

$$x(t) = \begin{cases} t+1 & -1 < t < 0 \\ 0 & \text{elsewhere} \end{cases}$$

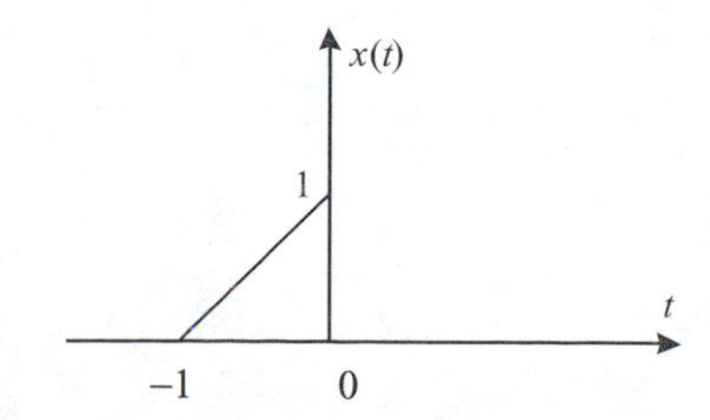

FIGURE 1.31
Signal for EOCE 1.1.

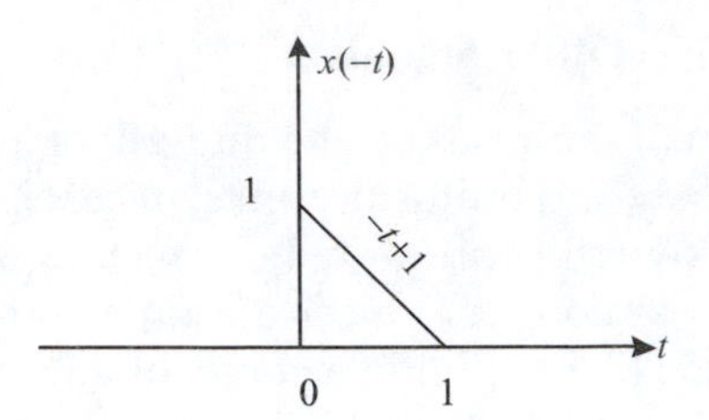

FIGURE 1.32
Signal for EOCE 1.1.

The reflection of $x(t)$ is $x(-t)$ and is written as

$$x(-t) = \begin{cases} -t+1 & -1 < -t < 0 \\ 0 & \text{elsewhere} \end{cases} = \begin{cases} -t+1 & 1 > t > 0 \\ 0 & \text{elsewhere} \end{cases}$$

or

$$x(-t) = \begin{cases} -t+1 & 0 < t < 1 \\ 0 & \text{elsewhere} \end{cases}$$

$x(-t)$ is sketched in Figure 1.32.

2. The signal

$$x(t) = \begin{cases} t+1 & -1 < t < 0 \\ 0 & \text{elsewhere} \end{cases}$$

reflected and translated three units is

$$x(-3-t) = \begin{cases} -3-t+1 & -1 < -3-t < 0 \\ 0 & \text{elsewhere} \end{cases}$$

Fixing the time scale we get

$$x(-3-t) = \begin{cases} -2-t & -1+3 < -t < 0+3 \\ 0 & \text{elsewhere} \end{cases}$$

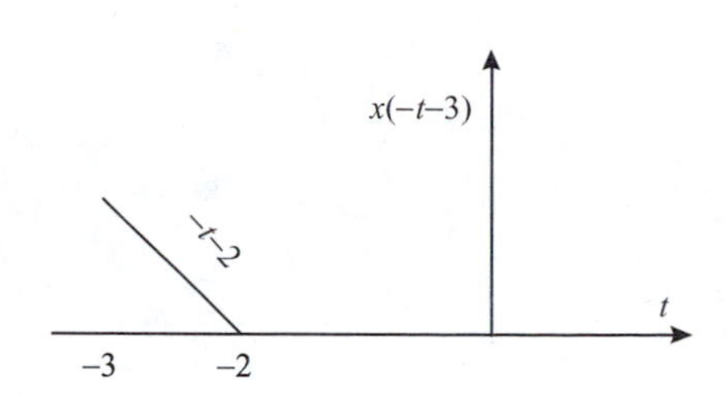

FIGURE 1.33
Signal for EOCE 1.1.

or

$$x(-3-t) = \begin{cases} -2-t & 2 < -t < 3 \\ 0 & \text{elsewhere} \end{cases}$$

$$x(-3-t) = \begin{cases} -2-t & -2 > t > -3 \\ 0 & \text{elsewhere} \end{cases}$$

and finally

$$x(-3-t) = \begin{cases} -2-t & -3 < t < -2 \\ 0 & \text{elsewhere} \end{cases}$$

This is a reflection and also a shifting operation involving $x(t)$. The sketch of $x(-3-t)$ is shown in Figure 1.33.

3. Rewriting $x(t)$ again we have

$$x(t) = \begin{cases} t+1 & -1 < t < 0 \\ 0 & \text{elsewhere} \end{cases}$$

Scaling with a factor of 3 gives

$$x(t/3) = \begin{cases} t/3+1 & -1 < t/3 < 0 \\ 0 & \text{elsewhere} \end{cases}$$

and finally

$$x(t/3) = \begin{cases} t/3+1 & -3 < t < 0 \\ 0 & \text{elsewhere} \end{cases}$$

The plot of $x(t/3)$ is shown in Figure 1.34.

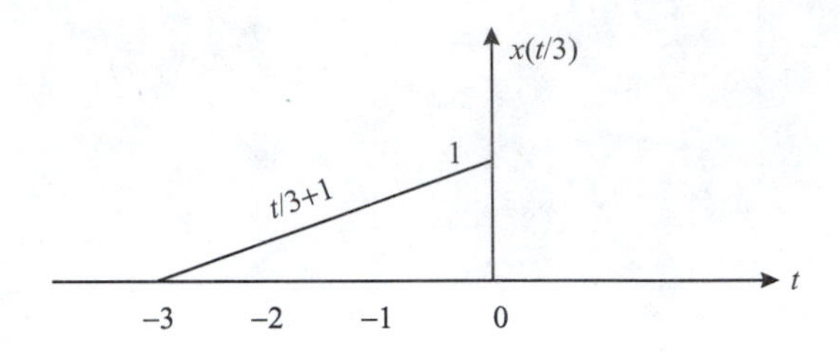

FIGURE 1.34
Signal for EOCE 1.1.

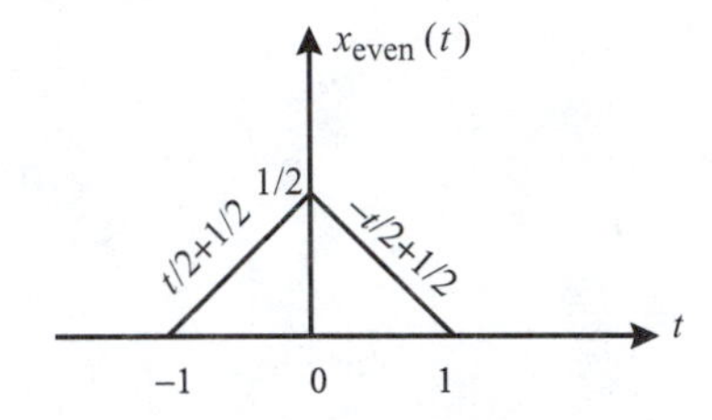

FIGURE 1.35
Signal for EOCE 1.1.

4. Any signal $x(t)$ can be represented as the sum of even and odd signals as

$$x(t) = x_{\text{even}}(t) + x_{\text{odd}}(t)$$

where

$$x_{\text{even}}(t) = 1/2[x(t) + x(-t)]$$

$$x_{\text{even}}(t) = \begin{cases} t/2 + 1/2 & -1 < t < 0 \\ 0 & \text{elsewhere} \end{cases} + \begin{cases} -t/2 + 1/2 & 0 < t < 1 \\ 0 & \text{elsewhere} \end{cases}$$

with the plot given in Figure 1.35.

$$x_{\text{odd}}(t) = 1/2[x(t) - x(-t)]$$

$$x_{\text{odd}}(t) = \begin{cases} t/2 + 1/2 & -1 < t < 0 \\ 0 & \text{elsewhere} \end{cases} - \begin{cases} -t/2 + 1/2 & 0 < t < 1 \\ 0 & \text{elsewhere} \end{cases}$$

with a plot as given in Figure 1.36.

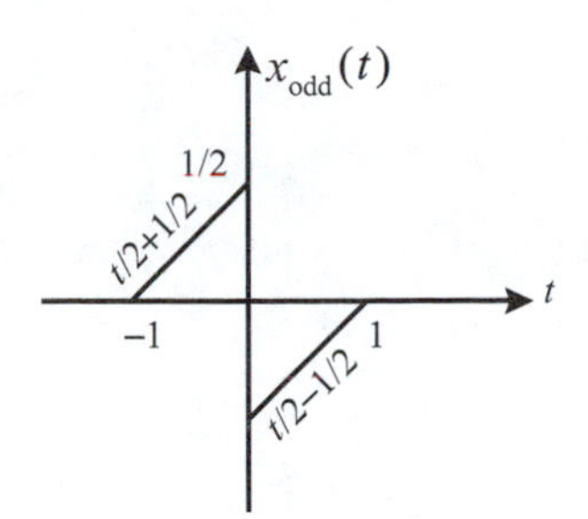

FIGURE 1.36
Signal for EOCE 1.1.

The sum of the odd and the even signals is obtained as shown in Figure 1.37. Using MATLAB, we can simulate EOCE 1.1 as in the following script.

```
clf %clear the plotting area
t = -1:0.05:0; %discrete time points
v = t + 1; %given plot
ft = -fliplr(t); %reversing time and changing sign
fv = fliplr(v); %reversing v
%generating a 3 × 2 plot and plotting v versus t
subplot(3, 2, 1), plot(t, v, '*')
ylabel('v(t)'); %labeling the y axis
subplot(3, 2, 2), plot (ft, fv, '*')
ylabel('v(-t)'); %labeling the next plot
st = ft - 3; %shifting by - 3
sv = fv;
subplot (3, 2, 3), plot(st, sv, '*')
ylabel('v(-3 - t)');
sct = 3 * t; %changing the range of t
scv = sct/3 + 1; %scaled v
subplot(3, 2, 4), plot (sct, scv, '*')
ylabel('v(t/3)');
veven =1/2* [v fv];
%row of t values extended to ft values
subplot(3, 2, 5), plot ([t ft], veven, '*')
ylabel('veven (t)'), xlabel ('Time (sec)')
vodd = 1/2* [v - fv];
subplot(3, 2, 6), plot([t ft], vodd,'*')
ylabel('vodd (t)'), xlabel('Time (sec)')
```

and the plots are given in Figure 1.38.

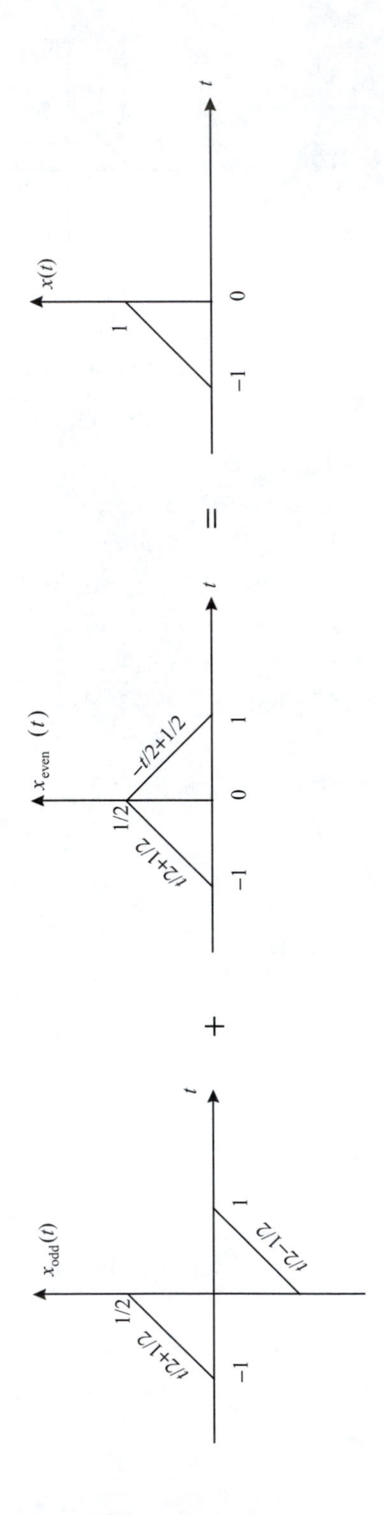

FIGURE 1.37
Signal for EOCE 1.1.

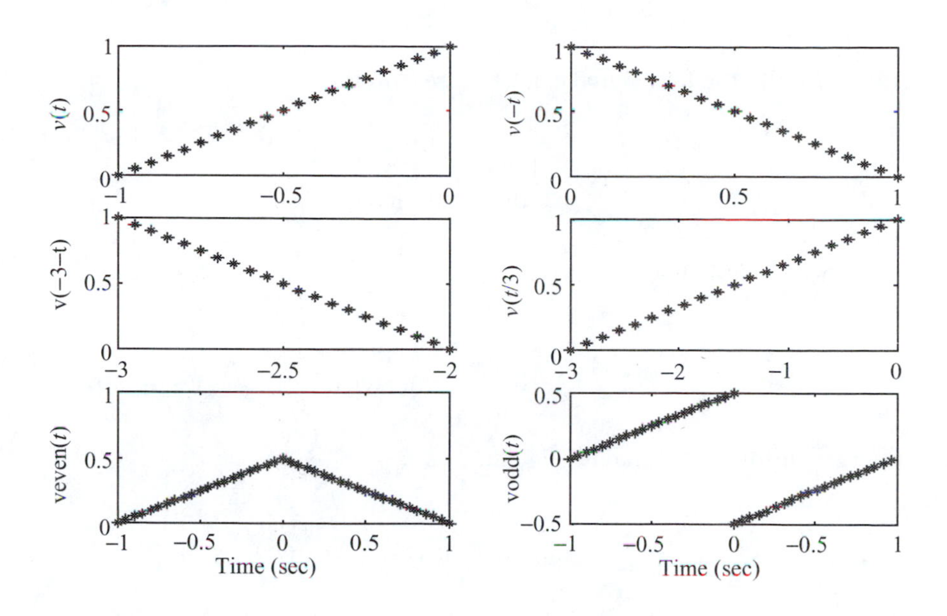

FIGURE 1.38
Signals for EOCE 1.1.

EOCE 1.2

Consider the two signals as given in Figures 1.39 and 1.40. Are these signals power or energy signals?

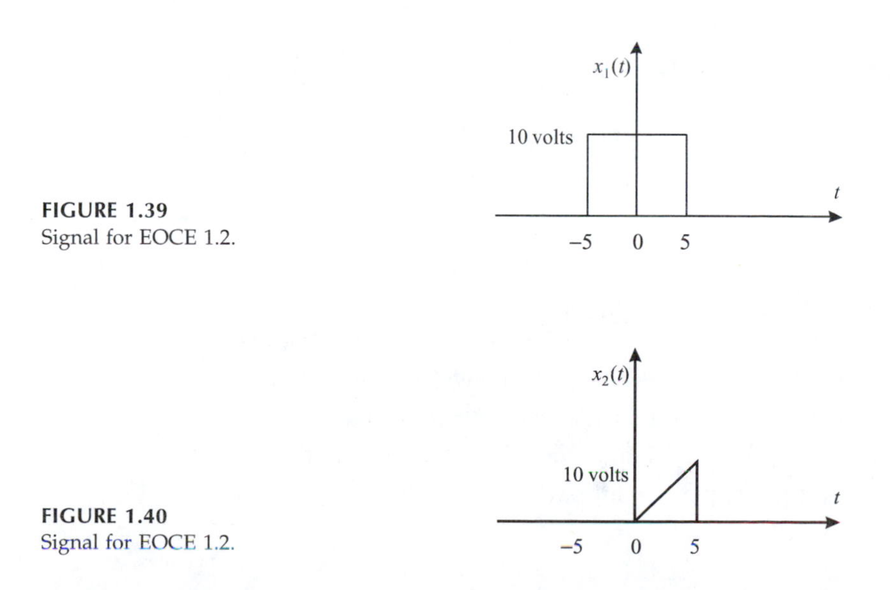

FIGURE 1.39
Signal for EOCE 1.2.

FIGURE 1.40
Signal for EOCE 1.2.

Solution

Mathematically the first signal can be represented as

$$x_1(t) = \begin{cases} 10 & -5 < t < 5 \\ 0 & \text{otherwise} \end{cases}$$

and the second signal as

$$x_2(t) = \begin{cases} 2t & 0 < t < 5 \\ 0 & \text{otherwise} \end{cases}$$

Energy for the first signal is calculated as

$$E_1 = \int_{-5}^{5} 100 dt = 100(5+5) = 1000$$

and energy for the second signal is calculated as

$$E_2 = \int_{0}^{5} 4t^2 dt = \frac{4}{3}(125-0) = \frac{500}{3}$$

Since E_1 and E_2 have finite energy, these signals are energy signals. Consequently, the average power is zero for both signals.

EOCE 1.3

Consider the following signals.

$$\begin{aligned} x_1(t) &= 10\cos(2t) \\ x_2(t) &= \cos(2\pi t) \\ x_3(t) &= (1/2)\cos((2/3)t + 10) \\ x_4(t) &= 10\cos(2t) + \sin(3\pi t) \end{aligned}$$

1. What are the periods of the first three signals given?
2. Is the sum of the first two signals periodic?
3. Is the sum of the last two signals periodic?
4. Is the sum of the first and the last signals periodic?
5. Plot the signals in all parts.

Solution

1. For the first signal: $w_1 = 2$ with $T_1 = \pi$ rad/sec.
   For the second signal: $w_2 = 2\pi$. Therefore $T_2 = 1$ rad/sec.
   For the third signal: $w_3 = 2/3$. Therefore $T_3 = 3\pi$ rad/sec.
2. $x_1(t) + x_2(t)$ is a periodic signal if $T_1/T_2 = m/k$ where m/k is a rational number and T_1 and T_2 are the periods of $x_1(t)$ and $x_2(t)$, respectively. In our case, $T_1 = \pi$ and $T_2 = 1$. $T_1/T_2 = \pi$ which is not a rational number and hence $x_1(t) + x_2(t)$ is not periodic.
3. The period for the third signal is $T_3 = 3\pi$. The last signal has two parts (signals). The first part has a period of π rad/sec and the second part has a period of 2/3 rad/sec. Again, $\pi/(2/3) = (3\pi)/2$ is not a rational number. Therefore, signal four is not periodic, and so the sum of the last two signals is not periodic.
4. Since the last signal is not periodic, the sum of the first and the last is not periodic.

We will use MATLAB to plot the signals. The MATLAB script is given below.

```
m1 = 10; % magnitude of signal 1
m3 = 1/2; % magnitude of signal 3
t = 0:0.05:10; %time range
x1 = m1*cos(2*t);
x2 = cos(2*pi*t);
x3 = m3*cos((2/3)*t + 10);
x4 = m1*cos(2*t) + sin(3*pi*t);
%A 2 × 2 plotting area, plotting x1 versus t
subplot (2, 2, 1), plot (t, x1), grid
title('x1(t)'), xlabel('Time (sec)')
%giving a title to the graph
subplot (2, 2, 2), plot (t, x2), grid
title('x2(t)'), xlabel ('Time (sec)')
subplot (2, 2, 3), plot (t, x3), grid
title ('x3(t)'), xlabel('Time (sec)')
subplot (2, 2, 4), plot (t, x4), grid
title('x4(t)'), xlabel('Time (sec)')
```

The plots are shown in Figure 1.41.

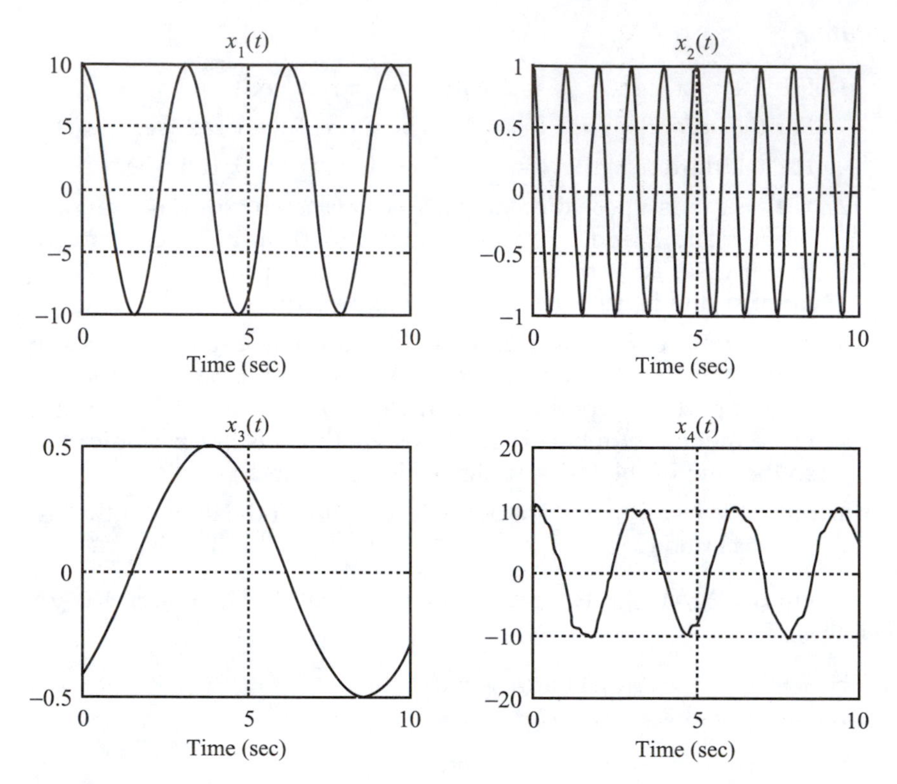

FIGURE 1.41
Signals for EOCE 1.3.

As you can see, the last signal is not periodic. We can write another script to plot the other signals as

```
%assuming x1 and x2 and t are defined in the work space
z1 = x1 + x2;
subplot(3, 1, 1), plot(t, z1), grid
title('x1(t) + x2(t)');
%assuming x3 and x4 are defined
z2 = x3 + x4;
subplot (3, 1, 2), plot (t, z2), grid
title ('x3(t) + x4(t)');
z3 = x1 + x4;
subplot(3, 1, 3), plot (t, z3), grid
title('x1(t) + x4(t)'), xlabel ('Time (sec)')
```

The plots are shown in Figure 1.42.

You also can see that the signals in Figure 1.42 are not periodic.

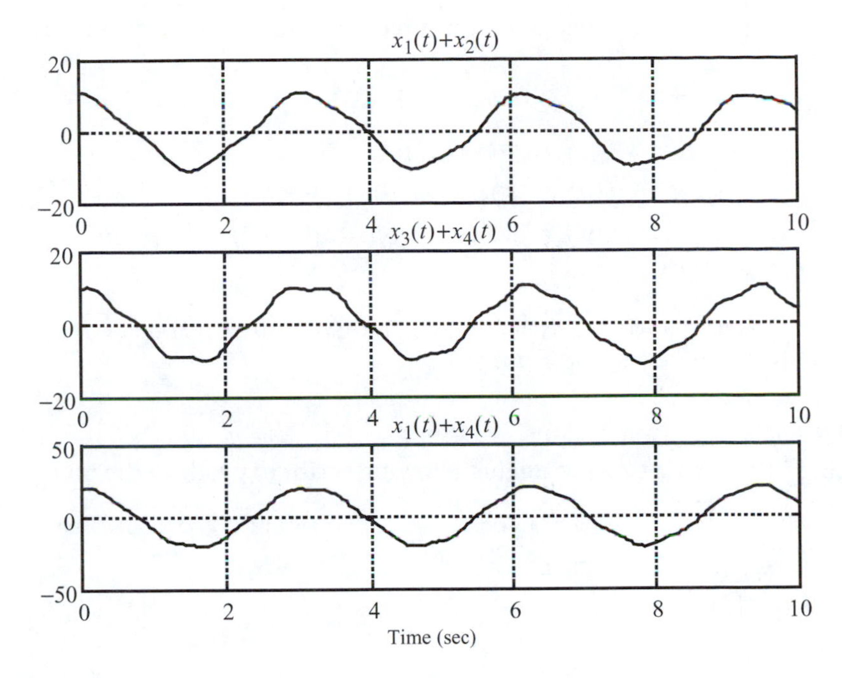

FIGURE 1.42
Signals for EOCE 1.3.

EOCE 1.4

Consider the single-phase circuit shown in Figure 1.43. Let the input voltage be given as

$$v(t) = V_m \cos(wt + \theta_v)$$

and the current be given as

$$i(t) = I_m \cos(wt + \theta_i)$$

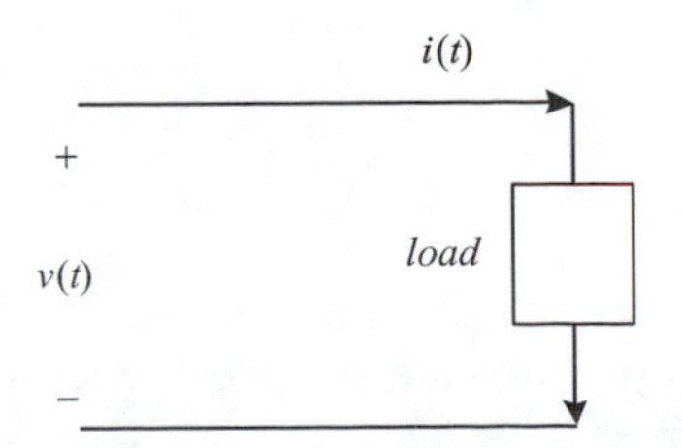

FIGURE 1.43
Circuit for EOCE 1.4.

For the purpose of calculations and plot, assume values for the variables used as

$$v(t) = 100\cos wt$$
$$i(t) = 80\cos(wt - 60)$$
$$p(t) = 8000\cos wt\cos(wt - 60)$$

Plot the curves for $i(t)$, $v(t)$, and $p(t)$, and show that $p(t)$ has a frequency twice the frequency of the input $v(t)$.

Solution

Using MATLAB we can write the following script to produce the plots.

```
%The range for wt from 0 to 3*pi with 0.05 increment
wt = 0:0.05:3*pi;
v = 100*cos(wt);
i = 80*cos(wt-60*pi/180);
%when we multiply element by element we use the •
%before *
p = v• * i;
subplot(3, 1, 1), plot(wt, v), grid;
ylabel('v(t)'),
subplot(3, 1, 2), plot(wt, i), grid;
ylabel('i(t)'),
subplot(3, 1, 3), plot(wt, p), grid;
ylabel('p(t)'), xlabel('wt in rad');
```

The plots are given in Figure 1.44.

Notice that the frequency of the power $p(t)$ is twice the frequency of the voltage $v(t)$. The instantaneous power delivered to the load is calculated as

$$p(t) = V_m\cos(wt + \theta_v)I_m\cos(wt + \theta_i)$$
$$p(t) = \frac{1}{2}V_mI_m[\cos(\theta_v - \theta_i) + \cos(2wt + \theta_v + \theta_i)]$$

You also can see mathematically that the frequency of the instantaneous power $p(t)$ is twice the frequency of the source voltage $v(t)$ in the term $\cos(2\text{wt} + \theta_v + \theta_i)$.

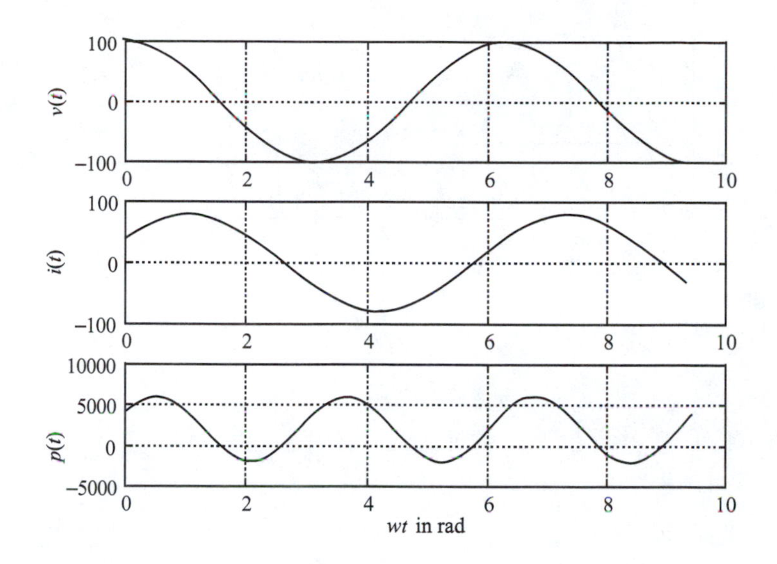

FIGURE 1.44
Signals for EOCE 1.4.

EOCE 1.5

Consider the signal shown in Figure 1.45.

1. Write the signal as a sum of unit step signals.
2. Knowing that the derivative of a unit step, $u(t)$, is the impulse $\delta(t)$ as shown in Figure 1.46, what is the derivative of the given signal?

Solution

1. The given signal can be written mathematically as

$$x(t) = 5u(t) - 11u(t-1) + 6u(t-2) + 8u(t-3)$$

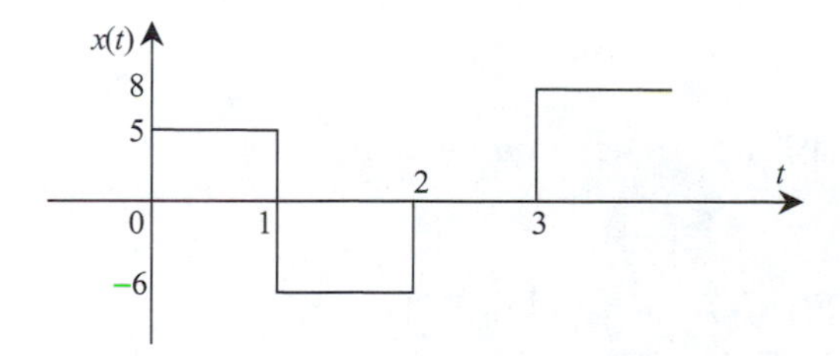

FIGURE 1.45
Signal for EOCE 1.5.

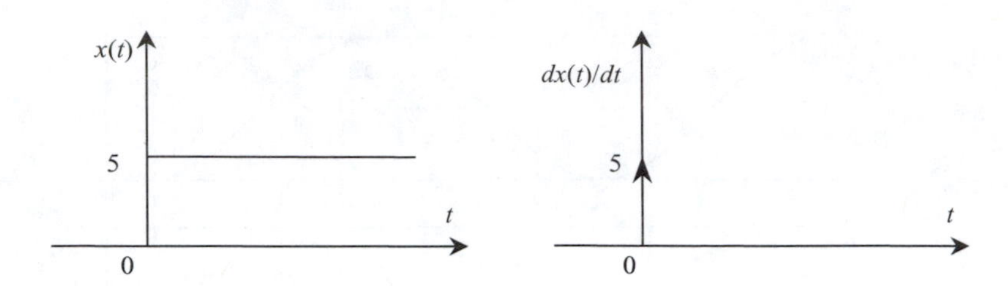

FIGURE 1.46
Signals for EOCE 1.5.

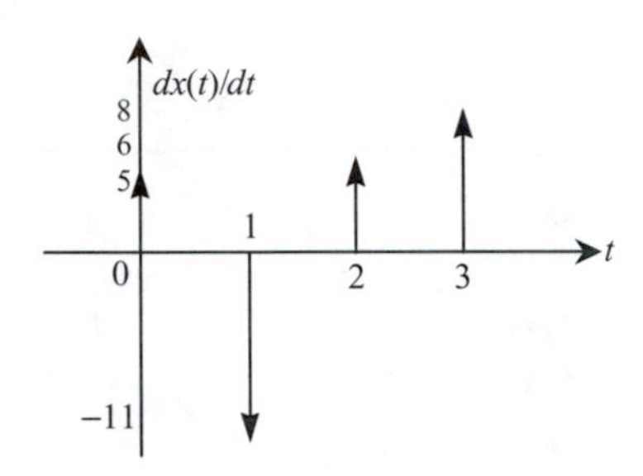

FIGURE 1.47
Signal for EOCE 1.5.

2. The derivative of the signal is

$$\frac{d}{dt}x(t) = 5\delta(t) - 11\delta(t-1) + 6\delta(t-2) + 8\delta(t-3)$$

and is shown in Figure 1.47.

EOCE 1.6

Consider the signal shown in Figure 1.48.

1. Write the signal as a sum of unit step signals.
2. Knowing that the derivative of a unit step, $u(t)$, is the impulse $\delta(t)$, what is the derivative of $x(t)$?

Solution

1. Mathematically $x(t)$ can be written as

$$\begin{aligned} x(t) = {} & 5t[u(t) - u(t-1)] - 6[u(t-1) - u(t-2)] \\ & + (5t-10)[u(t-2) - u(t-3)] \end{aligned}$$

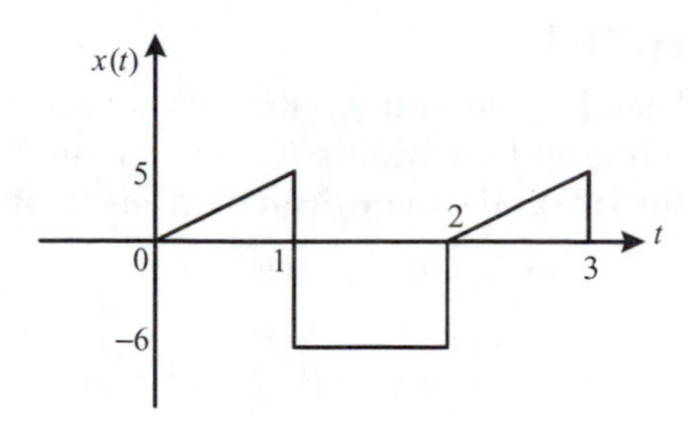

FIGURE 1.48
Signal for EOCE 1.6.

2. The derivative of $x(t)$ can be written as

$$
\begin{aligned}
\frac{d}{dt}x(t) = & [5u(t) + 5t\delta(t) - 5u(t-1) - 5t\delta(t-1)] \\
& + [-6\delta(t-1) + 6\delta(t-2)] \\
& + [5u(t-2) + 5t\delta(t-2) - 5u(t-3)] \\
& - 5t\delta(t-3)(-10\,\delta(t-2)) + 10\delta(t-3)]
\end{aligned}
$$

or

$$
\begin{aligned}
\frac{d}{dt}\,x(t) = & \; 5u(t) - 5u(t-1) - 5\delta(t-1) \\
& -6\delta(t-1) + (6\delta(t-2) + 5u(t-2)) \\
& +10\delta(t-2) - 5u(t-3) - 15\delta(t-3) \\
& -10\delta(t-2) + 10\delta(t-3)
\end{aligned}
$$

Finally

$$
\begin{aligned}
\frac{d}{dt}x(t) = & \; 5u(t) - 5u(t-1) - 11\delta(t-1) + 6\delta(t-2) \\
& + 5u(t-2) - 5u(t-3) - 5\delta(t-3)
\end{aligned}
$$

and is shown in Figure 1.49.

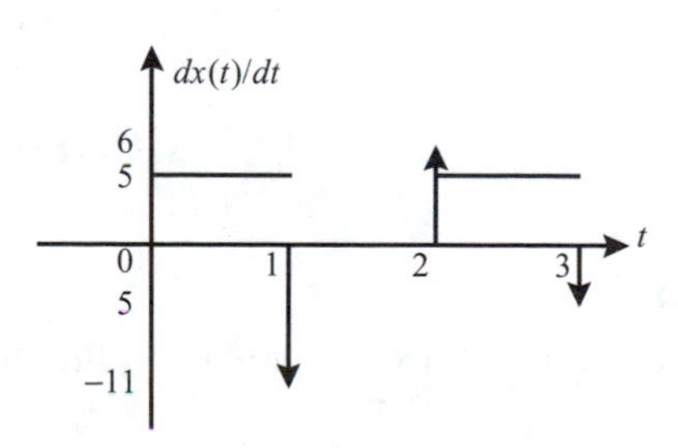

FIGURE 1.49
Signal for EOCE 1.6.

EOCE 1.7

Usually, evaluating integrals is somewhat difficult in many cases. When the integrand is a multiplication of a function, $x(t)$, and an impulse function, $\delta(t)$, the integral can be evaluated easily using the following formula.

$$\int_{t_1}^{t_2} x(t)\delta(t-t_0)dt = \begin{cases} x(t_0) & t_1 < t_0 < t_2 \\ 0 & \text{otherwise} \\ (1/2)x(t_0) & t_0 = t_1,\ t_0 = t_2 \end{cases}$$

Find the following integrals.

1. $\int_0^{10} e^{-t^2}\delta(t-6)dt$
2. $\int_0^{10} e^{-t^2}\cos(wt)\delta(t+6)dt$
3. $\int_0^{10} e^{-t^2}x(t)\delta(t-6)dt$

Solution

1. The impulse is applied at $t = 6$ units of time which is within the limits of the integral.
2. The impulse is applied at $t = -6$ units of time which is outside the limits of the integral.
3. The impulse is applied at $t = 6$ units of time which is within the limits of the integral.

Therefore,

$$\int_0^{10} e^{-t^2}\delta(t-6)dt = e^{-6^2} = e^{-36}$$

$$\int_0^{10} e^{-t^2}\cos(wt)\delta(t+6)dt = 0$$

$$\int_0^{10} e^{-t^2}x(t)\delta(t-6)dt = e^{-36}x(6)$$

EOCE 1.8

Use MATLAB to simulate the following signals.

1. The step signal, $u(t)$
2. The impulse signal, $\delta(t)$

3. The ramp signal, $r(t)$
4. A general sinusoidal signal

Solution

1. The following MATLAB script is used to simulate a unit step function with amplitude A that starts at time = 0.

```
clf % clearing latest plots
A = 1; % the amplitude of the step
t = 0:0.01:10; % generating points in time
% generating a row of length t of magnitude A
unitstep = A*ones(1, length(t));
plot(t, unitstep);
title('MATLAB simulated unit step signal');
xlabel('Time (sec)');
axis([0 10 0 1.2]); % fix the x and y axis
```

The plot for the unit step is shown in Figure 1.50.

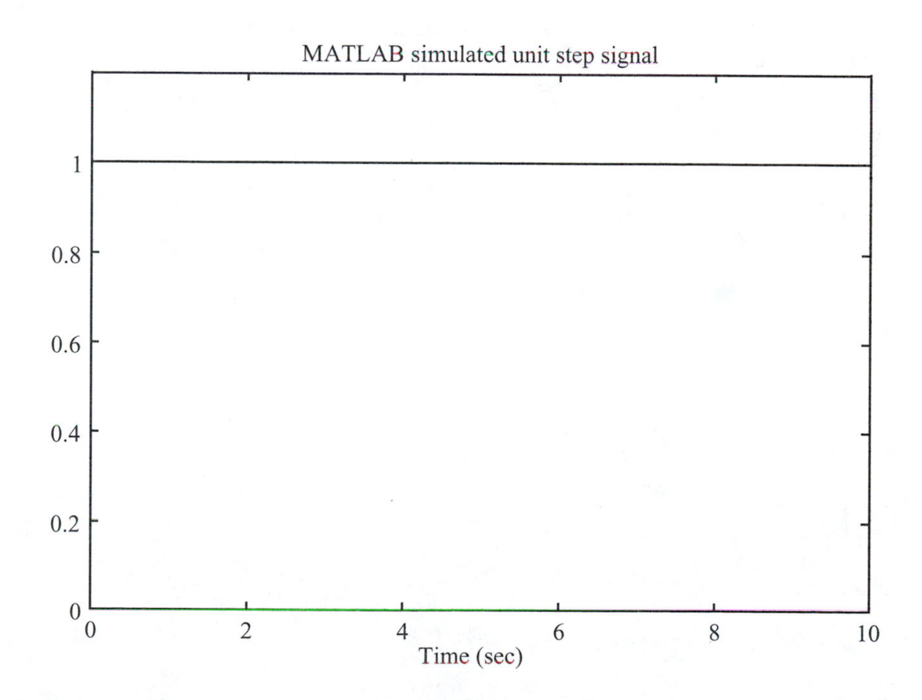

FIGURE 1.50
Signal for EOCE 1.8.

2. The following MATLAB script simulates the impulse signal.

```
clf % clearing latest plots
A = 1; % the amplitude of the impulse
t = 0:0.5:10; % generating points in time
%We generate a row of zeros of length t, magnitude A
unitimpulse = A*zeros(1, length(t));
unitimpulse(1) = A; % creating the impulse
stem(t, unitimpulse); % show plot as 'o'
title ('MATLAB simulated unit impulse signal');
xlabel('Time (sec)');
axis ([0 10 0 1.2]); % fix the x and y axis
```

The plot is shown in Figure 1.51.

3. The following MATLAB script simulates the ramp signal.

```
clf % clearing latest plots
A = 1; % the slope of the ramp
t = 0:0.5:10; % generating points in time
```

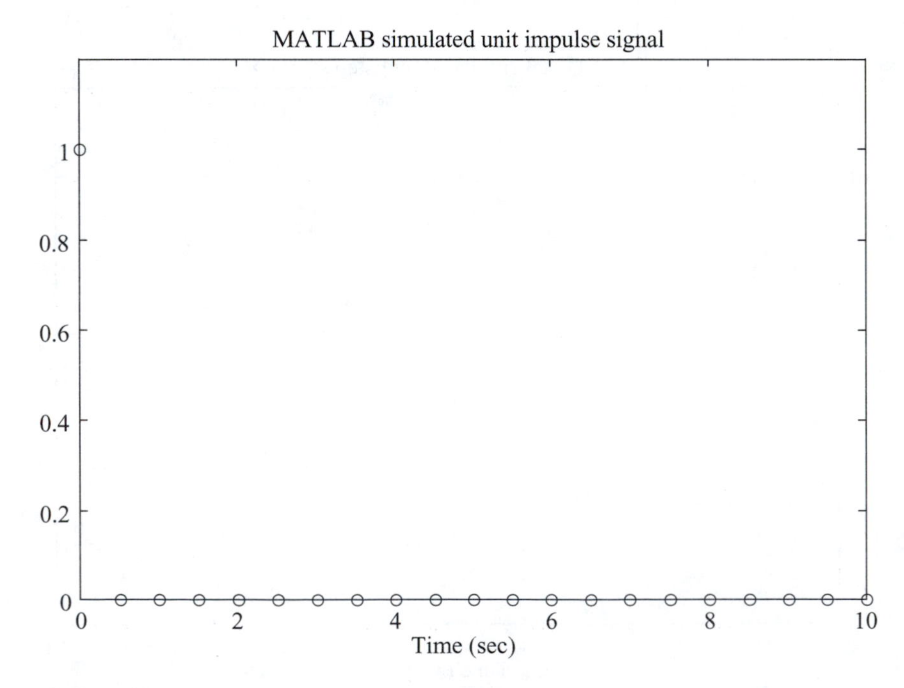

FIGURE 1.51
Signal for EOCE 1.8.

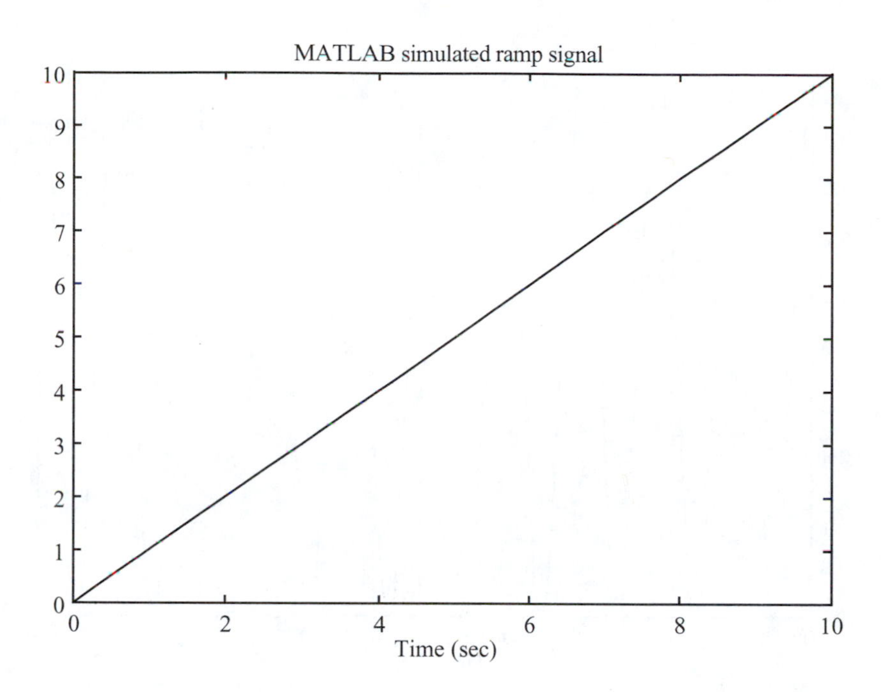

FIGURE 1.52
Signal for EOCE 1.8.

```
ramp = A*t; % generating the ramp signal
plot (t, ramp);
title ('MATLAB simulated ramp signal');
xlabel ('Time (sec)');
```

The plot is shown in Figure 1.52.

4. The following MATLAB script simulates the general sinusoidal signal.

```
clf % clearing latest plots
A = 1; % the amplitude of the sinusoidal signal
f = 1; % the frequency of the signal
phi = 0; % the phase of the signal in radians
% generating the sinusoidal signal
t = 0:0.01:10; % generating points in time
sinusoidal = A*cos(2*pi*t 1 phi);
plot(t, sinusoidal);
title('MATLAB simulated sinusoidal signal');
xlabel('Time (sec)');
```

The plot is shown in Figure 1.53.

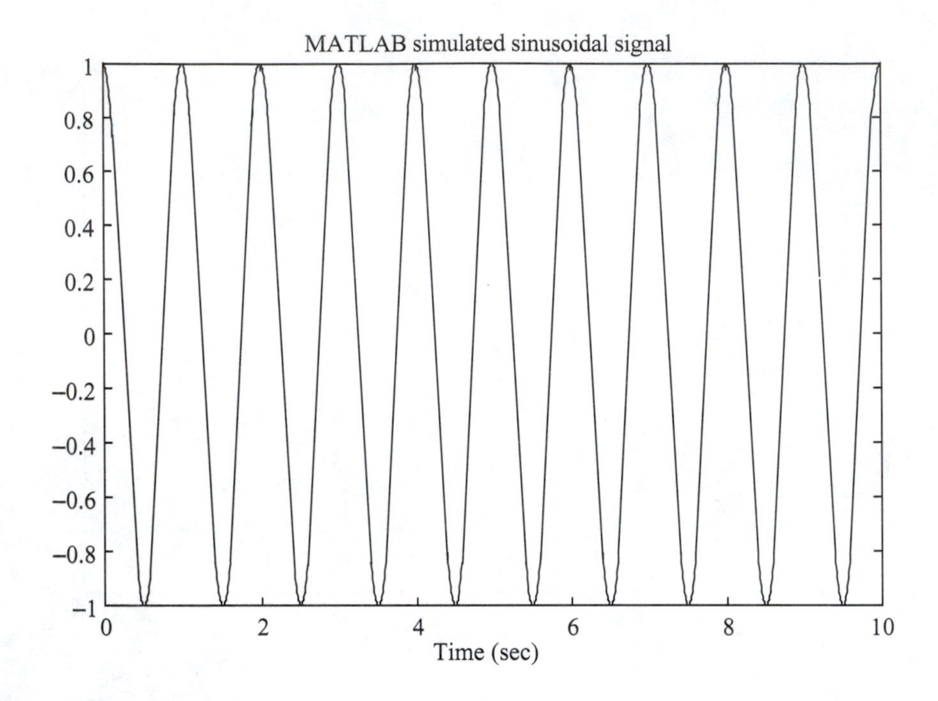

FIGURE 1.53
Signal for EOCE 1.8.

1.17 End-of-Chapter Problems

EOCP 1.1

Consider the following continuous signals.

1. $v(t) = 10t\cos(2t + 1)$
2. $v(t) = 10\cos(2t + 1) + 10$
3. $v(t) = \cos(5t + 1) + 10 + \sin(t)e^{-t}$
4. $v(t) = 100\cos(2t + 100) - 10 + \sin(3t - 3)$
5. $v(t) = 10\cos(2t - 1) - 100\sin(100t - 30) + \cos(t)$
6. $v(t) = 10\cos(2t + 1)e^{t} + \sin(t) - 1/2$
7. $v(t) = \cos(2t) + 10 + \dfrac{\sin(t)}{t}$
8. $v(t) = 1/t - 10\cos(2t + 1) + 10$
9. $v(t) = 1/2\cos(4t - 3) - 100\sin(100t) + \cos(t/2) - 1$
10. $v(t) = 1 + 5\cos(10t - 1) - 100\sin(100t - 30) + \cos(7t)$

If they are periodic, find the period, T, for each signal.

EOCP 1.2

Find the energy in each of the following signals.

1. $v(t) = e^{-3t}$ $\quad 0 \le t \le 100$
2. $v(t) = te^{-3t}$ $\quad 0 \le t \le 10$
3. $v(t) = e^{-3t} + 1$ $\quad 0 \le t \le 10$
4. $v(t) = \sin(t - 1)$ $\quad -1 \le t \le 10$
5. $v(t) = \sin(t)e^{-3t}$ $\quad 0 \le t \le 10$
6. $v(t) = \sin(t) + \cos(t)$ $\quad 0 \le t \le 1$
7. $v(t) = 10e^{-3t} + t$ $\quad -2 \le t \le 10$
8. $v(t) = 1 + \sin(t)$ $\quad 0 \le t \le 10$
9. $v(t) = 10$ $\quad -1 \le t \le 1$
10. $v(t) = 1 - \cos(2t)$ $\quad 0 \le t \le 1$

EOCP 1.3

Find the power in each of the following signals.

1. $v(t) = e^{-3t} + 1$ $\quad 0 \le t \le 100$
2. $v(t) = 1 + te^{-3t}$ $\quad 0 \le t \le 10$
3. $v(t) = e^{-3t} + t^2$ $\quad -10 \le t \le 10$
4. $v(t) = 1 - \sin(t - 1)$ $\quad -1 \le t \le 0$
5. $v(t) = \sin(t)e^{-3t} + 10$ $\quad 0 \le t \le 10$
6. $v(t) = \sin(t) + \cos(t)$ $\quad 0 \le t \le 1$
7. $v(t) = 10e^{-3t} + t$ $\quad -2 \le t \le 10$
8. $v(t) = 1 + \sin(t)/21$ $\quad 0 \le t \le 10$
9. $v(t) = 1$ $\quad -1 \le t \le 1$
10. $v(t) = \sin(2t)\cos(2t)$ $\quad 0 \le t \le 1$

EOCP 1.4

Consider the following signal.

$$v(t) = \begin{cases} t & 0 \le t \le 1 \\ 1 & 1 \le t \le 2 \\ 0 & t \ge 0 \end{cases}$$

a) Mathematically, find

   1. $v(-t)$
   2. $v(-t + 1)$
   3. $v(-t/2 + 1)$
   4. $2v(t/2 - 5)$

b) Sketch the above four versions of $v(t)$.

EOCP 1.5

Consider the signals shown in Figures 1.54a, 1.54b, and 1.54c. Plot and mathematically find for all signals

1. $v(-t - 1)$
2. $v(1 - t)$
3. $v(-1 - t/2)$
4. $v(-t/2 - 1/2)$

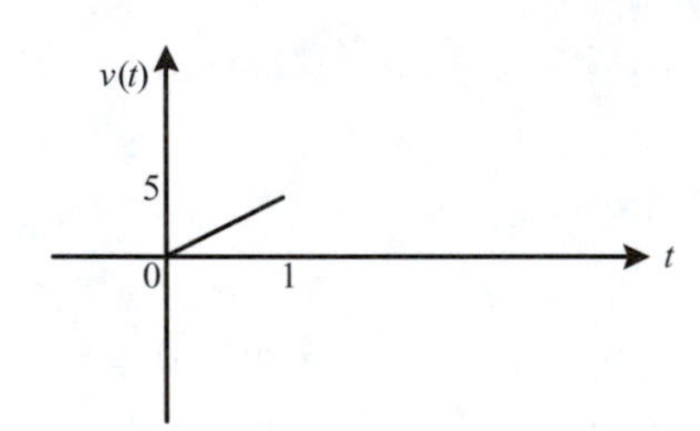

FIGURE 1.54a
Signal for EOCP 1.5.

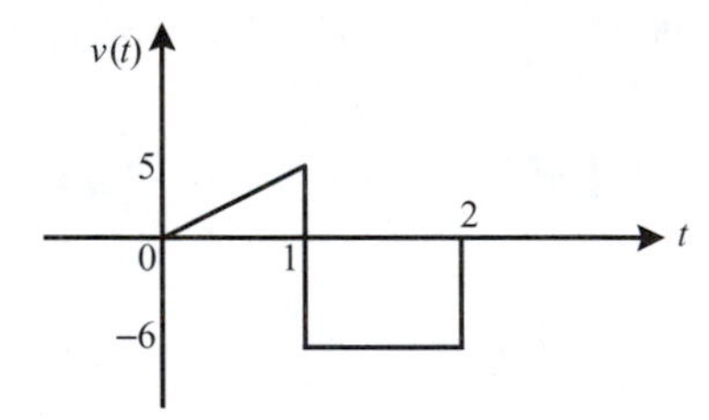

FIGURE 1.54b
Signal for EOCP 1.5.

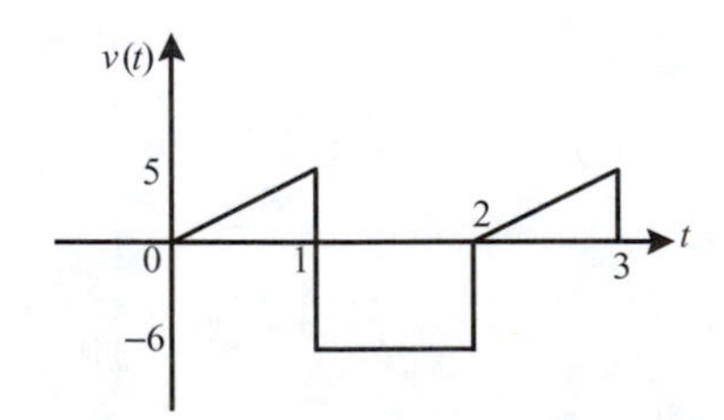

FIGURE 1.54c
Signal for EOCP 1.5.

EOCP 1.6

Express each of the signals shown in Figures 1.55, 1.56, and 1.57 as the sum of even and odd signals.

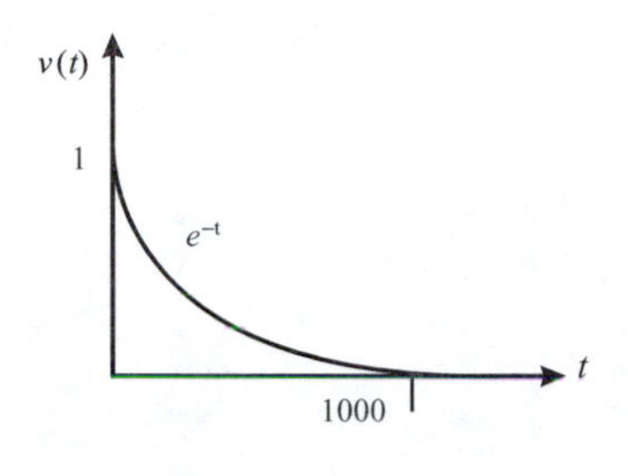

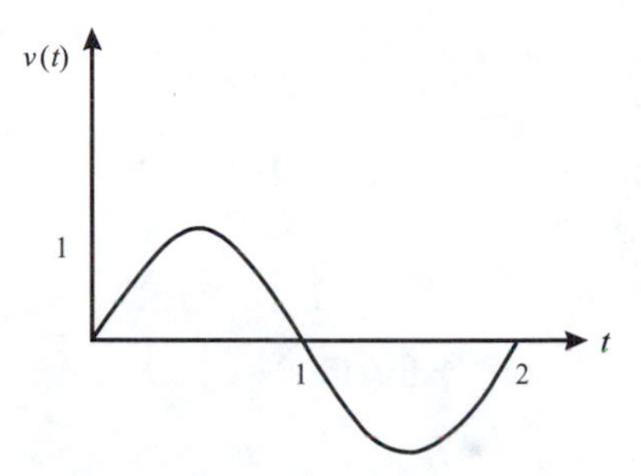

FIGURE 1.55
Signals for EOCP 1.6.

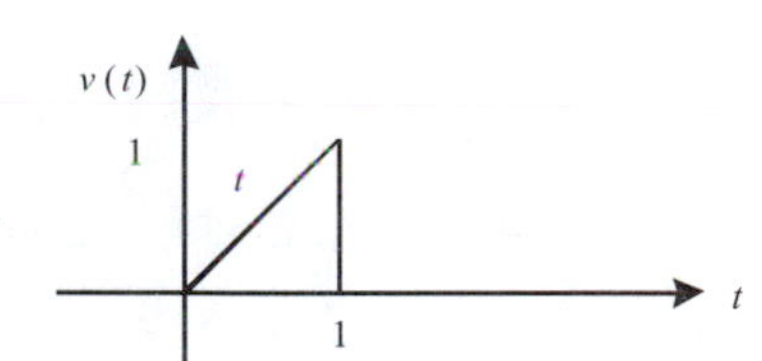

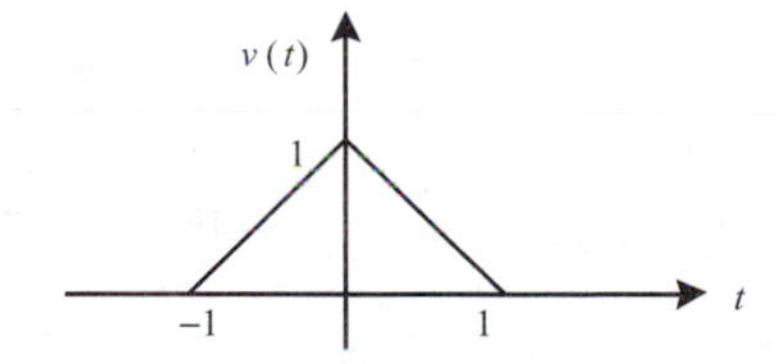

FIGURE 1.56
Signals for EOCP 1.6.

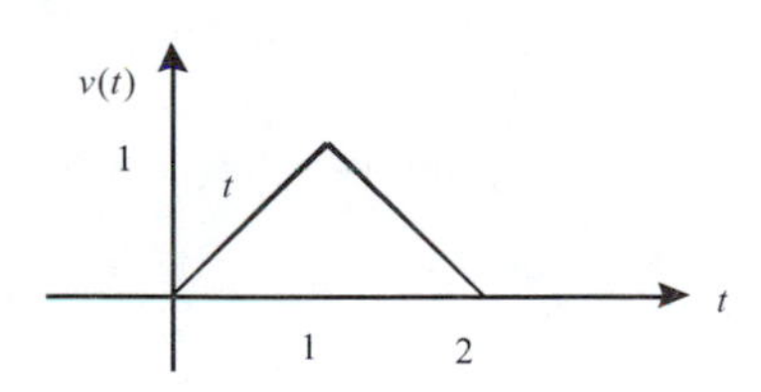

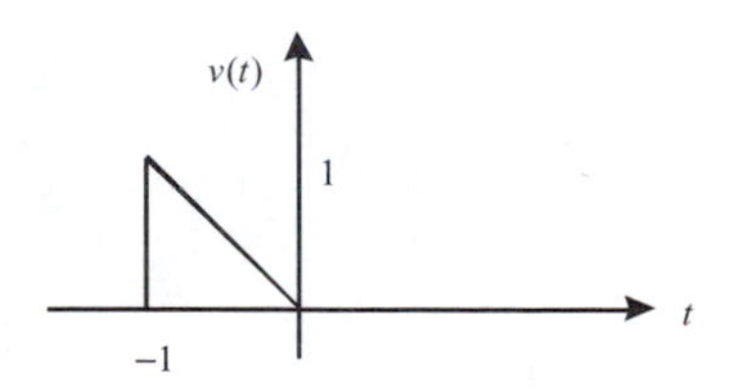

FIGURE 1.57
Signals for EOCP 1.6.

EOCP 1.7

Evaluate the following integrals using the sifting property of the delta function.

1. $\int_{-1}^{2}(t - \sin(t))\delta(t - 1)dt$
2. $\int_{-1}^{2}(t - \sin(t))t\delta(t)dt$
3. $\int_{-1}^{2}t\delta(t/2 + 3)dt$
4. $\int_{-1}^{2}(t - \sin(t))t\delta(2t + 1)dt$
5. $\int_{-\infty}^{2}(e^{-2t})\delta(t - 1)dt$
6. $\int_{-\infty}^{\infty}(e^{-2t})\sin(t)\delta(2t - 1)dt$
7. $\int_{-1}^{\infty}(e^{-2t})t\sin(t)\delta(t/3 + 1/2)dt$
8. $\int_{-\infty}^{0}(e^{-2t})(\sin(t)/t)\delta(t/3 + 1)dt$
9. $\int_{-\infty}^{2}(e^{-2t} + \cos(t - 20))\delta(t/2)dt$
10. $\int_{-\infty}^{10}(e^{-2t} + 1)\delta(t/2 - 20)dt$

EOCP 1.8

Express the signal shown in Figure 1.58 as a combination of unit step signals.

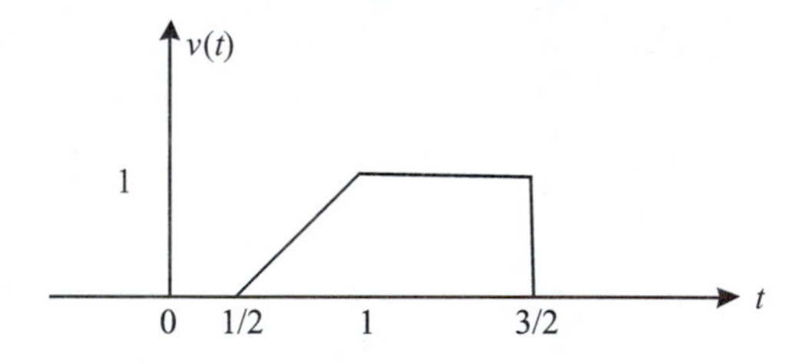

FIGURE 1.58
Signal for EOCP 1.8.

EOCP 1.9

Repeat the above problem considering the signal in Figure 1.59.

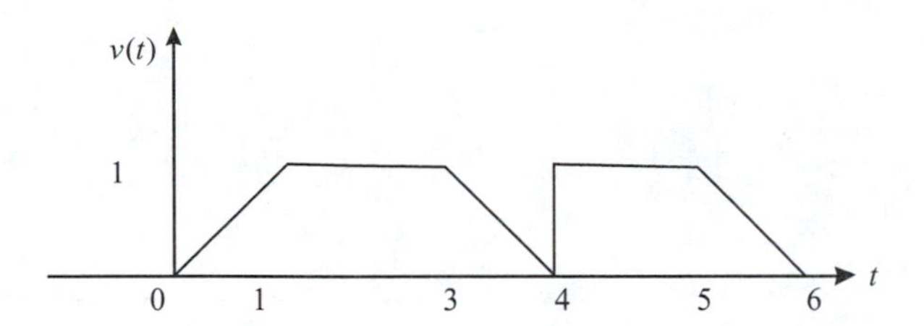

FIGURE 1.59
Signal for EOCP 1.9.

EOCP 1.10

Find the derivative of the signals shown in Figures 1.60a, 1.60b, 1.60c, 1.60d, and 1.60e.

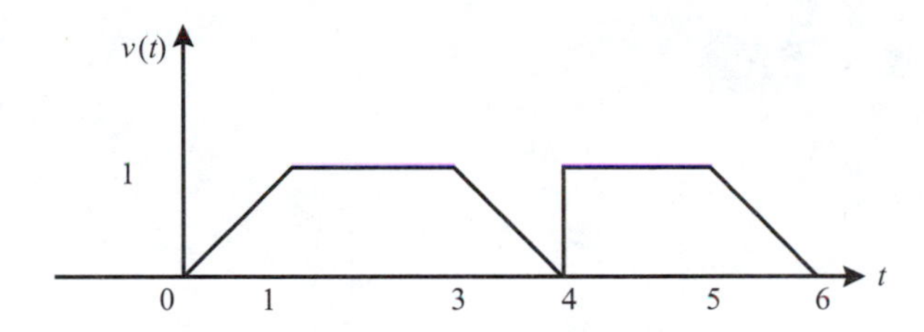

FIGURE 1.60a
Signal for EOCP 1.10.

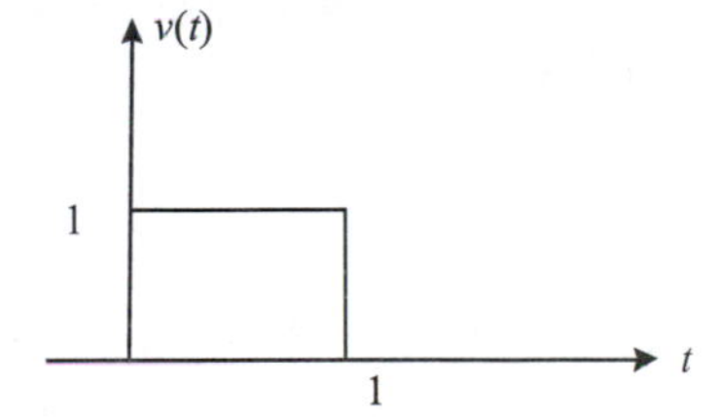

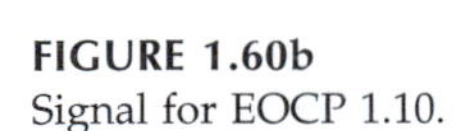
FIGURE 1.60b
Signal for EOCP 1.10.

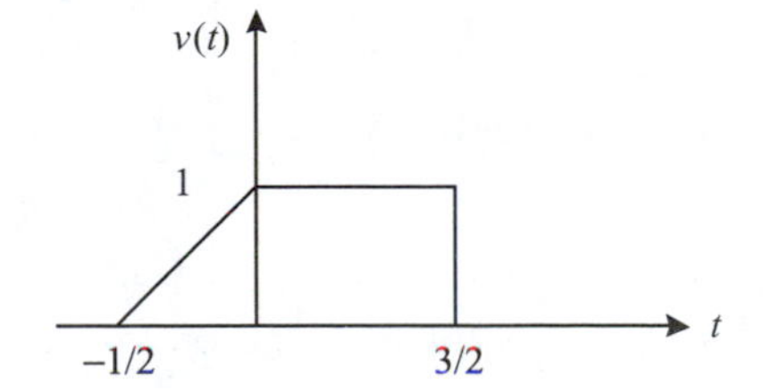

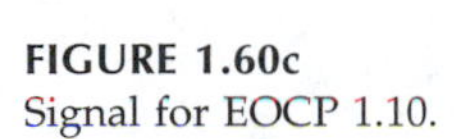
FIGURE 1.60c
Signal for EOCP 1.10.

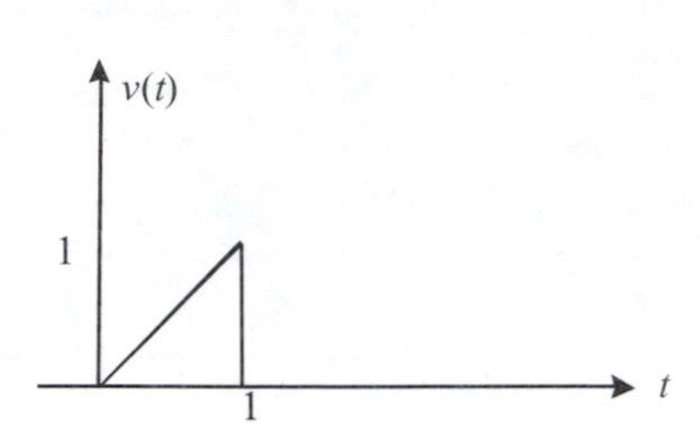

FIGURE 1.60d
Signal for EOCP 1.10.

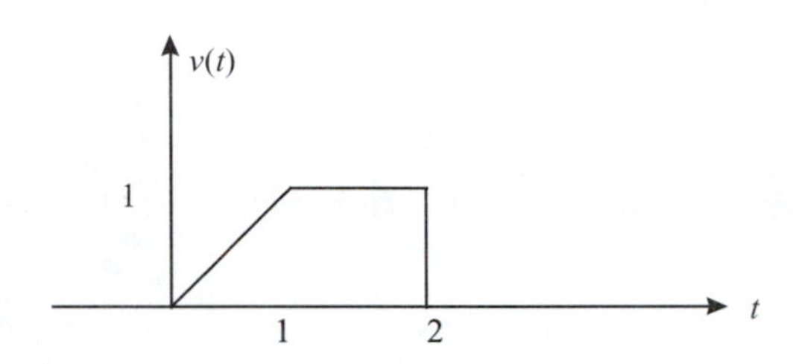

FIGURE 1.60e
Signal for EOCP 1.10.

EOCP 1.11

Consider that the signal $v(t) = 10\cos(100t - 100)$ volts represents the voltage across the terminals of a resistor and the current across the resistor is given by $i(t) = \cos(100t - 100)$ amps.

1. What is the instantaneous power consumed by the resistor?
2. Is the instantaneous power periodic?
3. If the instantaneous power signal is periodic, what is the period?

EOCP 1.12

An object of mass m is traveling a distance as given in Figure 1.61.

1. What is the energy in the signal that represents the distance?
2. What is the average power in the distance signal?
3. What is the energy in the force signal, F, acting on the object?
4. What is the average power in the force signal?
5. What is the velocity of the object? Sketch it.
6. What is the energy in the velocity signal?
7. What is the average power in the velocity signal?

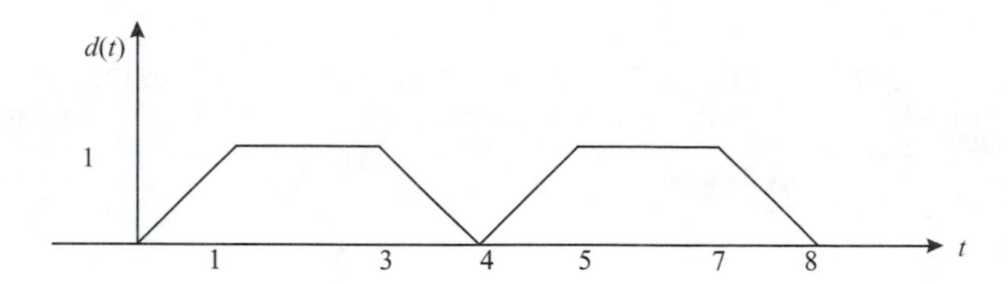

FIGURE 1.61
Signal for EOCP 1.12.

EOCP 1.13

Consider the circuit shown in Figure 1.62.

1. If $x(t)$ is an impulse of strength 10 that is applied at $t = 0$, what is the energy and average power in $x(t)$?
2. For $x(t)$ as given, what is the output $y(t)$ for $t > 0$?
3. What is the energy and the average power in $y(t)$?
4. What is the current in the resistor R?
5. What is the energy and the average power in the resistor's current signal?

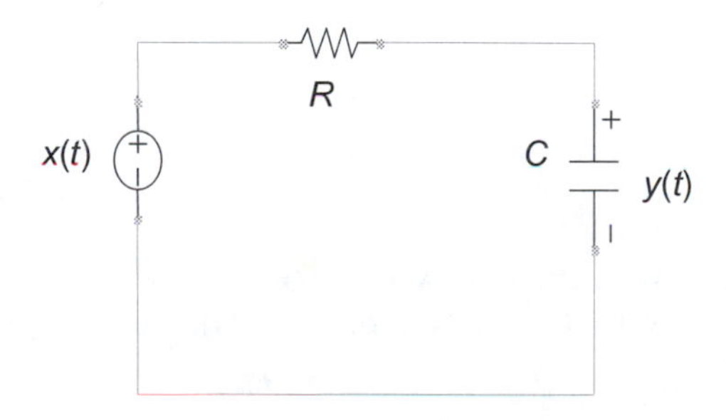

FIGURE 1.62
Circuit for EOCP 1.13.

EOCP 1.14

An object is hit by a hammer and starts to vibrate. If the hammer hit can be modeled as an impulse of unity strength and the output vibrations are modeled as

$$y(t) = e^{-t}\sin(t) \text{ for } t > 0$$

what is the energy and the average power in these vibrations?

EOCP 1.15

Consider the circuit shown in Figure 1.63.

1. If $x(t)$ is a unit step that is applied at $t = 0$, what is the energy and average power in $x(t)$?
2. For the given $x(t)$, what is the $y(t)$, the voltage in the capacitor if the inductance is 1 henry and the capacitance is 1 F?
3. What is the energy and the average power in $y(t)$?
4. If the capacitance is zero, what is the current in the inductor?
5. What is the energy and the average power in the inductor current?

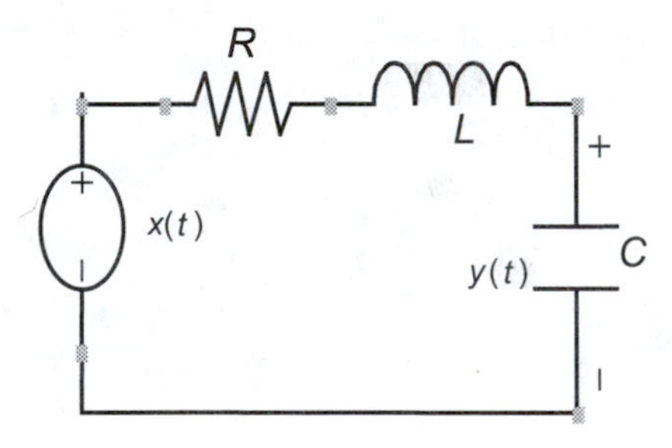

FIGURE 1.63
Circuit for EOCP 1.15.

EOCP 1.16

An object of mass M is pushed with a force F for a period of 10 seconds. Consider the two velocity signals over the same period of time as shown in Figure 1.64.

1. What is the force signal for each case? Sketch the force signals.
2. What is the power signal for each case? Sketch the power signals.
3. What is the energy signal for each case? Sketch the energy signals.

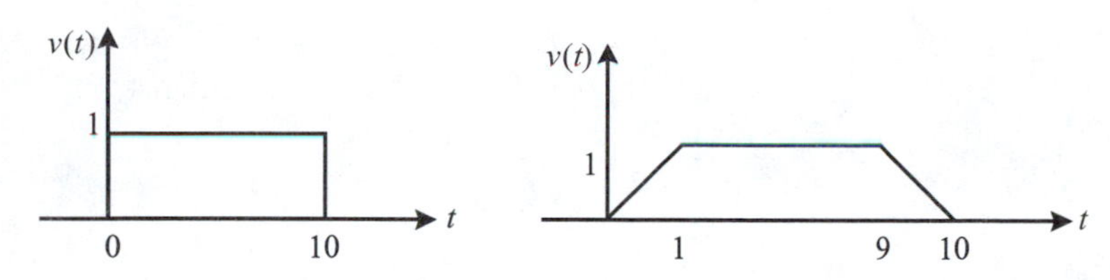

FIGURE 1.64
Signals for EOCP 1.16.

References

Denbigh, P. *System Analysis and Signal Processing,* Reading, MA: Addison-Wesley, 1998.

The MathWorks. *The Student Edition of MATLAB,* Englewood Cliffs, NJ: Prentice-Hall, 1997.

Phillips, C.L. and Parr, J.M. *Signals, Systems, and Transforms,* 2nd ed., Englewood Cliffs, NJ: Prentice-Hall, 1999.

Pratap, R. *Getting Started with MATLAB 5,* New York: Oxford University Press, 1999.

Strum, R.D. and Kirk, D.E. *Contemporary Linear Systems,* Boston: PWS, 1996.

Ziemer, R.E., Tranter, W.H., and Fannin, D.R. *Signals Systems Continuous and Discrete,* 4th ed., Englewood Cliffs, NJ: Prentice-Hall, 1998.

2

Continuous Systems

CONTENTS

2.1 Definition of a System

A system is an assemblage of things that are combined to form a complex whole. When we study systems we study the interactions and behavior of such an assemblage when subjected to certain conditions. These conditions are called inputs. In its simplest case a system can have one input and one output. The majority of this book will deal with linear systems. We will call the input $x(t)$ and the output $y(t)$ as depicted in Figure 2.1.

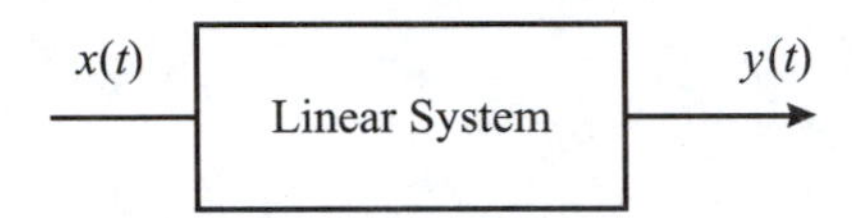

FIGURE 2.1
Linear system representation.

2.2 Input and Output

If an input signal $x(t)$ is available at the input of the linear system shown in Figure 2.1, the system will operate on the signal $x(t)$ and produce an output signal that we call $y(t)$. As an example, consider the case of an elevator where you push a button to go to the fifth floor. Pushing the button is the input $x(t)$. The elevator is the system under consideration here. The elevator system basically consists of the small room we ride in and the motor that drives the elevator belt. The input signal $x(t)$ "asks" the elevator to move to the fifth floor. The elevator system will process this request and move to the fifth floor. The motion of the elevator to the selected floor is called the output $y(t)$.

Pushing the button in this elevator case produces an electrical signal $x(t)$. This signal drives the motor of the elevator to produce a rotational motion which is transferred, via some gears, to a translational motion. This translational motion is the output $y(t)$. To summarize, when an electrical input signal or request $x(t)$ is applied to the elevator system, the elevator will operate on the signal and produce $y(t)$, which in this example can be thought of as a translational motion.

2.3 Linear Continuous System

A linear continuous system is one where if (a) the input is $\alpha x_1(t)$, the output is $\alpha y_1(t)$; and (b) the input is $x_1(t) + x_2(t)$, the output is $y_1(t) + y_2(t)$. By combining the two conditions, a system is considered linearly continuous if for the input $\alpha x_1(t) + \beta x_2(t)$, the output is $\alpha y_1(t) + \beta y_2(t)$, where α and β are constants.

Example 2.1

Consider the voltage divider circuit shown in Figure 2.2 where $x(t)$ is the input and $y(t)$ is the output voltage across the resistor R_2.

The output voltage $y(t)$ is

$$y(t) = x(t)R_2/(R_1 + R_2)$$

or

$$y(t) = [R_2/(R_1 + R_2)]x(t)$$

Is this system linear?

Solution

Using the definition of linearity introduced a while ago we can proceed as in the following. If the input is $x_1(t)$, the output is $y_1(t)$, and therefore

$$y_1(t) = [R_2/(R_1 + R_2)]x_1(t)$$

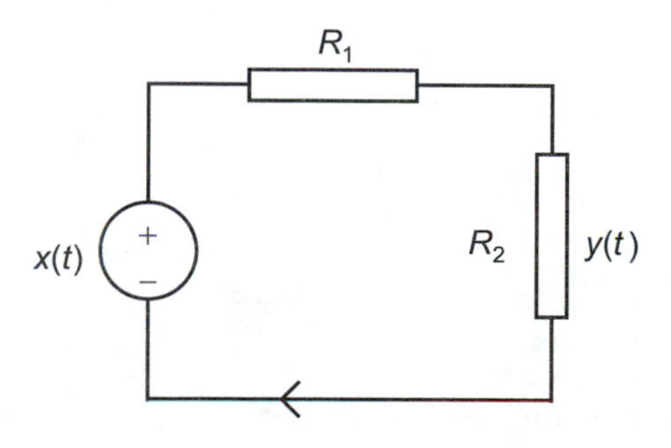

FIGURE 2.2
Circuit for Example 2.1.

If the input is $x_2(t)$, the output is $y_2(t)$, and therefore

$$y_2(t) = [R_2/(R_1 + R_2)]x_2(t)$$

If the input is $\alpha x_1(t) + \beta x_2(t)$, the output is $y(t)$, and therefore

$$y(t) = [R_2/(R_1 + R_2)](\alpha x_1(t) + \beta x_2(t))$$

or

$$y(t) = \alpha[R_2/(R_1 + R_2)]x_1(t) + \beta[R_2/(R_1 + R_2)]x_2(t)$$

which is

$$y(t) = \alpha y_1(t) + \beta y_2(t)$$

Therefore, the system is said to be linear.

Example 2.2

Consider the following system

$$y(t) = \sqrt{x(t)}$$

Is this system linear?

Solution

Consider two cases of the input signals, $x_1(t)$ and $x_2(t)$. The corresponding outputs are then given by

$$y_1(t) = \sqrt{x_1(t)}$$

for $x_1(t)$ and

$$y_2(t) = \sqrt{x_2(t)}$$

for $x_2(t)$. Now let us consider further that $\alpha x_1(t) + \beta x_2(t)$ has been applied to the system as its input. The corresponding output is then given by

$$y(t) = \sqrt{\alpha x_1(t) + \beta x_2(t)}$$

But

$$y(t) = \sqrt{\alpha x_1(t) + \beta x_2(t)} \neq \alpha y_1(t) + \beta y_2(t) = \alpha\sqrt{x_1(t)} + \beta\sqrt{x_2(t)}$$

Therefore, the system is nonlinear.

Example 2.3

Consider the following system

$$y(t) = (2/[2x(t) + 1])x(t)$$

Is this system linear?

Solution

As in Example 2.2, consider two cases of the input signals, $x_1(t)$ and $x_2(t)$. The corresponding outputs are then given by

$$y_1(t) = (2/[2x_1(t) + 1])x_1(t)$$

and

$$y_2(t) = (2/[2x_2(t) + 1])x_2(t)$$

Now let us apply $\alpha x_1(t) + \beta x_2(t)$ to the system as its input. The corresponding output is then given by

$$y(t) = (2/[2(\alpha x_1(t) + \beta x_2(t)) + 1])(\alpha x_1(t) + \beta x_2(t))$$

or

$$y(t) = [2\alpha x_1(t) + 2\beta x_2(t)]/[2\alpha x_1(t) + 2\beta x_2(t) + 1]$$

$$\neq \alpha y_1(t) + \beta y_2(t) = [2\alpha x_1(t)]/[2x_1(t) + 1] + [2\beta x_2(t)]/[2x_2(t) + 1]$$

Therefore, the system is nonlinear.

2.4 Time-Invariant System

A system is said to be time-invariant if for a shifted input $x(t-t_0)$, the output of the system is $y(t-t_0)$.

To see if a system is time-invariant or time-variant we do the following:

1. Find the output $y_1(t-t_0)$ that corresponds to the input $x_1(t)$.
2. Let $x_2(t) = x_1(t-t_0)$ and then find the corresponding output $y_2(t)$.
3. Check if $y_1(t-t_0) = y_2(t)$. If this is true then the system is time-invariant. Otherwise it is time-variant.

Example 2.4

Let

$$y(t) = \cos(x(t))$$

Find out if the system is time-variant or time-invariant.

Solution

According to the definition of time-invariant systems we write

$$y_1(t) = \cos(x_1(t))$$

$y_1(t)$ shifted by t_0 is

$$y_1(t-t_0) = \cos(x_1(t-t_0))$$

Also

$$y_2(t) = \cos(x_1(t-t_0))$$

Since

$$y_1(t-t_0) = y_2(t)$$

the system is time-invariant.

Example 2.5

Let

$$y(t) = x(t)\cos(t)$$

Find out if the system is time-variant or time-invariant.

Solution

Step One

$$y_1(t) = x_1(t)\cos(t)$$

Therefore,

$$y_1(t-t_0) = x_1(t-t_0)\cos(t-t_0)$$

Step Two

$$y_2(t) = x_1(t-t_0)\cos(t)$$

Step Three

$$y_1(t-t_0) \neq y_2(t)$$

Therefore, the system is time-variant.

Example 2.6

Let

$$y(t) = te^{-t}x(t)$$

Find out if the system is time-variant or time-invariant.

Solution

Step One

$$y_1(t) = te^{-t}x_1(t)$$

Therefore,

$$y_1(t-t_0) = (t-t_0)e^{-(t-t_0)}x_1(t-t_0)$$

Step Two

$$y_2(t) = te^{-t}x_1(t-t_0)$$

Step Three

$$y_1(t-t_0) \neq y_2(t)$$

Therefore, the system is time-variant.

2.5 Systems without Memory

Consider the system described by the following integral

$$y(t) = \int_{-\infty}^{t} x(\tau)\, d\tau$$

The value of $y(t)$, at a particular time t, depends on the time from $-\infty$ to t. In this case we say that the system has a memory. On the other hand, consider another system given by

$$y(t) = cx(t)$$

where c is a constant.

At any value of time, t, $y(t)$ depends totally on $x(t)$ at that point in time. In a case such as this we say the system is without a memory.

If the system is without memory we can write $y(t)$ as a function of $x(t)$ as

$$y(t) = F(x(t))$$

For linear systems

$$y(t) = c(t)x(t)$$

For time-invariant systems

$$y(t) = cx(t)$$

where c is a constant.

The system

$$y(t) = cx(t)$$

is linear, time-invariant, and has no memory.

2.6 Causal Systems

A causal system is a system where the output $y(t)$ at a certain time t_1, depends on the input $x(t)$ for $t < t_1$.

Example 2.7

Let $x(t)$ be an input as shown in Figure 2.3.

Let $y(t) = x(t+1)$. Is the system causal or non-causal? What happens to the system if $y(t) = x(t-1)$?

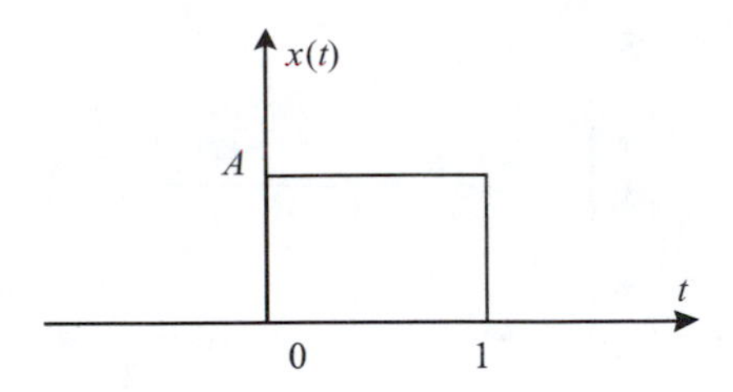

FIGURE 2.3
Signal for Example 2.7.

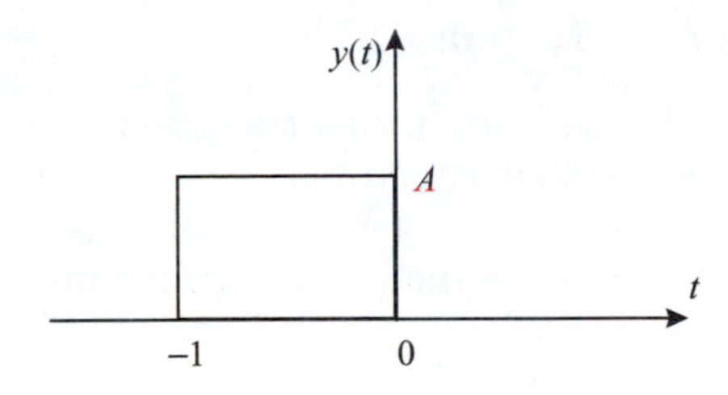

FIGURE 2.4
Signal for Example 2.7.

Solution

1. Let us look at the graph for the system

$$y(t) = x(t+1)$$

   as shown in Figure 2.4.
   You can observe that output $y(t)$ appears before the input has been introduced. Therefore, the given system is a non-causal system.
2. Let us look at the graph shown in Figure 2.5 for the system

$$y(t) = x(t-1)$$

   You can observe in this case that the output $y(t)$ appears only after the input has been introduced. Therefore, the given system is causal.

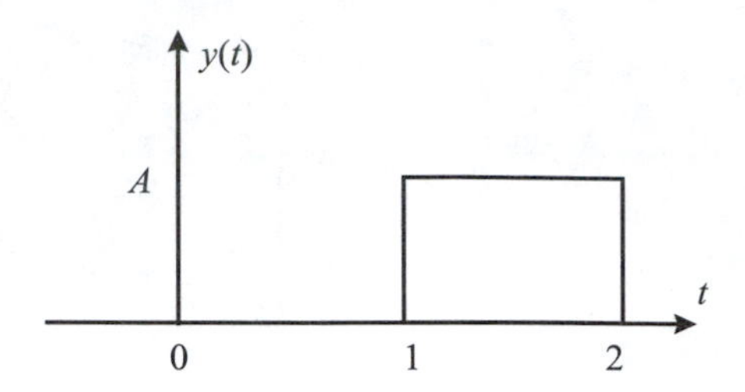

FIGURE 2.5
Signal for Example 2.7.

2.7 The Inverse of a System

If we can determine the input by measuring the output, then the system under consideration is said to be invertible.

Pictorially, an inverse system can be represented as shown in Figure 2.6. Notice that if two inputs give the same output then the system is not invertible.

Example 2.8

Consider the following systems.

1. $y(t) = x(t)$
2. $y(t) = 2x(t)$
3. $y(t) = a\cos(x(t))$
4. $y(t) = \int_{-\infty}^{t} x(\tau)\,d\tau$ such that $y(-\infty) = 0$

Are these systems invertible?

Solution

The first system is invertible. The corresponding pictorial representation is shown in Figure 2.7.

The second system is also invertible. Figure 2.8 shows the corresponding pictorial depiction.

The third system is not invertible. Why?

Let us consider two inputs, $x_1(t) = x(t)$ and $x_2(t) = x(t) + n\pi$, where n is an even integer.

For this system the output corresponding to $x_1(t) = x(t)$ is given by

$$y_1(t) = a\cos(x(t))$$

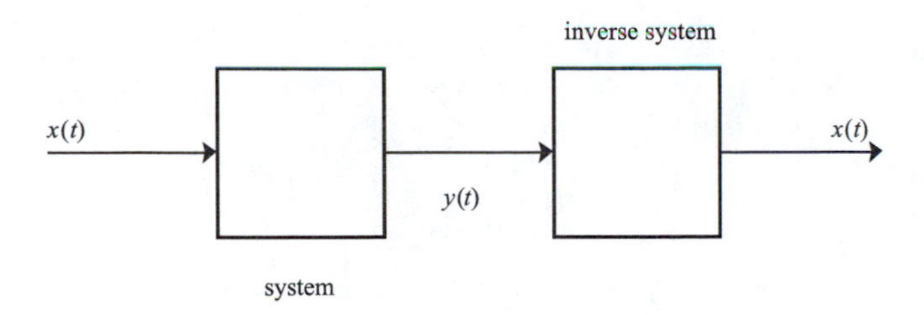

FIGURE 2.6
Inverse system.

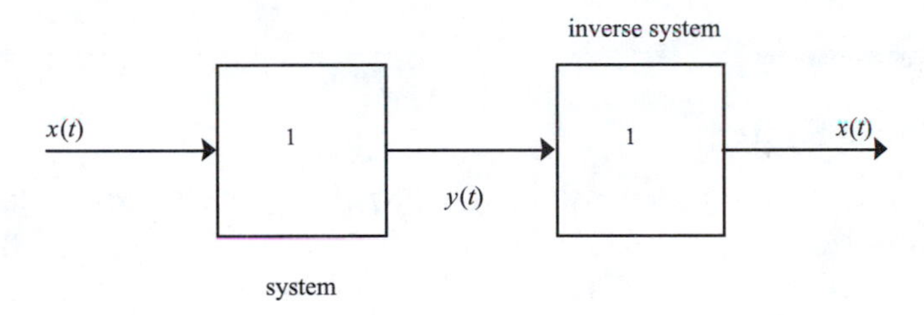

FIGURE 2.7
System for Example 2.8.

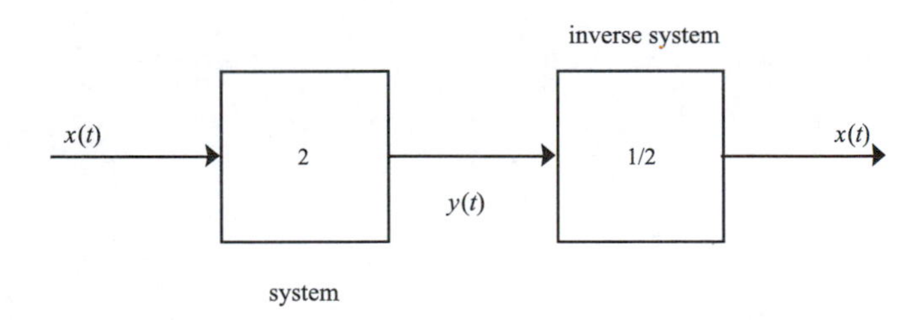

FIGURE 2.8
System for Example 2.8.

and the output corresponding to $x_2(t) = x(t) + n\pi$ is given by

$$y_2(t) = a \cos(x(t)+n\pi) = a \cos(x(t))$$

We can see here that two different inputs produced the same output. Therefore, the system is not invertible as claimed.

The fourth system is invertible. Notice that $\frac{d}{dt} y(t) = x(t)$ as shown in Figure 2.9.

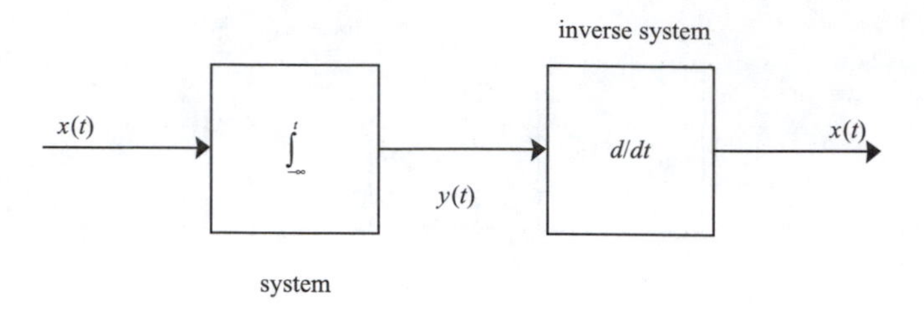

FIGURE 2.9
System for Example 2.8.

2.8 Stable Systems

The signal $x(t)$ is considered bounded if $|x(t)| < \beta < \infty$ for all t, where β is a real number. A system is said to be BIBO (bounded-input bounded-output) stable if and when the input is bounded the output is also bounded. The output, $y(t)$, is bounded if $|y(t)| < \beta < \infty$.

Example 2.9

Are the following systems bounded if it is known that the input $x(t)$ is bounded?

1. $y(t) = e^{x(t)}$
2. $y(t) = \int_{-\infty}^{t} x(\tau)\, d\tau$

Solution

1. Since $x(t)$ is bounded, then $|x(t)| < \beta < \infty$. Accordingly, the magnitude of the output is

$$|y(t)| = |e^{x(t)}| \leq |e^{\beta}| = e^{\beta} < \infty$$

Therefore, the system is BIBO stable.

2. Since

$$y(t) = \int_{-\infty}^{t} x(\tau) d\tau$$

The magnitude of the output is

$$|y(t)| = \left| \int_{-\infty}^{t} x(\tau)d\tau \right| \leq \int_{-\infty}^{t} |x(\tau)|d\tau \leq \beta \int_{-\infty}^{t} d\tau$$

Notice that

$$\int_{-\infty}^{t} d\tau = \tau\Big|_{-\infty}^{t} = t + \infty = \infty$$

Therefore, the system is not stable.

2.9 Convolution

As discussed earlier, by using the sifting property of the impulse signal we can write

$$\int_{t_1}^{t_2} x(\tau)\delta(\tau)\,dt = x(0) \quad \text{for } t_1 < 0 < t_2$$

Also

$$\int_{-\infty}^{+\infty} x(\tau)\delta(t-\tau)\,d\tau = x(t) \quad \text{for} -\infty < t < \infty$$

This last integral indicates that any signal can be represented as a sum of weighted impulses. For linear systems, if $x_1(t), x_2(t), x_3(t), x_4(t), \ldots, x_n(t)$ are the system inputs, the output of the system is obtained by superposition. It will be the sum of all outputs resulting from the individual inputs where each is presented alone. Mathematically we can write

$$y(t) = y_1(t) + y_2(t) + y_3(t) + y_4(t) + \cdots + y_n(t)$$

where $y(t)$ is the output due to the input $x(t)$.

The convolution integral is given by

$$y(t) = \int_{-\infty}^{+\infty} x(\tau)h(t,\tau)\,d\tau$$

where $h(t-\tau)$ is the output at time t for the input $\delta(t-\tau)$ applied at time $t=\tau$. If the system is time-invariant the convolution integral simplifies to

$$y(t) = \int_{-\infty}^{+\infty} x(\tau)h(t-\tau)\,d\tau \tag{2.1}$$

where $h(t)$ is the impulse response to the linear time-invariant system (LTI) and represents the output of the system at time t to a unit impulse signal when the system is relaxed (i.e., no initial conditions). As a matter of notation we write the equation above as

$$y(t) = x(t)^{*}h(t)$$

We read this equation as $y(t)$ equals $x(t)$ convolved with $h(t)$.

Example 2.10

What kinds of systems are the following?

1. $x(t)^{*}\delta(t)$
2. $x(t)^{*}u(t)$
3. $x(t)^{*}\frac{d}{dt}\delta(t)$

Solution

For the first system

$$x(t)^{*}\delta(t) = \int_{-\infty}^{+\infty} x(\tau)\delta(t-\tau)\,d\tau = x(t)$$

In the above equation $h(t)=\delta(t)$ and we say the system is a unity system. For the second system

$$x(t)^{*}u(t) = \int_{-\infty}^{+\infty} x(\tau)u(t-\tau)\,d\tau = \int_{-\infty}^{t} x(\tau)\,d\tau$$

In the above equation $h(t)=u(t)$ and the system is an integrator. For the third system

$$x(t)^{*}\frac{d}{dt}\delta(t) = \int_{-\infty}^{+\infty} x(\tau)\frac{d}{dt}\delta(t-\tau)d\tau = \frac{d}{dt}x(t)$$

In the above equation $h(t)=\frac{d}{dt}\delta(t)$ and the system is a differentiator.

2.10 Simple Block Diagrams

We have called $h(t)$ the impulse response of the system under consideration.

1. A block diagram representation for a system with $x(t)$ as the input, $y(t)$ as the output, and $h(t)$ as the impulse response can be represented as shown in Figure 2.10. Mathematically we write

$$y(t) = x(t)^{*}h(t)$$

2. If we have two systems connected in series as in Figure 2.11, then mathematically we write

$$z(t) = x(t)^{*}h_1(t)$$

and the output is

$$y(t) = z(t)^{*}h_2(t)$$

Therefore,

$$y(t) = [(x(t)^{*}h_1(t))^{*}h_2(t)]$$

This means that we convolve $x(t)$ with $h_1(t)$ to get $z(t)$ and then convolve $z(t)$ in turn with $h_2(t)$ to obtain $y(t)$.

3. If we have two systems connected in parallel as shown in Figure 2.12, then mathematically we write

$$z_1(t) = x(t)^{*}h_1(t)$$

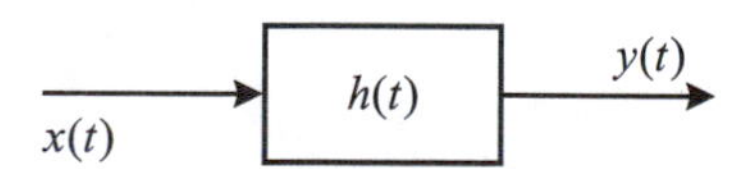

FIGURE 2.10
General block diagram representation.

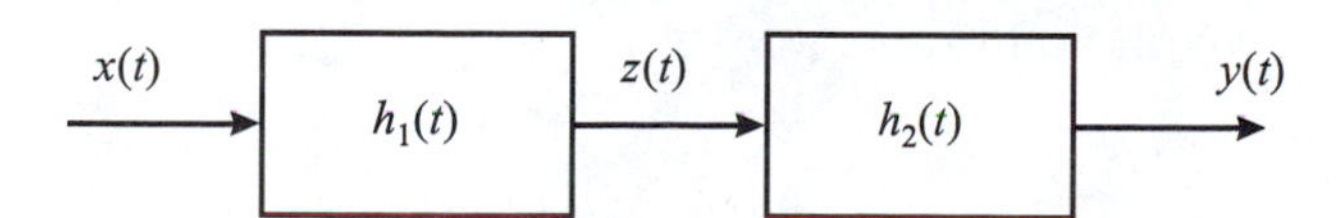

FIGURE 2.11
Series system.

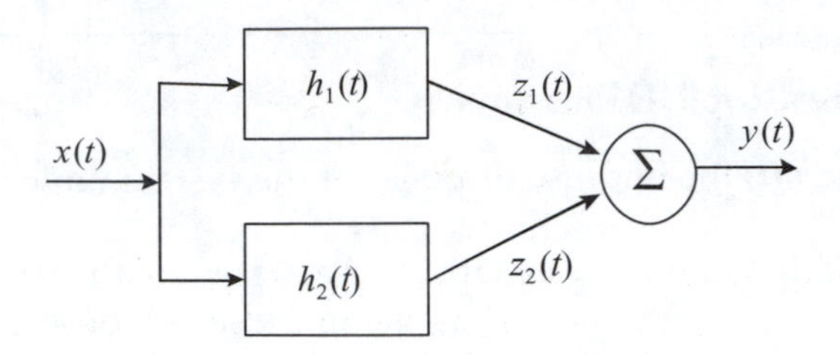

FIGURE 2.12
Parallel system.

and

$$z_2(t) = x(t)^* h_2(t)$$

The output $y(t)$ is

$$y(t) = z_1(t) + z_2(t)$$

Substituting for $z_1(t)$ and $z_2(t)$ we get

$$y(t) = [x(t)^* h_1(t)] + [x(t)^* h_2(t)]$$

Important properties:

1. Multiplication of the signal $x(t)$ by the shifted impulse signal

$$x(t)\delta(t-a) = x(a)\delta(t-a)$$

2. The sifting property

$$\int_{-\infty}^{\infty} x(t)\delta(t-a)\,dt = x(a)$$

3. The convolution property

$$x(t)^* \delta(t-a) = x(t-a)$$

Example 2.11

Let the impulse response for a linear time-invariant system be

$$h(t) = e^{-at}u(t)$$

where a is a positive constant. If the input to the system is $x(t) = u(t)$, find the output $y(t)$ using the convolution integral.

Solution

We can use the convolution integral to find the output $y(t)$ as

$$y(t) = \int_{-\infty}^{\infty} x(\tau)h(t-\tau)\,d\tau$$

Substituting for $x(t)$ and $h(t)$ we get

$$y(t) = \int_{-\infty}^{\infty} u(\tau)h(t-\tau)\,d\tau$$

which simplifies to

$$y(t) = \int_{-\infty}^{\infty} u(\tau)e^{-a(t-\tau)}u(t-\tau)\,d\tau$$

As we see in Figure 2.13

$$u(\tau)u(t-\tau) = 0 \qquad \text{for} \quad t < 0$$

and

$$u(\tau)u(t-\tau) = 1 \qquad \text{for} \quad t > 0$$

Therefore, the above convolution equation reduces to

$$y(t) = \int_0^t e^{-a(t-\tau)}\,d\tau = e^{-at}\int_0^t e^{a\tau}\,d\tau$$

Finally

$$y(t) = \frac{1}{a}[1 - e^{-at}] \qquad \text{for} \quad t > 0$$

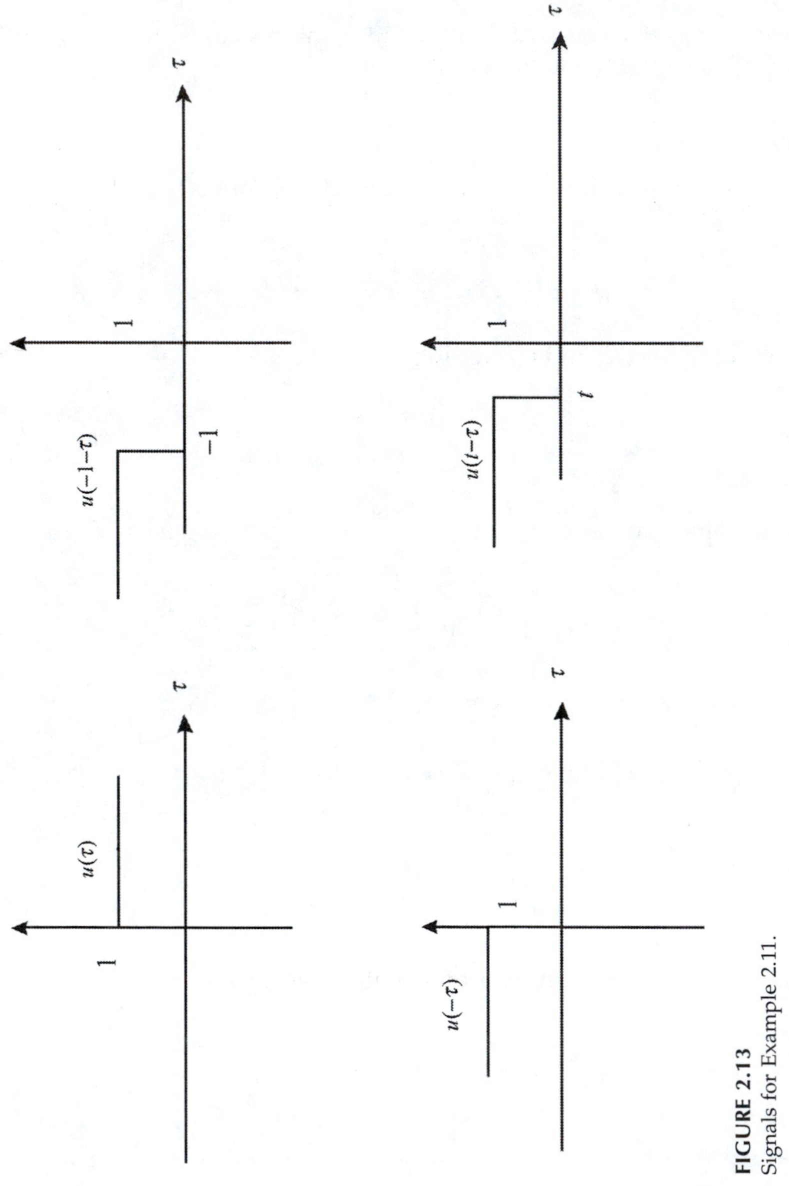

FIGURE 2.13
Signals for Example 2.11.

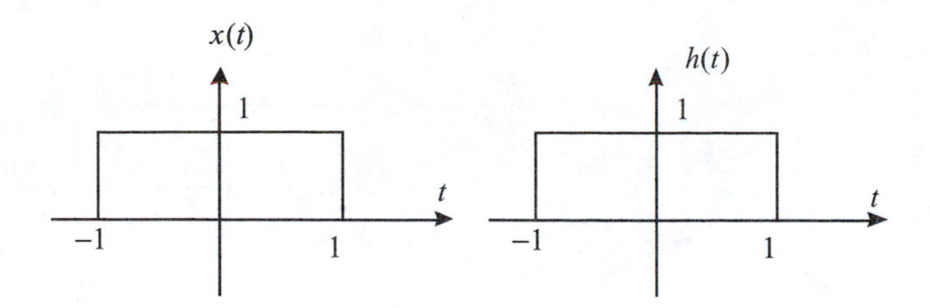

FIGURE 2.14
Signals for Example 2.12.

2.11 Graphical Convolution

Graphical convolution is best explained by an example.

Example 2.12

Let $x(t)$ and $h(t)$ be given as in Figure 2.14. Find $y(t)$ using graphical convolution.

Solution

Step One

Fix (but not slide) either $x(t)$ or $h(t)$ on the τ axis. Here we will fix $h(t)$.

Step Two

Sketch $h(\tau)$ by first writing the mathematical equation for $h(t)$ as

$$h(t) = \begin{cases} 1 & -1 < t < 1 \\ 0 & \text{otherwise} \end{cases}$$

and

$$h(\tau) = \begin{cases} 1 & -1 < \tau < 1 \\ 0 & \text{otherwise} \end{cases}$$

The sketch is given in Figure 2.15.

Step Three

We will sketch $x(t-\tau)$ by first writing its mathematical representation as

$$x(t-\tau) = \begin{cases} 1 & -1 < t-\tau < 1 \\ 0 & \text{otherwise} \end{cases}$$

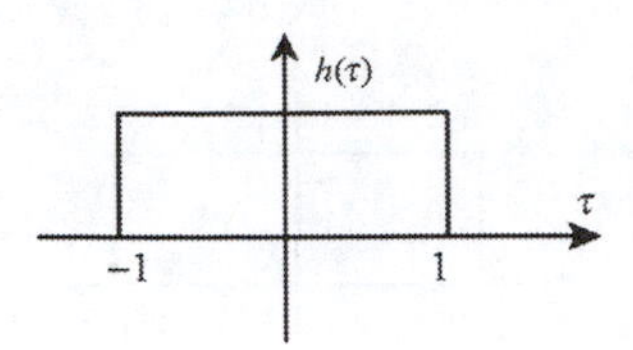

FIGURE 2.15
Signal for Example 2.12.

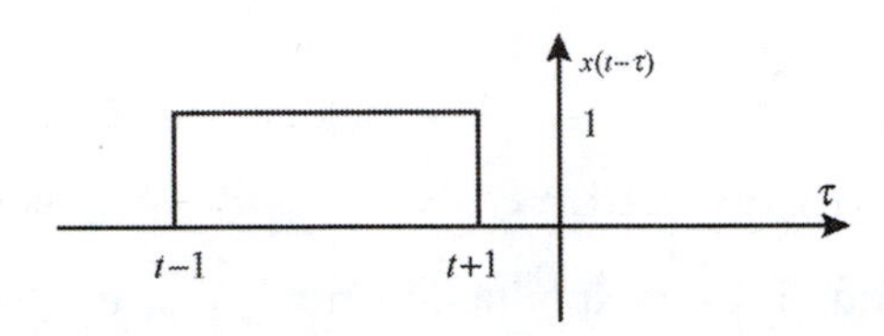

FIGURE 2.16
Signal for Example 2.12.

Fixing the time axis we get

$$x(t-\tau) = \begin{cases} 1 & -1-t < -\tau < -t+1 \\ 0 & \text{otherwise} \end{cases}$$

$$x(t-\tau) = \begin{cases} 1 & 1+t > \tau > t-1 \\ 0 & \text{otherwise} \end{cases}$$

and finally

$$x(t-\tau) = \begin{cases} 1 & t-1 < \tau < t+1 \\ 0 & \text{otherwise} \end{cases}$$

The sketch is given in Figure 2.16.

Step Four
Move $x(t-\tau)$ to the left of $h(\tau)$ first and then start moving it (sliding $x(t-\tau)$) to the right on the τ axis. This is shown in Figure 2.17.

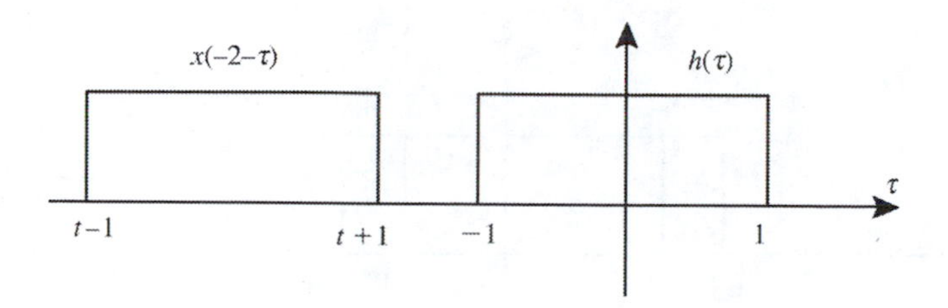

FIGURE 2.17
Signal for Example 2.12.

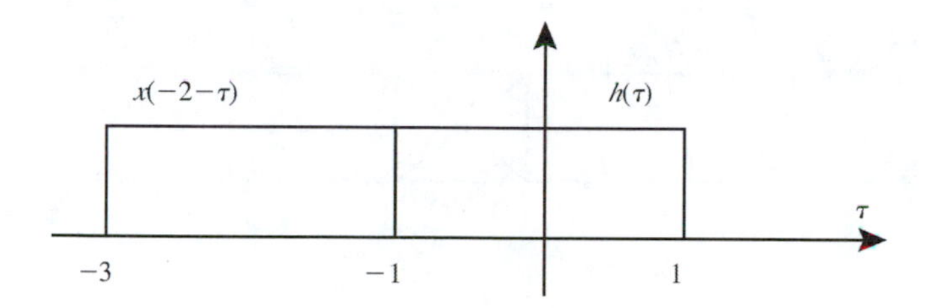

FIGURE 2.18
Signal for Example 2.12.

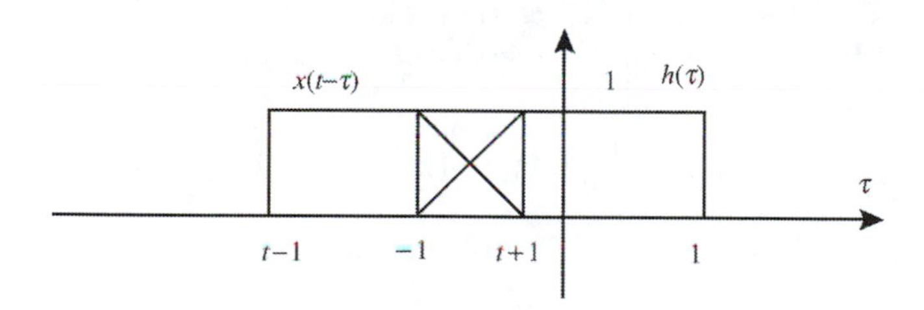

FIGURE 2.19
Signal for Example 2.12.

Now start moving $x(t - \tau)$ to the right to produce overlapping with $h(\tau)$ and determine the corresponding output.

1. When $t = -2$, we have Figure 2.18. The two functions are just overlapping and therefore

$$y(t) = 0 \qquad \text{for} \quad t < -2$$

2. When $-2 \leq t < -1$ we have the picture shown in Figure 2.19, where the X on the figure indicates the overlapping region.

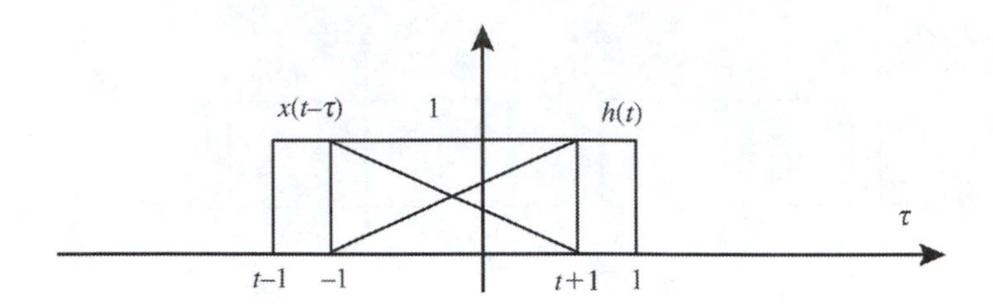

FIGURE 2.20
Signal for Example 2.12.

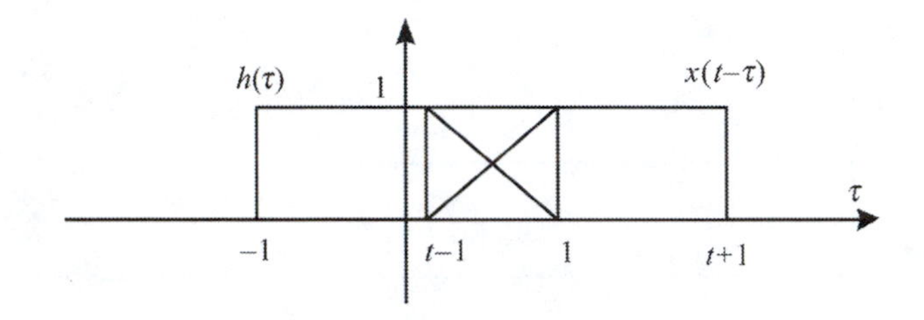

FIGURE 2.21
Signal for Example 2.12.

The limits for the corresponding convolution integral are the limits you see at the overlapping region and the magnitudes of the two functions are 1 and 1, respectively. Therefore

$$y(t) = \int_{-1}^{t+1} (1)(1)\, d\tau = (t+1) - (-1)$$

$$y(t) = t + 2 \qquad \text{for} \quad -2 \le t < -1$$

3. When $-1 \le t < 0$ we have the picture shown in Figure 2.20.

   The overlapping region is stretched now from -1 to $t + 1$. Therefore

$$y(t) = \int_{-1}^{t+1} (1)(1) d\tau = (t+1) - (-1)$$

$$y(t) = t+2 \qquad \text{for} \quad -1 \le t < 0$$

4. When $0 \le t < 1$ we have the picture shown in Figure 2.21.
   The overlapping extends now from $t - 1$ to 1. Therefore

$$y(t) = \int_{t-1}^{1} (1)(1)\, d\tau = (1) - (t-1)$$

$$y(t) = -t+2 \qquad \text{for} \quad 0 \le t < 1$$

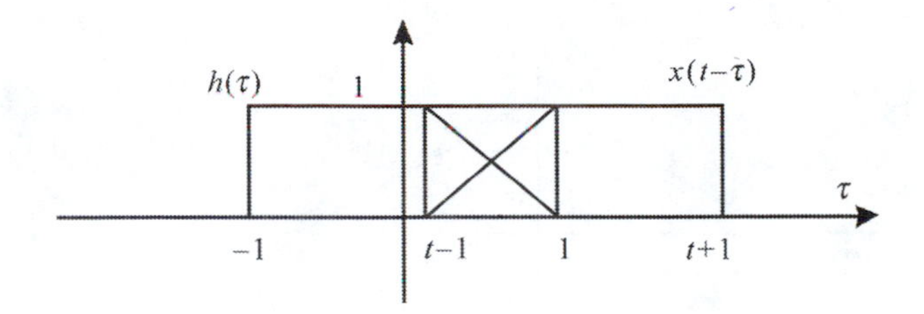

FIGURE 2.22
Signal for Example 2.12.

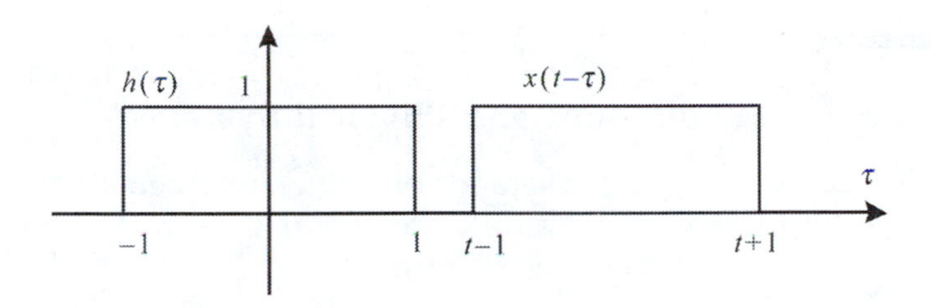

FIGURE 2.23
Signal for Example 2.12.

5. When $1 \le t < 2$ we have the picture shown in Figure 2.22, where in this case

$$y(t) = \int_{t-1}^{1}(1)(1)\,d\tau = (1) - (t-1)$$

$$y(t) = -t + 2 \qquad \text{for} \quad 1 \le t < 2$$

6. When $2 \le t < 3$ we have the picture shown in Figure 2.23.

   In this case, we see no overlap between the two graphs. Therefore

$$y(t) = 0 \qquad \text{for} \quad 2 \le t < 3$$

   Finally we can write the final results as

$$y(t) = \begin{cases} 0 & t < -2 \\ t + 2 & -2 \le t \le 0 \\ -t + 2 & 0 \le t \le 2 \\ 0 & t > 2 \end{cases}$$

   The output $y(t) = x(t)^{*}\ h(t)$ thus takes the graphical form such as that shown in Figure 2.24.

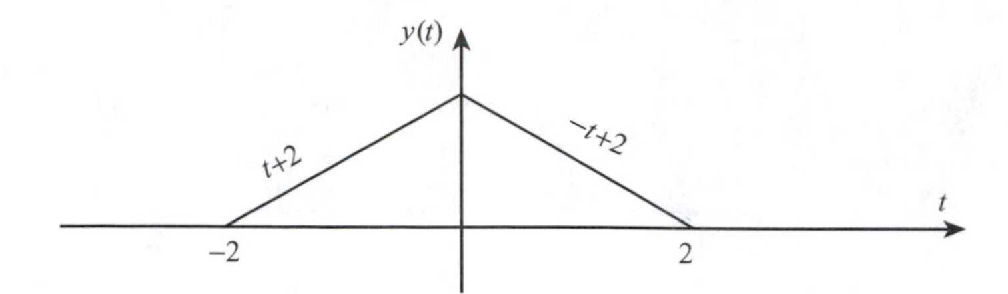

FIGURE 2.24
Signal for Example 2.12.

2.12 Differential Equations and Physical Systems

In most physical systems, $y(t)$ is a solution of a differential equation. Since we are dealing with LTI systems, our differential equations will be linear and have constant coefficients. The general form is

$$a_n\frac{d^n}{dt}y(t) + a_{n-1}\frac{d^{n-1}}{dt}y(t) + a_{n-2}\frac{d^{n-2}}{dt}y(t) + \cdots + a_0y(t) = b_m\frac{d^m}{dt}x(t)$$
$$+b_{m-1}\frac{d^{m-1}}{dt}x(t) + \cdots + b_0x(t) \tag{2.2}$$

where $x(t)$ is the input and $y(t)$ is the output of the system under consideration. In order to solve for the output $y(t)$, we must have all initial conditions needed. We will need n initial conditions where n is the order of the differential equation. These initial conditions are

$$\frac{d^{n-1}}{dt}y(0), \frac{d^{n-2}}{dt}y(0),\ldots, y(0) \tag{2.3}$$

2.13 Homogeneous Differential Equations and Their Solutions

If we set the left-hand side of Equation (2.2) to zero, we will have the homogeneous differential equation

$$a_n\frac{d^n}{dt}y(t) + a_{n-1}\frac{d^{n-1}}{dt}y(t) + a_{n-2}\frac{d^{n-2}}{dt}y(t) + \cdots + a_0y(t) = 0 \tag{2.4}$$

with the initial conditions

$$\frac{d^{n-1}}{dt}y(0), \frac{d^{n-2}}{dt}y(0),\ldots,y(0)$$

To solve the homogeneous differential equation we form the characteristic algebraic equation

$$a_n m^n + a_{n-1} m^{n-1} + \cdots + a_0 = 0 \tag{2.5}$$

The roots for this algebraic equation are called the poles or the eigenvalues of the system.

2.13.1 Case When the Roots Are All Distinct

If the roots are all distinct, the solution of the homogeneous equation is

$$y(t) = c_1 e^{m_1 t} + c_2 e^{m_2 t} + \cdots + c_n e^{m_n t} \tag{2.6}$$

where $m_1, m_2, \ldots, m_n$ are the n distinct roots of the characteristic equation describing the system. The constant c's are determined using the initial conditions which are usually given along with the description of the system.

2.13.2 Case When Two Roots Are Real and Equal

If we have a second order differential equation where m_1 and m_2 are real and equal, we are then tempted to write the solution as

$$y(t) = c_1 e^{m_1 t} + c_2 e^{m_2 t} = c_1 e^{m_1 t} + c_2 e^{m_1 t} = (c_1 + c_2) e^{m_1 t} = c e^{m_1 t} \tag{2.7}$$

Notice that this is a solution of a first order system with one energy storage element. In this case the solution is corrected and written as

$$y(t) = c_1 e^{m_1 t} + c_2 t e^{m_1 t}$$

where the two terms are independent. If the differential equation is of third order type, with m_3 as another root, then the solution is

$$y(t) = c_1 e^{m_1 t} + c_2 t e^{m_2 t} + c_3 e^{m_3 t}$$

2.13.3 Case When Two Roots Are Complex

If m_1 and m_2 are two complex roots of the characteristic equation, the two roots should appear in complex conjugate pair. In this case the roots are still distinct but complex. The solution is still of the form

$$y(t) = c_1 e^{m_1 t} + c_2 e^{m_2 t}$$

In general we can write m_1 and m_2 as $a+jb$ and $a-jb$. The solution becomes

$$y(t) = c_1 e^{(a+jb)t} + c_2 e^{(a-jb)t}$$

Using the fact that

$$e^{\pm jx} = \cos \pm j\sin x$$

and

$$e^{x+y} = e^x e^y$$

we can simplify the above equation as

$$y(t) = c_1 e^{(a+jb)t} + c_2 e^{(a-jb)t}$$

$$y(t) = c_1 e^{at}[\cos bt + j\sin bt] + c_2 e^{at}[\cos bt - j\sin bt]$$

Simplifying we get

$$y(t) = e^{at}[(c_1 + c_2)\cos bt + (c_1 - c_2)j\sin bt]$$

If we let $c_1 + c_2 = a_1$ and $(c_1 - c_2)j = a_2$ then

$$y(t) = e^{at}[a_1 \cos bt + a_2 \sin bt] \tag{2.8}$$

where a_1 and a_2 can be found using the given initial conditions.

2.14 Nonhomogeneous Differential Equations and Their Solutions

A differential equation is referred to as nonhomogeneous if it has an input (a forcing function) that appears to the right of the equal sign as in the following

$$a_n \frac{d^n}{dt} y(t) + a_{n-1} \frac{d^{n-1}}{dt} y(t) + a_{n-2} \frac{d^{n-2}}{dt} y(t) + \cdots + a_0 y(t) = x(t) \tag{2.9}$$

A solution of this equation has the form

$$y(t) = y_h(t) + y_p(t) \tag{2.10}$$

where $y_h(t)$ is the homogenous part of the solution and $y_p(t)$ is the particular part. The homogenous part has the form

$$y_h(t) = c_1 e^{m_1 t} + c_2 e^{m_2 t} + \cdots + c_n e^{m_n t}$$

In the homogenous solution we cannot use the initial conditions

$$\frac{d^{n-1}}{dt} y(0), \frac{d^{n-2}}{dt} y(0), \ldots, y(0)$$

to find the constants $c_1, \ldots, c_n$. These initial conditions should be applied to $y(t)$ not $y_h(t)$. After we find the combined solution

$$y(t) = y_h(t) + y_p(t)$$

we can use the initial condition to find the c's.

2.14.1 How Do We Find the Particular Solution?

Table 2.1 will help us in finding the form of the particular solution. The constants that correspond to the particular solution can be obtained by simply substituting $y_p(t)$ in the given nonhomogeneous equation and then equating coefficients. The constants corresponding to the homogeneous equation are determined using the initial conditions that come with the given nonhomogeneous differential equation after the two solutions, $y_h(t)$ and $y_p(t)$ with its constants have been determined, are added together to form the overall solution $y(t)$.

TABLE 2.1 Particular Solution Forms

Input	Particular Solution
constant A	constant B
Ae^{at}	Be^{at}
$A\cos(at)$ or $A\sin(at)$	$B_1\cos(at) + B_2\sin(at)$
At^n	$B_n t^n + B_{n-1}t^{n-1} + \cdots + B_0$

Example 2.13

Consider the following system described by the differential equation

$$\frac{d^2}{dt}y(t) + \frac{d}{dt}y(t) = 2t$$

with the initial conditions

$$y(0) = 0 \quad \text{and} \quad \frac{d}{dt}y(0) = 0$$

What is the solution $y(t)$ (the output)?

Solution

$y(t)$ will have two components: the homogeneous solution and the particular solution.

We find the homogeneous solution by first finding the characteristic equation as

$$m^2 + m = m(m+1) = 0$$

The two roots are real and distinct: $m = 0$ and $m = -1$. The corresponding homogeneous solution is therefore

$$y_h(t) = c_1 e^{-1t} + c_2 e^{0t} = c_1 e^{-t} + c_2$$

Here, the input $x(t)$ is $2t$. If we look at Table 2.1, we find out that the form of the particular solution is

$$y_p(t) = B_2 t + B_1$$

If we substitute this particular solution in the given differential equation we will get

$$0 + B_2 = 2t$$

By equating coefficients of $t(0t^0$ is compared with $B_2 t^0)$, B_2 is 0 and there is no way of calculating B_1. The terms in $y_p(t) + y_h(t)$ must be independent. Here we have the following combination as the total solution

$$y_h(t) + y_p(t) = c_1 e^{-t} + c_2 + B_2 t + B_1$$

c_2 and B_1 are dependents. In this situation we multiply B_1, the part of the particular solution that causes the dependency, by the minimum integer positive power of t to make the whole combination independent. This can be achieved by multiplying B_1 by t^2.

Therefore we now rewrite the particular solution as

$$y_p(t) = B_2 t + B_1 t^2$$

If we substitute in the differential equation given we will have

$$2B_1 + B_2 + 2B_1 t = 2t$$

By equating coefficients we have

$$B_1 = 1$$

and

$$2B_1 + B_2 = 0$$

which gives

$$B_2 = -2$$

The particular solution is then

$$y_p(t) = -2t + t^2$$

The overall solution is

$$y(t) = y_h(t) + y_p(t) = c_1 e^{-t} + c_2 + -2t + t^2$$

Using the initial conditions we can now find the constants as in the following.

$$y(0) = 0 = c_1 + c_2$$

$$\frac{d}{dt} y(t) = -c_1 e^{-t} - 2 + 2t$$

$$\frac{d}{dt} y(0) = -c_1 - 2 = 0$$

This indicates that $c_1 = -2$ and $c_2 = 2$. The final solution is

$$y(t) = y_h(t) + y_p(t) = -2e^{-t} + 2 - 2t + t^2$$

Example 2.14

Consider the following system described by the following differential equation

$$\frac{d^3}{dt}y(t) - y(t) = \sin t$$

with initial conditions

$$y(0) = 0, \frac{d}{dt}y(0) = 0, \text{ and } \frac{d^2}{dt}y(0) = 0$$

What is the output of the system?

Solution

The auxiliary equation is

$$m^3 - 1 = 0$$

with the roots $m_1 = 1$, $m_2 = -0.5 + 0.866j$, and $m_3 = -0.5 - 0.866j$.

The homogeneous solution is therefore

$$y_h(t) = c_1e^t + e^{-0.5t}[a_1\cos(0.866t) + a_2\sin(0.866t)]$$

To find the particular part of the solution, we look at the input to the system. The input is sin(t), and therefore, the form of the particular solution using Table 2.1 is

$$y_p(t) = B_1\cos t + B_2\sin t$$

Next we will find B_1 and B_2 by substituting the particular solution in the given differential equation to get

$$B_1\sin t - B_2\cos t - B_1\cos t - B_2\sin t = \sin t$$

By grouping like terms we arrive at

$$(B_1 - B_2)\sin t - (B_2 + B_1)\cos t = \sin t$$

and by equating coefficients we will have

$$B_1 - B_2 = 1$$

and

$$B_2 + B_1 = 0$$

where $B_1 = 1/2$ and $B_2 = -1/2$

The total solution now is

$$y(t) = c_1 e^t + e^{-0.5t}[a_1 \cos(0.866t) + a_2 \sin(0.866t)] + 1/2 \cos t - 1/2 \sin t$$

Using the initial conditions given we will get three algebraic equations in three unknowns as

$$c_1 + a_1 + 0a_2 = -0.5$$

$$c_1 - 0.5a_1 + 0.886a_2 = 0.5$$

$$c_1 - 0.535a_1 - 0.886 = 0.5$$

The solution to these equations is

$$c_1 = 0.159,\ a_1 = -0.659, \text{ and } a_2 = 0.013$$

The final solution is

$$y(t) = 0.159e^t + e^{-0.5t}[-0.659 \cos(0.866t) + 0.013 \sin(0.866t)] + 1/2 \cos t - 1/2 \sin t$$

2.15 The Stability of Linear Continuous Systems: The Characteristic Equation

If the coefficients of the characteristic equation that is derived from the differential equation have the same sign as the system the differential equation describes, then the system is stable. If there are sign changes in the coefficients of the characteristic equation, we can test stability by finding the roots of the characteristic equation analytically, or we can use the MATLAB function "roots" to find the roots of the characteristic equation. We can also find the solution to the given differential equation with zero input and nonzero initial conditions, and if the output settles at a constant value then we conclude that the system is stable. If the coefficients in the characteristic equation are not constants, then we can use the Routh test to check for stability. The Routh test can be used for any characteristic equation whether the coefficients are constants or not, but we will limit the use of this test for characteristic equations where some of the coefficients are not constant. In general consider the following characteristic equation

$$a_5m^5 + a_4m^4 + a_3m^3 + a_2m^2 + a_1m + a_0 = 0 \tag{2.11}$$

We form the Routh array as in the following:

m^5	a_5	a_3	a_1
m^4	a_4	a_2	a_0
m^3	$(a_4a_3 - a_5a_2)/a_4$ call this term x_1	$(a_4a_1 - a_5a_0)/a_4$ call this term x_2	0
m^2	$(x_1a_2 - a_4x_2)/x_1$ call this term x_3	$(x_1a_0 - a_4(0))/x_1 = a_0$	0
m^1	$(x_3x_2 - x_1a_0)/x_3$ call this term x_4	$(x_3(0) - x_1(0))/x_3 = 0$	0
m^0	$(x_4a_0 - x_3(0))/x_4 = a_0$	$(x_4(0) - x_3(0))/x_4 = 0$	0

Now we look at the first column of the Routh array and see that every term in this first column must be positive for stability. So for stability, a_5, a_4, x_1, x_3, x_4, and a_0 must all be positive.

Example 2.15

Consider the following systems.

1. $\frac{d^2}{dt}y + \frac{d}{dt}y + y = x(t)$
2. $\frac{d^2}{dt}y + \frac{d}{dt}y - y = x(t)$
3. $\frac{d^2}{dt}y + \frac{d}{dt}y + ky = x(t)$
4. $\frac{d^2}{dt}y + k\frac{d}{dt}y + y = x(t)$
5. $\frac{d^2}{dt}y + \frac{d}{dt}y + (1 + k)y = x(t)$
6. $\frac{d^2}{dt}y + k\frac{d}{dt}y + y = x(t)$

Are the above systems stable?

Solution

System 1

The characteristic equation is

$$m^2 + m + 1 = 0$$

We form the Routh array as

$$\begin{array}{cc} 1 & 1 \\ 1 & 0 \end{array}$$
$$(1(1) - 1(0))/1 = 1$$

The terms in the first column are all positive; therefore, the system is stable.

System 2

The characteristic equation is

$$m^2 + m - 1 = 0$$

We form the Routh array as

$$\begin{array}{cc} 1 & -1 \\ 1 & 0 \end{array}$$
$$(1(-1) - 1(0))/1 = -1$$

One term in the first column of the array is negative; therefore, the system is unstable.

System 3

The characteristic equation is

$$m^2 + m + k = 0$$

The Routh array is

$$\begin{array}{cc} 1 & k \\ 1 & 0 \end{array}$$
$$(1(k) - 1(0))/1 = k$$

All terms in the first column of the array are positive if k is positive. This is to say that the system is stable for all values of $k > 0$.

System 4

The characteristic equation is

$$m^2 + km + 1 = 0$$

We form the Routh array as

$$\begin{array}{cc} 1 & 1 \\ k & 0 \end{array}$$

$$(k(1) - 1(0))/k = 1$$

Here also k must be greater than zero for stability.

System 5

The characteristic equation is

$$m^2 + m + (1 + k) = 0$$

The Routh array is

$$\begin{array}{cc} 1 & 1 + k \\ 1 & 0 \end{array}$$

$$[1(1 + k) - 1(0)]/1 = 1 + k$$

For stability, $(k + 1)$ should be positive; therefore, the system is stable if $k > -1$.

System 6

The characteristic equation is

$$m^3 + km + 1 = 0$$

The Routh array is

$$\begin{array}{cc} 1 & k \\ 0 & 1 \end{array}$$

Before we continue and calculate the other terms in the Routh array, and since we have a leading zero in the first column, we substitute that zero with a very small positive number and we call it e. The Routh array becomes

$$\begin{array}{cc} 1 & k \\ e & 1 \end{array}$$

$$[e(k) - 1]/e = k - 1/e$$

For stability, $k - 1/e$ should be positive. This is equivalent to saying that $k > 1/e$. Notice that e is a very small positive number. So what is the range for k that must make this system stable? In theory we say $k > \infty$.

We will write this MATLAB script to examine this argument.

```
k = input('enter a value for k, a negative number
to exit:')
while k > 0
r = [1 0 k 1];
roots_of_system_6 = roots(r)
k = input(''enter a value for k, a negative number
to exit');
end
```

When we run this script the result is

```
enter a value for k, a negative number to exit: 1000
roots_of_system_6 =
0.0005 + 31.6228i
0.0005 - 31.6228i
-0.0010
enter a value for k, a negative number to exit: 10000
roots_of_system_6 =
1.0e + 002*
0.0000 + 1.0000i
0.0000 - 1.0000i
0.0000
enter a value for k, a negative number to exit: 1000000
roots_of_system_6 =
1.0e + 003*
0.0000 + 1.0000i
0.0000 - 1.0000i
0.0000
enter a value for k, a negative number to exit: 10000000000
roots_of_system_6 =
1.0e + 005*
0.0000 + 1.0000i
0.0000 - 1.0000i
0.0000
```

```
enter a value for k, a negative number to exit:
1000000000000000
roots_of_system_6 =
1.0e + 008*
0.0000 + 1.0000i
0.0000 - 1.0000i
0.0000
enter a value for k, a negative number to exit: -1
```

Notice that no matter how big k is, the real parts of the roots are either positive or zero, indicating instability.

2.16 Block Diagram Representation of Linear Systems

Block diagrams are representations of physical systems using blocks. Individual blocks can be put together to represent the physical system in block diagram form. Individual blocks are then considered.

2.16.1 Integrator

The integrator block diagram is shown in Figure 2.25.

The output $y(t)$ in this case is

$$y(t) = \int x(t)\,dt$$

2.16.2 Adder

The adder block diagram is shown in Figure 2.26.

The output $y(t)$ in this case is

$$y(t) = x_1(t) + x_2(t)$$

2.16.3 Subtractor

The subtractor block diagram is shown in Figure 2.27.

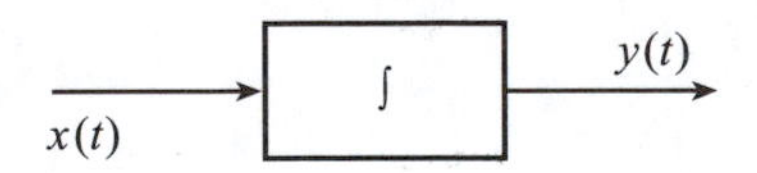

FIGURE 2.25
An integrator.

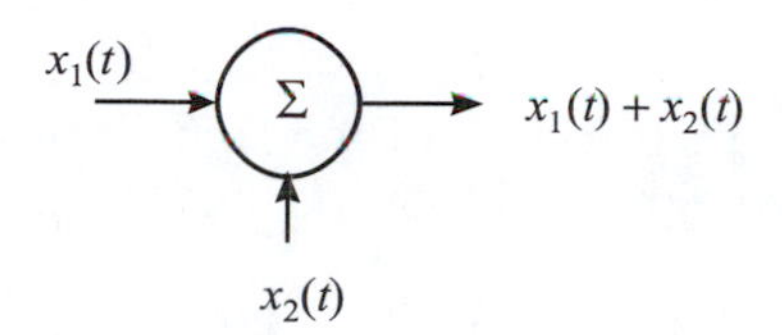

FIGURE 2.26
An adder.

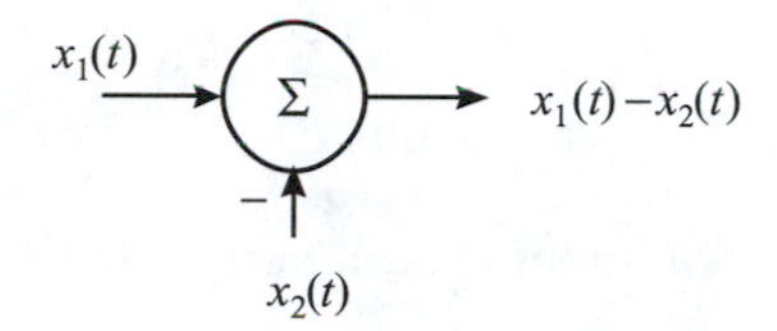

FIGURE 2.27
A subtractor.

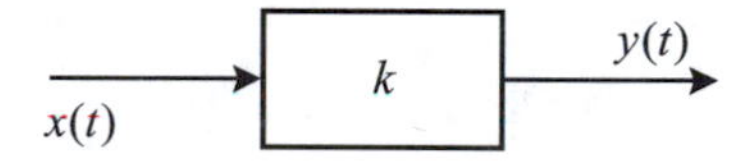

FIGURE 2.28
A multiplier.

The output is

$$y(t) = x_1(t) - x_2(t)$$

2.16.4 Multiplier

The multiplier block diagram is shown in Figure 2.28.

The output is

$$y(t) = k\,x(t)$$

2.17 From Block Diagrams to Differential Equations

This process is best understood by using examples.

Example 2.16

Consider the block diagram in Figure 2.29.

The arrows indicate the direction of the signal flow. What is the system described by this block diagram? Or what is the differential equation that this system represents?

Solution

Since $y(t)$ is the output of the integrator, the input to the integrator is $\frac{d}{dt}y(t)$.

The output of the summer is

$$\frac{d}{dt}y(t) = x(t) + ay(t)$$

Writing this first order differential equation in the standard form gives us the system differential equation

$$\frac{d}{dt}y(t) - ay(t) = x(t)$$

Example 2.17

Consider the block diagram in Figure 2.30.

What is the system described by this block diagram?

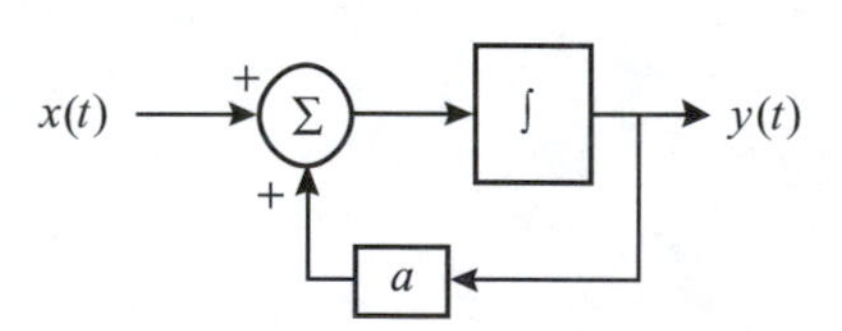

FIGURE 2.29
System for Example 2.16.

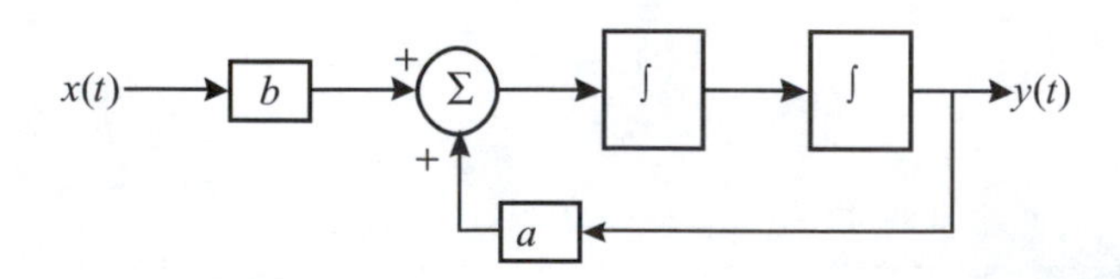

FIGURE 2.30
System for Example 2.17.

Solution

The input to the first integrator is the second derivative of $y(t)$. At the summer

$$\frac{d^2}{dt}y(t) = bx(t) + ay(t)$$

or

$$\frac{d^2}{dt}y(t) - ay(t) = bx(t)$$

which is the differential equation described by the block diagram given in Figure 2.30.

2.18 From Differential Equations to Block Diagrams

This method is also best understood by using examples.

Example 2.18

Consider the following differential equation.

$$\frac{d}{dt}y(t) - ay(t) = x(t).$$

What is the block diagram that represents the system or the differential equation?

Solution

Let D represent $\frac{d}{dt}$, D^2 represent $\frac{d^2}{dt}$ and so on.

Therefore, if we substitute in the differential equation given we get

$$Dy(t) - ay(t) = x(t)$$

Multiply by D^{-1}, where 1 is the highest power in the above differential equation, and the above equation becomes

$$y(t) - aD^{-1}y(t) = D^{-1}x(t)$$

We then solve for $y(t)$ to get

$$y(t) = D^{-1}[x(t) + ay(t)]$$

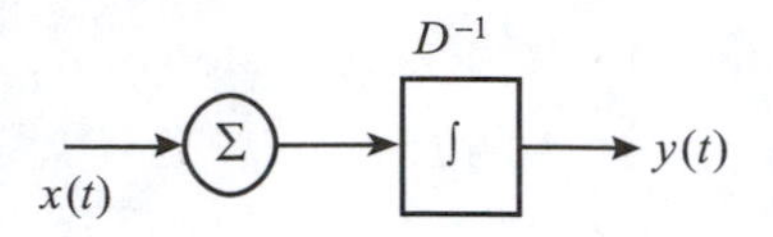

FIGURE 2.31
System for Example 2.18.

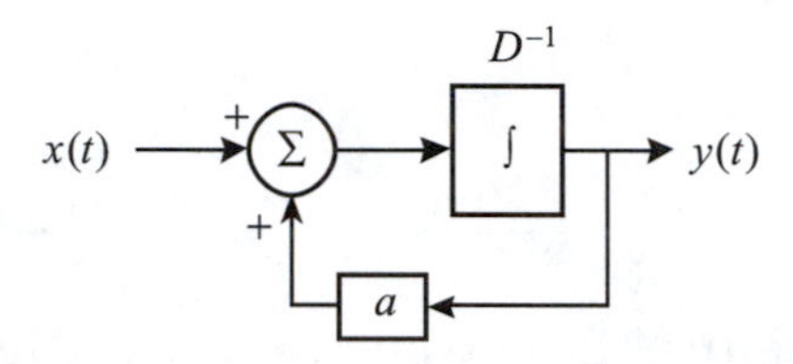

FIGURE 2.32
System for Example 2.18.

The differential equation we started with has only a first derivative and in the block diagram we will need one integrator, one input, $x(t)$, and one output, $y(t)$. We will also have a summer at each input of any integrator used in the block diagram and use that as a rule for constructing block diagrams.

We will take the following steps to draw the block diagram.

Step One

This step is represented graphically, according to the rule just explained, in Figure 2.31. The D on the integrator block in the figure is only to aid in understanding the procedure.

Step Two

The block in Figure 2.32 is a direct implementation of the equation

$$y(t) = D^{-1}[x(t) + ay(t)]$$

Notice that the terms $x(t)$ and $ay(t)$ are inputs to the summer that comes before the integrator.

Example 2.19

Consider the following differential equation

$$\frac{d^2}{dt}y(t) - ay(t) = bx(t)$$

What is the block diagram that represents this system?

Solution

Using the same procedure as explained before we can write the differential equation as

$$D^2y(t) - ay(t) = bx(t)$$

Multiply the above equation by D^{-2}, where 2 is the highest power in the equation above, to get

$$y(t) - aD^{-2}y(t) = bD^{-2}x(t)$$

We then solve for $y(t)$ to get

$$y(t) = D^{-2}[ay(t) + bx(t)]$$

Since we were given a second order differential equation, we will have two integrators in the block diagram. We will also have the input $x(t)$ and the output $y(t)$. Each integrator will have a summer at its input.

Step One

This step is accomplished by drawing the initial block in Figure 2.33.

Step Two

The terms $bx(t)$ and $ay(t)$ are associated with D^{-2} and are inputs to the summer which is the input of the first integrator from the left.

We do not see a D^{-1} term in the above equation and therefore the term associated with D^{-1} is zero. Zero means nothing is supplied as input to the second integrator other than the straight path from the first integrator. The final block representation is shown in Figure 2.34.

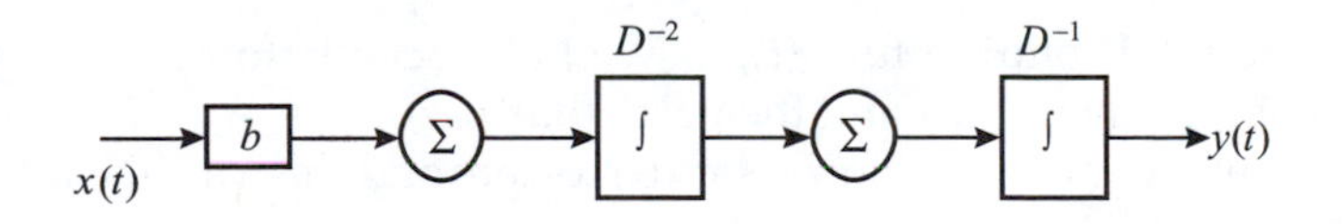

FIGURE 2.33
System for Example 2.19.

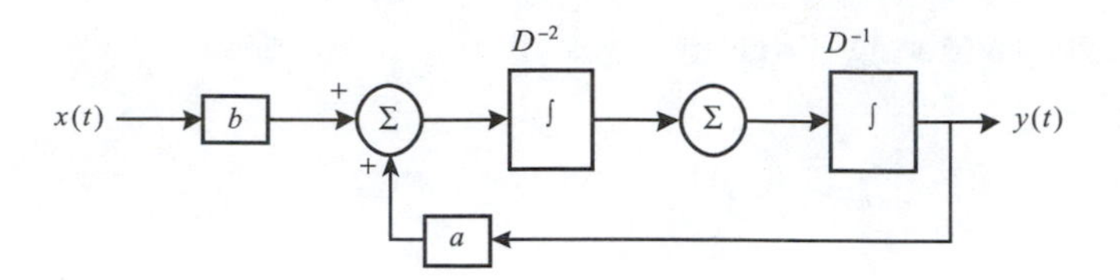

FIGURE 2.34
System for Example 2.19.

2.19 The Impulse Response

To find the response to an input $x(t)$, $y(t)$, the output of the system will have the following form

$$y(t) = \int_{-\infty}^{+\infty} x(\tau)h(t-\tau)\,d\tau$$

where $h(t)$ is the impulse response of the system when $x(t) = \delta(t)$. Therefore, we can say that if $x(t) = \delta(t)$, then $y(t) = h(t)$, the impulse response.

The following examples will show us how to find $h(t)$.

Example 2.20

Consider the first order differential equation

$$\frac{d}{dt}y(t) + 4y(t) = x(t)$$

Find the impulse response for this system.

Solution

To find the impulse response for the given system we set the input $x(t)$ equals to $\delta(t)$ and find $y(t)$ which is then $h(t)$, the impulse response. Therefore, the given equation representing the system in this case can be written as

$$\frac{d}{dt}h(t) + 4h(t) = \delta(t)$$

Again, the total solution for $h(t)$, as we have seen before, will have two parts: the homogenous part and the particular part.

For the homogenous solution we set $x(t)$ to zero and this kills the $\delta(t)$ signal. Therefore, we have

$$\frac{d}{dt}h(t) + 4h(t) = 0$$

The auxiliary equation is

$$m + 4 = 0$$

which indicates that $m = -4$ and the homogenous solution is

$$h_h(t) = c_1 e^{-4t} \quad t > 0$$

To find c_1 we need the total solution for $h(t)$ first.

For the particular solution we predict that the solution has the form

$$h_p(t) = 0$$

because this is always the case in most real systems where the order of the derivative in the output, $y(t)$, is greater than the order of the derivative in the input, $x(t)$.

Therefore, the total solution is

$$h(t) = h_h(t) + h_p(t)$$

$$h(t) = c_1 e^{-4t} \quad t > 0$$

or we can write instead

$$h(t) = c_1 e^{-4t} u(t)$$

where $u(t)$ is the unit step signal that when multiplied by other signal will force the other signal to start at $t = 0$ without any other alteration in the other signal.

Finally we have to find c_1. To do so we substitute the total solution for $h(t)$ in the original differential equation noting that the derivative of $u(t)$ is $\delta(t)$. The derivative of $h(t)$ is

$$\frac{d}{dt}h(t) = c_1(-4)e^{-4t}u(t) + c_1 e^{-4t}\delta(t)$$

where we used the product rule for derivatives.

Therefore, substituting back in the differential equation gives

$$[c_1(-4)e^{-4t}u(t) + c_1 e^{-4t}\delta(t)] + 4[c_1 e^{-4t}u(t)] = \delta(t)$$

Remember that the $\delta(t)$ signal is only valid at $t = 0$ and in this case any function multiplied by $\delta(t)$ is evaluated at $t = 0$. Therefore

$$[c_1(-4)e^{-4t}u(t) + c_1\delta(t)] + 4[c_1 e^{-4t}u(t)] = \delta(t)$$

The above equation reduces to

$$c_1\delta(t) = \delta(t)$$

Equating coefficients we find that $c_1 = 1$, and the final solution is

$$h(t) = y(t) = e^{-4t} \quad t > 0$$

2.20 Some Insights: Calculating $y(t)$

If we study carefully the solution of the system in Example 2.20, we see that the output, $y(t)$, is

$$y(t) = e^{-4t} \quad t > 0$$

As time approaches infinity, the output will approach the value zero. In this sense we say the output is stable for our particular input. For first order systems (that is described by first order differential equations) the output will have one exponential term at the most. This exponential term is called the transient of the system. For second order systems, the output will have two transients (two exponential terms) at the most. For third order systems we will have three exponential terms, and so on.

In many systems of order greater than two, and for the purpose of analysis and design, we can reduce the order of the system at hand to a second order system due to the fast decay of some transients. The solution for the output for these systems is in the following form

$$y(t) = c_1 e^{\alpha_1 t} + c_2 e^{\alpha_2 t}$$

The stability of the system is determined by the values of α_1 and α_2. If any of the α's is positive, the output $y(t)$ will grow wild as t approaches infinity. So if the α's are all negative then the output $y(t)$ will decay gradually and stays at a fixed value as time progresses. The α's are called the eigenvalues of the system. Therefore, we can say that a linear time-invariant system is stable if the eigenvalues of the system are all negative.

2.20.1 How Can We Find These Eigenvalues?

A linear time-invariant system can always be represented by a linear differential equation with constant coefficients

$$a_n \frac{d^n}{dt} y(t) + a_{n-1} \frac{d^{n-1}}{dt} y(t) + a_{n-2} \frac{d^{n-2}}{dt} y(t) + \cdots + a_0 y(t) = b_m \frac{d^m}{dt} x(t)$$

$$+ b_{m-1} \frac{d^{m-(1)}}{dt} x(t) + \cdots + b_0 x(t)$$

We can look at the auxiliary algebraic equation by setting the input, $x(t)$, to zero

$$a_n \frac{d^n}{dt} y(t) + a_{n-1} \frac{d^{n-1}}{dt} y(t) + \frac{d^{n-2}}{dt} y(t) + \cdots + a_0 y(t) = 0$$

and letting $\frac{d}{dt} = m$ to get

$$a_n m^n y(t) + a_{n-1} m^{n-1} y(t) + \cdots + a_0 y(t) = 0$$

We can factor out $y(t)$ as

$$y(t)[a_n m^n + a_{n-1} m^{n-1} + \cdots + a_0] = 0$$

$y(t)$ cannot be zero (in which case the output of the system would be zero at all times) and, therefore,

$$[a_n m^n + a_{n-1} m^{n-1} + \cdots + a_0] = 0$$

or

$$\left[m^n + \frac{a_{n-1}}{a_n} m^{n-1} + \cdots + \frac{a_0}{a_n}\right] = 0$$

This is an nth order algebraic equation with n roots. Let us call them the n α's. These are the eigenvalues of the system.

2.20.2 Stability and Eigenvalues

To summarize, any linear time-invariant system can be modeled by a linear differential equation with constant coefficients. The auxiliary algebraic equation that can be obtained from the differential equation will have a number of roots called the eigenvalues of the system. The stability of the system is determined by these roots. These roots may be real or complex. If **all** the real parts of the roots are negative then the system is stable. If **any** of the real parts of the roots is positive the system is unstable. The eigenvalues of the system are responsible for the shape of the output, $y(t)$. They dictate the shape of the transients of the system as well.

2.21 End-of-Chapter Examples

EOCE 2.1

Consider the following systems.

1. $y(t) = \frac{1}{x}$
2. $y(t) = t\sin(x(t))$
3. $y(t) = t\ln(x(t))$

Are the above systems linear?

Solution

For the first system

$$\begin{aligned} y_1(t) &= \frac{1}{x_1(t)} \\ y_2(t) &= \frac{1}{x_2(t)} \\ y(t) &= \frac{1}{\alpha x_1(t) + \beta x_2(t)} \end{aligned}$$

Is $y(t) = \alpha y_1(t) + \beta y_2(t)$?

$$\begin{aligned} \alpha y_1(t) &= \frac{\alpha}{x_1(t)} \\ \beta y_2(t) &= \frac{\beta}{x_2(t)} \\ y(t) &= \frac{1}{\alpha x_1(t) + \beta x_2(t)} \neq \frac{\alpha}{x_1(t)} + \frac{\beta}{x_2(t)} \end{aligned}$$

Therefore, the system is not linear.
For the second system

$$\begin{aligned} y_1(t) &= t\sin(x_1(t)) \\ y_2(t) &= t\sin(x_2(t)) \\ y(t) &= t\sin(\alpha x_1(t) + \beta x_2(t)) \end{aligned}$$

Is $y(t) = \alpha y_1(t) + \beta y_2(t)$?

$$\alpha y_1(t) = \alpha t \sin(x_1(t))$$

$$\beta y_2(t) = \beta t \sin(x_2(t))$$

$$y(t) = t \sin(\alpha x_1(t) + \beta x_2(t)) \neq \alpha t \sin(x_1(t)) + \beta t \sin(x_2(t))$$

Therefore, the system is not linear.
For the third system

$$\begin{aligned} y_1(t) &= t \ln(x_1(t)) \\ y_2(t) &= t \ln(x_2(t)) \\ y(t) &= t \ln(\alpha x_1(t) + \beta x_2(t)) \end{aligned}$$

Is $y(t) = \alpha y_1(t) + \beta y_2(t)$?

$$\begin{aligned} \alpha y_1(t) &= \alpha t \ln(x_1(t)) \\ \beta y_2(t) &= \beta t \ln(x_2(t)) \\ y(t) &= t \ln(\alpha x_1(t) + \beta x_2(t)) \neq \alpha t \ln(x_1(t)) + \beta t \ln(x_2(t)) \end{aligned}$$

Therefore, the system is not linear.

EOCE 2.2

Consider EOCE 2.1 again. Are the systems time-invariant?

Solution

For the first system

$$y_1(t) = \frac{1}{x_1(t)}$$

$$y_1(t - t_0) = \frac{1}{x_1(t - t_0)}$$

$$y_2(t) = \frac{1}{x_1(t - t_0)}$$

$$y_1(t - t_0) = \frac{1}{x_1(t - t_0)} = y_2(t) = \frac{1}{x_1(t - t_0)}$$

Therefore, the system is time-invariant.
For the second system

$$
\begin{aligned}
y_1(t) &= t\sin(x_1(t)) \\
y_1(t-t_0) &= (t-t_0)\sin(x_1(t-t_0)) \\
y_2(t) &= t\sin(x_1(t-t_0)) \\
y_1(t-t_0) &= (t-t_0)\sin(x_1(t-t_0)) \neq y_2(t) = t\sin(x_1(t-t_0))
\end{aligned}
$$

Therefore, the system is not time-invariant.
For the third system

$$y_1(t) = t\ln(x_1(t))$$

$$y_1(t-t_0) = (t-t_0)\ln(x_1(t-t_0))$$

$$y_2(t) = t\ln(x_1(t-t_0))$$

$$y_1(t-t_0) = (t-t_0)\ln(x_1(t-t_0)) \neq y_2(t) = t\ln(x_1(t-t_0))$$

Therefore, the system is not time-invariant.

EOCE 2.3

Consider the systems represented in block diagrams in Figure 2.35.
Find the output, $y(t)$, using convolution for both systems.

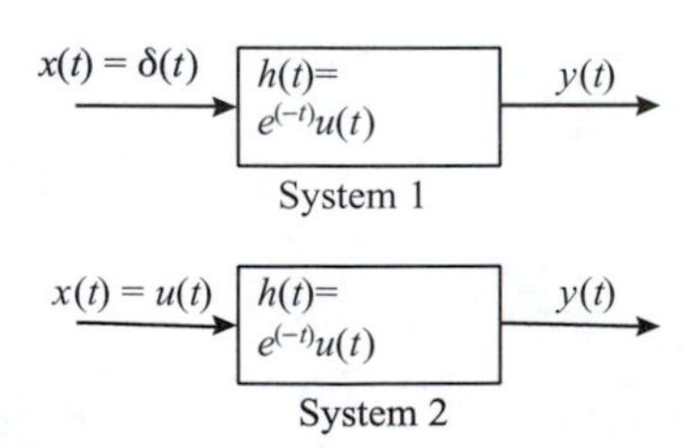

FIGURE 2.35
Systems for EOCE 2.3.

Solution

System 1

We can use the convolution integral to find the output $y(t)$ as

$$y(t) = \int_{-\infty}^{+\infty} x(\tau)h(t-\tau)\,d\tau = \int_{-\infty}^{+\infty} h(\tau)x(t-\tau)\,d\tau$$

$$y(t) = \int_{-\infty}^{+\infty} e^{-\tau}u(\tau)\delta(t-\tau)\,d\tau$$

Using the sifting property

$$\int_{-\infty}^{+\infty} x(\tau)\delta(t-\tau)\,d\tau = x(t)$$

and noticing that the impulse is applied at $\tau = t$, the output becomes

$$y(t) = e^{-t}u(t)$$

System 2

In this case we also use the convolution integral and write

$$y(t) = \int_{-\infty}^{+\infty} h(\tau)x(t-\tau)\,d\tau = \int_{-\infty}^{+\infty} e^{-\tau}u(\tau)u(t-\tau)\,d\tau$$

Notice that $u(t-\tau)$ is a unit step function that is reflected and shifted by t units. Also $h(\tau)$ is stationary with $u(t-\tau)$ moving to the right. Overlapping will start at $\tau = 0$ and stop at some $\tau = t$.

Remember also that $u(t-\tau)$ and $u(\tau)$ are multiplied under the integrand. Therefore,

$$y(t) = \int_{0}^{t} e^{-\tau}d\tau = [e^{-\tau}]_0^t$$

$$y(t) = -[e^{-t} - 1] = [1 - e^{-t}] \;\; t \geq 0$$

EOCE 2.4

Consider the block diagram in Figure 2.36.

Find the output $y(t)$.

Solution

This is the case of a parallel system and the output $y(t)$ is calculated as

$$y(t) = z_1(t) + z_2(t)$$

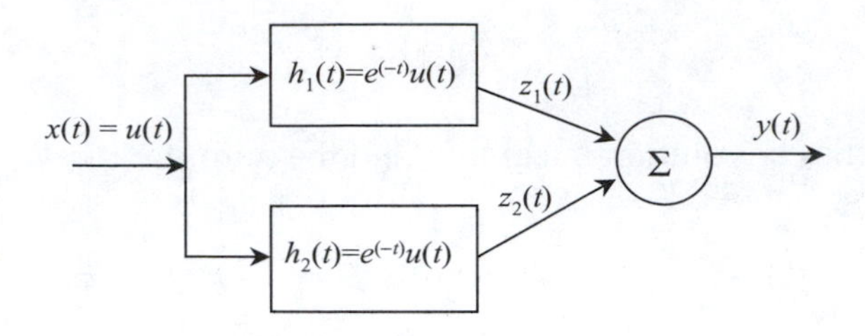

FIGURE 2.36
System for EOCE 2.4.

with

$$z_1(t) = x(t)*h_1(t)$$

and

$$z_2(t) = x(t)*h_2(t)$$

Using the convolution integral we write for $z_1(t)$

$$z_1(t) = \int_{-\infty}^{+\infty} h_1(\tau)x(t-\tau)\,d\tau = \int_{-\infty}^{+\infty} e^{-\tau}u(\tau)u(t-\tau)\,d\tau$$

$$z_1(t) = \int_0^t e^{-\tau}\,d\tau = -[e^{-\tau}]_0^t$$

and finally

$$z_1(t) = [1 - e^{-t}]\, t \geq 0$$

For $z_2(t)$ we write

$$z_2(t) = \int_{-\infty}^{+\infty} h_2(\tau)x(t-\tau)\,d\tau = \int_{-\infty}^{+\infty} e^{-\tau}u(\tau)u(t-\tau)\,d\tau$$

$$z_2(t) = \int_0^t e^{-\tau}\,d\tau = -[e^{-\tau}]_0^t$$

and

$$z_2(t) = [1 - e^{-t}]\, t \geq 0$$

The output is then

$$y(t) = z_1(t) + z_2(t)$$

$$y(t) = [1 - e^{-t}] + [1 - e^{-t}] = [2 - 2e^{-t}]\, t \geq 0$$

Using MATLAB we can plot $h_1(t)$, $h_2(t)$, $z_1(t)$, $z_2(t)$ and $y(t)$ for the impulse input $x(t)$ using the script

```
t = 0:0.05:10;  %establishing a range for time
h1 = exp(-t);
h2 = exp(-t);
z1 = 1-exp(-t);
z2 = 1-exp(-t);
y = 2-2*exp(-t);
subplot(5,1,1), plot(t,h1), grid
ylabel('h1(t)'),
%a plot of five rows with grids
subplot(5,1,2), plot(t,h2), grid
ylabel('h2(t)'),
subplot(5,1,3), plot(t,z1), grid
ylabel('z1(t)'),
subplot(5,1,4), plot(t,z2), grid
ylabel('z2(t)'),
subplot(5,1,5), plot(t,y), grid
xlabel('Time (sec)'), ylabel('y(t)');
```

The plots are shown in Figure 2.37.

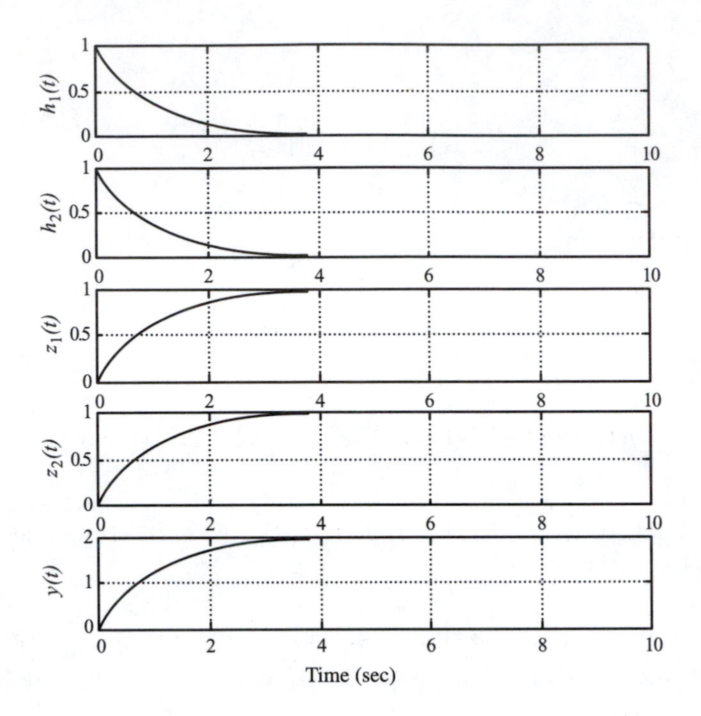

FIGURE 2.37
Plots for EOCE 2.4.

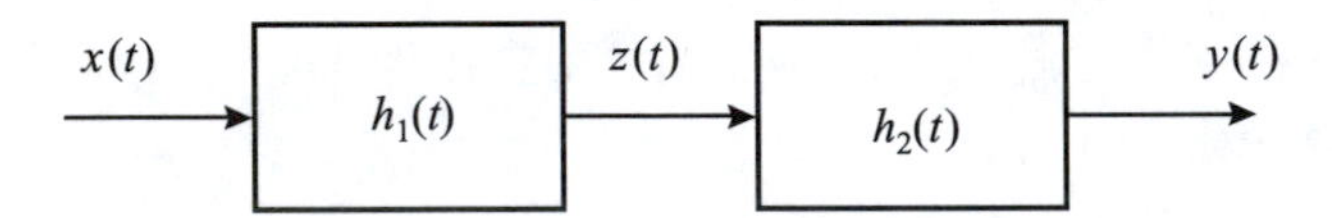

FIGURE 2.38
System for EOCE 2.5.

EOCE 2.5

Consider the block diagram in Figure 2.38.

Use $h_1(t)$ and $h_2(t)$ as given in EOCE 2.4 with $x(t)$ as a unit step input and find the output $y(t)$.

Solution

We start with the intermediate output $z(t)$ and write

$$z(t) = \int_{-\infty}^{+\infty} h_1(\tau)x(t-\tau)\,d\tau = \int_{-\infty}^{+\infty} e^{-\tau}u(\tau)u(t-\tau)\,d\tau$$

$$z(t) = \int_{0}^{t} e^{-\tau}\,d\tau = -[e^{-\tau}]_0^t$$

and finally

$$z(t) = [1 - e^{-t}]t \geq 0$$

But $y(t)$ is the convolution between $z(t)$ and $h_2(t)$

$$y(t) = z(t)^* h_2(t)$$

$$y(t) = \int_{-\infty}^{+\infty} h_2(\tau)z(t-\tau)d\tau = \int_{-\infty}^{+\infty} (e^{-\tau})u(\tau)(1 - e^{-(t-\tau)})u(t-\tau)\,d\tau$$

which simplifies to

$$y(t) = \int_0^t (e^{-\tau})(1 - e^{-(t-\tau)})\,d\tau = \int_0^t (e^{-\tau} - e^{-t})\,d\tau$$

$$y(t) = [1 - e^{-t}] - [e^{-t}\tau]_0^t$$

Finally

$$y(t) = [1 - e^{-t}] - [e^{-t}t] = [1 - e^{-t}(1+t)]\; t \geq 0$$

For this impulse input, we can use MATLAB to plot $z(t)$ and $y(t)$ by writing this script.

```
t = 0:0.05:10;
h1 = exp(-t);
h2 = exp(-t);
z = 1 - exp(-t);
y = 1 - exp(-t) - t.*exp(-t);
subplot(4,1,1), plot(t,h1), grid,
ylabel('h1(t)'),
subplot(4,1,2), plot(t,h2), grid,
ylabel('h2(t)'),
subplot(4,1,3), plot(t,z), grid,
ylabel('z(t)'),
subplot(4,1,4), plot(t,y), grid,
xlabel('Time (sec)'), ylabel('y(t)');
```

The plots are shown in Figure 2.39.

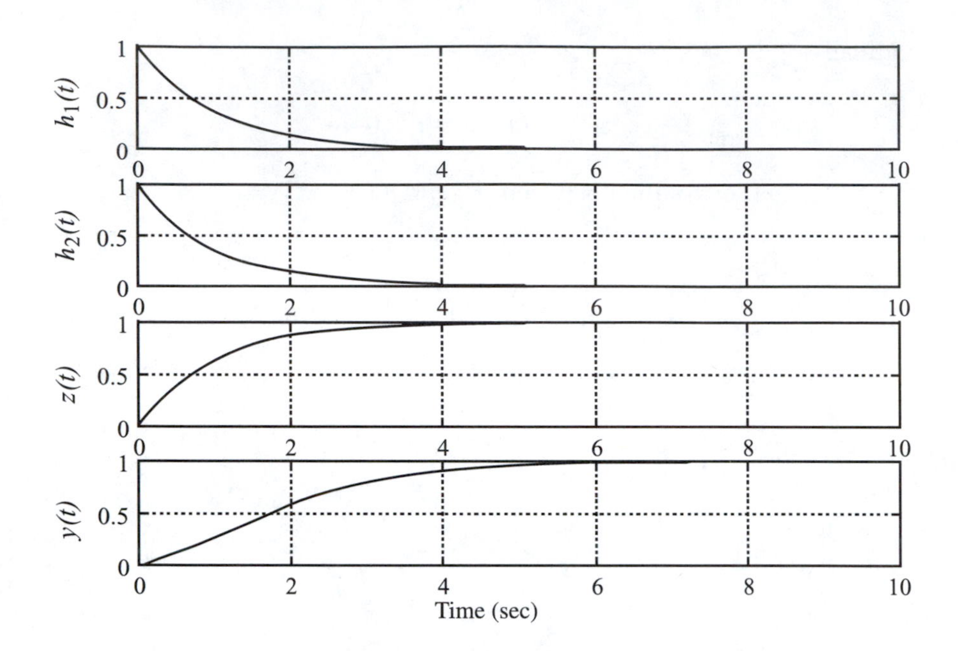

FIGURE 2.39
Plots for EOCE 2.5.

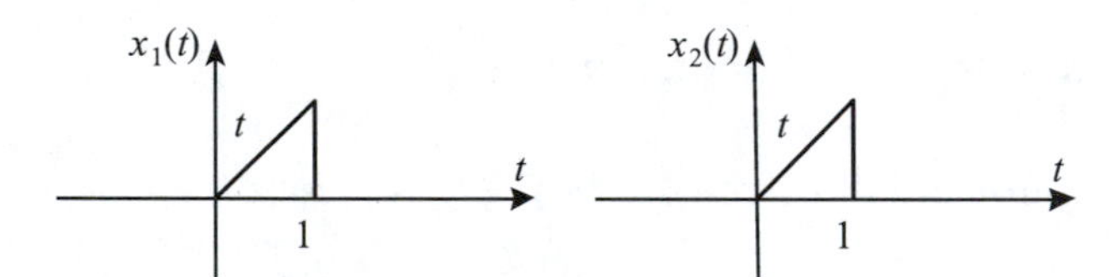

FIGURE 2.40
Signals for EOCE 2.6.

EOCE 2.6

Find the graphical convolution

$$y(t) = x_1(t)^* x_2(t)$$

where the signals $x_1(t)$ and $x_2(t)$ are shown in Figure 2.40.

Solution

We start by mathematically writing the two signals $x_1(t)$ and $x_2(t)$ as

$$x_1(t) = x_2(t) = \begin{cases} t & 0 < t < 1 \\ 0 & \text{otherwise} \end{cases}$$

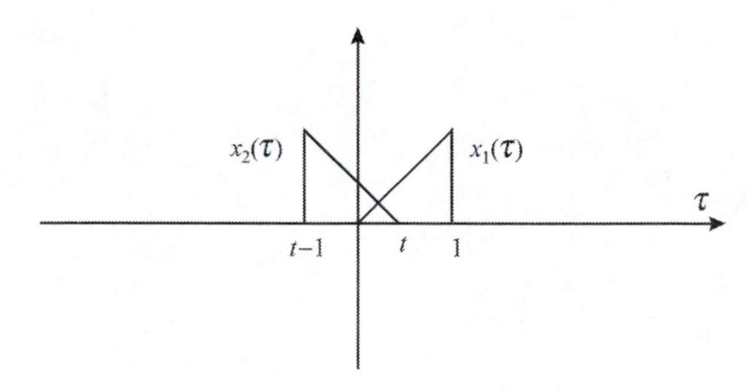

FIGURE 2.41
Signal for EOCE 2.6.

with

$$x_1(\tau) = \begin{cases} \tau & 0 < \tau < 1 \\ 0 & \text{otherwise} \end{cases}$$

We will fix $x_1(t)$ and slide $x_2(t)$. Therefore,

$$x_2(t - \tau) = \begin{cases} t - \tau & 0 < t - \tau < 1 \\ 0 & \text{otherwise} \end{cases}$$

or

$$x_2(t - \tau) = \begin{cases} t - \tau & t - 1 < \tau < t \\ 0 & \text{otherwise} \end{cases}$$

For $t < 2$, there is no overlap and

$$y(t) = 0$$

For $0 \leq t < 1$, the two graphs are shown in Figure 2.41 where

$$y(t) = \int_0^t \tau(t - \tau)\, d\tau$$

$$y(t) = \left(\frac{t\tau^2}{2} - \frac{\tau^3}{3} \right)_0^t$$

which results in

$$y(t) = \frac{t^3}{6}$$

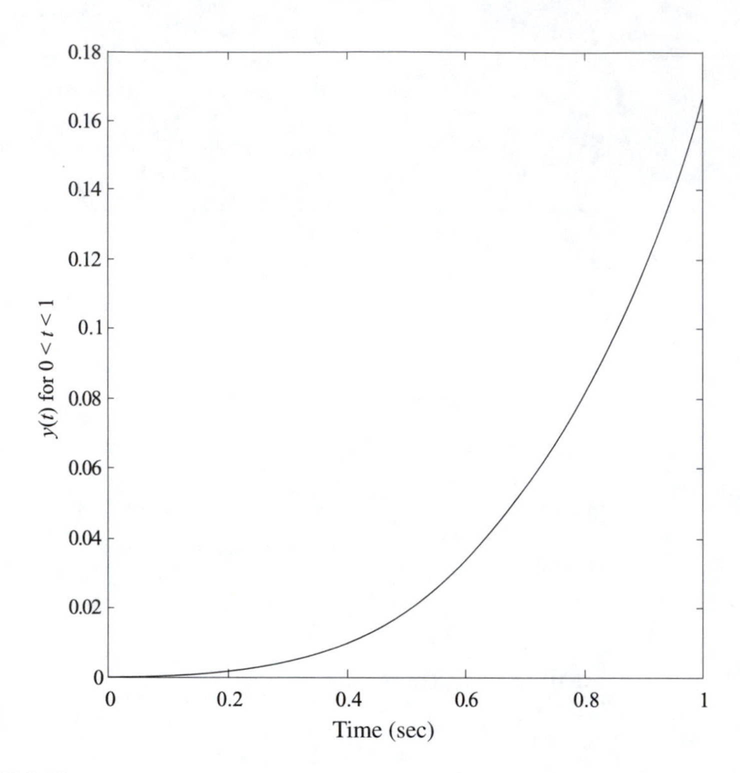

FIGURE 2.42
Plot for EOCE 2.6.

The plot of $y(t)$ for $0 < t < 1$, using the following MATLAB script, is shown in Figure 2.42.

```
t = 0:.001:1;
y = (t.^3)/6;
plot(t,y), xlabel('Time (sec)'),
ylabel('y(t) for 0 < t < 1');
```

For $1 \le t < 2$, the graphs for this range are now shown in Figure 2.43. The overlapping region is between $\tau = t-1$ and $\tau = 1$. The convolution results in

$$y(t) = \int_{t-1}^{1} \tau(t - \tau)\, d\tau$$

$$y(t) = \left(\frac{t\tau^2}{2} - \frac{\tau^3}{3}\right)_{t-1}^{1}$$

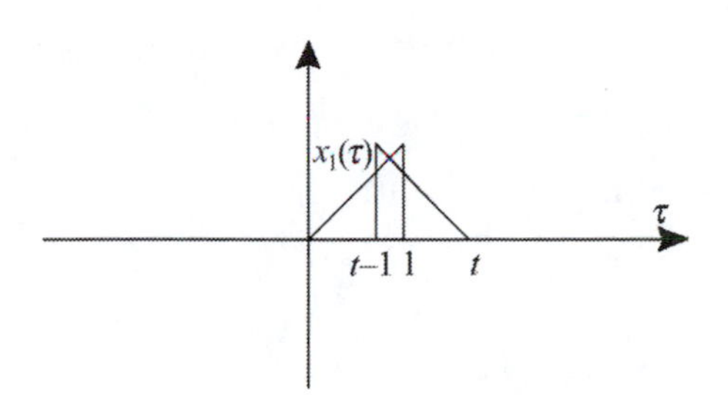

FIGURE 2.43
Signal for EOCE 2.6.

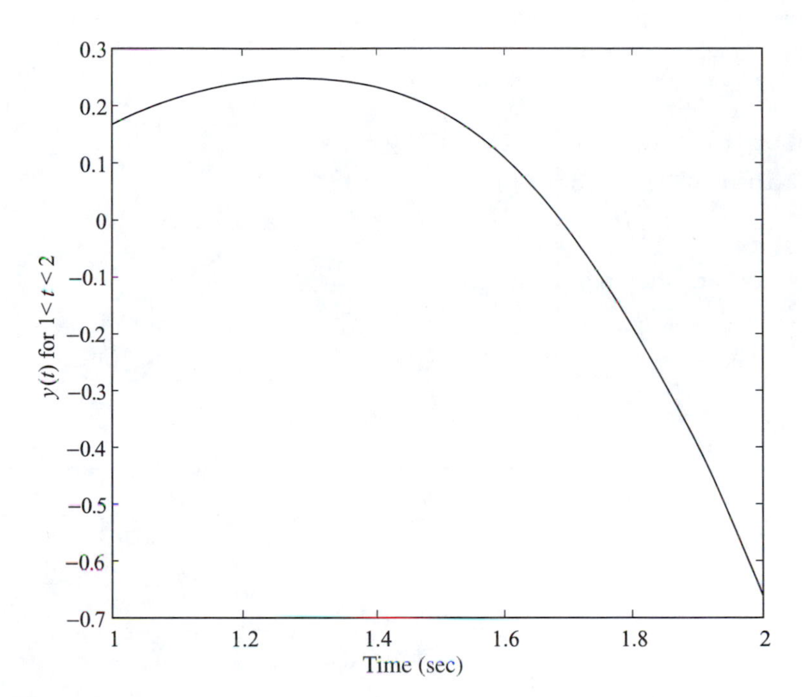

FIGURE 2.44
Plot for EOCE 2.6.

which results in

$$y(t) = \frac{-5t^3 + 12t^2 - 6t}{6}$$

Let us plot $y(t)$ for $1 < t < 2$ using the MATLAB script

```
t = 1:0.001:2;
y = (-5*t.^3 +12*t.^2 - 6*t)/6;
plot(t,y), xlabel('Time (sec)'),
ylabel('y(t) for 1 < t < 2');
```

The plot is shown in Figure 2.44.
For $t > 2$, $y(t) = 0$.

Now we can put all pieces of $y(t)$ together using the MATLAB script

```
t1 = 0:0.001:1;
t2 = 1:0.001:2;
t3 = -1:0.001:0;
t4 = 2:0.001:3;
y1 = (t1.^3)/6;
y2 = (-5*t2.^3 +12*2.^2 - 6*t2)/6;
y3 = 0*t3;
y4 = 0*t4;
plot(t1,y1,t2,y2,t3,y3,t4,y4);
xlabel('Time (sec)'),
ylabel('y(t) = x1(t) * x2(t) for all time')
```

The plots are shown in Figure 2.45.

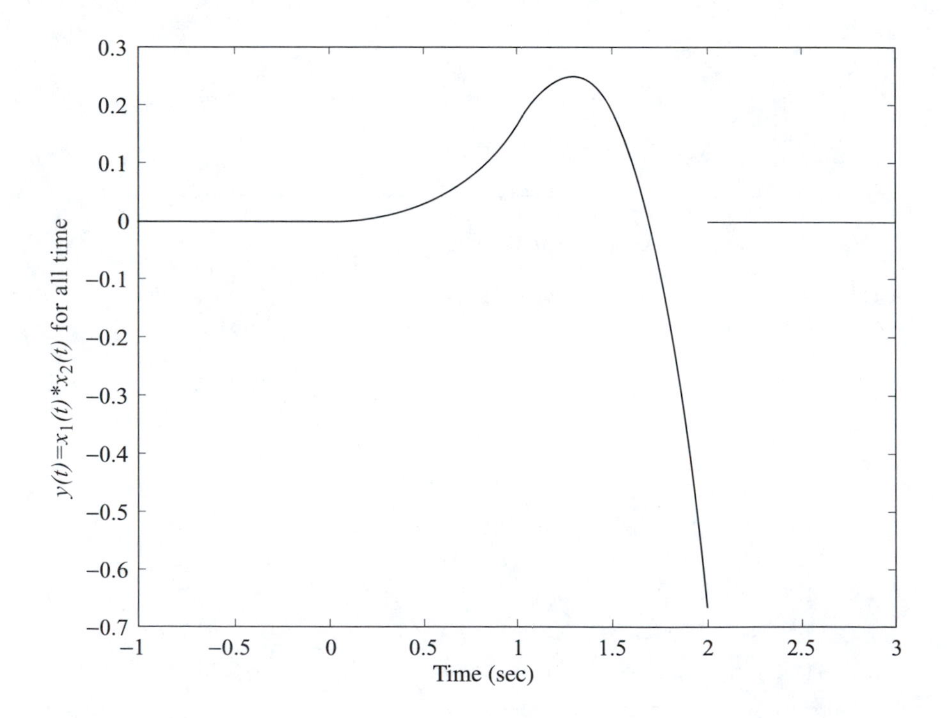

FIGURE 2.45
Plot for EOCE 2.6.

EOCE 2.7

Find the impulse response for the following systems.

1. $\frac{d}{dt}y(t) + 16y(t) = x(t)$

2. $\frac{d^2}{dt}y(t) + 8\frac{d}{dt}y(t) + 16y(t) = x(t)$

Solution

Let us consider the first system. With an impulsive input, the output of the system, $y(t)$, is the impulse response, $h(t)$. $h(t)$ has two parts: the homogenous part and the particular part. For most real physical systems, the order of the derivative involving $x(t)$, the input, is less than the order of the derivative of the output $y(t)$. This is the case here and hence the particular solution for $h(t)$ is zero. The homogenous part of the system given is

$$\frac{d}{dt}y(t) + 16y(t) = 0$$

The characteristic equation is

$$m + 16 = 0$$

for which m is -16.

Therefore, the homogenous part of $h(t)$ is

$$h_h(t) = c_1e^{(-16t)}u(t)$$

The total solution of $h(t)$ is the sum of the homogenous part and the particular part

$$h(t) = h_h(t) + h_p(t)$$

$$h(t) = c_1e^{(-16t)}u(t)$$

We need to calculate the constant c_1. To do that we substitute the solution $h(t)$ in the given system to get

$$[-16c_1e^{(-16t)}u(t) + c_1e^{(-16t)}\delta(t)] + 16[c_1e^{(-16t)}]u(t) = \delta(t)$$

Note that the middle term at the left part of the equal sign is multiplied by the impulse function that is applied at $t = 0$. At this point we use the relation $f(t)\delta(t) = f(0)\delta(t)$, where $f(t)$ is any function, to simplify the above equation. We will get

$$c_1\delta(t) = \delta(t)$$

Comparing coefficients, we clearly see that $c_1 = 1$. Therefore

$$h(t) = e^{(-16t)}u(t)$$

This response can be seen graphically using the MATLAB script as

```
t = 0:0.01:1;
h = exp(-16*t);
plot(t,h), xlabel('time in seconds'),
% ... is used for continuation in the next line
title('The impulse response for the first ...
system in EOCE 2.7');
```

The plot is shown in Figure 2.46.

For the second system, and using the same arguments as we did for the first system, $h_p(t)$ is zero. The homogenous equation is given as

$$\frac{d^2}{dt}y(t) + 8\frac{d}{dt}y(t) + 16y(t) = 0$$

Remember that when you calculate the impulse response, $y(t)$ is $h(t)$.

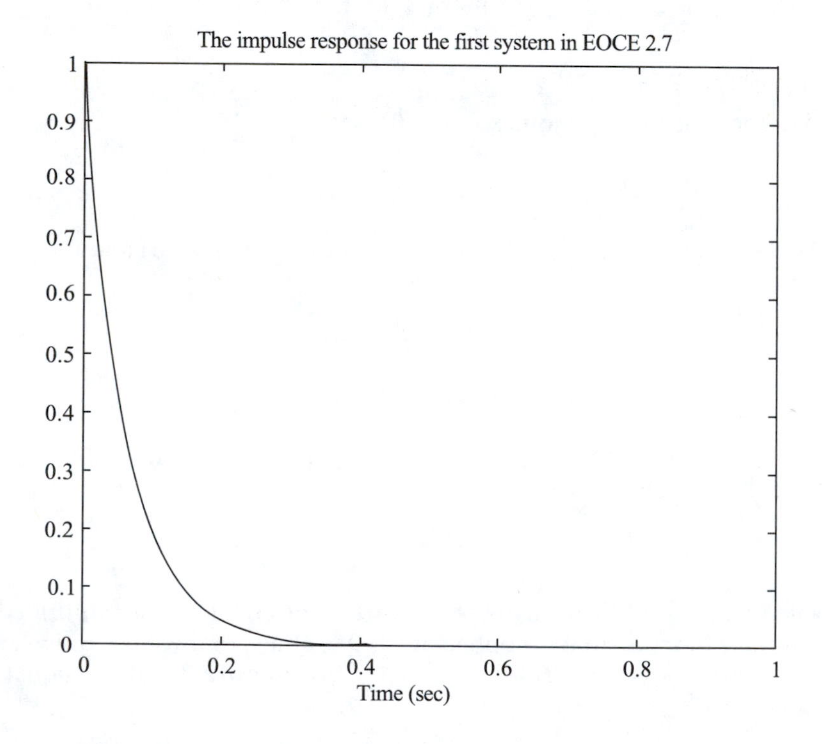

FIGURE 2.46
Signal for EOCE 2.7.

The characteristic equation is

$$m^2 + 8m + 16 = 0$$

with the double repeated roots at $m = -4$.

The homogenous solution in this case has the form

$$h_h(t) = c_1 e^{(-4t)} u(t) + c_2 t e^{(-4t)} u(t)$$

We next substitute in the original system to get

$$\begin{aligned}
&\left\{[16c_1 e^{(-4t)} u(t) - 4c_1 e^{(-4t)} \delta(t)] + \left[-4c_1 e^{(-4t)} \delta(t) + c_1 e^{(-4t)} \frac{d}{dt}\delta(t)\right]\right. \\
&\quad + [-4c_2 e^{(-4t)} u(t) + c_2 e^{(-4t)} \delta(t)] + [16c_2 t e^{(-4t)} u(t) - 4c_2 t e^{(-4t)} \delta(t)] \\
&\quad \left. + \left[-4c_2 e^{(-4t)} u(t) + c_2 e^{(-4t)} \delta(t) + c_2 t e^{(-4t)} \frac{d}{dt}\delta(t) - 4c_2 t e^{(-4t)} \delta(t)\right]\right\} \\
&\quad + \{-32c_1 e^{(-4t)} u(t) + 8c_1 e^{(-4t)} \delta(t) + 8c_2 e^{(-4t)} u(t) - 32c_2 t e^{(-4t)} u(t) \\
&\quad + 8c_2 t e^{(-4t)} \delta(t)\} + \{16c_1 e^{(-4t)} u(t) + 16c_2 t e^{(-4t)} u(t)\} = \delta(t)
\end{aligned}$$

We can clean the above to get

$$\begin{aligned}
&-4c_1 e^{(-4t)} \delta(t) - 4c_1 e^{(-4t)} \delta(t) + c_1 e^{(-4t)} \frac{d}{dt}\delta(t) + c_2 e^{(-4t)} \delta(t) - 4c_2 t e^{(-4t)} \delta(t) \\
&\quad + c_2 e^{(-4t)} \delta(t) + c_2 t e^{(-4t)} \frac{d}{dt}\delta(t) - 4c_2 t e^{(-4t)} \delta(t) + 8c_1 e^{(-4t)} \delta(t) \\
&\quad + 8c_2 t e^{(-4t)} \delta(t) = \delta(t)
\end{aligned}$$

With $f(t)\delta(t) = f(0)\delta(t)$, where $f(t)$ is any function, we can simplify further and write

$$-4c_1\delta(t) - 4c_1\delta(t) + c_1 \frac{d}{dt}\delta(t) + c_2\delta(t) + c_2\delta(t) + 8c_1\delta(t) = \delta(t)$$

By equating coefficients for $\delta(t)$ we get

$$-4c_1 - 4c_1 + c_2 + c_2 + 8c_1 = 1$$

and by equating coefficients for $\frac{d}{dt}\delta(t)$ we get

$$c_1 = 0$$

Therefore $c_2 = 1/2$ and the impulse response is

$$h(t) = (1/2) t e^{(-4t)} u(t)$$

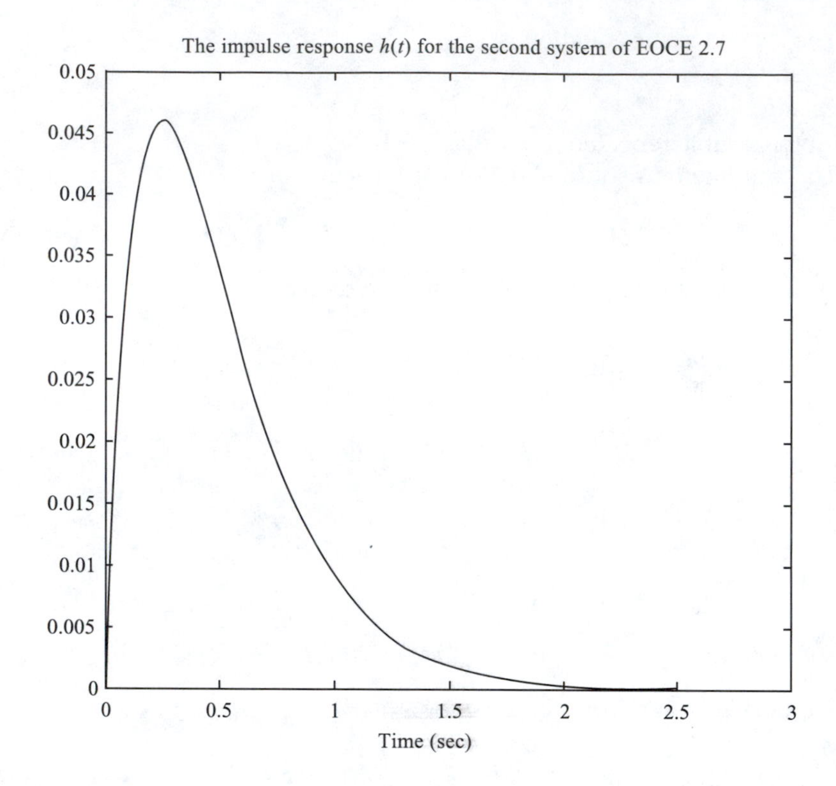

FIGURE 2.47
Signal for EOCE 2.7.

Using MATLAB we can plot this response as in Figure 2.47 by writing the following script.

```
t = 0:0.01:3;
h = (1/2)*t.*exp(-4*t);
plot(t,h), xlabel('Time (sec)');
% ... is used for continuation in the next line
title('The impulse response h(t) for the second ...
system of EOCE7');
```

EOCE 2.8

Consider the following systems.

1. $\frac{d^2}{dt}y(t) + 8\frac{d}{dt}y(t) + 16y(t) = x(t)$

2. $\frac{d^2}{dt}y(t) + \frac{d}{dt}y(t) + y(t) = x(t)$

3. $\frac{d^2}{dt}y(t) + 5\frac{d}{dt}y(t) + 6y(t) = x(t)$

4. $\frac{d^2}{dt}y(t) + 16y(t) = x(t)$

5. $\frac{d^2}{dt}y(t) + 16\frac{d}{dt}y(t) = x(t)$

For each system:

1. Discuss stability.
2. What is the general form of the solution for the homogenous part ?

Solution

System 1

The characteristic equation is

$$m^2 + 8m + 16 = 0$$

Using MATLAB we can find the roots by typing at the prompt

```
eigenvalues_for_system_1 = roots([1,8,16])
```

to get

```
eigenvalues_for_system_1 =
-4
-4
```

Since the eigenvalues are all negative, the system is stable. The general form of the solution is

$$y(t) = c_1e^{(-4t)} + c_2e^{(-4t)} \qquad \text{for} \quad t \geq 0$$

As you can see the solution, $y(t)$, converges to zero as $t \to \infty$. We can use MATLAB and type the following at the prompt

```
system_1_solution=dsolve('D2y+8*Dy+16*y=0','t')
```

to get

```
system_1_solution =
C1*exp(-4*t)+C2*exp(-4*t)*t
```

This solution agrees with the analytical solution presented previously.

System 2

The characteristic equation is

$$m^2 + m + 1 = 0$$

Using MATLAB we can find the roots as

```
eigenvalues_for_system_2 = roots([1,1,1])
```

to get

```
eigenvalues_for_system_2 =
-0.5000 + 0.8660i
-0.5000 - 0.8660i
```

Since the real parts of the eigenvalues are all negative, the system is stable. The general form of the solution is

$$y(t) = c_1 e^{[(-0.5+0.866j)t]} + c_2 e^{[(-0.5-0.866j)t]} \quad \text{for} \quad t \geq 0$$

$$y(t) = c_1 e^{(-0.5t)}[\cos(0.866t) + j\sin(0.866t)]$$
$$+ c_2 e^{(-0.5)t}[\cos(0.866)t - j\sin(0.866t)] \quad \text{for} \quad t \geq 0$$

$y(t)$ converges to zero as $t \rightarrow \infty$.

We can use MATLAB again to solve system two and type at the prompt

```
system_2_solution=dsolve('D2y+Dy+y=0','t')
```

to get

```
system_2_solution =
C1*exp(-1/2*t)*cos(1/2*3^(1/2)*t)+C2*exp(-1/2*t)
  *sin(1/2*3^(1/2)*t)
```

The two solutions are the same. Prove that.

System 3

The characteristic equation is

$$m^2 + 5m + 6 = 0$$

Using MATLAB we can find the roots as

```
eigenvalues_for_system_3 = roots([1,5,6])
```

to get

```
eigenvalues_for_system_3 =
-3
-2
```

Since all the roots are negative, the system is stable. The general form of the solution is

$$y(t) = c_1 e^{(-3t)} + c_2 e^{(-2t)} \qquad \text{for} \quad t \geq 0$$

$y(t)$ converges to zero as $t \to \infty$. This means stability.

With MATLAB we type at the prompt

```
system_3_solution=dsolve('D2y+5*Dy+6*y=0','t')
```

to get

```
system_3_solution =
C1*exp(-3*t)+C2*exp(-2*t)
```

The two solutions agree.

System 4

The characteristic equation is

$$m^2 + 16 = 0$$

Using MATLAB we can find the roots by typing the following at the prompt

```
eigenvalues_for_system_4 = roots([1,0,16])
```

to get

```
eigenvalues_for_system_4 =
0 + 4.0000i
0 - 4.0000i
```

Since there are no real parts for the roots, and since the roots are all imaginary, the system response is pure sinusoidal. In this case, the system is on the verge of instability and we consider it unstable. The general form of the solution is

$$y(t) \qquad c_1 e^{[(4j)t]} \qquad c_2 e^{[(-4j)t]} \qquad \text{for} \quad t$$

$y(t)$ does not converge to zero as $t \to \infty$. Pure sinusoidal signals are bounded and they do not die as time increases.

With MATLAB we type at the prompt

```
system_4_solution = dsolve('D2y+16*y=0','t')
```

to get

```
system_4_solution =
C1*sin(4*t)+C2*cos(4*t)
```

Try to prove that the two solutions are the same.

System 5

The characteristic equation is

$$m^2 + 16m = 0$$

Using MATLAB we can find the roots by typing the following at the MATLAB prompt.

```
eigenvalues_for_system_5 = roots([1,16,0])
```

to get

```
eigenvalues_for_system_5 =
  0
-16
```

Since one of the roots is at zero and the other is negative (no real roots in the positive side of the complex plane) the system is on the verge of stability. The general form of the solution is

$$y(t) = c_1 e^{(0t)} + c_2 e^{(-16t)} \qquad \text{for } t \geq 0$$

$$y(t) = c_1 + c_2 e^{(-16t)} \qquad \text{for } t \geq 0$$

As $t \to \infty$, $y(t)$ approaches $c_1 + c_2$, a constant. If the first root is one step in the right-half plane of the complex plane, the system would be unstable.

With MATLAB we type at the prompt

```
system_5_solution=dsolve('D2y+16*Dy=0','t')
```

to get

```
system_5_solution =
C1 + C2*exp(-16*t)
```

That is what we have analytically.

EOCE 2.9

Consider the following system described by the differential equation

$$\frac{d^2}{dt}y(t) + \frac{d}{dt}y(t) = 2t$$

with the initial conditions

$$y(0) = 0 \quad \text{and} \quad \frac{d}{dt}y(0) = 0$$

What is the solution $y(t)$?

Solution

The solution was found earlier in the chapter as

$$y(t) = y_h(t) + y_p(t) = -2e^{-t} + 2 - 2t + t^2$$

We can type at the MATLAB prompt the following command

```
y_total = dsolve('D2y+Dy=2*t','y(0)=0','Dy(0) = 0')
```

to get

```
y_total =
(-2*t*exp(t)+t^2*exp(t)+2*exp(t)-2)/exp(t)
```

which agrees with the analytical solution presented above and explained in Example 2.14.

EOCE 2.10

Consider the following system described by the following differential equation

$$\frac{d^3}{dt}y(t) - y(t) = \sin t$$

with initial conditions

$$y(0) = 0, \quad \frac{d}{dt}y(0) = 0, \quad \text{and} \quad \frac{d^2}{dt}y(0) = 0$$

What is the output of the system?

Solution

The solution was again derived in Example 2.15:

$$y(t) = 0.159e^{t} + e^{-0.5t}[-0.659\cos(0.866t) + 0.013\sin(0.866t)] + 1/2\cos t - 1/2\sin t$$

We can use MATLAB and type at the prompt

```
y_total=dsolve('D3y-y=sin(t)','y(0)=0',...
'Dy(0)=0','D2y(0)=0');
y_total=simple(y_total)% to simplify the result
```

to get

```
y_total =
1/6*exp(t)-1/2*sin(t)+1/2*cos(t)-2/3*exp
(-1/2*t)*cos(1/2*3^(1/2)*t)
```

The two solutions seem to be different as we look at them. Is that true? One way to know is to plot the two solutions using the following MATLAB script. The plots are shown in Figure 2.48.

```
t = 0:0.05:5;% … used for continuation in the
following line
y2 = 0.159*exp(t) + exp(-0.5*t).* (-0.659*cos(0.886*t)+...
0.013*sin(0.886*t))+0.5*cos(t)-0.5*sin(t);
y1=1/6*exp(t)-1/2*sin(t)+ ...
1/2*cos(t)-2/3*exp(-1/2*t).*cos(1/2*3^(1/2)*t);
plot(t,y1,'*',t,y2,'o')
title('Matlab and Analytical solution')
xlabel('Time (sec)')
```

The error you see in Figure 2.48 is due to truncation in the calculations.

You may also notice that the output is unstable. This is due to the presence of a positive root (a pole) at 1 as seen in the first term of the solution.

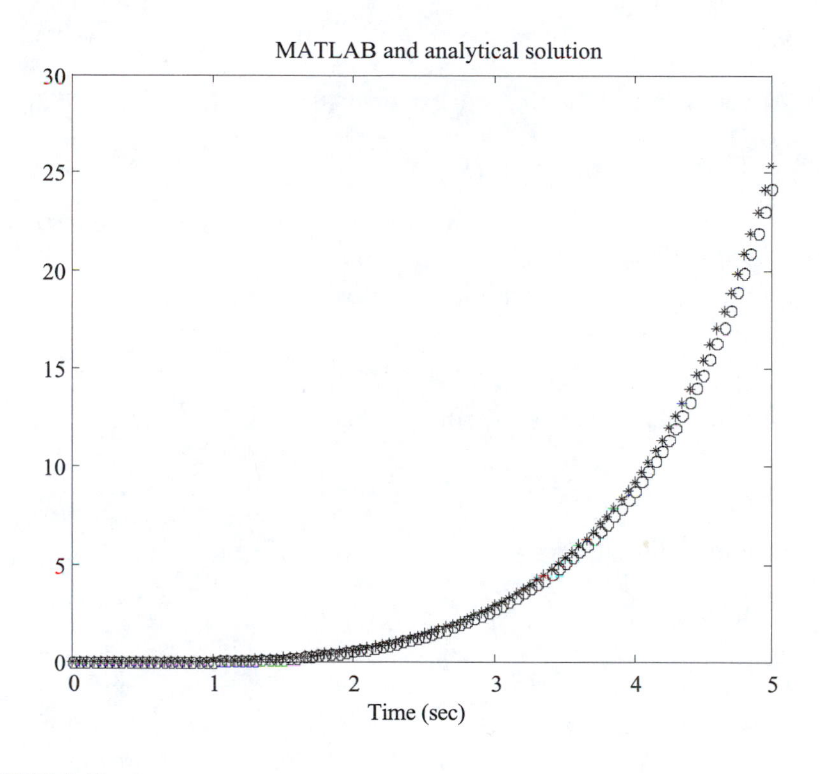

FIGURE 2.48
Plots for EOCE 2.10.

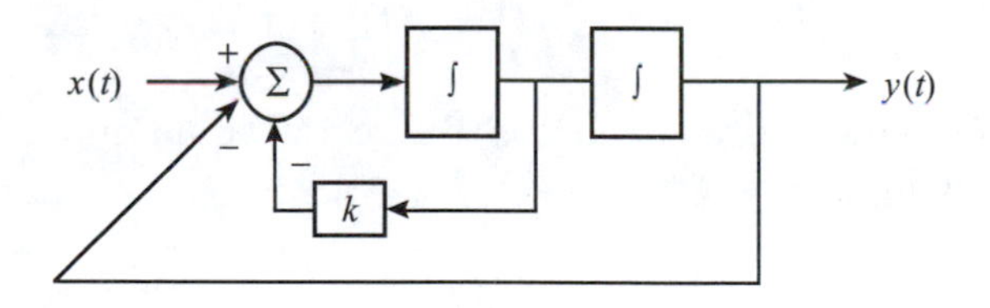

FIGURE 2.49
System for EOCE 2.11.

EOCE 2.11

Consider the system given in block diagram representation shown in Figure 2.49.

1. For what value(s) of k is the system stable?
2. Pick a value for k where the system is stable and find the impulse and the step responses.

Solution

1. At the output of the first integrator the signal is

$$y_1(t) = \int [x(t) - ky_1(t) - y(t)]\,dt$$

where $y_1(t)$ is the output of the first integrator. In this case $y_1(t)$ is the derivative of $y(t)$.

At the output of the second integrator

$$y(t) = \int y_1(t)\,dt$$

Differentiating the first equation we get

$$\frac{d}{dt}y_1(t) = x(t) - ky_1(t) - y(t)$$

Differentiating the second equation twice we get

$$\frac{d^2}{dt}y(t) = \frac{d}{dt}y_1(t)$$

or

$$\frac{d^2}{dt}y(t) = x(t) - ky_1(t) - y(t) = -k\frac{d}{dt}y(t) + x(t) - y(t)$$

Finally

$$\frac{d^2}{dt}y(t) + k\frac{d}{dt}y(t) + y(t) = x(t)$$

If we apply the Routh test to this differential equation we will have the following Routh array

$$\begin{array}{cc} 1 & 1 \\ k & 0 \\ 1 & \end{array}$$

which indicates that the system is stable for all positive values of k.

2. Using MATLAB we can find the step and the impulse responses for the system by writing the following script.

```
k=input('enter a value for k, a negative number
to exit:');
t=0:0.5:100;
while k > 0
```

```
n=[1];
d=[1 k 1];
ystep=step(n,d,t);
yimpulse=impulse(n,d,t);
plot(t,ystep,'*',t,yimpulse,'+')
title('The step and the impulse response')
gtext('step')
gtext('impulse')
% ... is used for continuation on the next line
k=input('enter a value for k, a negative number ...
to exit:');
end
xlabel('Time (sec)')
```

A sample run of the script is

```
enter a value for k, a negative number to exit: 1
```

and the plot is in Figure 2.50.

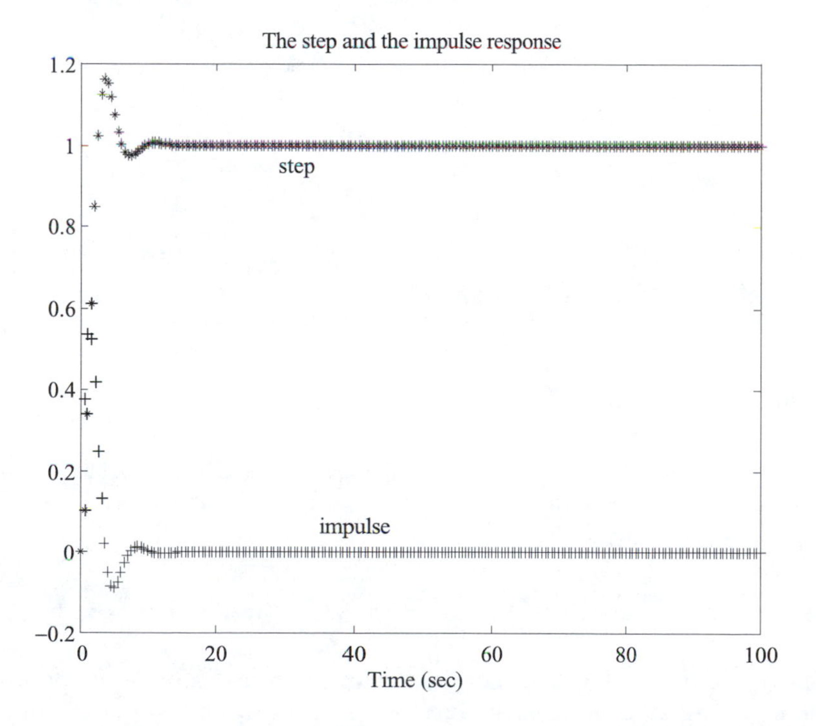

FIGURE 2.50
Plot for EOCE 2.11.

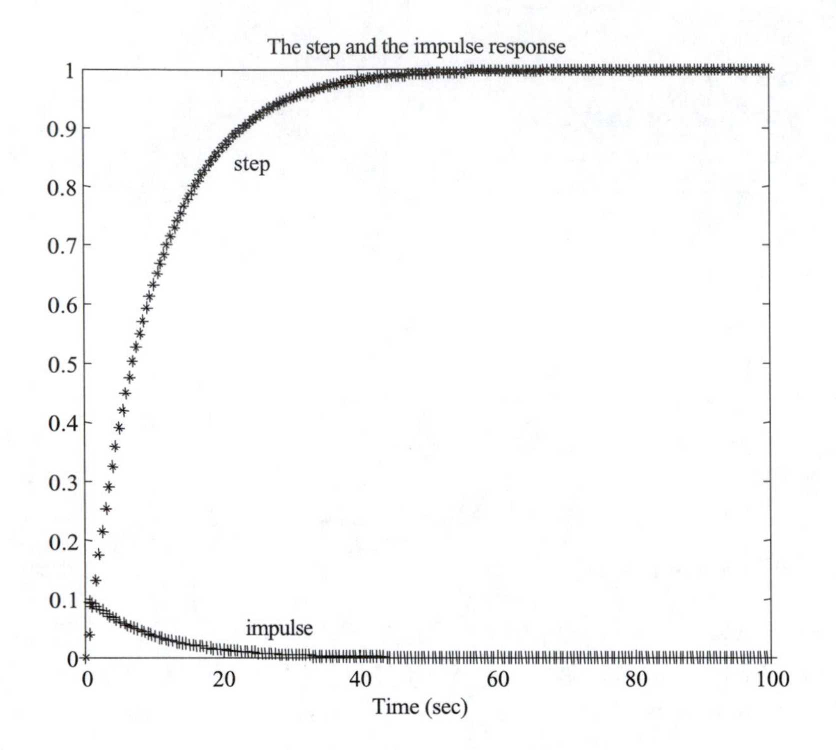

FIGURE 2.51
Plot for EOCE 2.11.

```
enter a value for k, a negative number to exit: 10
```

and the plot is in Figure 2.51.

```
enter a value for k, a negative number to exit: 100
```

and the plot is in Figure 2.52.

```
enter a value for k, a negative number to exit: -1
```

We can see that as k gets larger the system becomes slow in responding to the input signals. Notice also that for $k = 1$ the system is underdamped (has oscillations) and as k becomes larger it becomes undamped (no oscillations).

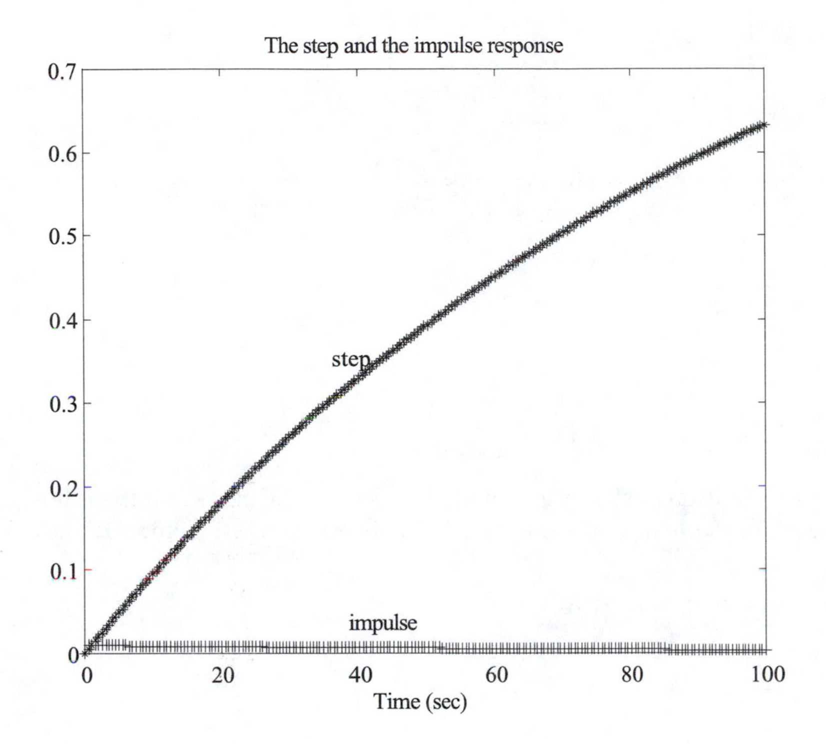

FIGURE 2.52
Plot for EOCE 2.11.

2.22 End-of-Chapter Problems

EOCP 2.1

Consider the following systems. State whether each is linear or nonlinear and give your reasons. Also check if each is time-variant and give reasons.

1. $y(t) = x(t)\sqrt{x(t)}$
2. $y(t) = x(t) + \sqrt{x(t)}$
3. $y(t) = \dfrac{1}{x(t)}$
4. $y(t) = tx(t) + \sqrt{x(t)}$
5. $y(t) = tx(t)$

6. $y(t) = \dfrac{1}{x(t)}$

7. $y(t) = tx(t) + e^{x(t)}$

8. $y(t) = tx(2t) + \sin(t)$

9. $y(t) = t + x(t)\sin(t)$

10. $y(t) = tx(2t) + x(t)$

EOCP 2.2

Use the definition of stability provided in Section 2.8 and see if the following systems are stable or not. Assume $x(t)$, the input to the system, is bounded.

1. $y(t) = \cos(x(t))$

2. $y(t) = t\cos(x(t))$

3. $y(t) = e^{t}\cos(x(t))$

4. $y(t) = e^{-t}\cos(x(t))$

5. $y(t) = te^{-t}\cos(x(t))$

6. $y(t) = te^{x(t)}\cos(x(t))$

7. $\int_{-\infty}^{t} 10x(\tau)\,d\tau$

8. $\int_{-\infty}^{t} x(\tau)e^{-\tau}\,d\tau$

9. $\int_{-\infty}^{t} x(\tau)e^{\tau}\,d\tau$

10. $\int_{-\infty}^{t} x(\tau)\tau e^{\tau}\,d\tau$

EOCP 2.3

Different systems are given below with specific inputs and the impulse response, $h(t)$. Find $y(t)$, the output, using the convolution integral.

1. $x(t) = \delta(t), \quad h(t) = e^{-t}u(t)$
2. $x(t) = \delta(t), \quad h(t) = (e^{-t} + e^{-2t})u(t)$
3. $x(t) = u(t), \quad h(t) = e^{-t}u(t)$
4. $x(t) = u(t), \quad h(t) = (e^{-t} + e^{-2t})u(t)$
5. $x(t) = \delta(t), \quad h(t) = te^{-t}u(t)$
6. $x(t) = \delta(t), \quad h(t) = (te^{-t} + e^{-t})u(t)$
7. $x(t) = u(t), \quad h(t) = te^{-t}u(t)$
8. $x(t) = u(t), \quad h(t) = (te^{-t} + e^{-t})u(t)$
9. $x(t) = e^{-t}u(t), \quad h(t) = e^{-t}u(t)$
10. $x(t) = e^{-t}u(t) + u(t), \quad h(t) = e^{-2t}u(t)$
11. $x(t) = e^{-t}u(t) + \delta(t), \quad h(t) = e^{-2t}u(t)$
12. $x(t) = e^{-t}u(t) + \delta(t) + u(t), \quad h(t) = e^{-2t}u(t)$

EOCP 2.4

Use the convolution integral to find $y(t)$ in each of the systems shown in Figures 2.53, 2.54, 2.55, 2.56, and 2.57.

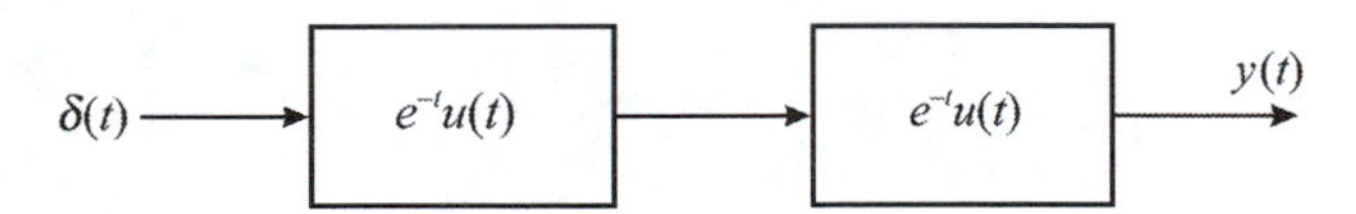

FIGURE 2.53
System for EOCP 2.4.

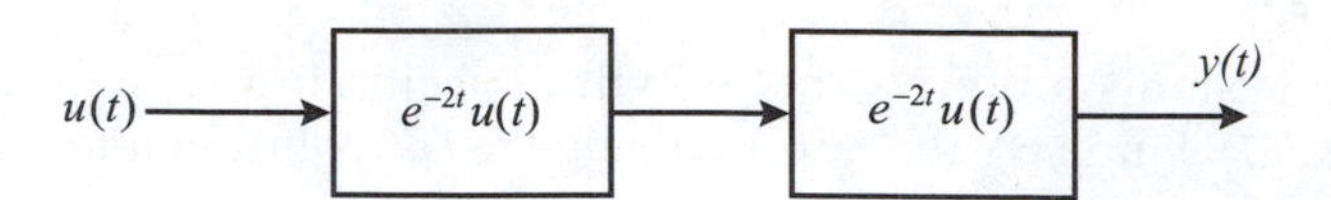

FIGURE 2.54
System for EOCP 2.4.

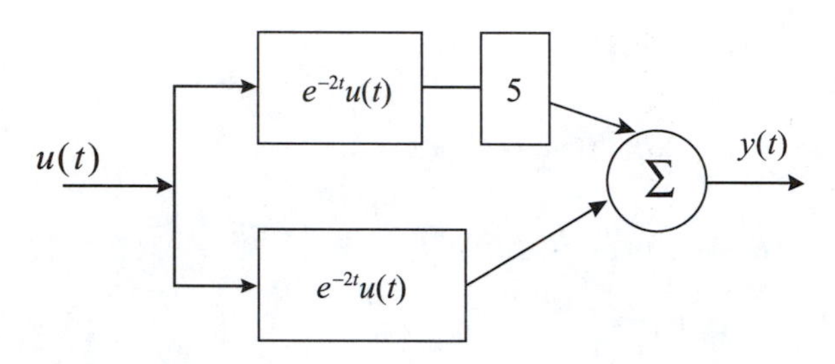

FIGURE 2.55
System for EOCP 2.4.

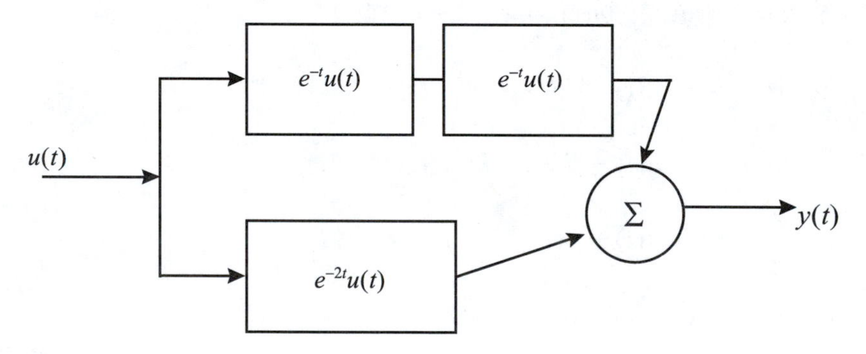

FIGURE 2.56
System for EOCP 2.4.

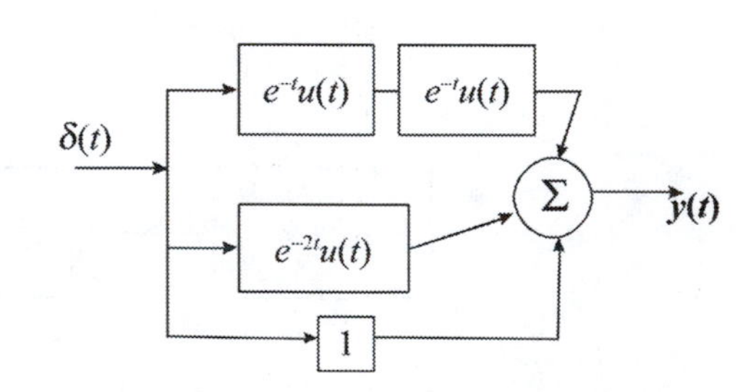

FIGURE 2.57
System for EOCP 2.4.

EOCP 2.5

Find the signal, $y(t)$, that results from convolving $x(t)$ and $h(t)$ in Figures 2.58, 2.59, 2.60, 2.61, 2.62, 2.63, and 2.64.

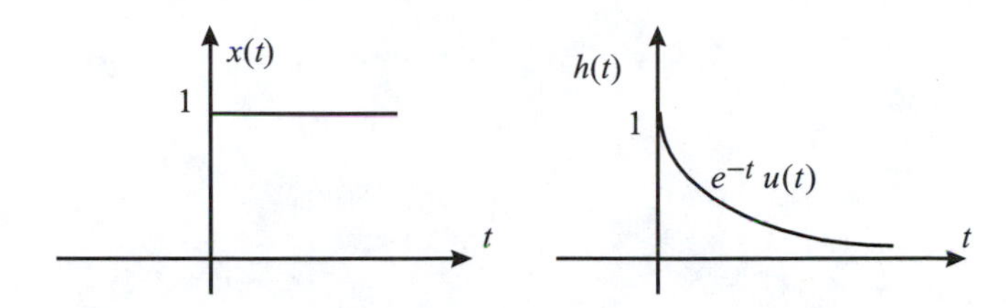

FIGURE 2.58
Signals for EOCP 2.5.

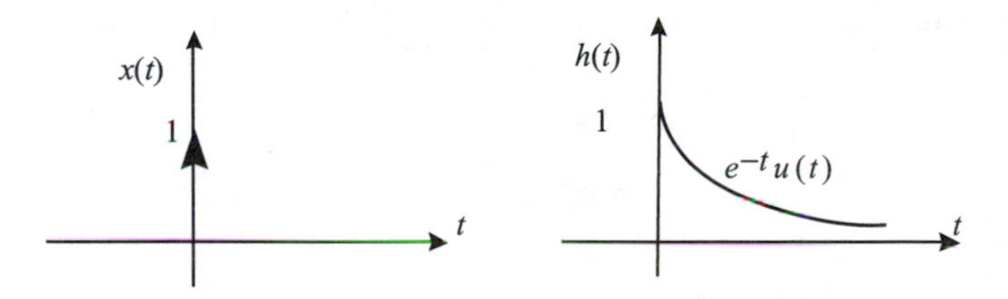

FIGURE 2.59
Signals for EOCP 2.5.

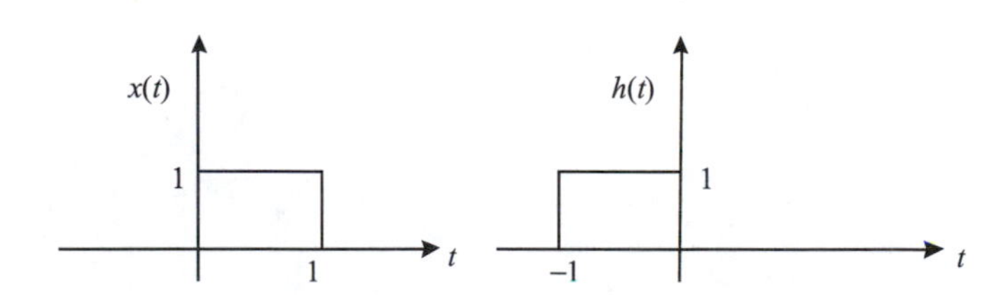

FIGURE 2.60
Signals for EOCP 2.5.

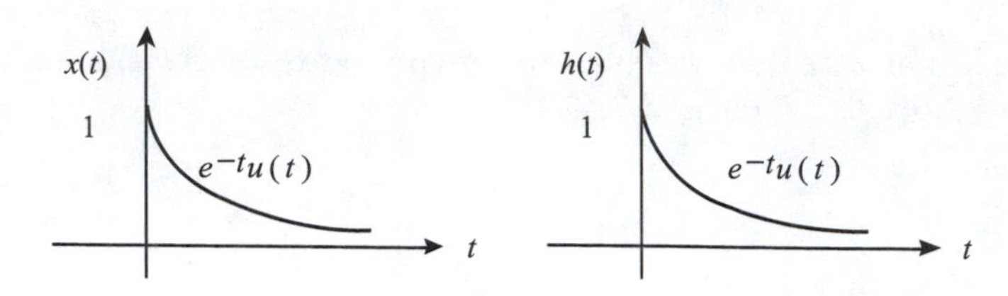

FIGURE 2.61
Signals for EOCP 2.5.

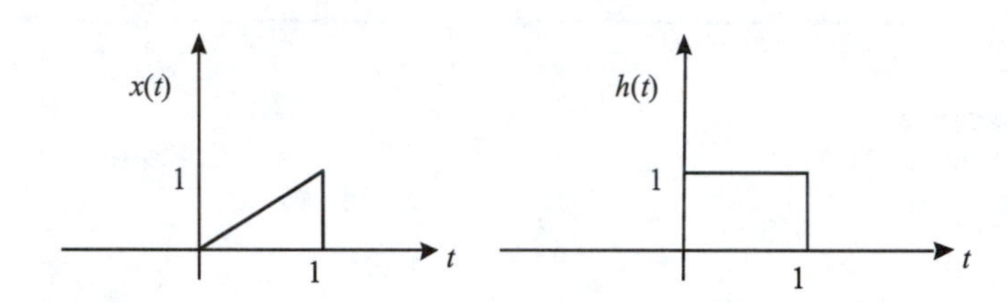

FIGURE 2.62
Signals for EOCP 2.5.

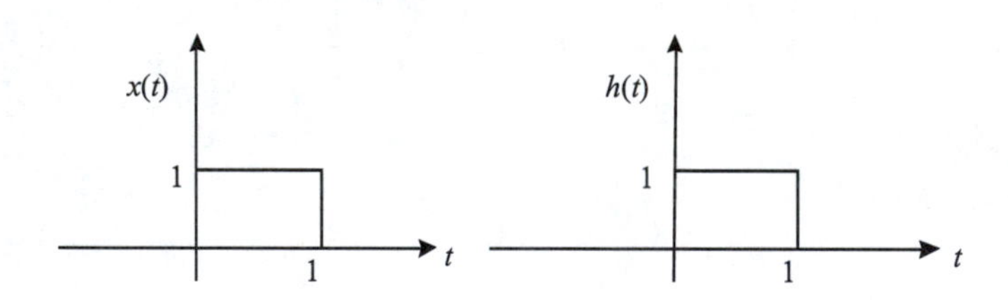

FIGURE 2.63
Signals for EOCP 2.5.

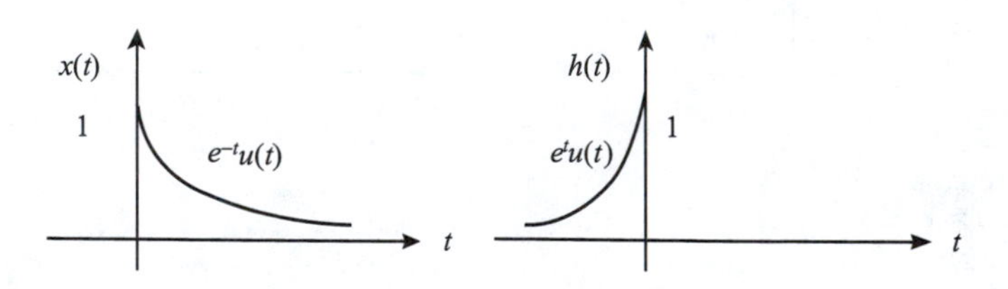

FIGURE 2.64
Signals for EOCP 2.5.

EOCP 2.6

Find the output, $y(t)$, for the following systems with the indicated initial conditions.

1. $\frac{d}{dt}y(t) + \frac{1}{2}y(t) = 0$ with $y(0) = -1$
2. $3\frac{d}{dt}y(t) - \frac{1}{2}y(t) = 0$ with $y(0) = 10$
3. $\frac{d^2}{dt}y(t) + \frac{1}{2}y(t) = 0$ with $y(0) = -1$ and $\frac{d}{dt}y(0) = 0$
4. $\frac{d^2}{dt}y(t) - \frac{1}{2}y(t) = 0$ with $y(0) = 1$ and $\frac{d}{dt}y(0) = -1$
5. $\frac{d^2}{dt}y(t) + \frac{1}{2}\frac{d}{dt}y(t) = 0$ with $y(0) = -1$ and $\frac{d}{dt}y(0) = 0$
6. $\frac{d^2}{dt}y(t) - \frac{1}{2}\frac{d}{dt}y(t) = 0$ with $y(0) = -1$ and $y(0) = 1$
7. $\frac{d^2}{dt}y(t) + \frac{1}{2}\frac{d}{dt}y(t) + y(t) = 0$ with $y(0) = -1$ and $\frac{d}{dt}y(0) = 0$
8. $\frac{d^2}{dt}y(t) + 10\frac{d}{dt}y(t) + 2y(t) = 0$ with $y(0) = -1$ and $\frac{d}{dt}y(0) = -1$
9. $2\frac{d^2}{dt}y(t) - 3\frac{d}{dt}y(t) = 0$ with $y(0) = 1$ and $\frac{d}{dt}y(0) = 0$
10. $2\frac{d^2}{dt}y(t) + 3\frac{d}{dt}y(t) - 3y(t) = 0$ with $y(0) = 1$ and $\frac{d}{dt}y(0) = 0$

EOCP 2.7

Use the roots of the characteristic equation to study the stability of the systems given in EOCP 2.6.

EOCP 2.8

Find the value(s) of k that makes each of the following systems stable.

1. $\frac{d}{dt}y(t) + \frac{1}{2}k\,y(t) = x(t)$
2. $3\frac{d}{dt}y(t) - \frac{1}{2}k\,y(t) = x(t)$

3. $\frac{d^2}{dt}y(t) + \frac{1}{2}k\,y(t) = x(t)$

4. $\frac{d^2}{dt}y(t) - \frac{1}{2}k\,y(t) = x(t)$

5. $\frac{d^2}{dt}y(t) + \frac{1}{2}(k+1)\frac{d}{dt}y(t) = x(t)$

6. $\frac{d^2}{dt}y(t) - \frac{1}{2}k\frac{d}{dt}y(t) = x(t)$

7. $\frac{d^2}{dt}y(t) + \frac{1}{2}\frac{d}{dt}y(t) + ky(t) = x(t)$

8. $\frac{d^2}{dt}y(t) + (k+1)\frac{d}{dt}y(t) + 2y(t) = x(t)$

9. $2\frac{d^2}{dt}y(t) - 3(k-1)\frac{d}{dt}y(t) = x(t)$

10. $2\frac{d^2}{dt}y(t) + (k+1)\frac{d}{dt}y(t) - 3ky(t) = x(t)$

EOCP 2.9

Find the total solution of the following systems. Assume zero initial conditions.

1. $y\frac{d^2}{dt}(t) - \frac{1}{2}\frac{d}{dt}y(t) = \sin(t)$

2. $\frac{d^2}{dt}y(t) + \frac{1}{2}\frac{d}{dt}y(t) + y(t) = u(t) + 1$

3. $\frac{d^2}{dt}y(t) + \frac{d}{dt}y(t) + 2y(t) = e^{-t}$

4. $2\frac{d^2}{dt}y(t) - 3\frac{d}{dt}y(t) = t + 10$

5. $2\frac{d^2}{dt}y(t) + \frac{d}{dt}y(t) - 3y(t) = 10$

EOCP 2.10

Find the impulse response for each of the following systems.

1. $\frac{d}{dt}y(t) + \frac{1}{2}y(t) = x(t)$

2. $3\frac{d}{dt}y(t) - \frac{1}{2}y(t) = x(t)$

3. $\frac{d^2}{dt}y(t) + \frac{1}{2}y(t) = x(t)$

4. $\frac{d^2}{dt}y(t) - \frac{1}{2}y(t) = x(t)$

5. $\frac{d^2}{dt}y(t) + \frac{1}{2}\frac{d}{dt}y(t) = x(t)$

EOCP 2.11

Draw the block diagrams for the following systems.

1. $\frac{d}{dt}y(t) + \frac{1}{2}y(t) = x(t) + \frac{d}{dt}x(t)$

2. $3\frac{d^2}{dt}y(t) - \frac{1}{2}\frac{d}{dt}y(t) = 2x(t)$

3. $\frac{d^3}{dt}y(t) + y(t) = \frac{d^2}{dt}x(t)$

4. $\frac{d^3}{dt}y(t) - \frac{1}{2}\frac{d}{dt}y(t) = x(t) + \frac{d}{dt}x(t)$

5. $\frac{d^3}{dt}y(t) + \frac{1}{2}\frac{d^2}{dt}y(t) = x(t)$

EOCP 2.12

For the block diagrams shown in Figures 2.65, 2.66, 2.67, 2.68, and 2.69, find the system that represents each.

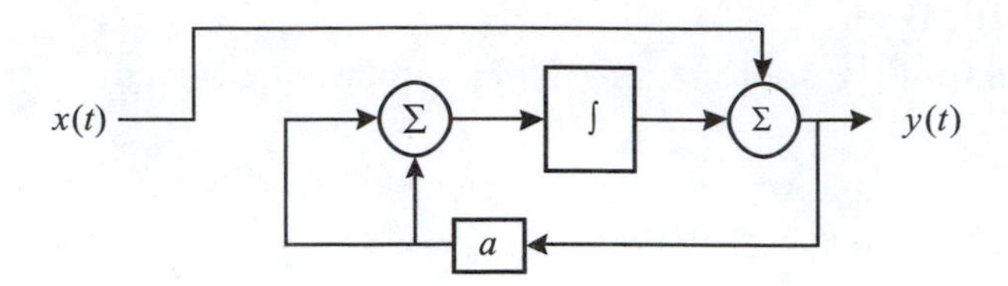

FIGURE 2.65
System for EOCP 2.12.

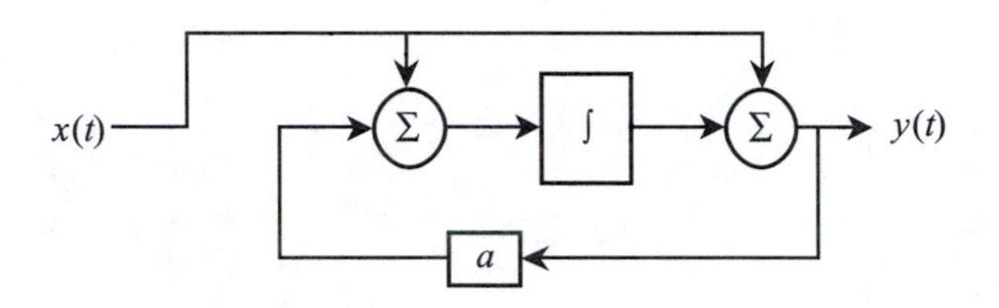

FIGURE 2.66
System for EOCP 2.12.

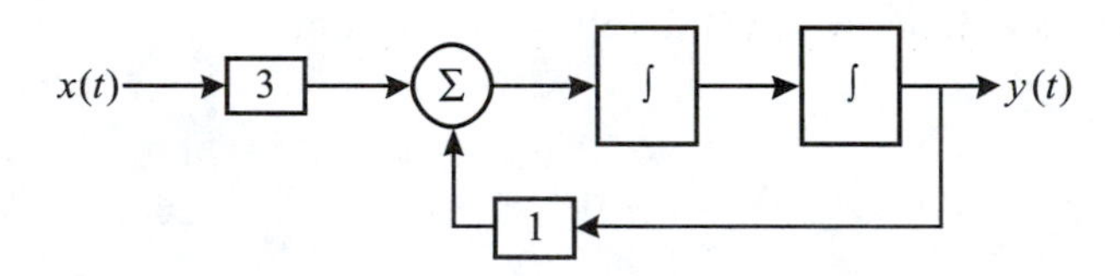

FIGURE 2.67
System for EOCP 2.12.

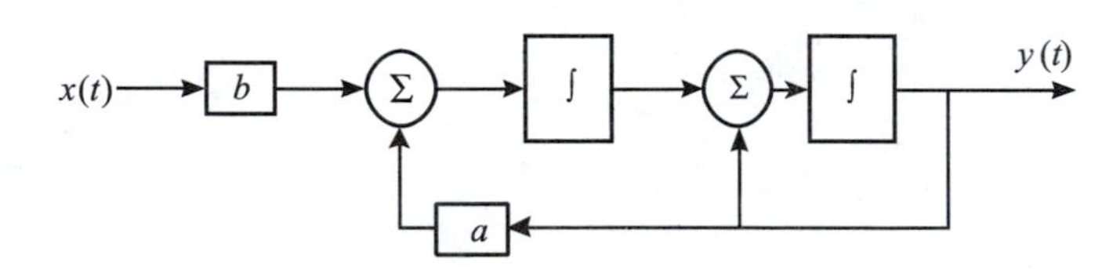

FIGURE 2.68
System for EOCP 2.12.

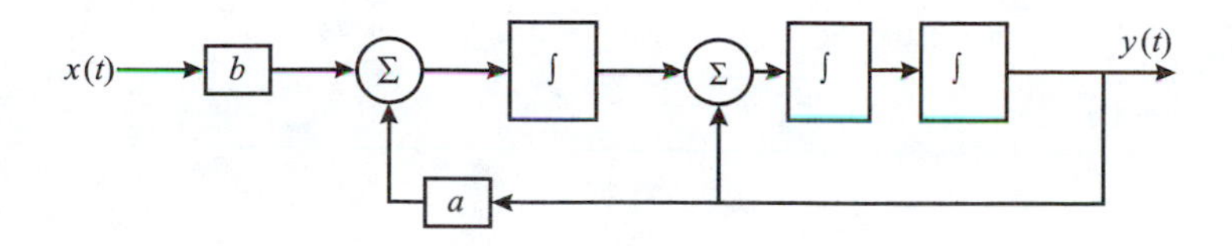

FIGURE 2.69
System for EOCP 2.12.

References

Cogdell, J.R. *Foundations of Electrical Engineering*, 2nd ed., Englewood Cliffs, NJ: Prentice-Hall, 1996.

Denbigh, P. *System Analysis and Signal Processing*, Reading, MA: Addison-Wesley, 1998.

Golubitsky, M. and Dellnitz, M. *Linear Algebra and Differential Equations Using MATLAB*, Stamford, CT: Brooks/Cole, 1999.

Harman, T.L., Dabney, J., and Richert, N. *Advanced Engineering Mathematics with MATLAB*, Stamford, CT: Brooks/Cole, 2000.

The Math Works. *The Student Edition of MATLAB*, Englewood Cliffs, NJ: Prentice-Hall, 1997.

Nilson, W.J. and Riedel, S.A. *Electrical Circuits*, 6th ed., Englewood Cliffs, NJ: Prentice-Hall, 2000.

Perko, L. *Differential Equations and Dynamical Systems*, New York: Springer-Verlag, 1991.

Phillips, C.L. and Parr, J.M. *Signals, Systems, and Transforms*, 2nd ed., Englewood Cliffs, NJ: Prentice-Hall, 1999.

Pratap, R. *Getting Started with MATLAB 5*, Oxford, 1999.

Strum, R.D. and Kirk, D.E. *Contemporary Linear Systems*, Boston: PWS, 1996.

Woods, R.L. and Lawrence, K.L. *Modeling and Simulation of Dynamic Systems*, Englewood Cliffs, NJ: Prentice-Hall, 1997.

Wylie, R.C. and Barrett, C.L. *Advanced Engineering Mathematics*, 6th ed., New York: McGraw-Hill, 1995.

Ziemer, R.E., Tranter, W.H., and Fannin, D.R. *Signals Systems Continuous and Discrete*, 4th ed., Englewood Cliffs, NJ: Prentice-Hall, 1998.

3

Fourier Series

CONTENTS

3.1 Review of Complex Numbers

This is a good point to give a brief review of complex numbers because this chapter and the chapters following are heavily involved with their arithmetic manipulation.

3.1.1 Definition

By definition, the complex number j is the square root of -1. In general, the complex number $C = P + jQ$ consists of two parts; the real part, P, and the imaginary part, Q. C can be represented in many forms as we will see later. If we are multiplying or dividing complex numbers we will prefer to use the polar form. If we are adding or subtracting complex numbers we will rather use the rectangular form. The reason for that is the ease each form provides in the corresponding calculation.

Consider two complex numbers C_1 and C_2 where $C_1 = P_1 + jQ_1$ and $C_2 = P_2 + jQ_2$. These complex numbers are given in the rectangular form.

3.1.2 Addition

When we add two complex numbers we add their real parts and their imaginary parts together to form the addition

$$C_1 + C_2 = (P_1 + P_2) + j(Q_1 + Q_2)$$

3.1.3 Subtraction

When we subtract two complex numbers we subtract their real parts and their imaginary parts to form the subtraction

$$C_1 - C_2 = (P_1 - P_2) + j(Q_1 - Q_2)$$

3.1.4 Multiplication

Let us now consider the two complex numbers in polar form. In polar form C_1 and C_2 are represented as

$$C_1 = M_1 e^{j\theta 1}$$

and

$$C_2 = M_2 e^{j\theta 2}$$

where

$$M_1 = \sqrt{P_1^2 + Q_1^2}$$
$$M_2 = \sqrt{P_2^2 + Q_2^2}$$

and

$$\theta_1 = \tan^{-1}\left[\frac{Q_1}{P_1}\right]$$

$$\theta_2 = \tan^{-1}\left[\frac{Q_2}{P_2}\right]$$

We can also represent the complex numbers C_1 and C_2 as

$$M_1\angle\theta_1$$

and

$$M_2\angle\theta_2$$

where M_1, M_2, θ_1, and θ_2 are as given above. The complex number $M_1\angle\theta_1$ is read as a complex number with magnitude M_1 and phase angle θ_1. Complex numbers are easily multiplied in polar form as

$$C_1C_2 = M_1 e^{j\theta 1} M_2 e^{j\theta 2}$$

$$C_1C_2 = M_1M_2 e^{j\theta 1} e^{j\theta 2} = M_1M_2 e^{j(\theta 1+\theta 2)}$$

Or we can write

$$C_1C_2 = M_1M_2 \angle(\theta_1 + \theta_2)$$

If we have more than two complex numbers in polar form to be multiplied we use the same procedure. We multiply their magnitudes and add their phase angles to form the new product.

3.1.5 Division

To divide two complex numbers we divide their magnitudes and subtract their phase angles.

$$C_1/C_2 = [M_1 e^{j\theta 1}]/[M_2 e^{j\theta 2}]$$

$$C_1/C_2 = [M_1/M_2]e^{j(\theta 1-\theta 2)}$$

Or we can represent the division as

$$C_1/C_2 = [M_1/M_2]\angle(\theta_1 - \theta 2)$$

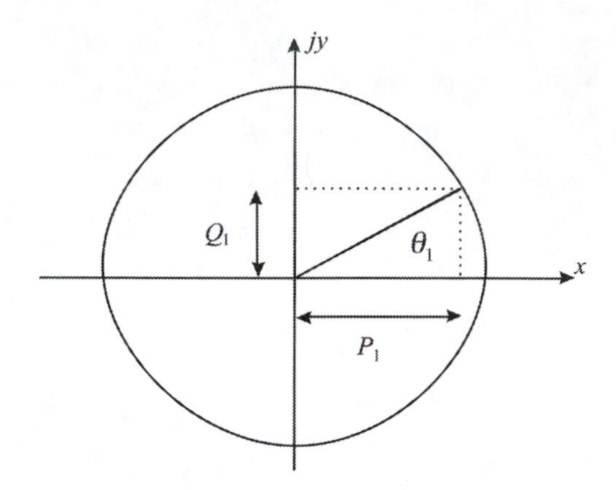

FIGURE 3.1
Rectangular form.

3.1.6 From Rectangular to Polar

Consider the complex number

$$C_1 = P_1 + jQ_1$$

This representation in the rectangular form is shown in Figure 3.1. To convert to polar form we write C_1 as

$$C_1 = M_1 e^{j\theta 1} = M_1 \angle \theta_1$$

where

$$M_1 = \sqrt{P_1^2 + Q_1^2}$$

and

$$\theta_1 = \tan^{-1}\left[\frac{Q_1}{P_1}\right]$$

3.1.7 From Polar to Rectangular

Consider the complex number

$$C_1 = M_1 e^{j\theta 1} = M_1 \angle \theta_1$$

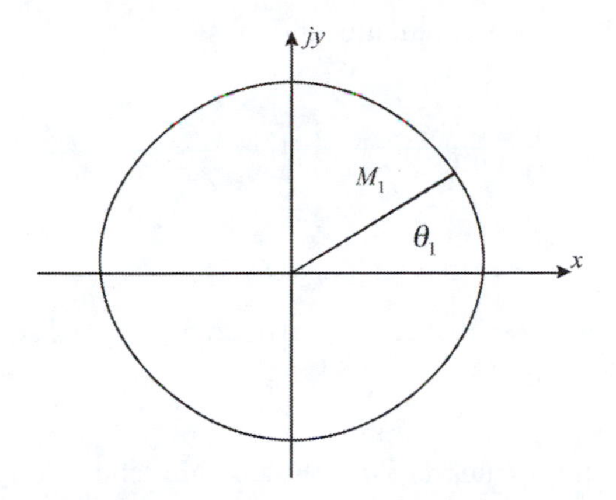

FIGURE 3.2
Polar form.

in polar form as seen in Figure 3.2.

To convert to its equivalent rectangular form we write C_1 as

$$C_1 = P_1 + jQ_1$$

where

$$M_1^2 = P_1^2 + Q_1^2$$

and

$$\tan(\theta_1) = \left[\frac{Q_1}{P_1}\right]$$

These two equations can be solved simultaneously for P_1 and Q_1.

3.2 Orthogonal Functions

A set of signals, Φ_i, $i = 0, \pm 1, \pm 2, \ldots$ are orthogonal over an interval (a, b) if

$$\int_a^b \Phi_l \Phi_k^* dt = \begin{cases} E_k & l = k \\ 0 & l \neq k \end{cases} = E_k \delta(l - k) \tag{3.1}$$

where * indicates complex conjugate, and

$$\delta(l-k) = \begin{cases} 1 & l = k \\ 0 & l \neq k \end{cases}$$

Example 3.1

Show that $\Phi_m = \sin(mt)$, $m = 1, 2, 3, \ldots$ form an orthogonal set on $(-\pi, \pi)$.

Solution

We will use the defining integral for orthogonal functions to get

$$\int_{-\pi}^{\pi} \Phi_m(t)\Phi_n^*(t)dt = \int_{-\pi}^{\pi} \sin(mt)\sin(nt)dt$$

$$= 1/2\int_{-\pi}^{\pi} \cos(m-n)tdt - 1/2\int_{-\pi}^{\pi} \cos(m+n)tdt$$

If $m = n$

$$\int_{-\pi}^{\pi} \Phi_m(t)\Phi_n^*(t)dt = 1/2\int_{-\pi}^{\pi} \cos(0)tdt - 1/2\int_{-\pi}^{\pi} \cos(2m)tdt$$

But cos(0) = 1 and

$$\int_{-\pi}^{\pi} \cos(2mt)dt = 0$$

Therefore,

$$\int_{-\pi}^{\pi} \Phi_m(t)\Phi_n^*(t)dt = 1/2\int_{-\pi}^{\pi} dt = 2\pi/2 = \pi$$

If $m \neq n$ then let $m - n = p$ and $m + n = q$, where p and q are integers. Then we will have

$$\int_{-\pi}^{\pi} \Phi_p(t)\Phi_q^*(t)dt = 1/2\int_{-\pi}^{\pi} \cos(pt)dt - 1/2\int_{-\pi}^{\pi} \cos(qt)dt$$

With the fact that

$$\int_{-\pi}^{\pi} \cos(2mt)dt = 0$$

we have

$$\int_{-\pi}^{\pi} \Phi_p(t)\Phi_q^*(t)dt = 0 - 0 = 0$$

Finally

$$\int_{-\pi}^{\pi} \Phi_m(t)\Phi_n^*(t)dt = \begin{cases} \pi & \text{for } n = m \\ 0 & \text{for } n \neq m \end{cases}$$

Therefore, the set $\Phi_m = \sin(mt)$ with $m = 1, 2, 3, \ldots$ is an orthogonal set.

3.3 Periodic Signals

A signal, $v(t)$, is periodic with a period T, if $v(t) = v(t \pm nT)$ where $n = 1, 2, 3, \ldots$. A periodic signal repeats itself every $\pm\, nT$ units of time. This is illustrated in Figure 3.3.

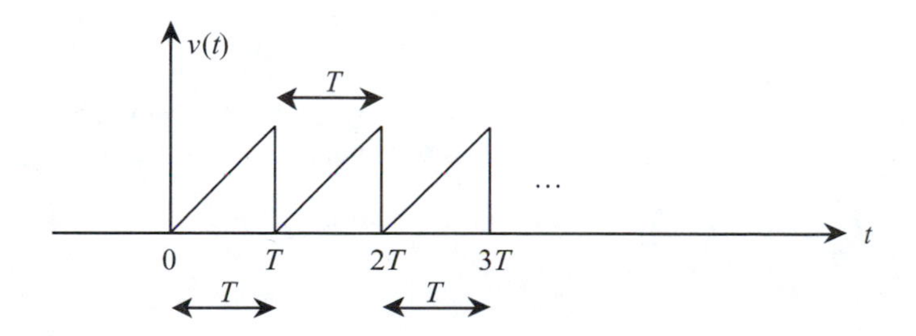

FIGURE 3.3
Periodic signal.

3.4 Conditions for Writing a Signal as a Fourier Series Sum

The following conditions must be satisfied for any signal to be written as a Fourier series sum.

1. $v(t)$ must be a single-valued signal, a function of one variable.
2. There must be a finite number of discontinuities in the periodic interval.
3. There must be a finite number of maximum and minimum in the periodic interval.
4. The integral $\int_{t_0}^{t_0+T} |v(t)|dt$ must exist.

Any periodic function that satisfies the four conditions above can be written as a sum of basis functions.

3.5 Basis Functions

In the case of Fourier series representation, we will consider the set

$$\left\{ e^{\frac{j2\pi n}{T}t} \right\}_{n=-\infty}^{n=+\infty}$$

where n is an integer, as the set of basis functions.

Therefore, any signal, $x(t)$, that satisfies the four conditions given earlier can be represented as a sum of these basis functions. Mathematically we write

$$x(t) = \sum_{-\infty}^{+\infty} c_n e^{\frac{j2\pi n}{T}t} \tag{3.2}$$

with $n = 0 \pm 1, \pm 2, \ldots$

Our goal is to find c_n and $x(t)$ will be completely represented as a sum of these basis functions. Remember that the basis functions are orthogonal and

$$\int_0^T e^{\frac{j2\pi n}{T}t} e^{\frac{-j2\pi m}{T}t} dt = \begin{cases} T & \text{if } n = m \\ 0 & \text{if } n \neq m \end{cases}$$

By multiplying (Equation 3.2) by $e^{\frac{-j2\pi m}{T}t}$ we get

$$e^{\frac{-j2\pi m}{T}t} x(t) = \sum_{-\infty}^{+\infty} c_n e^{\frac{-j2\pi m}{T}t} e^{\frac{j2\pi n}{T}t}$$

Now we can integrate over one period to get

$$\int_0^T e^{\frac{-j2\pi m}{T}t} x(t)dt = \int_0^T \sum_{-\infty}^{+\infty} c_n e^{\frac{-j2\pi m}{T}t} e^{\frac{j2\pi n}{T}t} dt$$

$$\int_0^T e^{\frac{-j2\pi m}{T}t} x(t)dt = \sum_{-\infty}^{+\infty} c_n \int_0^T e^{\frac{-j2\pi m}{T}t} e^{\frac{j2\pi n}{T}t} dt$$

Remember also that

$$\int_0^T e^{\frac{j2\pi n}{T}t} e^{\frac{-j2\pi m}{T}t} dt = \begin{cases} T & \text{if } n = m \\ 0 & \text{if } n \neq m \end{cases}$$

Therefore,

$$\int_0^T e^{\frac{-j2\pi m}{T}t} x(t)\, dt = Tc_n$$

and

$$c_n = 1/T \int_0^T x(t) e^{\frac{-j2\pi n}{T}t} dt \tag{3.3}$$

where in (Equation 3.3) we set $m = n$ where $n = 0, \pm 1, \pm 2, \ldots$.

3.6 The Magnitude and the Phase Spectra

We will denote $|c_n|$ as the magnitude spectra of $x(t)$ and $\langle c_n$ as the angle spectra of $x(t)$ or we call it the phase of $x(t)$. Notice that

$$C_{-n} = 1/T \int_0^T x(t) e^{\frac{j2\pi n}{T}t} dt$$

and

$$|c_{-n}| = |c_n| = \left| 1/T \int_0^T x(t) e^{\frac{j2\pi n}{T}t} dt \right|$$

since the magnitude of e^{jz} and the magnitude of e^{-jz} are always unity. We also see that $|c_n|$ has even symmetry and in the same way we can deduce that $\langle c_n$ has odd symmetry.

3.7 Fourier Series and the Sin-Cos Notation

In terms of sin and cos functions, $x(t)$ can be expressed as

$$x(t) = a_0 + \sum_{n=1}^{\infty} a_n \cos\left(\frac{2\pi nt}{T}\right) + \sum_{n=1}^{\infty} b_n \sin\left(\frac{2\pi nt}{T}\right) \tag{3.4}$$

where

$$a_0 = c_0 = 1/T \int_0^T x(t)dt \tag{3.5}$$

$$a_n = 2/T \int_0^T x(t)\cos\left(\frac{2\pi nt}{T}\right)dt \tag{3.6}$$

$$b_n = 2/T \int_0^T x(t)\sin\left(\frac{2\pi nt}{T}\right)dt \tag{3.7}$$

$$c_n = 1/T \int_0^T x(t)e^{\frac{-j2\pi n}{T}t} dt \tag{3.8}$$

where

$$a_n = 2\text{Real}(c_n)$$

and

$$b_n = -2\text{Imag}(c_n)$$

while Real stands for real part and Imag stands for imaginary part.

Example 3.2

Express the signal, $v(t)$, shown in Figure 3.4, as a sum of exponential Fourier series and then in terms of the sin-cos notations.

Solution

The four conditions needed to represent $v(t)$ in terms of the Fourier series are satisfied.

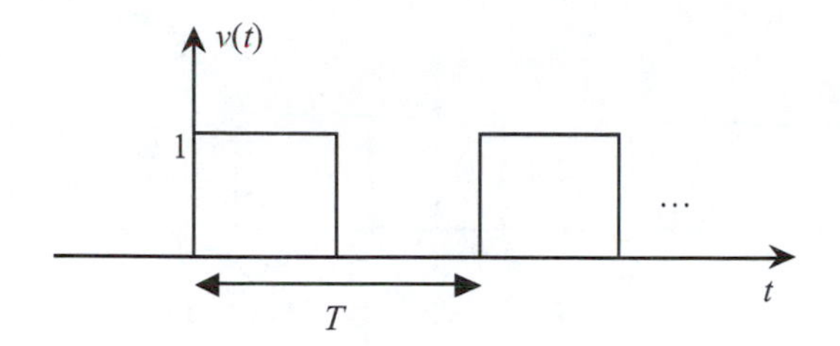

FIGURE 3.4
Signal for Example 3.2.

Therefore,

$$c_n = 1/T \int_0^T v(t) e^{\frac{-j2\pi n}{T}t} dt$$

Substituting for $v(t)$ we get

$$c_n = 1/T \int_0^{T/2} (1) e^{\frac{-j2\pi n}{T}t} dt + 1/T \int_{T/2}^{T/2} (0) e^{\frac{-j2\pi n}{T}t} dt$$

$$c_n = 1/T \left[\frac{e^{\frac{-j2\pi n}{T}t}}{-j2\pi n/T} \right]_0^{T/2}$$

$$c_n = \left[\frac{e^{\frac{-j2\pi n}{2T}T} - 1}{-j2\pi n} \right] = \left[\frac{e^{\frac{-j2\pi n}{2}} - 1}{-j2\pi n} \right]$$

and finally

$$c_n = \left[\frac{e^{-jn\pi} - 1}{-j2n\pi} \right]$$

Using Euler's identity we can simplify the above equation and write

$$c_n = [\cos(n\pi) - j\sin(n\pi) - 1]/[-j2n\pi]$$

But $\sin(n\pi)$ is always zero for any integer n. Therefore c_n reduces to

$$c_n = [\cos(n\pi) - 1]/[-j2n\pi] = \begin{cases} \dfrac{1}{jn\pi} & \text{for } n \text{ odd} \\ 0 & \text{for } n \text{ even and } n \neq 0 \\ 1/2 & \text{for } n = 0 \end{cases}$$

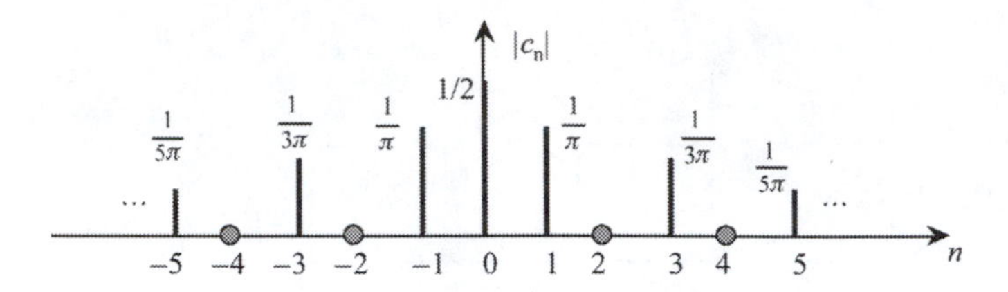

FIGURE 3.5
Signal for Example 3.2.

and the magnitude is

$$|c_n| = \begin{cases} \dfrac{1}{|n|\pi} & \text{for } n \text{ odd} \\ 0 & \text{for } n \text{ even and } n \neq 0 \\ 1/2 & \text{for } n = 0 \end{cases}$$

The phase of c_n is calculated as

$$\langle c_n = \begin{cases} \dfrac{-\pi}{2} & \text{for } n \text{ odd and negative} \\ \dfrac{\pi}{2} & \text{for } n \text{ odd and positive} \\ 0 & \text{for } n \text{ even and zero} \end{cases}$$

The plots in Figures 3.5 and 3.6 illustrate these results.

Now let us express $v(t)$ in terms of the sin-cos representation. We will evaluate c_n at $n = 0$ to get $a_0 = c_0 = 1/2$. We can also calculate c_0 using its defining integral as

$$\begin{aligned} c_0 &= 1/T \int_0^T v(t)dt \\ &= \int_0^{T/2} (1)dt + 1/T \int_{T/2}^T (0)dt = 1/2 \end{aligned}$$

$$\begin{aligned} a_n &= 2\,\mathrm{Re}\,al(c_n) \\ &= 2\mathrm{Real}\{[\cos(n\pi) - 1]/[-j2n\pi]\} \\ &= 2\mathrm{Real}\{[j\cos(n\pi) - j]/[2n\pi]\} = 0 \end{aligned}$$

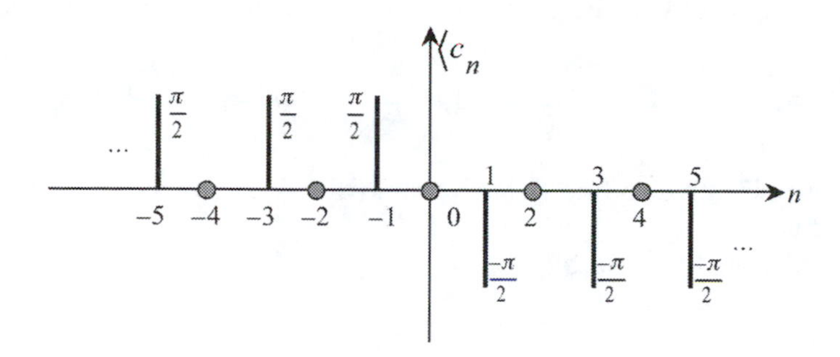

FIGURE 3.6
Signal for Example 3.2.

$$b_n = -2\mathrm{Im}\,ag\{[\cos(n\pi) - 1]/[-j2n\pi]\}$$

$$= -2[\cos(n\pi) - 1]/[2n\pi] = \begin{cases} \dfrac{4}{2n\pi} & \text{for } n \text{ odd} \\ 0 & \text{for } n \text{ even} \end{cases}$$

Finally we substitute in the general form

$$x(t) = a_0 + \sum_{n=1}^{\infty} a_n \cos\left(\frac{2\pi nt}{T}\right) + \sum_{n=1}^{\infty} b_n \sin\left(\frac{2\pi nt}{T}\right)$$

to get the approximation to the signal $v(t)$ as

$$x(t) = 1/2 + \sum_{n \text{ odd}} \left(\frac{2}{n\pi} \sin\left(\frac{2n\pi}{T} t\right)\right)$$

3.8 Fourier Series Approximation and the Resulting Error

In the Fourier series representation of $x(t)$, the integer n runs from one to infinity. As n approaches infinity, the series converges to the true function representation, $x(t)$. If we fix the value for the integer n, we will be truncating in effect the Fourier series and we will have an error in the approximation of $x(t)$, the given signal. We will take a look at the mean squared error which uses the average difference between the truncated $x(t)$ and the original $x(t)$ over one period and square the difference.

Let the truncated $x(t)$ be given by $X_N(t)$ (we truncate at $n = N$), where

$$X_N(t) = \sum_{n=-N}^{N} d_n\, e^{(jnwt)}$$

It is our objective to find d_n such that $x(t) - X_N(t)$ has the minimum mean squared value.

Let us denote the error by

$$E(t) = x(t) - x_N(t)$$

By substituting we get

$$E(t) = \sum_{n=-\infty}^{\infty} c_n\, e^{(jnwt)} - \sum_{n=-N}^{N} d_n\, e^{(jnwt)}$$

Let

$$z_n = \begin{cases} c_n & |n| > N \\ c_n - d_n & -N < n < N \end{cases}$$

Therefore,

$$E(t) = \sum_{n=-\infty}^{\infty} z_n\, e^{(jnwt)} \tag{3.9}$$

Notice that $E(t)$ is periodic with period $T = 2\pi/w$ and the mean squared error (MSE) is

$$\text{MSE} = \frac{1}{T}\int_{\langle T\rangle} |E(t)|^2 dt$$

where here we are integrating over one period. Substituting for $E(t)$ in the last integral and after some mathematical manipulations we get

$$\text{MSE} = \sum_{n=-\infty}^{\infty} |z_n|^2$$

$$\text{MSE} = \sum_{n=-N}^{N} |c_n - d_n|^2 + \sum_{|n|>N} |c_n|^2 \tag{3.10}$$

The above equation has two terms and both are positive. Therefore, to minimize the MSE we need to select $c_n = d_n$. The MSE finally is given as

$$\text{MSE} = \sum_{|n|>N} |c_n|^2 \tag{3.11}$$

3.9 The Theorem of Parseval

Let us take a look again at the formula we use to calculate the average power of a periodic signal

$$p_{\text{average}} = \frac{1}{T}\int_{\langle T\rangle} |x(t)|^2 dt \tag{3.12}$$

The root-mean-square (*rms*) of the average power, on the other hand, can be calculated as

$$p_{\text{average}}(rms) = \sqrt{\frac{1}{T}\int_{\langle T\rangle} |x(t)|^2 dt} \tag{3.13}$$

If $x(t) = c_n e^{jnwt}$ then

$$p_{\text{average}} = \frac{1}{T}\int_{\langle T\rangle} |c_n e^{jnwt}|^2 dt$$

$$p_{\text{average}} = \frac{1}{T}\int_{\langle T\rangle} |c_n|^2 dt = |c_n|^2 \tag{3.14}$$

This last equation is the Parseval equation for periodic signals.

In Example 3.1 there are two ways to calculate the average power and these two ways are presented here.

1. Using the average power formula

   The average power in this case is calculated as

$$p_{\text{average}} = \frac{1}{T}\int_{\langle T\rangle} |v(t)|^2 dt = \frac{1}{T}\int_0^{T/2} |(1)|^2 dt = 1/2$$

2. Using the theorem of Parseval

$$p_{\text{average}} = \sum_{n=-\infty}^{\infty} |c_n|^2$$

$$= \underbrace{(1/2)^2}_{\text{for } n=0} + \underbrace{(1/\pi)^2 + (1/3\pi)^2 + \dots}_{\text{right side of magnitude spectra}} + \underbrace{(1/\pi)^2 + (1/3\pi)^2 + \dots}_{\text{left side of magnitude spectra}}$$

$$p_{\text{average}} = (1/2)^2 + 2[(1/\pi)^2 + (1/3\pi)^2 + (1/5\pi)^2 + (1/7\pi)^2 + \dots]$$

But for large enough n, p_{average} will approach 1/2.

3.10 Systems with Periodic Inputs

In general and in terms of time domain representation, the output of any LTI system is given by the convolution equation

$$y(t) = \int_{-\infty}^{\infty} h(\tau)x(t-\tau)d\tau$$

If $x(t)$ is periodic, then it can be written as

$$x(t) = e^{jwt}$$

In this case we will have $y(t)$ rewritten as

$$y(t) = \int_{-\infty}^{\infty} h(\tau)e^{jw(t-\tau)}d\tau = e^{jwt}\int_{-\infty}^{\infty} h(\tau)e^{-jw\tau}d\tau$$

Let us define $H(w)$ as

$$H(w) = \int_{-\infty}^{\infty} h(\tau)e^{-jw\tau}d\tau$$

Therefore

$$y(t) = H(w)e^{jwt} \tag{3.15}$$

$H(w)$ is called the system transfer function. When we use the Fourier series representation, Equation (3.15) can be written as

$$y(t) = \sum_{-\infty}^{\infty} H(nw)c_n e^{jnwt} \tag{3.16}$$

In the above equation we considered the input in its Fourier series representation

$$x(t) = \sum_{-\infty}^{\infty} c_n e^{jnwt}$$

Example 3.3

Consider the following system.

$$\frac{d}{dt}y(t) + y(t) = x(t)$$

Let $x(t)$ be a periodic signal where $x(t) = e^{jwt}$. Find $H(w)$.

Solution

With $x(t) = e^{jwt}$ and $y(t) = H(w)e^{jwt}$ we have

$$\frac{d}{dt}y(t) = jwH(w)e^{jwt}$$

Therefore, the differential equation can be written as

$$jwH(w)e^{jwt} + H(w)e^{jwt} = e^{jwt}$$

We can drop the e^{jwt} term from both sides of the equation above (dropping the e^{jwt} term by dividing by e^{jwt} for e^{jwt} is always possible) to get

$$jwH(w) + H(w) = 1$$

from which

$$H(w) = 1/[1 + jw]$$

which is the system transfer function of the system described by the differential equation given in this example. We also have

$$H(nw) = 1/[1 + njw]$$

Example 3.4

Given $x(t)$ in Figure 3.7.

With $H(nw)$ as given in Example 3.3, find $y(t)$.

Solution

We found earlier that c_n for this signal is

$$c_n = [\cos(n\pi) - 1]/[-j2n\pi]$$

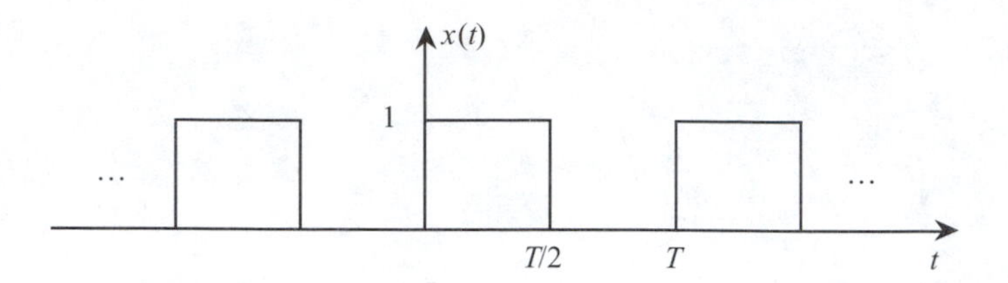

FIGURE 3.7
Signal for Example 3.4.

and

$$H(nw) = 1/[1 + njw]$$

Therefore

$$y(t) = \sum_{-\infty}^{\infty}\{1/[1 + njw]\}\{[\cos(n\pi) - 1]/[-j2n\pi]\}e^{jnwt}$$

3.11 A Formula for Finding $y(t)$ When $x(t)$ Is Periodic: The Steady-State Response

Let

$$x(t) = A\cos(wt + \phi)$$

and let us assume that we have the system transfer function $H(w)$. $y(t)$ in this case can be computed as

$$y(t) = |x(t)||H(w)|\cos(wt + \phi + \angle H(w)) \tag{3.17}$$

where $|\ \ |$ indicates the magnitude and $\angle$ indicates the phase. Our reference in this equation is the cos angle where ϕ is the phase angle of the input $x(t)$ and A is its amplitude.

Example 3.5

Let

$$H(w) = 1/[1 + jw]$$

and consider the input $x(t)$ to be

$$x(t) = \cos(t) - \cos(2t)$$

Find the output $y(t)$.

Solution

We can use the superposition principle here and let

$$x_1(t) = \cos(t)$$

and

$$x_2(t) = \cos(2t)$$

and find $y_1(t)$ that corresponds to $x_1(t)$ and $y_2(t)$ that corresponds to $x_2(t)$. After finding $y_1(t)$ and $y_2(t)$ we can add the two outputs and get $y(t)$.

For $x_1(t)$, $w = 1$ rad/sec and we have

$$H(w = 1) = 1/[1 + j(1)]$$

and $|x_1(t)| = 1$, $|H(1)| = \frac{1}{\sqrt{2}}$, $\phi_1 = 0$, and $\angle H(1) = -45°$

Therefore, the output $y_1(t)$ is

$$y_1(t) = |x_1(t)||H(1)|\cos(t + \phi_1 + \angle H(1))$$

$$y_1(t) = (1)\left(\frac{1}{\sqrt{2}}\right)\cos(t + 0 - 45°)$$

and finally

$$y_1(t) = \frac{1}{\sqrt{2}}\cos(t - 45°)$$

For $x_2(t)$, $w = 2$ rad/sec and we have

$$H(w = 2) = 1/[1 + j(2)]$$

and $|x_2(t)| = 1$, $|H(2)| = \frac{1}{\sqrt{5}}$, $\phi_2 = 0$, and $\angle H(2) = -63°$. Therefore, we write

$$y_2(t) = |x_2(t)||H(2)|\cos(2t + \phi 2 + \angle H(2))$$

$$y_2(t) = (1)\left(\frac{1}{\sqrt{5}}\right)\cos(2t + 0 - 63°)$$

and finally

$$y_2(t) = \frac{1}{\sqrt{5}}\cos(2t - 63°)$$

Now we can add the two solutions $y_1(t)$ and $y_2(t)$ (notice the minus sign before $x_2(t)$) to get

$$y(t) = \frac{1}{\sqrt{2}}\cos(t - 45°) - \frac{1}{\sqrt{5}}\cos(2t - 63°)$$

3.12 Some Insight: Why the Fourier Series?

We will discuss some of the many uses of the Fourier series.

3.12.1 No Exact Mathematical Form for $x(t)$

As we saw in this chapter, the Fourier series representation can be obtained for periodic signals with finite discontinuities and bounded magnitudes. In cases where the inputs to the systems that we study in real life have such shapes, periodicity, and magnitudes, it is very difficult to express these signals mathematically in order for us to evaluate the output, $y(t)$, due to such input signals.

Let us look at the signal in Figure 3.8.

We can start somewhere in time and proceed to the right to write

$$x(t) = \begin{cases} \vdots & \\ 1 & 0 < t < T/2 \\ 0 & T/2 < t < T \\ 1 & T < t < 3T/2 \\ \vdots & \end{cases}$$

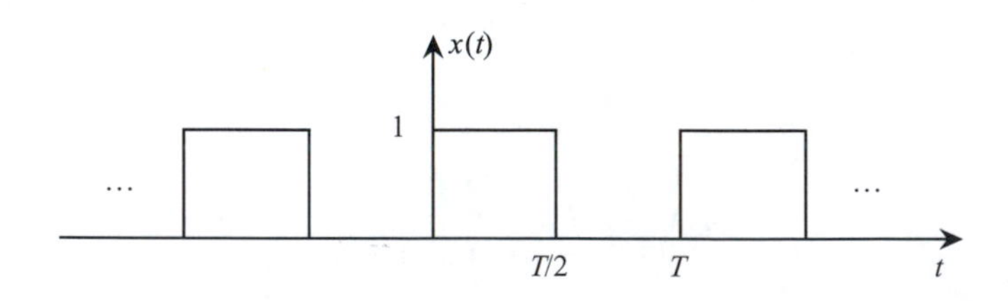

FIGURE 3.8
Periodic signal.

Assume that $x(t)$ is an input to a certain linear time-invariant system. We can use the method of superposition and use every interval for $x(t)$ to find its corresponding $y(t)$ and sum all the outputs to get the final output for the system. But how many mathematically defined inputs do you have? How many outputs will you have?

If we can write $x(t)$ as a sum of sin and cos terms in a compact mathematical, yet closely approximate the actual $x(t)$ form, we can find the corresponding $y(t)$ which should closely match the true output. This is one of the advantages the Fourier series offers.

3.12.2 The Frequency Components

The other advantage is the information we get from c_n, the Fourier series coefficient. We are able to identify the frequency contents of the input signal $x(t)$, both the magnitude and the phase spectrum. Knowing the frequency contents of the input signal gives us more insights into the way we analyze and design linear systems; we will know what frequencies will pass and what frequencies will not pass through a particular system.

3.13 End-of-Chapter Examples

EOCE 3.1

Evaluate each of the following expressions.

1. $\dfrac{1+j}{1-j}+\dfrac{1+2j}{1-2j}$

2. $\dfrac{1+j}{1-j}-\dfrac{3+j}{1-j}+\dfrac{1+j}{2-j}$

3. $\dfrac{1+j}{1-j}+\dfrac{1+2j}{1-2j}-1+j$

Use MATLAB in the calculations.

Solution

Every complex number has a magnitude and an angle. At the MATLAB prompt we type the following commands for expression 1.

```
Expression_1 = (1 + j)/(1 - j) + (1 + 2*j)/(1 - 2*j)
```

to get

```
Expression_1 =
-0.6000 + 1.8000i
Magnitude_of_Expression_1 = abs(Expression_1)
```

to get

```
Magnitude_of_Expression_1 =
1.8974
Angle_of_Expression_1 = angle(Expression_1)
```

to get

```
Angle_of_Expression_1 =
1.8925
```

For expression 2 we type the commands

```
Expression_2=(1+j)/(1-j)-(3+j)/(1-j)+(1+j)/(2-j)
```

to get

```
Expression_2 =
-0.8000 - 0.4000i
Magnitude_of_Expression_2 = abs(Expression_2)
```

to get

```
Magnitude_of_Expression_2 =
0.8944
Angle_of_Expression_2 = angle(Expression_2)
```

to get

```
Angle_of_Expression_2 =
-2.6779
```

For expression 3 we type the commands at the prompt

```
Expression_3=(1+j)/(1-j)+(1+2*j)/(1-2*j)-1+j
```

to get

```
Expression_3 =
-1.6000 + 2.8000i
Magnitude_of_Expression_3 = abs(Expression_3)
```

to get

```
Magnitude_of_Expression_3 =
3.2249
Angle_of_Expression_3 = angle(Expression_3)
```

to get

```
Angle_of_Expression_3 =
2.0899
```

EOCE 3.2

In Example 3.2 of this chapter we had the following problem to work: Express the signal, $v(t)$, shown again in Figure 3.9, as a sum of exponential Fourier series and then use the sin-cos notations to express the same signal. The final calculation for c_n was

$$c_n = [\cos(n\pi) - 1]/[-j2n\pi]$$

which is a complex number that has a magnitude and an angle.

Use MATLAB to show that the plots obtained before for the magnitude and the angle for c_n are accurate.

Solution

Here is the MATLAB script.

```
n = -10:10; % we fixed the range for n
cn = (cos(n*pi) -1) ./ (-2*j*n*pi);
cn(11) = 0.5; % cn = 1/2 at n = 0
```

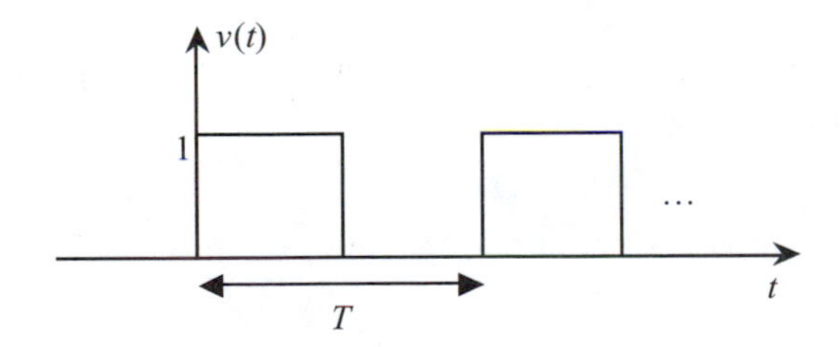

FIGURE 3.9
Signal for EOCE 3.2.

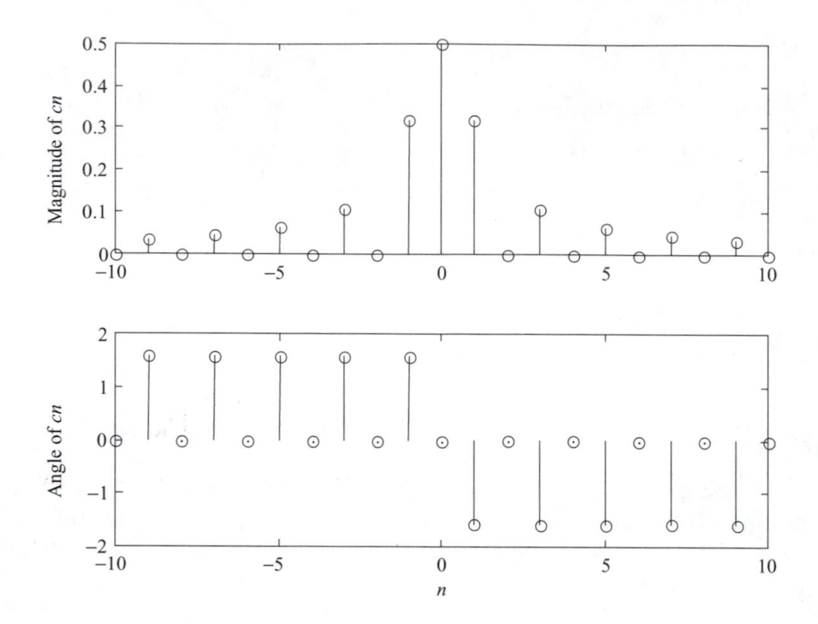

FIGURE 3.10
Plots for EOCE 3.2.

```
subplot (2, 1, 1), stem(n, abs(cn)),
ylabel('Magnitude of cn'),
subplot(2,1,2),stem(n,angle(cn)),ylabel('Angleofcn'),
xlabel('n');
```

The plots are shown in Figure 3.10.

For $T = 1$, we can plot approximation to the given signal. In Example 3.2 we have calculated $x(t)$ as

$$x(t) = 1/2 + \sum_{n \text{ odd}} \left(\frac{2}{n\pi} \sin\left(\frac{2n\pi}{T} t \right) \right)$$

Using MATLAB we can plot $x(t)$ to see how it can approximate $v(t)$ by writing the following script

```
clf % clear the screen
t = 0:0.001:1; % the interval where v(t) is defined
% (T = 1)
x = 0.5;
for n = 1:2:7, % considering 4 terms (n odd)
x = x + (2/(n*pi))*sin(2*n*pi*t);
end
```

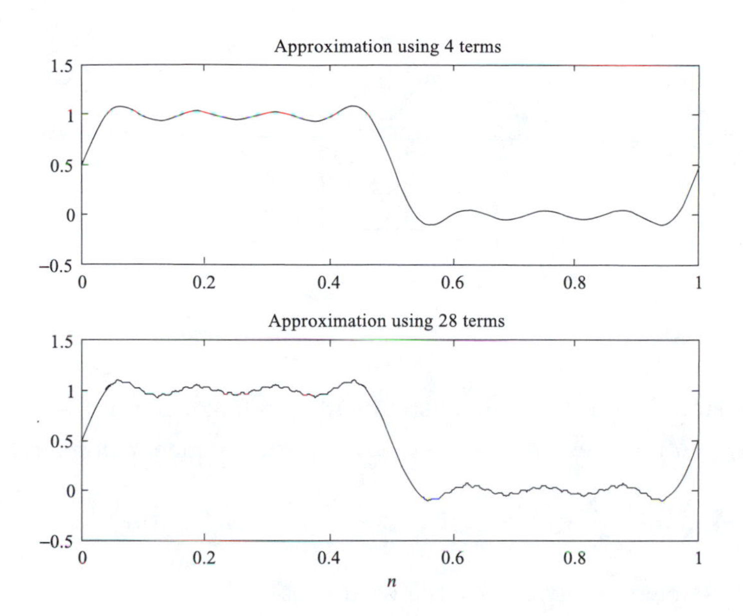

FIGURE 3.11
Plots for EOCE 3.2.

```
for n = 9:2:55, % considering 28 terms (n odd)
y = x + (2/(n*pi))*sin(2*n*pi*t);
end
subplot(2, 1, 1),plot(t, x),
title('Approximation using 4 terms'),
subplot(2, 1, 2),plot(t, y),
title('Approximation using 28 terms'),
xlabel('n');
```

The plots are shown in Figure 3.11.

EOCE 3.3

Consider the signal $x(t)$ shown in Figure 3.12 as an input to the linear first order system described by the differential equation

$$\frac{d}{dt}y + y = x(t)$$

1. Find an approximation to $x(t)$.
2. Plot the magnitude and phase of c_n.
3. Plot the approximation to $x(t)$.

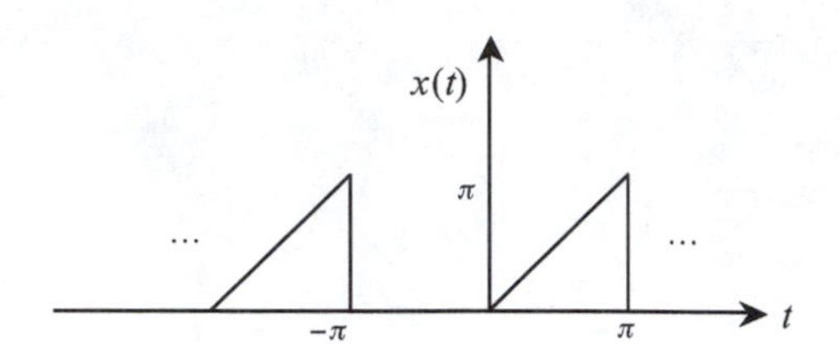

FIGURE 3.12
Signal for EOCE 3.3.

4. Find the output to the system with $n = 2$ and plot it.
5. Find the average power in the time and frequency domains.

Solution

1. Mathematically, $x(t)$ can be written as

$$x(t) = \begin{cases} 0 & -\pi < t < 0 \\ t & 0 < t < \pi \end{cases}$$

with period 2π.

The Fourier coefficients are determined as in the following.

$$a_0 = \frac{1}{2\pi}\int_{-\pi}^{\pi} x(t)dt$$

By substituting for $x(t)$ we arrive at

$$a_0 = \frac{1}{2\pi}\int_{0}^{\pi} tdt = \frac{\pi}{4}$$

and

$$a_n = \frac{1}{\pi}\int_{-\pi}^{\pi} x(t)\cos(nt)dt$$

Substituting again for $x(t)$ we obtain

$$a_n = \frac{1}{\pi}\int_{-\pi}^{0} 0\cos(nt)dt + \frac{1}{\pi}\int_{0}^{\pi} t\cos(nt)dt$$

Using the method of Integrating by Parts, we will get

$$a_n = \frac{1}{\pi}\left[\frac{t}{n}\sin(nt) + \frac{1}{n^2}\cos(nt)\right]_0^{\pi} = \frac{1}{n^2\pi}[\cos(n\pi) - 1]$$

for $n = 1, 2, 3\ldots$

We can find the other coefficient as in the following.

$$b_n = \frac{1}{\pi}\int_{-\pi}^{\pi} x(t)\sin(nt)dt$$

With $x(t)$ substituted we get

$$b_n = \frac{1}{\pi}\int_{-\pi}^{0} 0\sin(nt)dt + \frac{1}{\pi}\int_{0}^{\pi} t\sin(nt)dt$$

By carrying out the integration we get

$$b_n = \frac{1}{\pi}\left[-\frac{t}{n}\cos(nt) + \frac{1}{n^2}\sin(nt)\right]_0^{\pi} = -\frac{1}{n}[\cos(n\pi)]$$

for $n = 1, 2, 3\ldots$

By observing that $(-1)^n = \cos(n\pi)$ and that $a_n = 0$ for even n, we can write the approximation for $x(t)$ as

$$x(t) = \frac{\pi}{4} - \frac{2}{\pi}\sum_{1}^{\infty}\frac{\cos(2n-1)t}{(2n-1)^2} - \sum_{1}^{\infty}(-1)^n\frac{\sin(nt)}{n}$$

2. We can plot the magnitude and the phase of

$$c_n = \frac{1}{2n^2\pi}[(-1)^n - 1] + j\frac{1}{2n}(-1)^n$$

for $n = 0, \pm 1, \pm 2, \ldots$ by writing the following script.

```
n = -10:10;
cn = (1./(2*n.^2*pi)).*((-1).^n-1)+j*(1./(2*n)).*
(-1.^n);
cn(11) = pi/4; % cn = pi/4 at n = 0
```

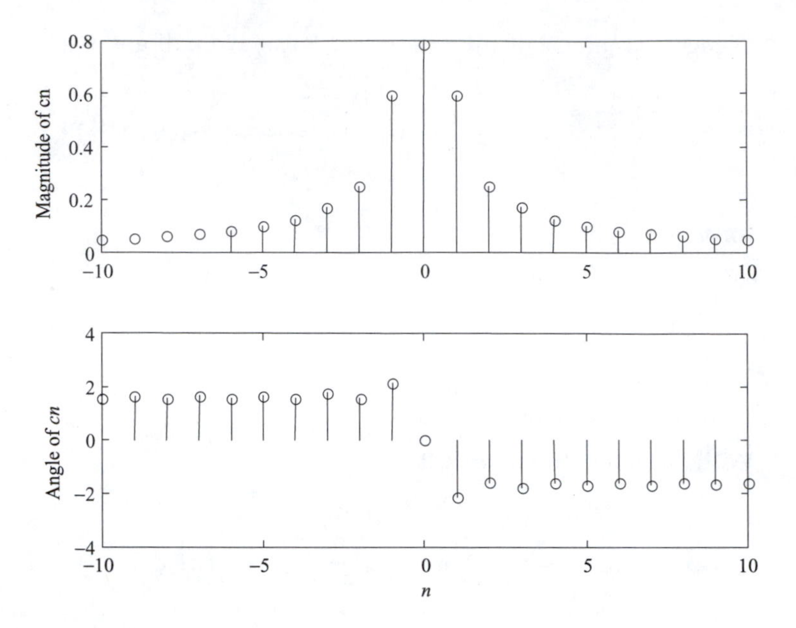

FIGURE 3.13
Plots for EOCE 3.3.

```
subplot(2, 1, 1), stem(n, abs(cn)),
ylabel('Magnitude of cn'),
subplot(2, 1, 2), stem(n, angle(cn)),
ylabel('Angle of cn'),
xlabel('n');
```

The plots are shown in Figure 3.13.

3. Let us now use MATLAB to plot the approximation, $x(t)$, to the triangular wave given using the script

```
% We first generate points in time for plotting
t = [2pi: .05:pi];
size_of_t = size(t);
F = (pi/4)*ones(size_of_t); % for Fourier series
for n = 1:6 % considering 6 points
F = F + (1/pi)*(-2*cos((2*n -1)*t)/(2*n -1)^2)- ...
((-1)^n*sin(n*t)/n);
end
% now we create x(t)
xplot = zeros(size_of_t);
```

```
for k = 1:length(t)
if t(k) < 0
xplot(k) = 0;
else
xplot(k) = t(k);
end
end
subplot(2, 1, 1), plot(t,F,t,xplot);
ylabel('x(t)')
title('Approximation to x(t), n = 6')
% we can add more point: add 15 more
for n = 7:21
F = F + (1/pi)*(-2*cos((2*n-1)t)/(2*n-1)^2)-...
((-1)^n*sin(n*t)/n);
end
subplot(2, 1, 2), plot(t, F, t, xplot);
xlabel('Time (sec)'),
ylabel('x(t)'),
title('Approximation to x(t), n = 21')
```

The plots are shown in Figure 3.14.

4. If we take $n = 2$, the input $x(t)$ will be

$$x(t) = \frac{\pi}{4} - \frac{2\cos(t)}{\pi} - \frac{2\cos(3t)}{9\pi} + \sin(t) - \frac{\sin(2t)}{2}$$

Writing terms as cos terms only so that our reference will be the cos function, $x(t)$ becomes

$$x(t) = \frac{\pi}{4} - \frac{2\cos(t)}{\pi} - \frac{2\cos(3t)}{9\pi} + \cos\left(t - \frac{\pi}{4}\right) - \frac{\cos\left(2t - \frac{\pi}{2}\right)}{2}$$

This input can be divided into 5 terms, $x_1(t), \ldots, x_5(t)$.

The transfer function of the system, $H(jw)$, is determined as

$$H(jw) = \frac{1}{jw + 1}$$

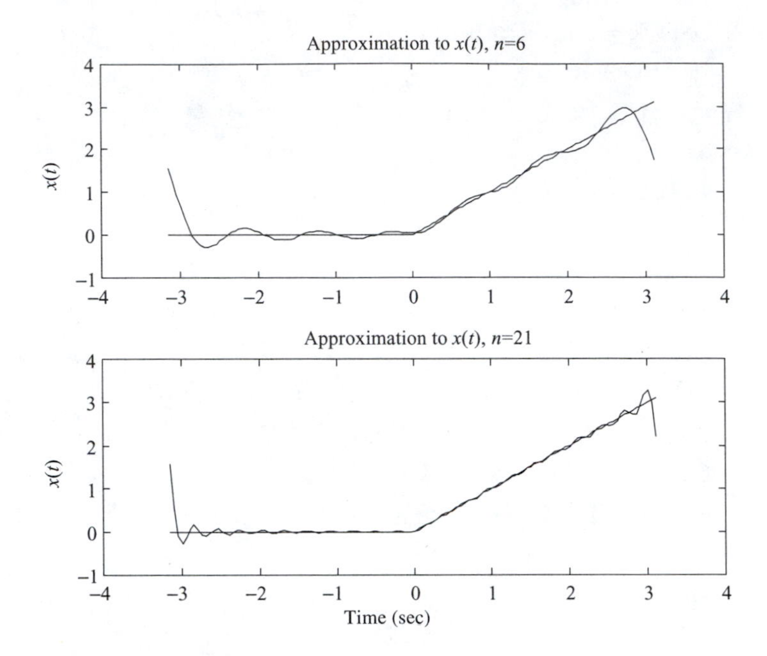

FIGURE 3.14
Plots for EOCE 3.3

We will use MATLAB commands plus the formula for calculating $y(t)$ if the input is sinusoidal in the following.

For the first term, $x_1(t) = \pi/4$, $w = 0$, and $H_1(jw) = 1$.

```
magnitude_of_x1 = abs(pi/4)
```

to get

```
magnitude_of_x1 =
0.7854
phase_of_x1 = angle(pi/4)
```

to get

```
phase_of_x1 =
0
magnitude_of_H1 = abs(1)
```

to get

```
magnitude_of_H1 = 1
```

```
phase_of_H1 = angle(1)
```

to get

```
phase_of_H1 =
0
```

Therefore,

$$y_1(t) = .7854$$

For the second term, $x_2(t) = 2(\cos(t))/\pi$, $w = 1$, and $H_2(jw) = 1/(1 + j)$. The magnitude of $x2(t) = 2/\pi$ and the phase is 0.

```
magnitude_of_H2 = abs(1/(1 + j))
```

to get

```
magnitude_of_H2 =
0.7071
```

```
phase_of_H2 = angle(1/(1 + j))
```

to get

```
phase_of_H2 =
-0.7854
```

Therefore,

$$y_2(t) = (2/\pi)(.7071)\cos(t - .7854)$$

For the third term, $x_3(t) = -2/(9\pi)\cos(3t)$, $w = 3$, and $H_3(jw) = 1/(1+3j)$. The magnitude of $x_3(t) = 2/(9\pi)$ and the phase is $= 0$.

```
magnitude_of_H3 = abs(1/(1 = 3*j))
```

to get

```
magnitude_of_H3 =
0.3162
```

```
phase_of_H3 = angle(1/(1 + 3*j))
```

to get

```
phase_of_H3 =
-1.2490
```

Therefore,

$$y_3(t) = [2/(9\pi)][.3162]\cos(3t - 1.249)$$

For the fourth term, $x_4(t) = \cos(t - \pi/2)$. The phase of $x_4(t) = -\pi/2$, $w = 1$, $|H_4(jw)| = |H_2(jw)| = .7071$, and phase of $H_4(jw)$ is the same as the phase of $H_2(jw) = -.7854$ rad.
Therefore,

$$y_4(t) = .7071\cos(t - \pi/2)$$

For the fifth term, $x_5(t) = \cos(2t - \pi/2)$, $w = 2$, and $H_4(jw) = 1/(1 + 2j)$. The magnitude of $x_5(t) = 1$ and the phase is $-\pi/2$.

```
magnitude_of_x5 = abs(1/(1 + 2*j))
```

to get

```
magnitude_of_x5 =
0.4472

phase_of_H5 = angle(1/(1 + 2*j))
```

to get

```
phase_of_H5 =
-1.1071
```

Therefore,

$$y_5(t) = (.4472)(1)\cos(2t - \pi/2 - 1.1071)$$

The total solution is

$$y(t) = y_1(t) + y_2(t) + y_3(t) + y_4(t) + y_5(t)$$

By substituting for the individual outputs we arrive at

$$\begin{aligned} y(t) = {} & .7854 - (2/\pi)(.7071)\cos(t - .7854) - [2/(9\pi)][.3162] \\ & \times \cos(3t - 1.249) + .7071\cos(t - \pi/2) \\ & - (.4472)(1)\cos(2t - \pi/2 - 1.1071) \end{aligned}$$

Let us now plot the response for $n = 1$ and $n = 2$. We use MATLAB to do that by writing the script

```
clf % to clear the screen
t=-pi:.05:pi;
y_for_n_1 = .7854 - (2/pi)*(.7071)*cos(t-.7854)+...
.7071*cos(t-pi/2);
y_for_n_2=.7854-(2/pi)*(.7071)*cos(t-.7854)...
-(2/(9*pi))*.3162*cos(3*t - 1.249)+...
.7071*cos(t-pi/2)-.4472*cos(2*t-pi/2-1.1071);
plot(t, y_for_n_1,'-',t, y_for_n_2, 'go');
xlabel('Time (sec)');
ylabel('y(t)');
legend('n = 1', 'n = 2', 4);
```

The plot is shown in Figure 3.15.

5. The average power in the signal is calculated as in the following. In time domain the average power is

$$P_{\text{average}} = \frac{1}{T}\int_{\langle T\rangle} |x(t)|^2 dt$$

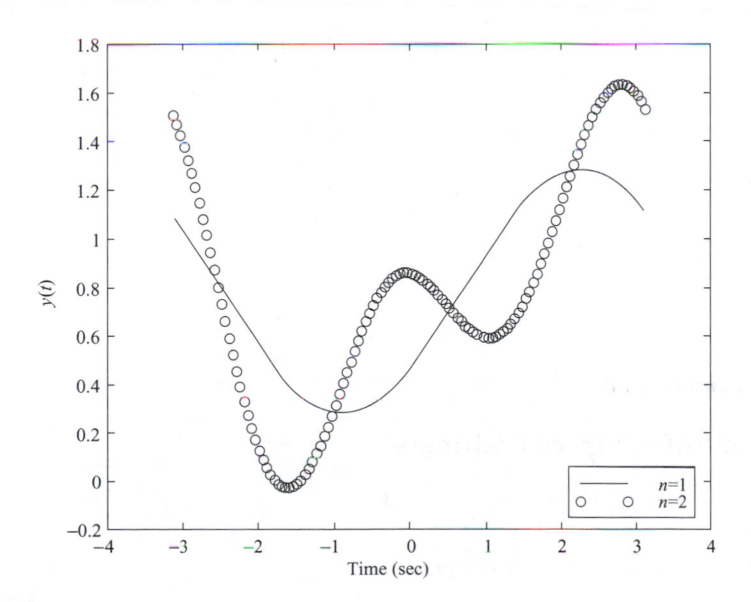

FIGURE 3.15
Plots for EOCE 3.3.

By substituting for $x(t)$ we get

$$P_{\text{average}} = \frac{1}{2\pi}\int_{-\pi}^{\pi}|x(t)|^2 dt = \frac{1}{2\pi}\int_{-\pi}^{0} 0dt + \frac{1}{2\pi}\int_{0}^{\pi} t^2 dt$$

By carrying out the integration we get

$$P_{\text{average}} = \frac{1}{2\pi}\int_{0}^{\pi} t^2 dt = \frac{1}{2\pi}\left[\frac{t^3}{3}\right]_0^{\pi} = \frac{\pi^2}{6}$$

In the frequency domain the average power is

$$P_{\text{average}} = \sum_{-\infty}^{\infty}|c_n|^2$$

and

$$c_n = \frac{1}{2n^2\pi}[\cos(n\pi) - 1] + j\left[\frac{1}{2n}\cos(n\pi)\right]$$

Remember that $|c_n|$ is even symmetric and $c_0 = \pi/4$. The average power becomes

$$P_{\text{average}} = \frac{\frac{\pi^2}{16}}{\text{term for } n = 0} + \frac{2}{2 - \text{sided spectra}} \times \left[\frac{1}{\pi^2} + \frac{1}{4} + \frac{1}{16} + \frac{1}{3^4\pi^2} + \frac{1}{4(9)} + \frac{1}{64} + \cdots\right]$$

Using the above equation, $p_{\text{average}} = 1.5338$ and is obtained by just considering five terms only. The actual average power is 1.6449. We know that the Parseval theorem, for large n, will result in an average power that is very close to 1.6449. The error in this example is about 11%.

3.14 End-of-Chapter Problems

EOCP 3.1

Perform the following complex arithmetic.

1. $\dfrac{1 + 2j}{1 - j} + 1$

2. $\dfrac{-2j}{1-j} - \dfrac{1}{2j} + j$

3. $\dfrac{1+2j}{j} + 1 - j$

4. $1 + 10j - \dfrac{1+10j}{1-j}$

5. $\dfrac{1+10j}{1-10j} + \dfrac{1+10j}{1-j} - 10$

EOCP 3.2

Put the following complex numbers in polar form.

1. $\dfrac{1-j}{1+j} + 2$

2. $\dfrac{-j}{1+j} + j + 1$

3. $\dfrac{1-2j}{-j} - 1 + j$

4. $-j + \dfrac{-j}{1-j}$

5. $\dfrac{-1+3j}{-1+10j} + \dfrac{1-j}{-1-j}$

EOCP 3.3

The angles are given in degrees. Put the following complex numbers in rectangular form.

1. $10\langle 90$

2. $10\langle 90 + \dfrac{\langle 120}{\langle -30}$

3. $10\langle -90 + \dfrac{(2\langle 0)(4\langle 120)}{\langle -30} - \langle -120$

4. $1\langle -60 - \dfrac{5\langle 45}{(12\langle -45)(2\langle -60)} - 3\langle -120 + \dfrac{5\langle -45}{1\langle -45}$

5. $10\langle 0 + \dfrac{\langle -120}{\langle -30} + (1\langle 180)(10\langle 90)$

EOCP 3.4

Consider the signals shown in Figures 3.16, 3.17, and 3.18. In each case find the mathematical approximation to $x(t)$.

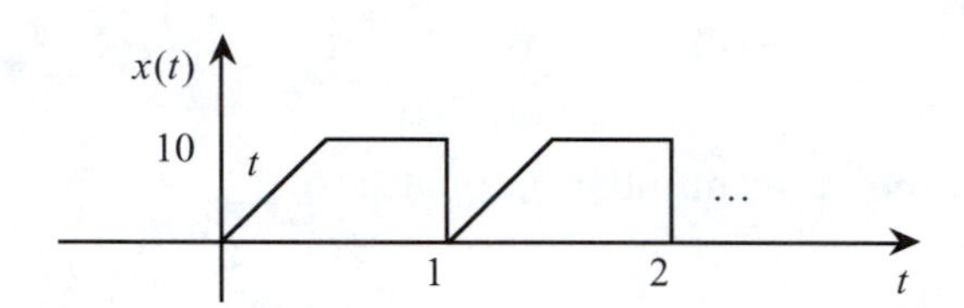

FIGURE 3.16
Signal for EOCP 3.4.

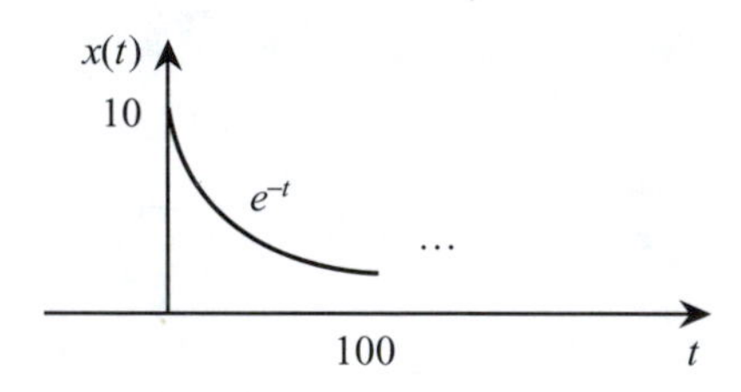

FIGURE 3.17
Signal for EOCP 3.4.

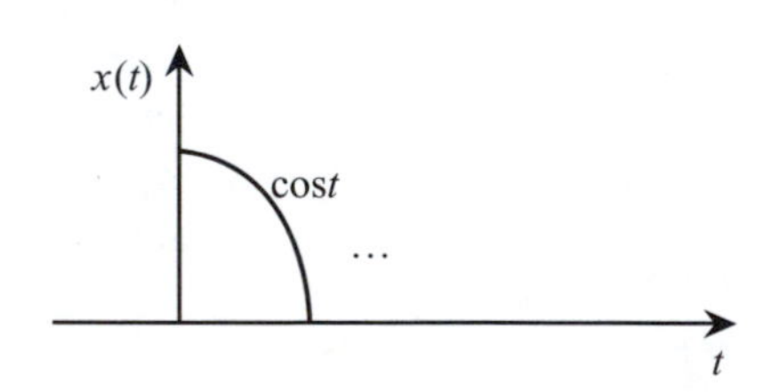

FIGURE 3.18
Signal for EOCP 3.4.

EOCP 3.5

Find the magnitude and the phase spectra for the signals in EOCP 3.4. Plot the results.

EOCP 3.6

Use the time-domain formula for the average power to find the average power in each signal in EOCP 3.4.

EOCP 3.7

Use the magnitude spectra to find the average power in each signal in EOCP 3.4.

EOCP 3.8

Find the system transfer function for each of the following systems.

1. $$\frac{d}{dt}y(t) + 5y(t) = x(t)$$

2. $$3\frac{d}{dt}y(t) + 3y(t) = 3x(t)$$

3. $$3\frac{d}{dt}y(t) + 3y(t) = x(t) + 3\frac{d}{dt}x(t)$$

4. $$\frac{d^2}{dt}y(t) - 1/2\,y(t) = x(t)$$

5. $$\frac{d^2}{dt}y(t) + 1/2\frac{d}{dt}y(t) = x(t) + \frac{d}{dt}x(t)$$

6. $$\frac{d^2}{dt}y(t) - 1/2\frac{d}{dt}y(t) = \frac{d}{dt}x(t)$$

7. $$\frac{d^2}{dt}y(t) + 1/2\frac{d}{dt}y(t) + y(t) = x(t)$$

8. $$\frac{d^2}{dt}y(t) + 10\frac{d}{dt}y(t) + 2\,y(t) = 10x(t)$$

9. $$2\frac{d^2}{dt}y(t) - 3\frac{d}{dt}y(t) = x(t)$$

10. $$2\frac{d^2}{dt}y(t) + 3\frac{d}{dt}y(t) - 3y(t) = x(t)$$

EOCP 3.9

Consider the systems with the input $x(t)$ as represented in Figure 3.19.

1. $$\frac{d}{dt}y(t) + 5y(t) = x(t)$$

2. $$3\frac{d}{dt}y(t) + 3y(t) = 3x(t)$$

3. $$3\frac{d}{dt}y(t) + 3y(t) = x(t) + 3\frac{d}{dt}x(t)$$

4. $$\frac{d^2}{dt}y(t) - 1/2\,y(t) = x(t)$$

5. $$\frac{d^2}{dt}y(t) + 1/2\frac{d}{dt}y(t) = x(t) + \frac{d}{dt}x(t)$$

Take only three terms from the approximated $x(t)$ to find the output for the first three systems. Take only two terms from the approximated $x(t)$ to find the output for the last two systems.

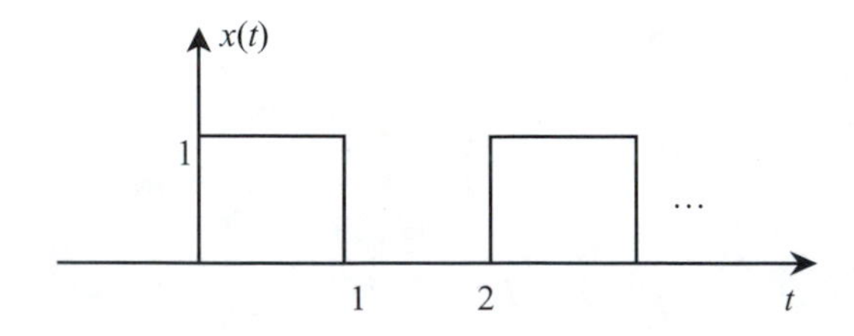

FIGURE 3.19
Signal for EOCP 3.9.

References

Cogdell, J.R. *Foundations of Electrical Engineering,* 2nd ed., Englewood Cliffs, NJ: Prentice-Hall, 1996.

Denbigh, P. *System Analysis and Signal Processing,* Reading, MA: Addison-Wesley, 1998.

Golubitsky, M. and Dellnitz, M. *Linear Algebra and Differential Equations Using MATLAB,* Stamford, CT: Brooks/Cole, 1999.

Harman, T.L., Dabney, J., and Richert, N. *Advanced Engineering Mathematics with MATLAB,* Stamford, CT: Brooks/Cole, 2000.

The MathWorks. *The Student Edition of MATLAB,* Englewood Cliffs, NJ: Prentice-Hall, 1997.

Nilson, W.J. and Riedel, S.A. *Electrical Circuits,* 6th ed., Englewood Cliffs, NJ: Prentice-Hall, 2000.

Phillips, C.L. and Parr, J.M. *Signals, Systems, and Transforms,* 2nd ed., Englewood Cliffs, NJ: Prentice-Hall, 1999.

Pratap, R. *Getting Started with MATLAB 5,* Oxford, 1995.

Strum, R.D. and Kirk, D.E. *Contemporary Linear Systems,* Boston: PWS, 1996.

Wylie, R.C. and Barrett, C.L. *Advanced Engineering Mathematics,* 6th ed., New York: McGraw-Hill, 1995.

Ziemer, R.E., Tranter, W.H., and Fannin, D.R. *Signals Systems Continuous and Discrete,* 4th ed., Englewood Cliffs, NJ: Prentice-Hall, 1998.

4

The Fourier Transform and Linear Systems

CONTENTS

4.1 Definition

The Fourier transform is a frequency domain transform that makes solution, design, and analysis of linear systems simpler. It also gives some insights about the frequency contents of signals where these insights are harder to see in real-time systems. There are other important uses for the Fourier transform but we will concentrate only on the issues introduced in this definition.

4.2 Introduction

We needed the Fourier series expansion to approximate signals that are periodic in time. In this chapter we will utilize Fourier series to try to express non-periodic signals using what we call the Fourier transform. The Fourier transform is the frequency-domain representation of non-periodic time-domain signals. Consider the time-domain periodic signal $x_p(t)$ in Figure 4.1.

Let us fix the size of the pulse (the height and the width) as seen in the graph in Figure 4.2. The height of the pulse is a, and the width of the pulse is

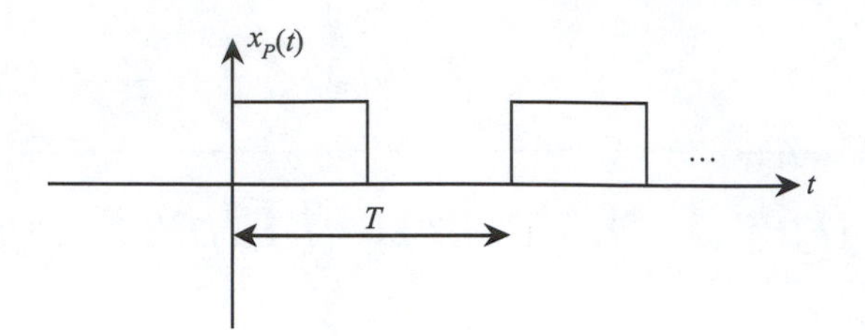

FIGURE 4.1
Periodic signal.

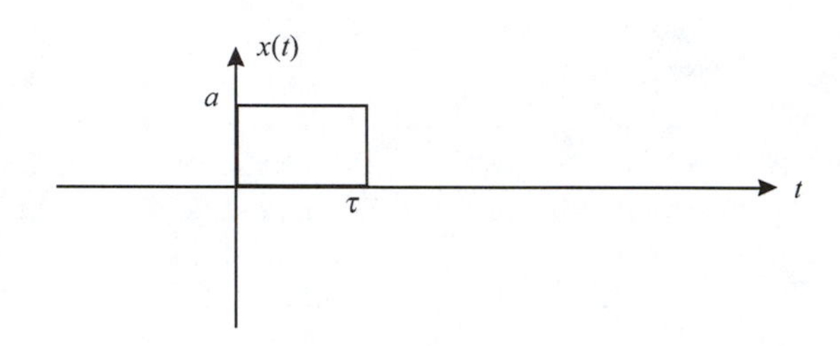

FIGURE 4.2
Signal when period is infinite.

τ units of time. Imagine now that the period T is extended to infinity. In this case our signal in Figure 4.1 will look like the signal in Figure 4.2.

You can see that this signal looks like a non-periodic signal. We also know that $w = 2\pi/T$. As $T \to \infty$, $w \to 0$. As w becomes very small, $\frac{w \to 0}{2\pi}$ can be written as $\frac{dw}{2\pi}$ where dw is a very small differential. In terms of Fourier series we write

$$x_p(t) = \sum_{n=-\infty}^{\infty} c_n e^{jnwt} \tag{4.1}$$

and the coefficients are

$$c_n = \frac{1}{T} \int_{\frac{-T}{2}}^{\frac{T}{2}} x_p(t) e^{-jnwt} dt \tag{4.2}$$

As $T \to \infty$, nw will be a continuous variable (w will be small). Substitute $dw/2\pi$ for $1/T$ in the equation for c_n above to get

$$\frac{c_n}{dw} = \frac{1}{2\pi} \int_{-\infty}^{+\infty} x_p(t) e^{-jwt} dt \tag{4.3}$$

In the equation for $x_p(t)$ above and in the limit as seen in the graph for $x(t)$, $x_p(t)$ becomes $x(t)$ and the summation becomes an integration. In this case

$$x(t) = \int_{-\infty}^{+\infty} \left[\underbrace{\left[\int_{-\infty}^{+\infty} x(t)e^{-jwt}dt \right]}_{\text{this whole term is a function of } w\text{, call it } X(w)} \right] e^{jwt} \frac{dw}{2\pi} \tag{4.4}$$

Therefore,

$$x(t) = \frac{1}{2\pi} \int_{-\infty}^{\infty} X(w)e^{jwt}dw \tag{4.5}$$

4.3 The Fourier Transform Pairs

In general, for any input $x(t)$ that satisfies the conditions

1. $\int_{-\infty}^{\infty} |x(t)|\, dt < \infty$
2. $x(t)$ has bounded variations.

$x(t)$ can be transformed into a domain called the Fourier transform domain. The transform pairs that take $x(t)$ from the time domain to $X(w)$ in the Fourier frequency domain, and vice versa, are

$$X(w) = \int_{-\infty}^{\infty} x(t)e^{-jwt}dt \tag{4.6}$$

$$x(t) = \frac{1}{2\pi} \int_{-\infty}^{\infty} X(w)e^{jwt}dw \tag{4.7}$$

These are referred to as the Fourier transform pairs. If $X(w)$ is the Fourier transform of $x(t)$ and $x(t)$ is the inverse transform of $X(w)$, then we can use the notation

$$x(t) \Leftrightarrow X(w)$$

to indicate the two-way transform from time to frequency and from frequency back to time.

Example 4.1

Find the Fourier transform of $x(t) = \delta(t)$.

Solution

Using the defining integral, the Fourier transform equation that takes $x(t)$ from its time domain and puts it in its frequency domain is

$$X(w) = \int_{-\infty}^{\infty} x(t)e^{-jwt}dt = \int_{-\infty}^{\infty} \delta(t)e^{-jwt}dt$$

Using the sifting property of the impulse signal we can simplify to get

$$X(w) = e^{-jw(0)} = 1$$

Therefore, we can write using the established notation

$$x(t) = \delta(t) \Leftrightarrow X(w) = 1$$

Example 4.2

Find the Fourier transform of $sgn(t)$.

Solution

Let us take the derivative of $sgn(t)$, the signal that was presented in Chapter 2, and write

$$\frac{d}{dt}sgn(t) = 2\delta(t)$$

Let us take the Fourier transform on both sides and use the Fourier transform derivative property as shown in Table 4.2 with $x(t) = sgn(t)$ to get

$$jwX(w) \Leftrightarrow 2(1)$$

This last expression says that the Fourier transform of $sgn(t)$ is $2/jw$ or

$$sgn(t) \Leftrightarrow 2/jw$$

Example 4.3

Find the Fourier transform of $x(t) = e^{jw_0t}$ and then deduce the Fourier transform of the constant signal $x(t) = k$.

Solution

Consider an impulse signal in the frequency domain as

$$X(w) = \delta(w - w_0)$$

The inverse transform of $X(w)$ is

$$x(t) = \frac{1}{2\pi}\int_{-\infty}^{\infty}\delta(w - w_0)e^{jwt}dt$$

By using the sifting property of the impulse function as explained in Chapter 2, we can write

$$x(t) = \frac{1}{2\pi}e^{jw_0t}$$

But $x(t)$ is the inverse transform of $\delta(w - w_0)$, therefore

$$e^{jw_0t} \Leftrightarrow 2\pi\delta(w - w_0)$$

Now let $w_0 = 0$. The above transform becomes

$$1 \Leftrightarrow 2\pi\delta(w)$$

Now we can use the linearity property and deduce

$$k \Leftrightarrow 2\pi k\delta(w)$$

Example 4.4

Consider the following signal $x(t)$ of period T, where $w = 2\pi/T$.

$$x(t) = \sum_{-\infty}^{\infty} c_n e^{jnw_0t}$$

Let $\Im$ represent the Fourier transform operator. What is the $\Im(x(t))$?

Solution

The Fourier transform of $x(t)$ is

$$X(w) = \sum_{-\infty}^{\infty} c_n \Im[e^{jnw_0t}]$$

which simplifies to

$$X(w) = \sum_{-\infty}^{\infty} c_n[2\pi\delta(w - nw_0)]$$

$X(w)$ can be viewed as a train of impulses on the w axis having values only at nw_0 where $n = 0, \pm1, \pm2, \pm3,\ldots$. These impulses are separated from each other by w_0.

We can also find the magnitude of $X(w)$ as

$$|X(w)| = |\sum_{-\infty}^{\infty} c_n \Im[e^{jnw_0t}]|$$

or

$$|X(w)| = 2\pi|c_n|$$

Example 4.5

Consider the following signal with period T.

$$x(t) = \sum_{-\infty}^{\infty} \delta(t - nT)$$

Find the Fourier transform of the given signal.

Solution

Let us first write the given signal as a sum of Fourier series. We can write the coefficients first as

$$c_n = \frac{1}{T}\int_{\langle T \rangle} x(t)e^{-j\frac{2\pi nt}{T}}\,dt$$

With $x(t)$ as the given train of impulses we can utilize the sifting property for integrals involving impulses to get

$$c_n = \frac{1}{T}\int_{\langle T \rangle} x(t)e^{-j\frac{2\pi nt}{T}}\,dt = \frac{1}{T}$$

With $x(t) = \Sigma_{-\infty}^{\infty}\delta(t - nT)$, the signal can be written as

$$x(t) = \sum_{-\infty}^{\infty} \frac{1}{T}\left[e^{\frac{j2\pi nt}{T}}\right]$$

from which

$$X(w) = \sum_{-\infty}^{\infty} \frac{1}{T} \Im\left[e^{\frac{j2\pi nt}{T}} \right]$$

or finally

$$X(w) = \frac{2\pi}{T} \sum_{n=-\infty}^{\infty} \delta\left(w - \frac{2n\pi}{T}\right)$$

In this last equation we used entry number 4 in Table 4.1 for the Fourier transform pairs.

TABLE 4.1 Fourier Transform Table: Selected Pairs

$x(t)$	$X(w)$
1. $u(t)$	$\pi\delta(w) + \frac{1}{jw}$
2. $\delta(t)$	1
3. 1	$2\pi\delta(w)$
4. e^{jw_0t}	$2\pi\delta(w - w_0)$
5. $\cos(w_0t)$	$\pi[\delta(w - w_0) + \delta(w + w_0)]$
6. $\sin(w_0t)$	$\pi/j[\delta(w - w_0) - \delta(w + w_0)]$
7. $\cos(w_0t)u(t)$	$\pi/2[\delta(w - w_0) + \delta(w + w_0)] + \frac{jw}{w_0^2 - w^2}$
8. $\sin(w_0t)u(t)$	$\pi/2j[\delta(w - w_0) + \delta(w + w_0)] + \frac{w_0}{w_0^2 - w^2}$
9. $e^{-at}u(t)\ a > 0$	$\frac{1}{a + jw}$
10. $te^{-at}u(t)\ a > 0$	$\left(\frac{1}{a + jw}\right)^2$
11. $\frac{t^{n-1}}{(n-1)!}e^{-at}u(t)\ a > 0$	$\left(\frac{1}{a + jw}\right)^n$
12. $e^{-a\|t\|}u(t)\ a > 0$	$\frac{2a}{a^2 + w^2}$
13. $e^{-at^2}u(t)\ a > 0$	$\sqrt{\frac{\pi}{a}}e^{\frac{-w^2}{4a}}$
14. $\sum_{n=-\infty}^{\infty} \delta(t - nT)$	$\frac{2\pi}{T}\sum_{n=-\infty}^{\infty} \delta\left(w - \frac{2n\pi}{T}\right)$

TABLE 4.2 Properties of the Fourier Transform

Time Signal	Fourier Transform Signal
1. $x(t)$	$X(w)$
2. $\sum_{n=1}^{N} a_n x_n(t)$	$\sum_{n=1}^{N} a_n X_n(w)$
3. $x(-t)$	$X(-w)$
4. $x(t - t_0)$	$X(w)e^{-jwt_0}$
5. $x(\alpha t)$	$\frac{1}{\lvert\alpha\rvert}X(w/\alpha)$
6. $\frac{d^n}{dt}x(t)$	$(jw)^n X(w)$
7. $x(t) * h(t)$	$X(w)H(w)$
8. $x(t)h(t)$	$\frac{1}{2\pi}X(w) * H(w)$
9. $x(t)e^{jw_0 t}$	$X(w - w_0)$
10. $\int_{-\infty}^{\infty} x(\tau)d\tau$	$\frac{X(w)}{jw} + \pi X(0)\delta(w)$
11. $x(t)/a$	$\frac{1}{\lvert a\rvert}X(w/a)$

Example 4.6

Let $x_1(t) = \delta(t)$ and $x_2(t) = e^{jw_0 t}$. Find the Fourier transform of $(3x_1(t) + 5x_2(t))$.

Solution

We can use the first entry in Table 4.2 to transform the indicated sum. $x_1(t)$ can be transformed using Table 4.1 as

$$x_1(1) \Leftrightarrow 1$$

With the help of Table 4.1 we can also write

$$x_2(t) \Leftrightarrow 2\pi\delta(w - w_0)$$

and therefore,

$$3x_1(t) + 5x_2(t) \Leftrightarrow 3(1) + 5(2)\pi\delta(w - w_0) = 3 + 10\pi\delta(w - w_0)$$

Example 4.7

Find the Fourier transform of $x(t) = 1 + 2u(t)$.

Solution

From entries 3 and 2 in Table 4.1 and entry 2 in Table 4.2 (the linearity property) we see that

$$1 \Leftrightarrow 2\pi\delta(w)$$

and

$$2u(t) \Leftrightarrow 2[\pi\delta(w) + 1/jw]$$

Therefore,

$$x(t) \Leftrightarrow X(w) = 4\pi\delta(w) + 2/jw$$

Example 4.8

Find the Fourier transform of $x(t - 2)$ where $x(t)$ is as given in Example 4.7 above.

Solution

Since in Example 4.7 we had $x(t) = 1 + 2u(t)$, therefore,

$$x(t-2) = 1 + 2u(t-2)$$

Using the time-shifting property, entry 4 in Table 4.2, we write

$$x(t-2) \Leftrightarrow X(w)e^{-2jw}$$

Therefore,

$$x(t-2) = 1 + 2u(t-2) \Leftrightarrow X(w)e^{-2jw} = [4\pi\delta(w) + 2/jw]e^{-2jw}$$

Example 4.9

Find the Fourier transform of $\frac{d}{dt}x(t)$ where $x(t)$ is as given in Example 4.7.

Solution

First Method

From entry 6, Table 4.2, with $n = 1$ we have

$$\frac{d}{dt}x(t) \Leftrightarrow jwX(w) = jw[4\pi\delta(w) + 2/jw]$$

Remember that with any function $f(t)$, $f(t)\delta(t) = \delta(t)f(t = 0)$

Therefore,

$$\frac{d}{dt}x(t) \Leftrightarrow jwX(w) = jw[4\pi\delta(w) + 2/jw] = 2$$

Second Method

Let us differentiate $x(t)$ first and write

$$\frac{d}{dt}x(t) = \frac{d}{dt}[1 + 2u(t)] = 2\delta(t)$$

where the derivative of $u(t)$ is $\delta(t)$.

Therefore,

$$2\delta(t) \Leftrightarrow 2(1) = 2$$

Example 4.10

Find the Fourier transform of $x(t/3)$ where $x(t)$ is as in Example 4.7.

Solution

Using entry 5, Table 4.2, we have

$$x(t/3) \Leftrightarrow [1/|1/3|]X(w/1/3)$$

By substituting for $x(t/3)$ and $X(w/(1/3))$ we arrive at

$$\begin{aligned}[1 + 2u(t/3)] &\Leftrightarrow [1/|1/3|][4\pi\delta(w/(1/3)) + 2/j(w/(1/3))] \\ &= 3[4\pi\delta(3w) + 6/jw]\end{aligned}$$

Example 4.11

Find the Fourier transform of $x(t)*x(t-2)$, where $*$ means convolution and $x(t)$ as given in Example 4.7.

Solution

From entry 7, Table 4.2, we have

$$x(t)*x(t-2) \Leftrightarrow X(w)X(w)e^{-2jw}$$

If we substitute for $x(t)$ and $x(t-2)$ we have

$$[1 + 2u(t)]*[1 + 2u(t-2)] \Leftrightarrow [4\pi\delta(w) + 2/jw][4\pi\delta(w) + 2/jw]e^{-2jw}$$

Example 4.12

Let $x(t) = e^{-t}u(t)$, and the system's impulse response, $h(t) = \delta(t)$.

1. Find the Fourier transform of $x(t)$ and $h(t)$.
2. Let the output be $y(t)$, where $y(t)$ is the output of a linear time-invariant system, with $x(t)$ and $h(t)$ as given in this example. Find the output $y(t)$.
3. Let $x(t) = u(t)$ and $h(t) = e^{-t}u(t)$. Find $y(t)$, the output, for this new system.

Solution

1. From Table 4.1, entry 9, we have

$$x(t) = e^{-t}u(t) \Leftrightarrow X(w) = 1/[1 + jw]$$

and from Table 4.1, entry 2, we also have

$$h(t) = \delta(t) \Leftrightarrow H(w) = 1$$

2. From Table 4.2, entry 7, we have

$$y(t) = x(t)*h(t) \Leftrightarrow Y(w) = X(w)H(w)$$

Therefore,

$$Y(w) = (1)[1/(1 + jw)] = 1/[1 + jw]$$

We can also look at Table 4.1 again and see the inverse transform of $Y(w)$ which is

$$y(t) = e^{-t}u(t)$$

3. With $x(t) = u(t)$ and $h(t) = e^{-t}u(t)$, the output $y(t)$ is

$$y(t) = x(t)*h(t) = u(t)*e^{-t}u(t)$$

We can take the Fourier transform of this last equation to get

$$Y(w) = \left[\frac{1}{jw} + \pi\delta(w)\right]\left[\frac{1}{1 + jw}\right]$$

$$Y(w) = \left[\frac{\pi\delta(w)}{1 + jw}\right] + \left[\frac{1}{1 + jw}\right]\left[\frac{1}{jw}\right]$$

$$Y(w) = \left[\frac{\pi\delta(w)}{1 + jw}\right] + \left[\frac{1}{w(j - w)}\right]$$

Always remember that with any function $f(t)$, $f(t)\delta(t) = \delta(t)f(t = 0)$. With this note we write

$$Y(w) = \left[\frac{\pi\delta(w)}{1}\right] + \left[\frac{1}{w(j - w)}\right]$$

We can do partial fraction expansion on $\frac{1}{w(j-w)}$ and write

$$Y(w) = \left[\frac{\pi\delta(w)}{1}\right] + \left[\frac{1}{jw}\right] - \left[\frac{1}{1 + jw}\right]$$

Let us define $\mathfrak{F}^{-1}$ to be the inverse Fourier transform, then by using Table 4.1 again we get

$$y(t) = \mathfrak{F}^{-1}\left[\frac{\pi\delta(w)}{1} + \frac{1}{jw}\right] - \mathfrak{F}^{-1}\left[\frac{1}{1 + jw}\right]$$

Finally

$$y(t) = u(t) - e^{-t}u(t)$$

Example 4.13

Let $x_1(t) = e^{-t}u(t)$ and $x_2(t) = \cos(2t)$. Find $y(t) = x_1(t)^*x_2(t)$.

Solution

Using the convolution property we write

$$y(t) = x_1(t)^*x_2(t) \Leftrightarrow Y(w) = X_1(w)X_2(w)$$

With the values for $X_1(w)$ and $X_2(w)$ substituted using the tables, we have

$$Y(w) = \left[\frac{1}{1 + jw}\right][\pi\delta(w - 2) + \pi\delta(w + 2)]$$

$$Y(w) = \left[\frac{\pi\delta(w - 2)}{1 + jw}\right] + \left[\frac{\pi\delta(w + 2)}{1 + jw}\right]$$

Using the fact that

$$F(w)\delta(w - w_0) = F(w_0)\delta(w - w_0)$$

we write $Y(w)$ as

$$Y(w) = \left[\frac{\pi\delta(w - 2)}{1 + j2}\right] + \left[\frac{\pi\delta(w + 2)}{1 - j2}\right]$$

Notice that $\frac{\pi}{1+j2}$ and $\frac{\pi}{1-j2}$ are constant numbers. Therefore by utilizing Table 4.1 again we can inverse transform $Y(w)$ to write

$$y(t) = \left[\frac{\pi}{1+j2}\right]\frac{1}{2\pi}e^{2jt} + \left[\frac{\pi}{1-j2}\right]\frac{1}{2\pi}e^{-2jt}$$

In the above equation for $y(t)$, the terms

$$\left[\frac{\pi}{1+j2}\right]\frac{1}{2\pi}e^{2jt}$$

and

$$\left[\frac{\pi}{1-j2}\right]\frac{1}{2\pi}e^{-2jt}$$

are complex conjugate functions. If you add two complex conjugate functions, the sum is twice the real part of the complex number. Using this fact we can write $y(t)$ as

$$y(t) = 2\text{Real}\left\{\left[\frac{\pi}{1+j2}\right]\frac{1}{2\pi}e^{2jt}\right\}$$

$$y(t) = 2\text{Real}\left[\frac{1}{2}\left(\frac{1-2j}{5}\right)(\cos)(2t) + j\sin(2t)\right]$$

and finally

$$y(t) = [1/10][2\cos(2t) + 4\sin(2t)]$$

Example 4.14

Before we send an information signal through a communication medium, the information signal is converted to another signal via a process called modulation. Let us look at the block diagram in Figure 4.3.

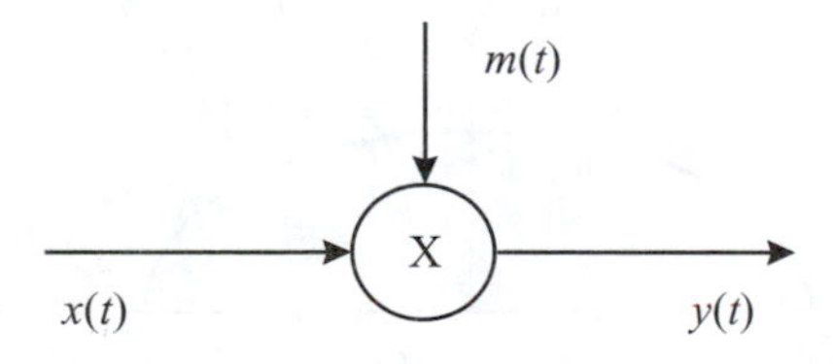

FIGURE 4.3
Sending end.

$x(t)$ is the signal that carries the information, $m(t)$ is the signal that we call carrier, and $y(t)$ is the signal that the receiver gets as input. Let $m(t) = \cos(w_0 t)$ where w_0 is the carrier frequency. As shown in the multiplier block diagram in Figure 4.3

$$y(t) = x(t)m(t) = x(t)\cos(w_0 t).$$

From Table 4.2 we see that multiplication in the time domain is convolution in the frequency domain. With the help of Table 4.1 we can write the Fourier transform of $y(t)$ as

$$Y(w) = \frac{1}{2\pi}[X(w)]^{*}\pi[\delta(w - w_0) + \delta(w + w_0)]$$

with

$$M(w) = \pi[\delta(w - w_0) + \delta(w + w_0)]$$

$M(w)$ will have the shape shown in Figure 4.4.

Let us look at the band limited signal $x(t)$ in the frequency domain. Assume $X(w)$ has the form as shown in Figure 4.5.

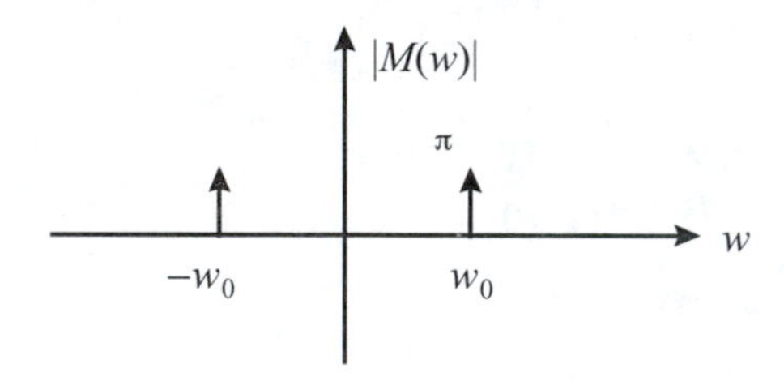

FIGURE 4.4
$m(t)$ Magnitude spectra.

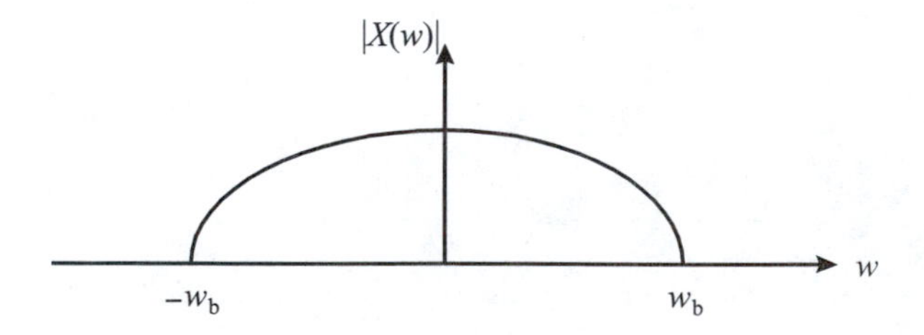

FIGURE 4.5
$x(t)$ Magnitude spectra.

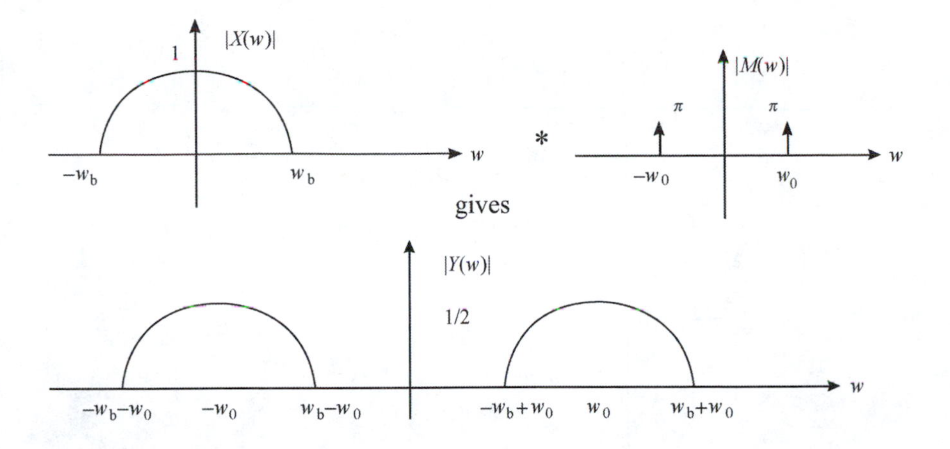

FIGURE 4.6
The convolution of $x(t)$ and $m(t)$ in the frequency domain.

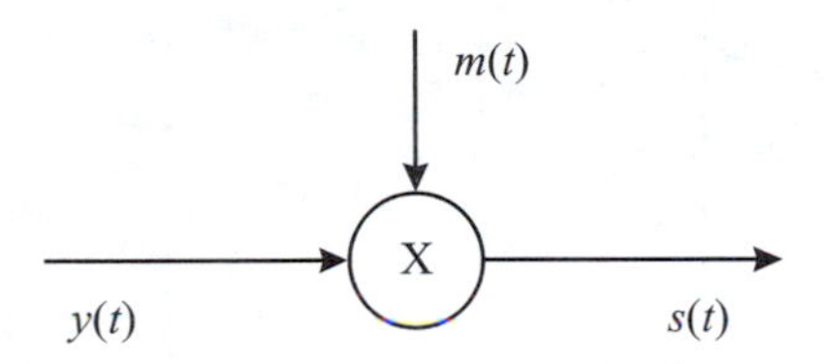

FIGURE 4.7
The end of transmission.

The convolution of $X(w)$ and $M(w)$ is presented by the graph in Figure 4.6. Where $|X(w)|$ is shifted twice (due to the presence of the two impulses) to the left and to the right by w_0. But at the output, $y(t)$, we are interested in our transmitted signal $x(t)$.

Let us introduce another signal and call it $s(t)$ as seen in Figure 4.7. From Figure 4.7 we can see that

$$s(t) = y(t)\cos(w_0 t)$$

Taking the Fourier transform of this equation for $s(t)$ we will have

$$s(t) = y(t)\cos(w_0 t) \Leftrightarrow S(w) = \frac{1}{2\pi}[Y(w)] * \pi[\delta(w - w_0) + \delta(w + w_0)]$$

This process now is called demodulation and it takes place at the receiving end of the transmission media.

Let us look now at the convolution process graphically as seen in Figure 4.8. To get $x(t)$ back again, we pass $s(t)$ through a low-pass filter with $H(w)$ as shown

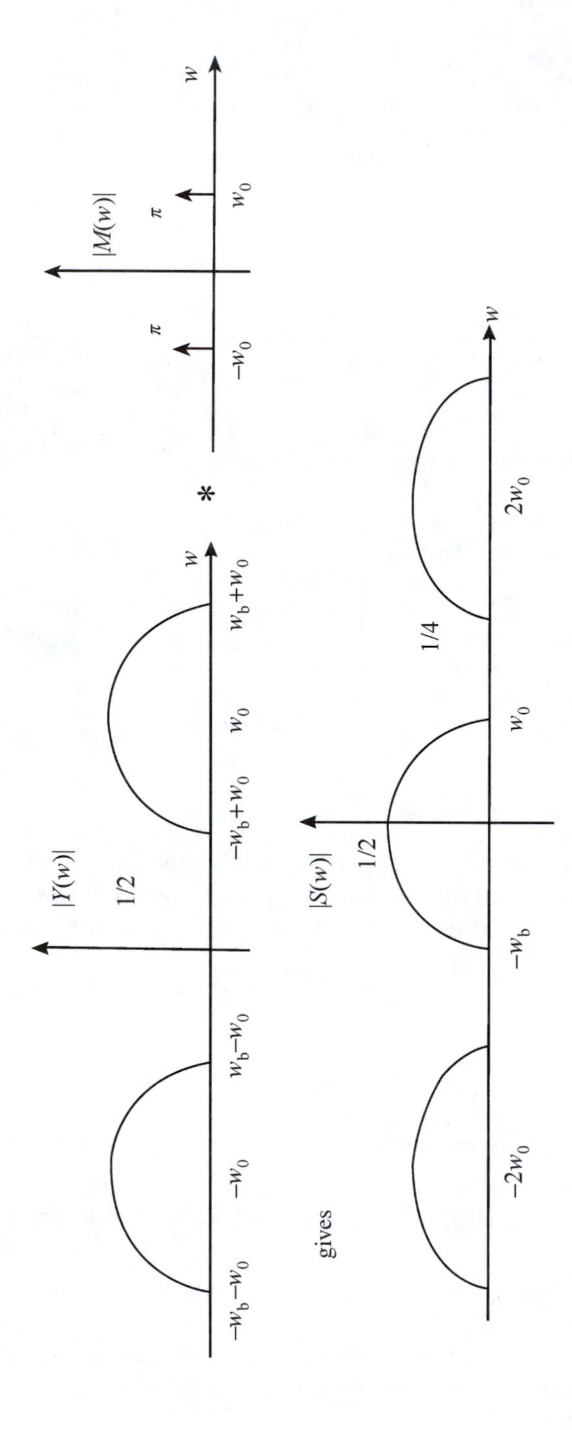

FIGURE 4.8
Convolution of $y(t)$ and $m(t)$ in the frequency domain.

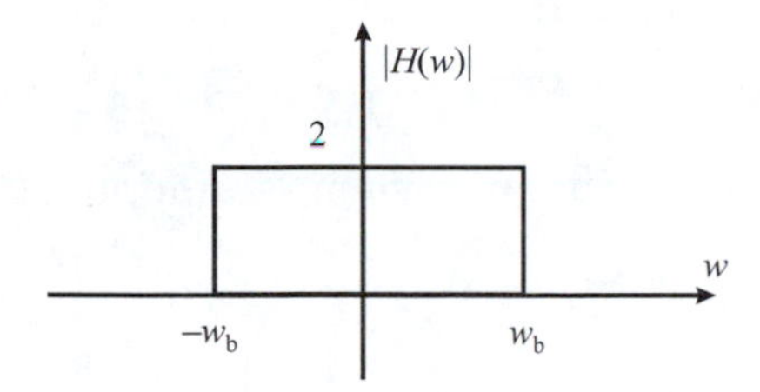

FIGURE 4.9
Low-pass filter.

in Figure 4.9, to get our desired signal $x(t)$ at the receiving end of the transmission media.

4.4 Energy of Non-Periodic Signals

The energy of the signal $x(t)$ is defined as

$$E = \int_{-\infty}^{\infty} |x(t)|^2 dt = \int_{-\infty}^{\infty} x(t)x^*(t)dt$$

where $*$ means conjugate.

Taking the Fourier transform of $x^*(t)$ we get

$$E = \int_{-\infty}^{\infty} x(t)\left[\frac{1}{2\pi}\int_{-\infty}^{\infty} X^*(w)e^{-jwt}dw\right]dt$$

$$E = \frac{1}{2\pi}\int_{-\infty}^{\infty} X^*(w)\left[\int_{-\infty}^{\infty} x(t)e^{-jwt}dt\right]dw$$

The last integral term in the above equation is the Fourier transform of $x(t)$, $X(w)$. Therefore,

$$E = \frac{1}{2\pi}\int_{-\infty}^{\infty} X^*(w)X(w)dw = \frac{1}{2\pi}\int_{-\infty}^{\infty} |X(w)|^2 dw$$

that is, what we call the theorem of Parseval. Since $|X(w)|$ is an even function we have

$$E = \frac{1}{\pi}\int_{0}^{\infty} |X(w)|^2 dw \tag{4.9}$$

4.5 The Energy Spectral Density of a Linear System

Consider a linear time-invariant system with input $x(t)$, impulse response function $h(t)$, and output $y(t)$. In real time we write the output as

$$y(t) = x(t)^{*}h(t)$$

where $*$ means convolution. In the Fourier transform language we write the output $y(t)$ as

$$Y(w) = X(w)H(w)$$

and by taking magnitudes we get

$$|Y(w)|^2 = |X(w)H(w)|^2$$

which simplifies to

$$|Y(w)|^2 = |X(w)|^2|H(w)|^2 \tag{4.10}$$

Therefore, the energy spectral density for the output of the system is the product of the energy spectral density of the input times the squared magnitude of the system's transfer function $H(w)$.

4.6 Some Insights: Notes and a Useful Formula

Fourier series are limited as to their use. They only represent signals of periodic nature. Many important functions like the isolated pulse and the decaying exponential are nonperiodic and therefore cannot be represented by the Fourier series. Using the Fourier transform we were able to represent such functions.

Fourier series coefficients, c_n, in the previous chapter were used to study and examine the frequency components of the signal under investigation. The Fourier transform allows us to look at the Fourier transform of the impulse function, $h(t)$, and thus, study the frequency response of the system $h(t)$ represents. From $H(jw)$ we can deduce the magnitude and the phase of the system. If the system under investigation is to pass a signal that has many unknown frequencies, plotting $H(jw)$ vs. w will tell us what frequencies the system will pass and what frequencies it will block or attenuate.

Knowing the transfer function, $H(jw)$, we can find its magnitude and phase at a particular frequency. If the system $H(jw)$ represents is subject to a sinusoidal input signal at that particular frequency of $H(jw)$, the steady-state output, $y_{ss}(t)$, of the system can be evaluated as

$$y_{ss}(t) = |x(t)||H(jw)|\cos(wt + \theta + \varphi) \tag{4.11}$$

where

$$x(t) = A\cos(wt + \theta)$$

and φ is the phase of $H(jw)$ at a particular given w.

If

$$x(t) = A\sin(wt + \theta)$$

then

$$y_{ss}(t) = |x(t)||H(jw)|\sin(wt + \theta + \varphi) \tag{4.12}$$

This says that the output of a linear time-invariant system, if subject to a sinusoidal input, will have a steady-state solution equal to the magnitude of the input signal multiplied by the magnitude of the transfer function of the system evaluated at the frequency of the input signal and shifted by the phase angle of the transfer function evaluated at the input frequency as well. Some notes on Fourier transform are worth mentioning:

1. The magnitude spectrum of the Fourier transform of a signal $x(t)$ is even and the phase spectrum is odd.
2. Shifting the signal $x(t)$ in time does not alter its magnitude.
3. Time compression of a signal $x(t)$ corresponds to frequency expansion of $X(w)$.
4. Time expansion of a signal $x(t)$ corresponds to frequency compression of $X(w)$.
5. A time-limited signal $x(t)$ has a Fourier transform, $X(w)$, which is not band limited (by band limited we mean the frequency band).
6. If $X(w)$ is band limited then $x(t)$ is not time limited.

4.7 End-of-Chapter Examples

EOCE 4.1

Let the Fourier transforms of $x_1(t)$ and $x_2(t)$ be $X_1(w)$ and $X_2(w)$, respectively. Let $\Im$ denote 'the Fourier transform of.' Is $\Im(ax_1(t) + bx_2(t)) = a\,\Im(x_1(t)) + b\,\Im(x_2(t))$? Assume a and b are constants.

Solution

Using the defining integral of the Fourier transform and properties of integrals we have

$$\Im(ax_1(t) + bx_2(t)) = \int_{-\infty}^{\infty}(ax_1(t) + bx_2(t))e^{-jwt}dt$$

$$\Im(ax_1(t) + bx_2(t)) = a\int_{-\infty}^{\infty}x_1(t)e^{-jwt}dt + b\int_{-\infty}^{\infty}x_2(t)e^{-jwt}dt$$

$$\Im(ax_1(t) + bx_2(t)) = aX(w) + bX(w) = a\Im(x_1(t)) + b\Im(x_2(t))$$

This is what we call the linearity property of the Fourier transform.

EOCE 4.2

If $X(w)$ is the Fourier transform of $x(t)$, what is the relation between $X(t)$ and $2\pi x(-w)$?

Solution

From the defining integral

$$X(w) = \int_{-\infty}^{\infty}x(t)e^{-jwt}\,dt$$

and

$$x(t) = \frac{1}{2\pi}\int_{-\infty}^{\infty}X(w)^{jwt}\,dw$$

If we *set* $t = -t$, then the last integral becomes

$$x(-t) = \frac{1}{2\pi}\int_{-\infty}^{\infty}X(w)e^{-jwt}dw$$

or

$$2\pi x(-t) = \int_{-\infty}^{\infty} X(w)e^{-jwt}dw$$

Now replace w by t to get

$$2\pi x(-w) = \int_{-\infty}^{\infty} X(t)e^{-jwt}dt$$

This last equation says that the Fourier transform of $X(t)$ is $2\pi x(-w)$ where $X(t)$ is $X(w)$ evaluated at $w = t$.

EOCE 4.3

If a is a positive constant, what is the Fourier transform of $x(at)$?

Solution

Let us look at the familiar integral

$$\int_{-\infty}^{\infty} x(at)e^{-jwt}dt$$

If we let $at = m$, then $dt = dm/a$. The integral becomes

$$1/a\int_{-\infty}^{\infty} x(m)e^{-j(m/a)w}dm$$

If we now let $m = t$, then $dm = dt$ and the above integral becomes

$$1/a\int_{-\infty}^{\infty} x(t)e^{-j\frac{w}{a}t}dt = \frac{1}{a}X(w/a)$$

Therefore,

$$\Im(x(at)) = \frac{1}{a}X(w/a)$$

EOCE 4.4

What is the Fourier transform of the shifted signal $x(t - a)$?

Solution

Using the defining integral we have

$$\Im(x(t-a)) = \int_{-\infty}^{\infty} x(t-a)e^{-jwt}dt$$

Let $t - a = m$, then $dt = dm$. The above integral becomes

$$\Im(x(t-a)) = \int_{-\infty}^{\infty} x(m)e^{-jw(m+a)}dm$$

$$\Im(x(t-a)) = e^{-jwa}\int_{-\infty}^{\infty} x(m)e^{-jwm}dm$$

Now it is clear from the last step that

$$\Im(x(t-a)) = e^{-jwa}X(w)$$

EOCE 4.5

What is the Fourier transform of $e^{jat}x(t)$?

Solution

By using the defining integral again we can write

$$\Im(e^{jat}x(t)) = \int_{-\infty}^{\infty} e^{-jat}x(t)e^{jwt}dt$$

$$\Im(e^{jat}x(t)) = \int_{-\infty}^{\infty} x(t)e^{-j(w-a)t}dt = X(w-a)$$

EOCE 4.6

What is the Fourier transform of $x(t)\cos(at)$?

Solution

$\cos(at)$ can be written in exponential form as

$$\cos(at) = \frac{e^{jat} + e^{-jat}}{2}$$

Taking the Fourier transform we get

$$\begin{aligned}\Im(x(t)\cos(at)) &= \Im\left(x(t)\frac{e^{jat}+e^{-jat}}{2}\right)\\ &= \Im\left(x(t)\frac{e^{jat}}{2}\right)+\Im\left(x(t)\frac{e^{-jat}}{2}\right)\end{aligned}$$

By simplification we arrive at

$$\begin{aligned}\Im(x(t)\cos(at)) &= \frac{1}{2}\Im(x(t)e^{jat})+\frac{1}{2}\Im(x(t)e^{-jat})\\ &= \frac{1}{2}X(w-a)+\frac{1}{2}X(w+a)\end{aligned}$$

EOCE 4.7

What is the Fourier transform of the derivative of $x(t)$?

Solution

$$\Im\left(\frac{d}{dt}x(t)\right) = \int_{-\infty}^{\infty}\left(\frac{d}{dt}x(t)\right)e^{-jwt}dt$$

If we let $u = e^{-jwt}$ and $dv = (\frac{d}{dt}x(t))\,dt$ then

$$\Im\left(\frac{d}{dt}x(t)\right) = e^{-jwt}x(t)\Big|_{-\infty}^{\infty} + jw\int_{-\infty}^{\infty}x(t)e^{-jwt}dt$$

The above equation was obtained by integrating by parts.

$$\Im\left(\frac{d}{dt}x(t)\right) = jw\int_{-\infty}^{\infty}x(t)e^{-jwt}dt$$

$$\Im\left(\frac{d}{dt}x(t)\right) = jwX(W)$$

EOCE 4.8

What is the Fourier transform of the second and the nth derivative of $x(t)$?

Solution

By using the properties of the Fourier transform we write

$$\Im\left(\frac{d^2}{dt}x(t)\right) = jw\Im\left(\frac{d}{dt}x(t)\right) = jw(jw)X(w)$$

$$\Im\left(\frac{d^2}{dt}x(t)\right) = (jw)^2X(w) = -w^2X(w)$$

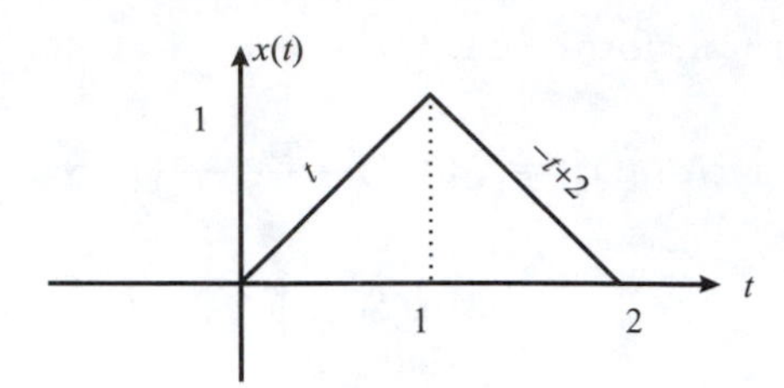

FIGURE 4.10
Signal for EOCE 4.9.

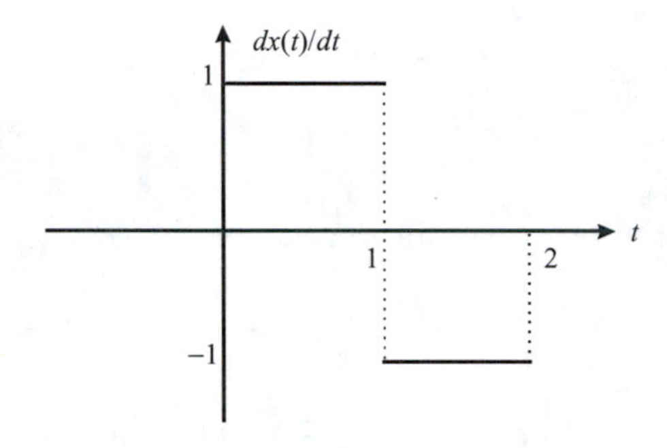

FIGURE 4.11
Signal for EOCE 4.9.

Similarly,

$$\Im\left(\frac{d^2}{dt}x(t)\right) = (jw)^n X(w)$$

EOCE 4.9

Consider the signal in Figure 4.10 and its derivative in Figure 4.11. What is the Fourier transform of $x(t)$?

Solution

Let us start by taking the Fourier transform of $\frac{d}{dt}x(t)$. We can write $x(t)$ mathematically as

$$x(t) = \begin{cases} 0 & t < 0 \\ t & 0 \le t < 1 \\ -t+2 & 1 \le t < 2 \\ 0 & t \ge 2 \end{cases}$$

Its derivative can also be written mathematically as

$$\frac{d}{dt}x(t) = \begin{cases} 1 & 0 \le t < 1 \\ -1 & 1 \le t < 2 \\ 0 & \text{otherwise} \end{cases}$$

Taking the Fourier transform of $\frac{d}{dt}x(t)$ piece by piece we get

$$\Im\left(\frac{d}{dt}x(t)\right) = \begin{cases} \int_0^1 e^{-jwt}dt = \frac{1}{-jw}[e^{-jw} - 1] \\ \int_1^2 -e^{-jwt}dt = \frac{1}{jw}[e^{-2jw} - e^{-jw}] \end{cases}$$

Therefore, the Fourier transform of $\frac{d}{dt}x(t)$ is

$$\Im\left(\frac{d}{dt}x(t)\right) = \frac{1}{jw}[e^{-2jw} - 2e^{-jw} + 1]$$

Now we can use entry 6, Table 4.2, to find the Fourier transform of $x(t)$ as

$$\Im(x(t)) = X(w) = \Im(dx(t)/dt)jw$$

$$\Im(x(t)) = X(w) = \frac{e^{-2jw} - 2e^{-jw} + 1}{(jw)(jw)} = \frac{-e^{-2jw} + 2e^{-jw} - 1}{w^2}$$

Note that in many situations taking the Fourier transform of the derivative of a signal to calculate the Fourier transform of the signal itself is easier when the signal is hard to integrate using the defining integral directly.

EOCE 4.10

We have seen in this chapter that convolution in time domain is multiplication in frequency domain. If

$$\Im(a(t)^*b(t)) = \int_0^t a(\tau)b(t-\tau)d\tau = A(w)B(w)$$

what is $x(t)$ if

$$X(w) = \frac{1}{5 + jw}\frac{1}{6 + jw}$$

Solution

We can use partial fraction expansion to find $x(t)$, but instead we will do the following.

Let

$$A(w) = \frac{1}{5 + jw}$$

and

$$B(w) = \frac{1}{6 + jw}$$

Therefore,

$$a(t) = e^{-5t}u(t)$$

and

$$b(t) = e^{-6t}u(t)$$

But according to what is given in this example

$$x(t) = \int_0^t a(\tau)b(t - \tau)d\tau$$

$$x(t) = \int_0^t e^{-5\tau}e^{-6(t - \tau)}\,d\tau = e^{-6t}\int_0^t e^{\tau}d\tau$$

and finally

$$x(t) = (e^{-5t} - e^{-6t})u(t)$$

EOCE 4.11

Consider the following differential equation

$$\frac{d^2}{dt}y - 5\frac{d}{dt}y + 6y = x(t)$$

where $x(t)$ is the input and $y(t)$ is the output.

If $h(t)$ is the impulse response (the transfer function of the system as seen in Chapter 3), $H(w)$ is the Fourier transform of $h(t)$. $H(w)$ is the ratio of $Y(w)$ to $X(w)$. For a particular input $x(t)$,

$$\Im(y(t)) = \Im((x(t)^{*}h(t)) = \Im\left(\int_0^t x(\tau)h(t-\tau)d\tau\right) = X(w)H(w)$$

We can see that if $x(t)$ is given we can use the Fourier transform to find $y(t)$. Let $x(t) = e^{-2t}$ for $t > 0$. Find $y(t)$ using the Fourier techniques.

Solution

We start by finding $H(w)$. We do that by taking the Fourier transform of the differential equation with the input as $x(t)$ to get

$$(-w^2)Y(w) - 5jwY(w) + 6Y(w) = X(w)$$

Taking $Y(w)$ common we get

$$Y(w)[-w^2 - 5jw + 6] = X(w)$$
$$H(w) = Y(w)/X(w)$$
$$H(w) = \frac{1}{-w^2 - 5jw + 6} = \frac{1}{(jw)^2 - 5jw + 6} = \frac{1}{jw-3} - \frac{1}{jw-2}$$

But

$$Y(w) = H(w)X(w)$$

Therefore

$$Y(w) = \left[\frac{1}{jw-3} - \frac{1}{jw-2}\right]\left[\frac{1}{jw+2}\right]$$

Multiply out to get

$$Y(w) = \left[\frac{1}{jw-3}\frac{1}{jw+2} - \frac{1}{jw-2}\frac{1}{jw+2}\right]$$
$$Y(w) = \frac{1/5}{jw-3} - \frac{1/5}{jw+2} - \frac{1/4}{jw-2} + \frac{1/4}{jw+2}$$

and $y(t)$ is

$$y(t) = 1/5[e^{3t} - e^{-2t}]u(t) + 1/4[-e^{2t} - e^{-2t}]u(t)$$

EOCE 4.12

Let $y(t)$ be the output of a linear system with $Y(w)$ as its Fourier transform. If $Y(w)$ can be written in polar form as

$$Y(w) = M(w)e^{j\theta(w)}$$

where $M(w)$ and $\theta(w)$ are the magnitude and the angle of $Y(w)$, respectively, is the equation

$$\int_{-\infty}^{\infty} y^2(t)\,dt = \frac{1}{2\pi}\int_{-\infty}^{\infty} M^2(w)\,dw$$

true?

Solution

Let us look at the defining integral of the Fourier transform

$$Y(w) = \int_{-\infty}^{\infty} y(t)e^{-jwt}\,dt$$

$Y(w)$ conjugated and reflected is

$$\overset{*}{Y}(-w) = Y(w) = \int_{-\infty}^{\infty} y(t)e^{-jwt}\,dt$$

where * represents complex conjugate. Similarly,

$$Y(w - \tau) = \overset{*}{Y}(\tau - w)$$

If we convolve $Y(w)$ with $Y(w)$, we will have

$$Y(w)^*Y(w) = \int_{-\infty}^{\infty} Y(\tau)Y(w - \tau)\,d\tau = \int_{-\infty}^{\infty} Y(\tau)\overset{*}{Y}(\tau - w)\,d\tau$$

We know that

$$\Im(y^2(t)) = \frac{1}{2\pi}Y(w)^*Y(w) = \frac{1}{2\pi}\int_{-\infty}^{\infty} Y(\tau)\overset{*}{Y}(\tau - w)\,d\tau$$

Also

$$\Im(y^2(t)) = \int_{-\infty}^{\infty} y^2(t)e^{-jwt}dt$$

If we let $w = 0$ in the above two equations then

$$\Im(y^2(t)) = \frac{1}{2\pi}\int_{-\infty}^{\infty} y(\tau)\overset{*}{Y}(\tau)d\tau$$

and

$$\Im(y^2(t)) = \int_{-\infty}^{\infty} y^2(t)dt$$

Now let $w = \tau$ to get

$$\Im(y^2(t)) = \frac{1}{2\pi}\int_{-\infty}^{\infty} Y(w)\overset{*}{Y}(w)dw = \frac{1}{2\pi}\int_{-\infty}^{\infty} M^2(w)dw$$

But

$$\Im(y^2(t)) = \int_{-\infty}^{\infty} y^2(t)\,dt$$

Therefore

$$\int_{-\infty}^{\infty} y^2(t)dt = \frac{1}{2\pi}\int_{-\infty}^{\infty} M^2(w)\,dw$$

which is the Parseval theorem.

$M^2(w)$ is the energy spectrum and the integral from $-\propto$ to $+\propto$ is the energy in the signal.

EOCE 4.13

If $y(t) = e^{-t}u(t)$, use the Parseval theorem to find the total energy in $y(t)$.

Solution

We will use the time domain and the frequency domain to find the energy in the signal $y(t)$.

In time domain the energy is given as

$$\int_{-\infty}^{\infty} y^2(t)dt = \int_0^{\infty} e^{-2t}dt = -\frac{1}{2}e^{-2t}\Big|_0^{\infty} = \frac{1}{2}$$

In frequency domain the energy is given as

$$\frac{1}{2\pi}\int_{-\infty}^{\infty} M^2(w)\, dw$$

with $M(w)$ the magnitude of $Y(w)$.

$$Y(w) = \frac{1}{jw+1}$$

and

$$M(w) = \sqrt{\frac{1}{w^2+1}}$$

Therefore, the energy is given as

$$\frac{1}{2\pi}\int_{-\infty}^{\infty} M^2(w)dw = \frac{1}{2\pi}\int_{-\infty}^{\infty} \frac{1}{w^2+1}dw$$

or

$$\frac{1}{\pi}\int_0^{\infty} M^2(w)dw = \frac{1}{\pi}\tan^{-1}(w)\Big|_0^{\infty} = \frac{1}{2}$$

EOCE 4.14

Consider the circuit in Figure 4.12 with the input as $x(t)$ and the output is the voltage across the capacitor, $y(t)$. If $x(t) = \cos(10t)$, what is the capacitor steady-state voltage?

Solution

The differential equation describing the circuit is

$$\frac{d}{dt}yt + y(t)/RC = x(t)/RC$$

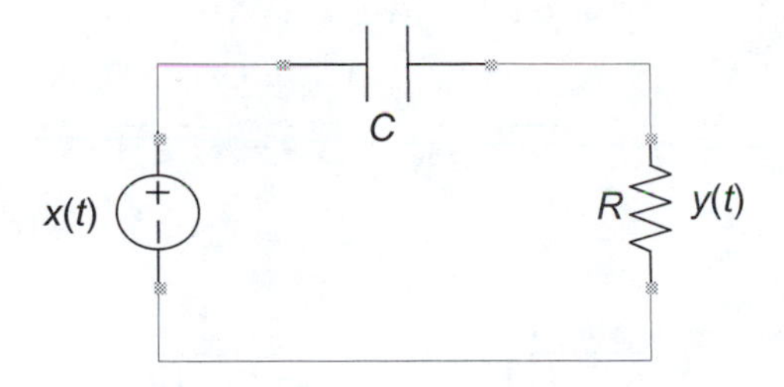

FIGURE 4.12
Circuit for EOCE 4.14.

Taking the Fourier transform of the differential equation above we get

$$jwY(w) + Y(w)/RC = X(w)/RC$$

with $x(t) = \cos(10t)$

$$X(w) = \frac{\pi}{2}[\delta(w - 10) + \delta(w + 10)]$$

Therefore

$$Y(w)[jw + 1/RC] = \frac{\frac{\pi}{2}[\delta(w - 10) + \delta(w + 10)]}{RC}$$

Simplifying the above equation we get

$$Y(w) = \frac{\frac{\pi}{2}[\delta(w - 10) + \delta(w + 10)]}{(jw + 1/RC)RC}$$

$$Y(w) = \frac{\frac{\pi}{2}[\delta(w - 10)]}{(10j + 1/RC)RC} + \frac{\frac{\pi}{2}[\delta(w + 10)]}{(-j10 + 1/RC)RC}$$

which can be written as

$$Y(w) = \frac{\frac{\pi}{2}[\delta(w - 10)]}{(RC10j + 1)} + \frac{\frac{\pi}{2}[\delta(w + 10)]}{(-RCj10 + 1)}$$

Using the Fourier transform table, we can write $y(t)$ as

$$y(t) = \frac{\pi}{2\pi 2(RCj10 + 1)}e^{j10t} + \frac{\pi}{2\pi 2(RCj10 + 1)}e^{-j10t}$$

or

$$y(t) = \frac{1/4}{(RCj10 + 1)}e^{j10t} + \frac{1/4}{(-RCj10 + 1)}e^{-j10t}$$

Let $(1 + RCj10)$ be written in polar form as $Me^{j\theta}$ then

$$(1 - RCj10) = Me^{-j\theta}$$

The output $y(t)$ becomes

$$y(t) = \frac{1}{4M}e^{-j\theta}e^{j10t} + \frac{1}{4M}e^{j\theta}e^{-j10t}$$

$$y(t) = \frac{1}{4M}[e^{j(10t - \theta)} + e^{-j(10t - \theta)}]$$

Finally

$$y(t) = \frac{1}{2M}\cos(10t - \theta)$$

EOCE 4.15

What is the Fourier transform of $u(t) = \begin{cases} 1 & t > 0 \\ 0 & t < 0 \end{cases}$?

Solution

We have seen that $u(t)$ can be written as

$$u(t) = 1/2 + \text{sgn}(t)/2$$

where $sgn(t)$ is as described in Chapter 2.

$$\Im(u(t)) = \Im(1/2) + \frac{1}{2}\Im(sgn(t)) = \pi\delta(w) + \frac{1}{jw}$$

EOCE 4.16

What is $x(t)$ if its Fourier transform is $X(w) = sgn(w)$ where

$$\text{sgn}(w) = \begin{cases} -1 & w < 0 \\ 1 & w > 0 \end{cases}$$

Solution

From EOCE 4.2, we saw that the Fourier transform of $X(t)$ is $2\pi x(-w)$ where $X(t)$ is $X(w)$ evaluated at $w = t$.

$$\Im(X(t)) = \Im(sgn(t)) = \frac{2}{jw}$$

Therefore,

$$2/(jw) = 2\pi x(-w)$$

We can solve for $x(-w)$ to get

$$x(-w) = \frac{2}{2jw\pi}$$

By letting $-w = t$ we get

$$x(t) = \frac{1}{-jt\pi}$$

4.8 End-of-Chapter Problems

EOCP 4.1

Find the Fourier transform for each of the following signals.

1. $2u(t - 10) + 10\delta(t - 1)$
2. $\sin(2(t - 1))u(t - 1)$
3. $\cos(3(t - 5))u(t - 5)$
4. $e^{-a(t-b)}u(t - b) + u(t - 2)$
5. $(t - 1)e^{-a(t-1)}u(t - 1)$
6. $10e^{-2(t-2)^2}u(t - 2)$
7. $2\delta(t - 2) + 3\delta(t - 4)$
8. $e^{-2|t-1|}u(t - 1) + 10\delta(t - 3)$
9. $e^{-3|t-5|}u(t - 5) + \frac{t^2}{2!}e^{-3t}u(t)$
10. $e^{-3t}u(t) + \frac{t^5}{5!}e^{-10t}u(t) + u(t - 5)$

EOCP 4.2

Given the following signals, find the Fourier transform of the derivative for each.

1. $tu(t)$
2. $t\sin(t)u(t)$

3. $(e^{-3t} + \sin(t))u(t)$
4. $(\cos wt + \sin wt)u(t) + 10u(t)$
5. $e^{jwt}\, u(t)$

EOCP 4.3

Find the Fourier transform for each of the following signals, where $*$ indicates convolution.

1. $e^{-t}u(t)^* e^{-t}u(t)$
2. $[\sin(t)u(t)^* u(t)]^* u(t)$
3. $[e^{-t}u(t) + e^{-2t}u(t)]^* \sin(t)u(t)$
4. $[e^{-3t}u(t)^* u(t)] + u(t)^* u(t)$
5. $\delta(t)^* \sin(10t)u(t) - \delta(t-1)^* u(t)$

EOCP 4.4

Consider the following systems represented by the differential equations. $x(t)$ is the input, and $y(t)$ is the output. Find a particular solution to each system.

1. $\frac{d}{dt}y + 2y = x(t) = \sin(t)u(t)$
2. $\frac{d^2}{dt}y + 3\frac{d}{dt}y + 2y = x(t) = u(t)$
3. $\frac{d^2}{dt}y + 3\frac{d}{dt}y = x(t) = e^{-t}(t)$
4. $\frac{d^2}{dt}y + 5y = x(t) = e^{-t}u(t)$
5. $\frac{d^2}{dt}y + \frac{d}{dt}y + y = x(t) = \delta(t)$

EOCP 4.5

Given are the Fourier transforms for some signals. What are the time-domain signals?

1. $\frac{1}{jw+2} + \frac{1}{jw+3}$
2. $\frac{1}{-w^2 + 7jw + 10}$
3. $\frac{1}{-w^2 + 5jw + 6}$
4. $\frac{1}{-w^2 + 4jw + 4}$
5. $\frac{2}{jw} + 4\pi\delta(w) + \frac{2\pi\delta(w)}{5}$

EOCP 4.6

Use the Fourier transform to find $y(t)$, a particular solution, in each of the systems shown in Figures 4.13, 4.14, 4.15, 4.16, and 4.17.

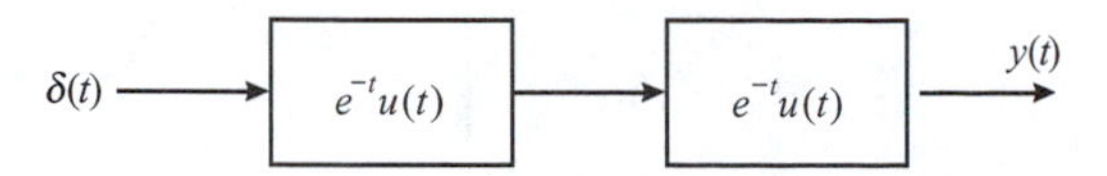

FIGURE 4.13
System for EOCP 4.6.

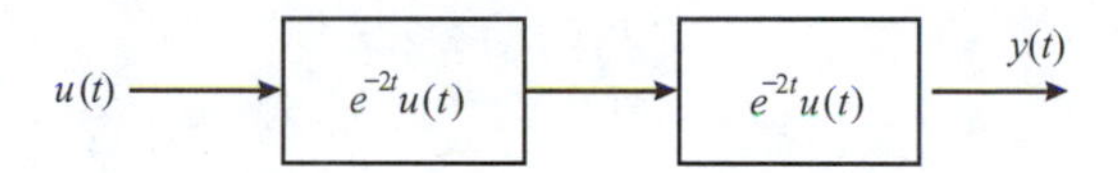

FIGURE 4.14
System for EOCP 4.6.

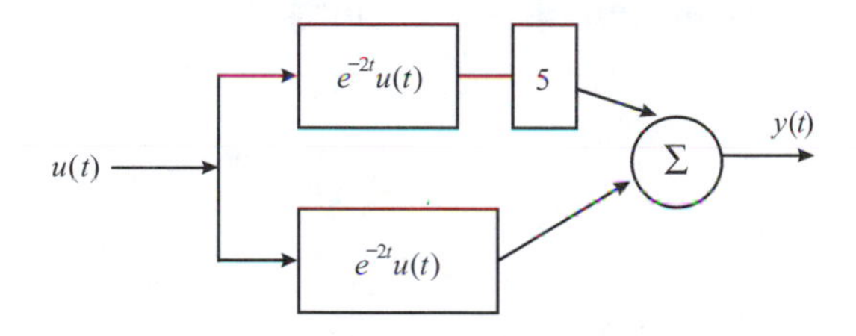

FIGURE 4.15
System for EOCP 4.6.

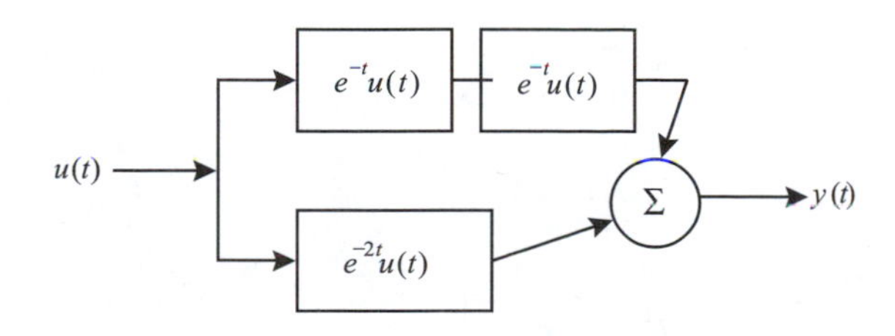

FIGURE 4.16
System for EOCP 4.6.

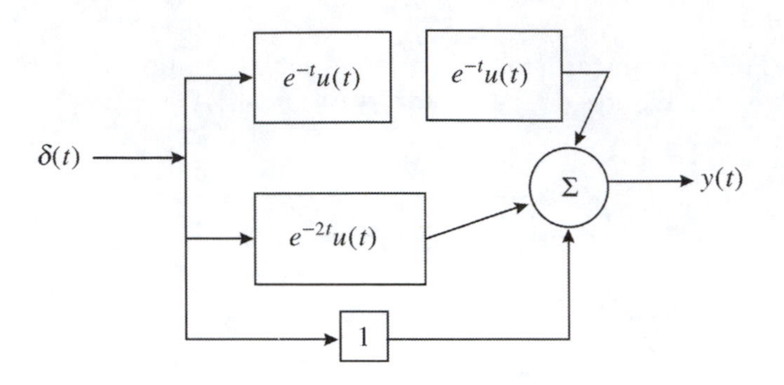

FIGURE 4.17
System for EOCP 4.6.

EOCP 4.7

Given a band-limited signal $x(t)$ where its magnitude spectra is shown in Figure 4.18.

In each of the block diagrams shown in Figures 4.19 and 4.20, let $m(t) = \cos(w_0 t)$, and plot the magnitude spectra for $y(t)$.

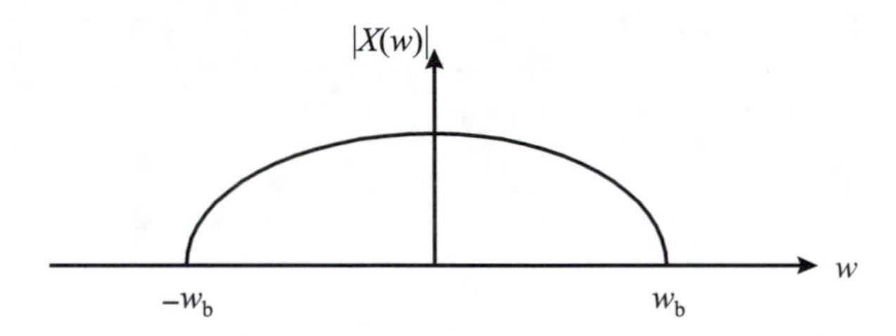

FIGURE 4.18
Signal for EOCP 4.7.

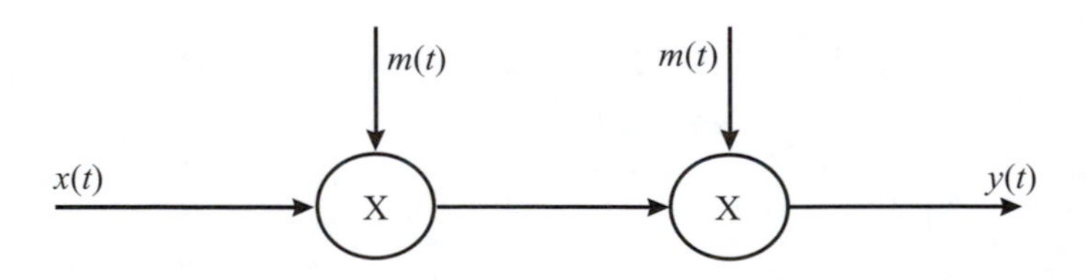

FIGURE 4.19
Block for EOCP 4.7.

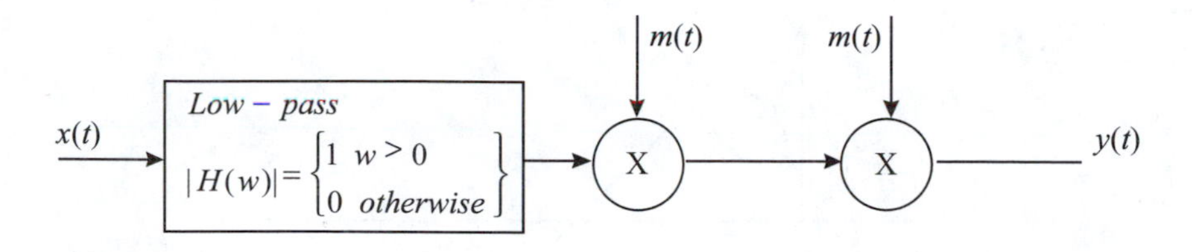

FIGURE 4.20
Block for EOCP 4.7.

EOCP 4.8

Consider the following signals.

1. $\cos(w_0 t)$
2. $\sin(w_0 t)$
3. $e^{-3t}u(t)$
4. $e^{-3t}u(t)$

Use the Fourier transform to find the energy in each signal in the frequency band between 0 and 10 rad/sec.

EOCP 4.9

What is the Fourier transform for each of the signals in Figures 4.21, 4.22, 4.23, and 4.24?

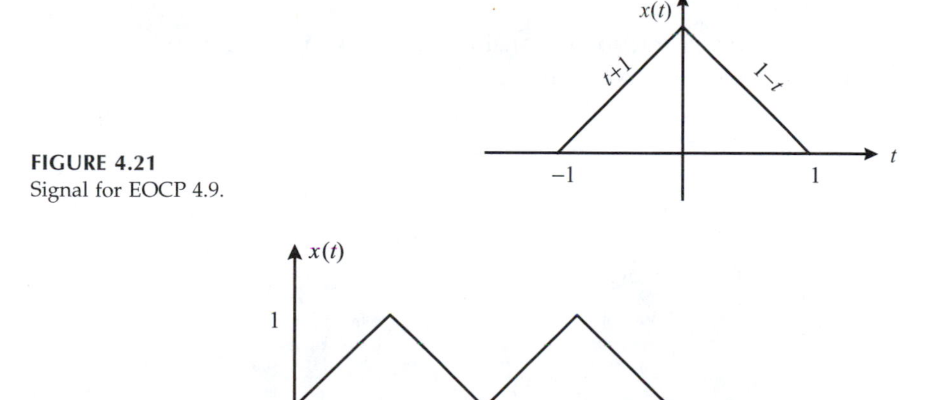

FIGURE 4.21
Signal for EOCP 4.9.

FIGURE 4.22
Signal for EOCP 4.9.

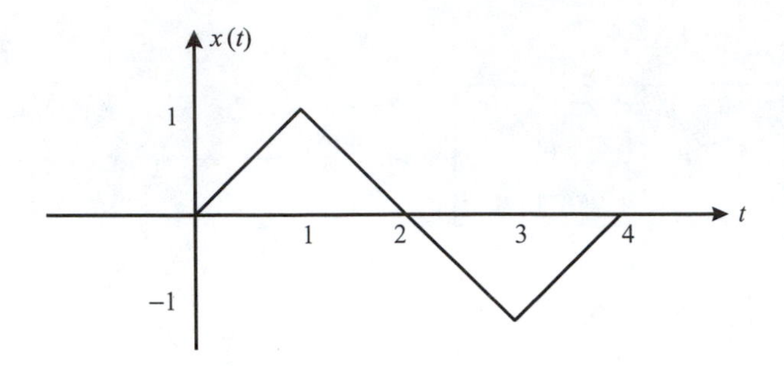

FIGURE 4.23
Signal for EOCP 4.9.

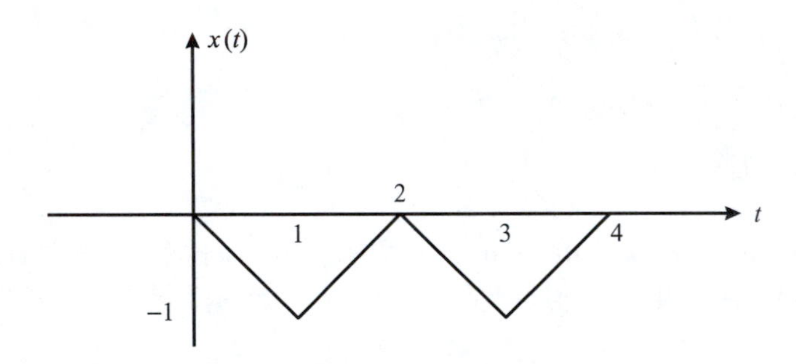

FIGURE 4.24
Signal for EOCP 4.9.

EOCP 4.10

Do not use partial fraction expansion and refer to EOCE 4.10 to find $x(t)$ in each of the following if $X(w)$ is

1. $\dfrac{10}{2 + jw}\,\dfrac{1}{4 + jw}$
2. $2\pi\delta(w - 1)\dfrac{1}{1 + jw}$
3. $\dfrac{\pi}{j}[\delta(w - 1) - \delta(w + 1)]\dfrac{1}{1 + jw}$
4. $2\pi\delta(w - 10)\dfrac{1}{(1 + jw)^2}$

EOCP 4.11

Consider the circuit in Figure 4.25.

1. Find the output voltage $y(t)$, for $t > 0$, using the Fourier transform method.

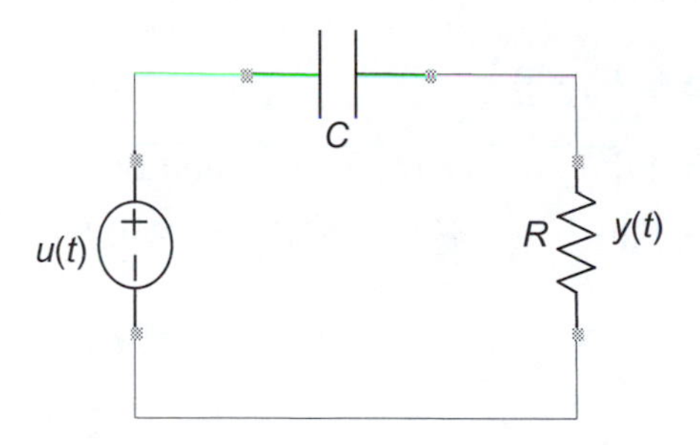

FIGURE 4.25
Circuit for EOCP 4.11.

2. Find the current in the resistor R using the Fourier method.
3. Find the energy in $y(t)$ and the current calculated in 2.

EOCP 4.12

Consider the mechanical system in Figure 4.26. The rod to the left of the mass, M, can be modeled by a translational spring with spring constant k, in parallel with a translational damper of constant B.

1. If $x(t)$ is an impulse of strength 10, use the Fourier transform method to find the displacement $y(t)$. Assume no friction.
2. Repeat if $x(t)$ is a sinusoidal shaking of magnitude 10 and period 1 second.

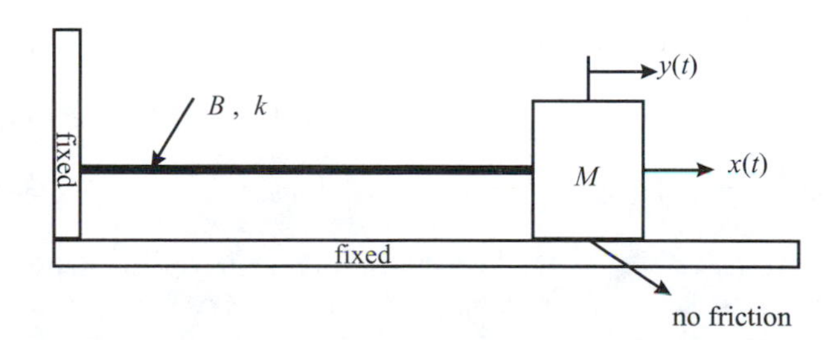

FIGURE 4.26
System for EOCP 4.12.

EOCP 4.13

Consider the circuit shown in Figure 4.27.

1. Find $y(t)$ using the Fourier transform method.
2. Find the current in the capacitor using the Fourier transform method.

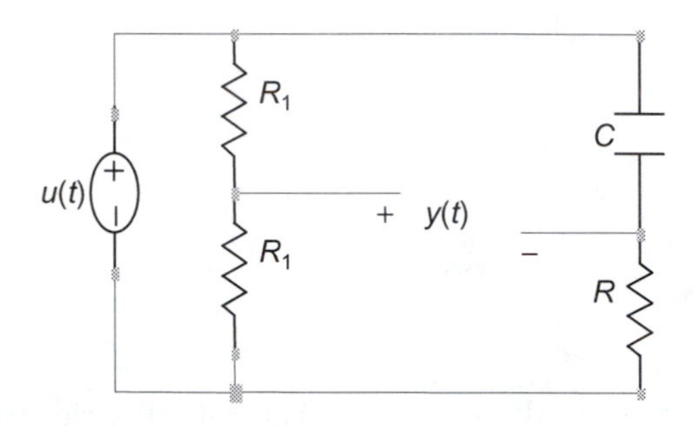

FIGURE 4.27
Circuit for EOCP 4.13.

References

Cogdell, J.R. *Foundations of Electrical Engineering,* 2nd ed., Englewood Cliffs, NJ: Prentice-Hall, 1996.

Denbigh, P. *System Analysis and Signal Processing,* Reading, MA: Addison-Wesley, 1998.

Golubitsky, M. and Dellnitz, M. *Linear Algebra and Differential Equations Using MATLAB,* Stamford, CT: Brooks/Cole, 1999.

Harman, T.L., Dabney, J., and Richert, N. *Advanced Engineering Mathematics with MATLAB,* Stamford, CT: Brooks/Cole, 2000.

The MathWorks. *The Student Edition of MATLAB,* Englewood Cliffs, NJ: Prentice- Hall, 1997.

Nilson, W.J. and Riedel, S.A. *Electrical Circuits,* 6th ed., Englewood Cliffs, NJ: Prentice-Hall, 2000.

Phillips, C.L. and Parr, J.M. *Signals, Systems, and Transforms,* 2nd ed., Englewood Cliffs, NJ: Prentice-Hall, 1999.

Pratap, R. *Getting Started with MATLAB 5,* New York: Oxford University Press, 1999.

Strum, R.D. and Kirk, D.E. *Contemporary Linear Systems,* Boston: PWS, 1996.

Wylie, R.C. and Barrett, C.L. *Advanced Engineering Mathematics,* 6th ed., New York: McGraw-Hill, 1995.

Ziemer, R.E., Tranter, W.H., and Fannin, D.R. *Signals Systems Continuous and Discrete,* 4th ed., Englewood Cliffs, NJ: Prentice-Hall, 1998.

5

The Laplace Transform and Linear Systems

CONTENTS

5.1 Definition

The Laplace transform, like the Fourier transform, is a frequency domain representation that makes solution, design, and analysis of linear systems simpler. It also gives some insights about the frequency contents of signals where these insights are hard to see in real-time systems. There are other important uses for the Laplace transform but we will concentrate only on the issues described in this definition.

5.2 The Bilateral Laplace Transform

The bilateral Laplace transform of a signal $x(t)$ is defined as

$$X(s) = \int_{-\infty}^{\infty} x(t)e^{-st}dt \tag{5.1}$$

where $s = \sigma + jw$.

If $s = jw$ and $\sigma = 0$, the bilateral Laplace transform is then

$$X(s) = \int_{-\infty}^{\infty} x(t)e^{-jwt}dt \tag{5.2}$$

which is nothing but the Fourier transform of $x(t)$, $X(w)$. If $\sigma \neq 0$ then

$$X(s) = \int_{-\infty}^{\infty} x(t)e^{-(\sigma + jwt)}dt \tag{5.3}$$

$$X(s) = \int_{-\infty}^{\infty} x(t)e^{-\sigma t}e^{-jwt}dt \tag{5.4}$$

You can easily see in the above equation that $X(s)$ is, in this case, the Fourier transform of $x(t)e^{-\sigma t}$.

5.3 The Unilateral Laplace Transform

If $x(t) = 0$ for $t < 0$, and $\sigma = 0$, then

$$X(s) = \int_{0^-}^{\infty} x(t)e^{-jwt}dt$$

We will be interested in the unilateral Laplace transform because, in real life, most systems start at a specific time which we call 0^-, where 0^- indicates that time starts just before 0.

Again, if $x(t)$ is an input signal to a linear system then the Laplace transform of $x(t)$ is $X(s)$ and we write the defining integral as

$$X(s) = \int_{0^-}^{\infty} x(t)e^{-st}dt \tag{5.5}$$

TABLE 5.1 Laplace Transform Table: Selected Pairs

$x(t)$	$X(s)$
1. $\delta(t)$	1
2. $\delta(t-a)$	e^{-as}
3. $u(t)$	$\frac{1}{s}$
4. $u(t-a)$	$\frac{e^{-as}}{s}$
5. $e^{-at}u(t)$	$\frac{1}{s+a}$
6. $t^n u(t)$ $n = 1, 2, \ldots$	$\frac{n!}{s^{n+1}}$
7. $t^n e^{-at} u(t)$ $n = 1, 2, \ldots$	$\frac{n!}{(s+a)^{n+1}}$
8. $\cos(wt)u(t)$	$\frac{s}{s^2+w^2}$
9. $\sin(wt)u(t)$	$\frac{w}{s^2+w^2}$
10. $e^{-at}\cos(wt)u(t)$	$\frac{s+a}{(s+a)^2+w^2}$
11. $e^{-at}\sin(wt)u(t)$	$\frac{w}{(s+a)^2+w^2}$
12. $t\sin(wt)u(t)$	$\frac{2ws}{(s^2+w^2)^2}$
13. $t\cos(wt)u(t)$	$\frac{s^2-w^2}{(s^2+w^2)^2}$

and

$$x(t) \Leftrightarrow X(s)$$

says that the transform goes both sides.

We will present two tables: Table 5.1 for the Laplace transform pairs and Table 5.2 for the properties of the Laplace transform.

5.4 The Inverse Laplace Transform

The inverse Laplace transform can be found using Tables 5.1 and 5.2. It can also be found using the following equation

$$x(t) = \frac{1}{2\pi j}\int_{\sigma-jw}^{\sigma+jw} X(s)e^{st}ds \tag{5.6}$$

where $s = \sigma + jw$.

In the next few examples, we will look at some entries in Table 5.1 and see how they were developed. You need to know that each entry in Table 5.1 is derived using the defining integral

$$X(s) = \int_{-\infty}^{\infty} x(t)e^{-st}dt$$

TABLE 5.2 Properties of the Laplace Transform

Time Signal	LaSignal
1. $x(t)$	$X(s)$
2. $\sum_{n=1}^{N} a_n x_n(t)$	$\sum_{n=1}^{N} a_n X_n(s)$
3. $x(t - t_0)$	$X(s)e^{-St_0}$
4. $e^{(at)}x(t)$	$X(s - a)$
5. $x(\alpha t)$	$\frac{1}{\lvert\alpha\rvert}X(s/\alpha), \alpha > 0$
6. $\frac{d}{dt}x(t)$	$sX(s) - x(0^-)$
7. $\frac{d^2}{dt}x(t)$	$s^2X(s) - x(0^-) - \frac{dx(t)}{dt}(0^-)$
8. $x(t) * h(t)$	$X(s)H(s)$
9. $\int_{0^-}^{\infty} x(\tau)d\tau$	$\frac{1}{s}X(s)$
10. $tx(t)$	$\frac{-dX(s)}{ds}$
11. $x(t)\sin(w_0 t)$	$\frac{1}{2j}[X(s - jw_0) - X(s + jw_0)]$
12. $x(t)\cos(w_0 t)$	$\frac{1}{2}[X(s - jw_0) + X(s + jw_0)]$
13. $x(o^+)$	$\text{limit}_{s\to\infty} sX(s)$
14. $\text{limit}_{t\to\infty} x(t)$	$\text{limit}_{s\to\infty} sX(s)$

Example 5.1

Find the Laplace transform of the impulse signal $\delta(t)$.

Solution

Using the defining integral, we can find the Laplace transform of $\delta(t)$ as

$$X(s) = \int_{0^-}^{\infty} x(t)e^{-st}dt$$

By substituting for $x(t)$ we get

$$X(s) = \int_{0^-}^{\infty} \delta(t)e^{-st}dt = e^{-s(0)} = 1$$

Therefore,

$$x(t) = \delta(t) \Leftrightarrow X(s) = 1$$

Example 5.2

Find the Laplace transform of the signal $u(t)$.

Solution

Using the defining integral, we can find the Laplace transform of $u(t)$ as

$$X(s) = \int_{0^-}^{\infty} x(t)e^{-st}dt$$

By substituting for $x(t)$ we have

$$X(s) = \int_{0^-}^{\infty} u(t)e^{-st}dt$$

and finally

$$X(s) = \int_{0}^{\infty} e^{-st}dt = \left.\frac{e^{-st}}{-s}\right|_0^{\infty} = \frac{1}{s}$$

Therefore, we write

$$x(t) = u(t) \Leftrightarrow x(s) = \frac{1}{s}$$

Example 5.3

Find the Laplace transform of the signal $x(t) = e^{-at}u(t)$.

Solution

We can find the Laplace transform of $e^{-at}u(t)$ using the defining integral and write

$$X(s) = \int_{0^-}^{\infty} x(t)e^{-st}dt$$

$$X(s) = \int_{0^-}^{\infty} e^{-at}u(t)e^{-st}dt$$

$$X(s) = \int_{0}^{\infty} e^{-(a+s)t}dt$$

and finally

$$X(s) = \left.\frac{e^{-(s+a)t}}{-(s+a)}\right|_0^{\infty} = \frac{1}{s+a}$$

Therefore, we have the pairs

$$x(t) = e^{-at}u(t) \Leftrightarrow X(s) = \frac{1}{s+a}$$

The other entries in Table 5.1 can be evaluated in the same way.

Example 5.4

Find the inverse Laplace transform of

$$X(s) = 1 - \frac{5}{s+1}$$

Solution

As Table 5.2 shows, we can use the first entry to inverse transform each term separately. Using Table 5.1 we get

$$\delta(t) \Leftrightarrow 1$$

and

$$\frac{5}{s+1} \Leftrightarrow 5e^{-t}$$

Therefore,

$$X(s) = 1 - \frac{5}{s+1} \Leftrightarrow x(t) = \delta(t) - 5e^{-t}$$

Example 5.5

Find the inverse Laplace transform of

$$X(s) = \frac{1}{s^2 + s + 1}$$

Solution

We can see that the roots of the denominator in the above expression for $X(s)$ are complex. In this case, we can try to avoid using partial fraction expansion and try to complete the squares in the denominator in $X(s)$ and take advantage of the entries in Table 5.1.

Therefore, we can write

$$X(s) = \frac{1}{\left(s+\frac{1}{2}\right)^2 + \frac{3}{4}} = \frac{1}{\left(s+\frac{1}{2}\right)^2 + \left(\sqrt{\frac{3}{4}}\right)^2}$$

And by putting the above expression in a form similar to the form in Table 5.1, entry 11, we arrive at

$$X(s) = \frac{1}{\left(\sqrt{\frac{3}{4}}\right)} \frac{\left(\sqrt{\frac{3}{4}}\right)}{\left(s+\frac{1}{2}\right)^2 + \left(\sqrt{\frac{3}{4}}\right)^2}$$

We can now look at entry 11 in Table 5.1 and write

$$X(s) = \frac{1}{\left(\sqrt{\frac{3}{4}}\right)} \frac{\left(\sqrt{\frac{3}{4}}\right)}{\left(s+\frac{1}{2}\right)^2 + \left(\sqrt{\frac{3}{4}}\right)^2} \Leftrightarrow x(t) = \sqrt{\frac{4}{3}} e^{-\frac{1}{2}t} \sin\left(\sqrt{\frac{3}{4}}t\right) u(t)$$

Note that when the roots of the denominator in the s-plane (frequency domain) are complex, use entries 10 and 11 in Table 5.1, depending on the problem at hand.

Example 5.6

Find the inverse Laplace transform of

$$X(s) = \frac{s}{s^2 - 1}$$

Solution

In this case, the roots are not complex. This $X(s)$ is not shown in Table 5.1. We will use partial fraction expansion as in the following.

$$X(s) = \frac{s}{s^2 - 1} = \frac{A}{s-1} + \frac{B}{s+1}$$

where

$$A = (s-1)\left[\frac{s}{(s-1)(s+1)}\right]\Bigg|_{s=1}$$

Notice here that we can cancel the $(s - 1)$ terms before evaluating at $s = 1$. Therefore, we write

$$A = \left[\frac{s}{(s+1)}\right]\Bigg|_{s=1} = \frac{1}{2}$$

B can also be calculated in a similar way as

$$B = (s+1)\left[\frac{s}{(s-1)(s+1)}\right]\bigg|_{s=-1}$$

or

$$B = \left[\frac{s}{(s-1)}\right]\bigg|_{s=-1} = \frac{1}{2}$$

Therefore, we now write

$$X(s) = \frac{s}{s^2-1} = \frac{A}{s-1} + \frac{B}{s+1} = \frac{\frac{1}{2}}{s-1} + \frac{\frac{1}{2}}{s+1}$$

Using Table 5.1 we can inverse transform $X(s)$ and write

$$X(S) = \frac{\frac{1}{2}}{s-1} + \frac{\frac{1}{2}}{s+1} \Leftrightarrow x(t) = \left[\frac{1}{2}e^{t} + \frac{1}{2}e^{-t}\right]u(t)$$

5.5 Block Diagrams Using the Laplace Transform

Any transfer function in the s-domain can be written in terms of a fraction: the numerator as a function of s and the denominator as a function of s, too. We will denote such a transfer function as $H(s)$, the numerator as $N(s)$, and the denominator as $D(s)$. In this case, we can write

$$H(s) = \frac{N(s)}{D(s)}$$

where $N(s)$ is a polynomial in s and $D(s)$ is a polynomial in s, as well.

The transfer function relates the input to an LTI (linear time-invariant) system, $x(t)$, to the output of the same system, $y(t)$. The transfer function can also be written as

$$H(s) = \frac{Y(s)}{X(s)} = \frac{N(s)}{D(s)}$$

From the above equation we can write

$$Y(s) = X(s)H(s)$$

As a block diagram representation, we represent the system in the above equation as in Figure 5.1.

5.5.1 Parallel Systems

Consider three systems represented by their transfer functions $H_1(s)$, $H_2(s)$, and $H_3(s)$, with the single input $X(s)$ and the single output $Y(s)$. These can be represented in a block diagram as shown in Figure 5.2.

In Figure 5.2, the output is

$$Y(s) = (H_1(s) + H_2(s) + H_3(s))X(s)$$

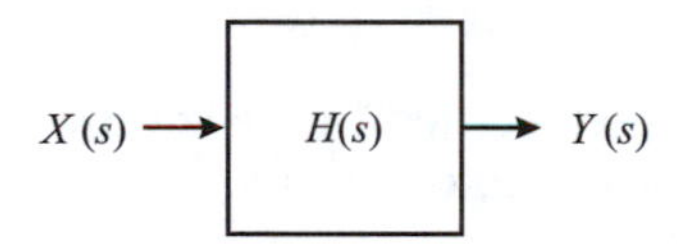

FIGURE 5.1
General block diagram.

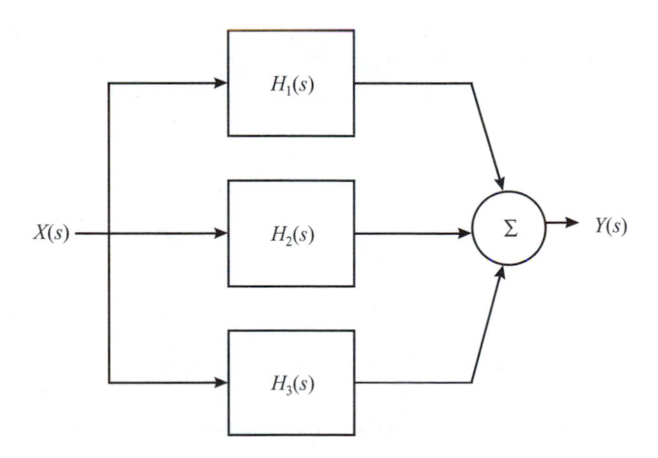

FIGURE 5.2
Parallel system.

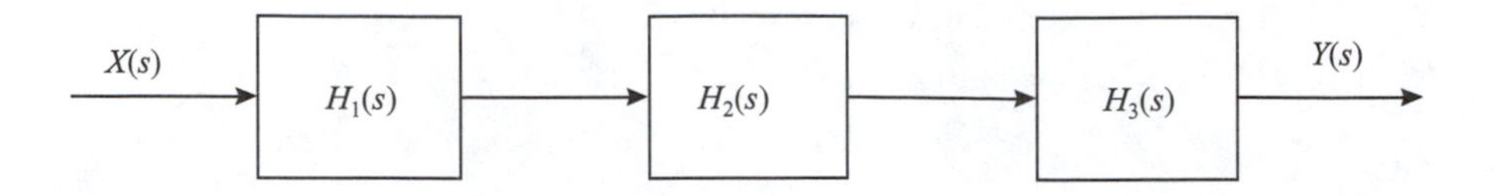

FIGURE 5.3
Series system.

If we have more transfer functions, the procedure is the same where you add all the individual transfer functions and multiply by the input $X(s)$ to get $Y(s)$. You can find the equivalent transfer function for the whole system as

$$H(s) = \frac{Y(s)}{X(s)} = H_1(s) + H_2(s) + H_3(s)$$

5.5.2 Series Systems

Consider three systems represented by their transfer functions $H_1(s)$, $H_2(s)$, and $H_3(s)$, with the single input $X(s)$ and the single output $Y(s)$. These can be represented in a block diagram as shown in Figure 5.3.

In Figure 5.3, the output is

$$Y(s) = (H_1(s)H_2(s)H_3(s))X(s)$$

If we have more transfer functions, the procedure is the same where you multiply all the individual transfer functions first and then multiply the result by the input $X(s)$ to get $Y(s)$. You can find the equivalent transfer function for the whole system as

$$H(s) = \frac{Y(s)}{X(s)} = H_1(s)H_2(s)H_3(s)$$

5.6 Representation of Transfer Functions as Block Diagrams

Block diagram representation is not unique, as you will see in the following examples, which illustrate the procedure.

Example 5.7

Represent the transfer function

$$H(s) = \frac{1}{s^2 - 1}$$

using block diagrams.

Solution

The representation can be accomplished in many ways.

1. With the relation

$$Y(s) = H(s)X(s)$$

   the procedure is straightforward, as shown in Figure 5.4.
2. We can write $H(s)$ as the product of two terms

$$H(s) = \frac{1}{s^2 - 1} = \left(\frac{1}{s - 1}\right)\left(\frac{1}{s + 1}\right)$$

   This is the case of multiplying two transfer functions and the block is shown in Figure 5.5.
3. We can write $H(s)$ as the sum of two terms

$$H(s) = \frac{1}{s^2 - 1} = \left(\frac{\frac{1}{2}}{s - 1}\right) + \left(\frac{-\frac{1}{2}}{s + 1}\right)$$

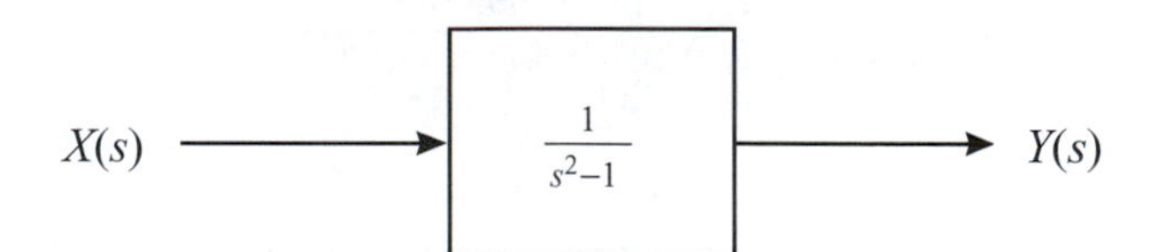

FIGURE 5.4
Block for Example 5.7.

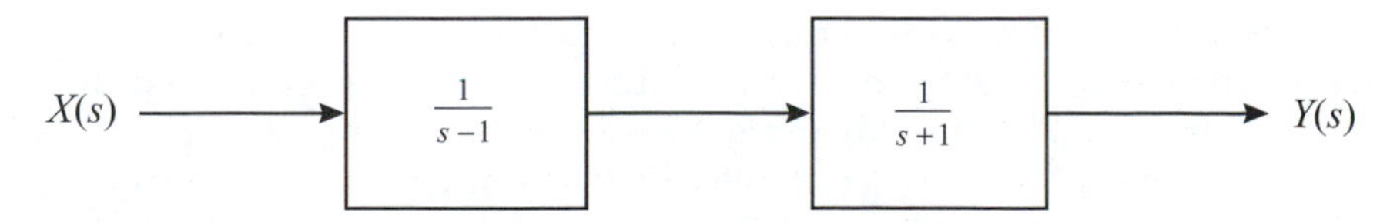

FIGURE 5.5
Block for Example 5.7.

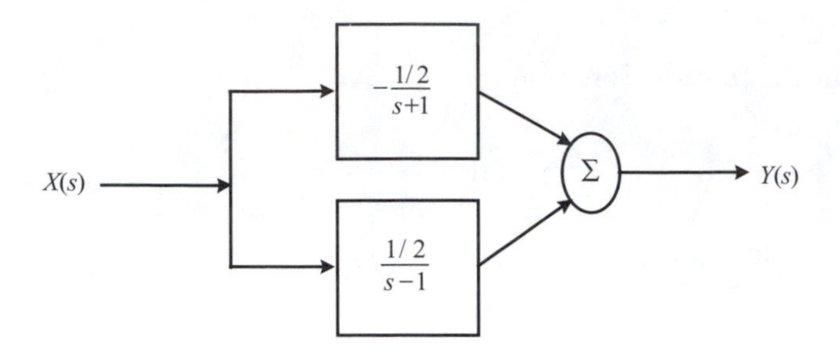

FIGURE 5.6
Block for Example 5.7.

This is the case of adding two transfer functions. The block is shown in Figure 5.6.

5.7 Procedure for Drawing the Block Diagram from the Transfer Function

We will learn this procedure by example.

Consider the following third order system (we say third order because the denominator polynomial is of degree three) or third order differential equation represented as the transfer function

$$H(s) = \frac{as^3 + bs^2 + cs + d}{s^3 + es^2 + fs + g}$$

where a, b, c, d, e, f, and g are constants.

It is required that we represent this transfer function in block diagram form.

The first thing we do here is to make the coefficient of s^3 in the denominator unity. Our transfer function has that characteristic.

Because the order of the denominator in this expression for $H(s)$ is 3, we will need three integrators. Let us precede each integrator by a summer and follow the last integrator by a summer, too. The initial step in the development of drawing this block diagram is as shown in Figure 5.7.

Next, we will feedback $Y(s)$ multiplied by the negative of each of the constants e, f, and g, in the denominator of $H(s)$. If we proceed from left to right, the signal $[Y(s)(-g)]$ will be fed to the summer that precedes the first integrator, the signal $[Y(s)(-f)]$ will be fed to the summer that precedes the second integrator, and the signal $[Y(s)(-e)]$ will be fed to the summer that precedes the third integrator. The diagram becomes as shown in Figure 5.8.

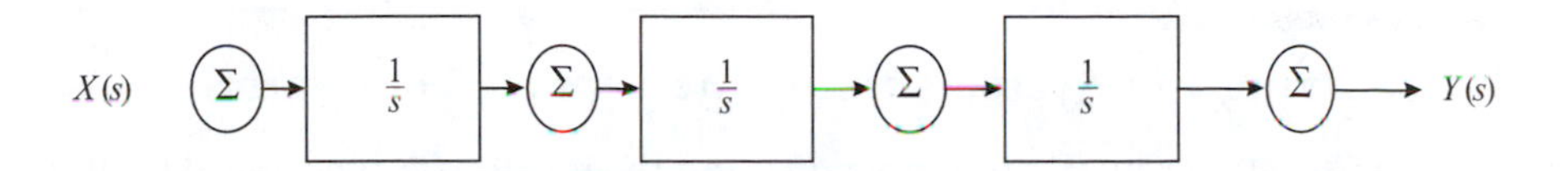

FIGURE 5.7
Three-integrators block diagram.

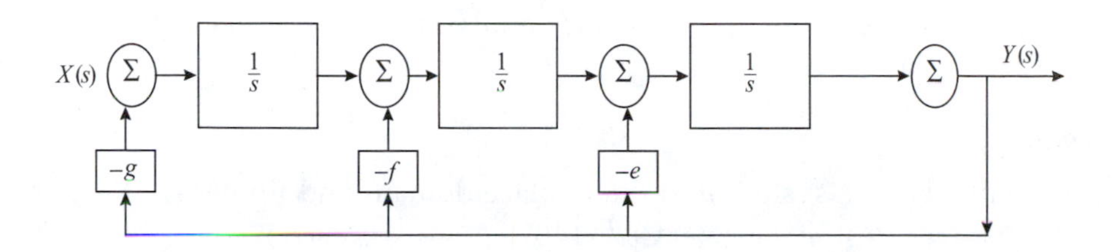

FIGURE 5.8
Three-integrators block diagram.

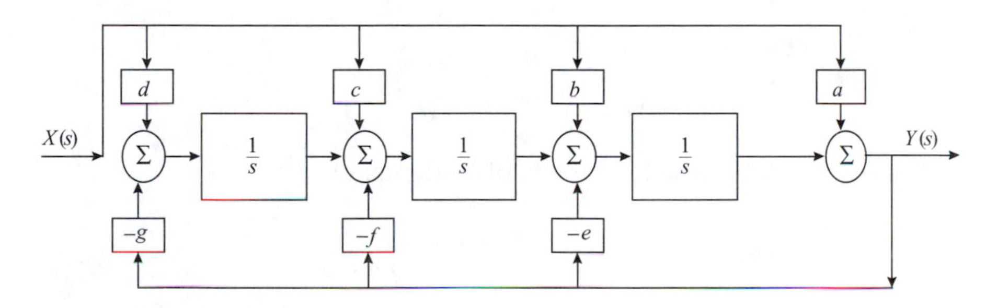

FIGURE 5.9
Three-integrators block diagram.

Finally, we will feed forward the input signal $[X(s)(d)]$ to the summer that precedes the first integrator, $[X(s)(c)]$ to the summer that precedes the second integrator, $[X(s)(b)]$ to the summer that precedes the third integrator, and $[X(s)(a)]$ to the summer that follows the third integrator. The final block diagram is as shown in Figure 5.9.

Notes on the above example:

1. If the order of the numerator in the example at hand is 2, then the feed forward to the summer after the third integrator is zero, and we represent that without connection.
2. If the coefficient of s^2 in the numerator in the example at hand is zero, then there will be no feed forward signal to the summer that precedes the third integrator.
3. If there is no constant term in the numerator (if $d = 0$) then there will be no feed forward from $x(t)$ to the first summer before the first integrator.

We will consider further examples to illustrate these notes.

5.8 Solving LTI Systems Using the Laplace Transform

Given the system transfer function, $h(t)$, and the input, $x(t)$, we should be able to find the output, $y(t)$, easily, using the Laplace transform techniques.

Example 5.8

For a certain system let $h(t) = e^{-t}u(t)$ and $x(t) = \delta(t)$. What is the output of the system?

Solution

We will take this system into the Laplace domain and find $Y(s)$. We then inverse transform $Y(s)$ to get $y(t)$. For the impulse signal

$$h(t) = e^{-t}u(t) \Leftrightarrow H(s) = \frac{1}{s+1}$$

and for the input

$$x(t) = \delta(t) \Leftrightarrow X(s) = 1$$

The output of the system in the Laplace domain is

$$Y(s) = X(s)H(s)$$

or

$$Y(s) = (1)\left(\frac{1}{s+1}\right)$$

Using Table 5.1, we can see that the inverse transform of $Y(s)$ is

$$y(t) = e^{-t}u(t)$$

Let us look at the system graphically. In the Laplace domain we have the block in Figure 5.10.

In the time domain, we have the signals as shown in Figure 5.11.

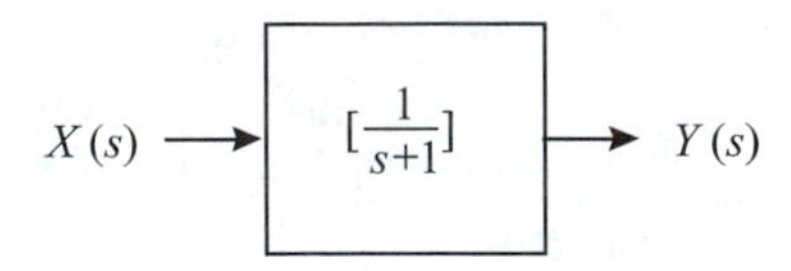

FIGURE 5.10
System for Example 5.8.

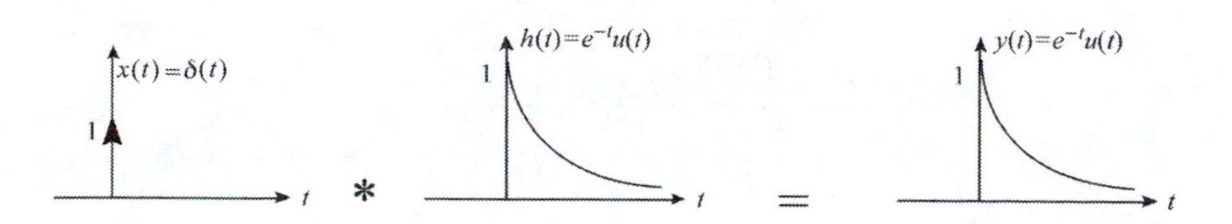

FIGURE 5.11
System for Example 5.8.

Example 5.9

For a certain system let

$$h(t) = e^{-t}u(t)$$

and

$$x(t) = u(t)$$

What is the output of the system?

Solution

We will take this system into the Laplace domain and find $Y(s)$. We then inverse transform $Y(s)$ to get $y(t)$. For the impulse signal

$$h(t) = e^{-t}u(t) \Leftrightarrow H(s) = \frac{1}{s+1}$$

and for the input

$$x(t) = u(t) \Leftrightarrow X(s) = \frac{1}{s}$$

The output is then

$$Y(s) = X(s)H(s)$$

$$Y(s) = \left(\frac{1}{s}\right)\left(\frac{1}{s+1}\right)$$

If we look at Table 5.1, we cannot see an entry similar to $Y(s)$ as given in this example. We will do partial fraction expansion on $Y(s)$ and write

$$Y(s) = \left(\frac{1}{s}\right)\left(\frac{1}{s+1}\right) = \frac{1}{s} - \frac{1}{s+1}$$

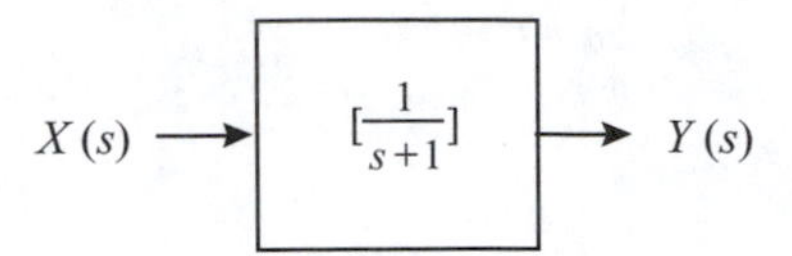

FIGURE 5.12
System for Example 5.9.

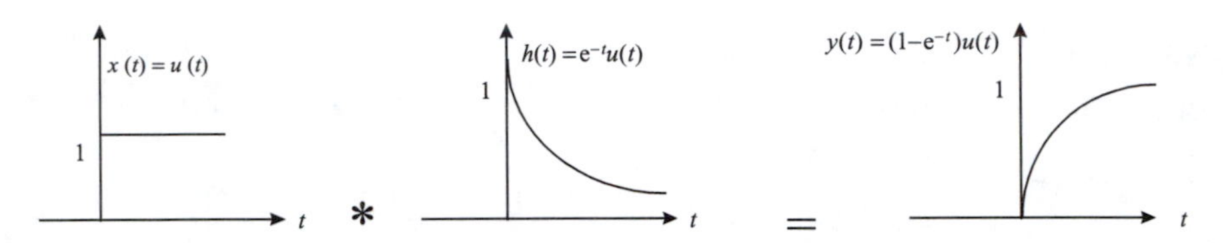

FIGURE 5.13
System for Example 5.9.

We can now use Table 5.1 and inverse transform each term in $Y(s)$ to get

$$y(t) = u(t) - e^{-t}u(t) = (1 - e^{-t})u(t)$$

Graphically, and in the Laplace domain we have the system shown in Figure 5.12.

In the time domain we have the picture in Figure 5.13.

5.9 Solving Differential Equations Using the Laplace Transform

We need to remember that the systems we are dealing with here are nothing but dynamical systems that can be represented as differential equations. The differential equation

$$\frac{d}{dt}y(t) + y(t) = x(t)$$

is a first order differential equation that describes a linear system with $x(t)$ as an input and $y(t)$ as an output. Using Tables 5.1 and 5.2, we can solve differential equations easily.

Example 5.10

Find the output of the following system represented by the differential equation

$$\frac{d}{dt}y(t) + 9y(t) = \delta(t)$$

with the initial condition $y(0^-) = 0$.

Solution

In this equation $x(t)$ is given as $\delta(t)$. We need to find $y(t)$. Using Table 5.2, we can write the Laplace equivalent of the differential equation given. We will transform the differential equation term by term and write

$$[sY(s) - y(0^-)] + 9Y(s) = 1$$

With the initial condition set to zero as given, we can solve for $Y(s)$ as

$$Y(s) = \frac{1}{s+9}$$

We now take the inverse transform of $Y(s)$ using Table 5.1 to get

$$y(t) = e^{-9t}u(t)$$

Example 5.11

Suppose we have the same differential equation as in Example 5.10 with non-zero initial condition $y(0^-) = 1$. What is the output $y(t)$ of this system?

Solution

We proceed with the solution as we did in the previous example and write the Laplace transform of the system this time as

$$[sY(s) - y(0^-)] + 9Y(s) = 1$$

$$[sY(s) - 1] + 9Y(s) = 1$$

Finally

$$Y(s) = \frac{2}{s+9}$$

Next, we take the inverse transform to get

$$y(t) = 2e^{-9t}u(t)$$

Example 5.12

Let us now consider a second order system described by the differential equation

$$\frac{d^2}{dt}y(t) + 2\frac{d}{dt}y(t) = \delta(t)$$

The input is applied at $t = 0$ and the initial conditions are

$$y(0^-) = \left.\frac{d}{dt}y(t)\right|_{t=0^-} = 0$$

Find the output of this LTI system.

Solution

We can look again at Table 5.2 to transform the given differential equation term by term into its Laplace domain and write

$$\left[s^2Y(s) - s\left.\frac{d}{dt}y(t)\right|_{t=0^-} - y(0^-)\right] + 2[sY(s) - y(0^-)] = 1$$

Solving for $Y(s)$ we get

$$Y(s) = \frac{1}{s^2 + 2s} = \frac{1}{s(s+2)}$$

Using partial fractions we write $Y(s)$ as

$$Y(s) = \frac{\frac{1}{2}}{s} + \frac{-\frac{1}{2}}{s+2}$$

The inverse transform of $Y(s)$, $y(t)$, is obtained using Table 5.1 as

$$y(t) = (1/2)u(t) - (1/2)e^{-2t}u(t)$$

and finally,

$$y(t) = (1/2)[1 - e^{-2t}]u(t)$$

5.10 The Final Value Theorem

Sometimes we are interested in the final value of the output $y(t)$ as $t \rightarrow \infty$. Instead of first solving for $y(t)$ in time domain, which can be very tedious in some cases, we can use the Laplace transform and write

$$\text{limit}_{t \rightarrow \infty}\, y(t) = \text{limit}_{s \rightarrow 0}[sY(s)]$$

To use the theorem above the roots of the denominator of $sY(s)$ must not be positive.

5.11 The Initial Value Theorem

For the initial value of $y(t)$ we have

$$\text{limit}_{t \rightarrow 0^+} y(t) = y(0^+) = \text{limit}_{s \rightarrow \infty}[sY(s)]$$

Example 5.13

If $y(t) = 2e^{-9t}u(t)$, what is $\text{limit}_{t \rightarrow \infty} y(t)$?

Solution

In this example, the solution for the output $y(t)$ was given for us. As we said earlier, this solution may not be available but what we can find in such a case is its transform $Y(s)$ from which we can find

$$\text{limit}_{t \rightarrow \infty} y(t) = \text{limit}_{s \rightarrow 0}[sY(s)]$$

In this example, we are lucky and have $y(t)$. We can use this $y(t)$ for checking our answer. Since $u(\infty) = 1$ and $e^{-\infty} = 0$, $\text{limit}_{t \rightarrow \infty} y(t) = e^{-9(\infty)}u(\infty)$ $= (0)(1) = 0$. According to the final value theorem

$$\text{limit}_{t \rightarrow \infty} y(t) = \text{limit}_{s \rightarrow 0}[sY(s)] = \text{limit}_{s \rightarrow 0}\left[s\frac{2}{s+9}\right] = \left[\frac{0}{0+9}\right] = 0$$

Here we see that the two approaches agree.

5.12 Some Insights: Poles and Zeros

The transfer function, $H(s)$, of a linear time-invariant system is a very important representation. It tells us many things about the stability of the system, the poles, the zeros, and the shape of the transients of the output of the system. Using $H(s)$, we can find the steady-state response of the system and the particular solution of the system, all in one shot.

5.12.1 The Poles of the System

The poles of the system are the roots of the denominator, the algebraic equation in the variable s, of the transfer function $H(s)$

$$H(s) = \frac{N(s)}{D(s)}$$

$D(s)$ is a polynomial in s of order equal to the order of the system. The roots of the denominator $D(s)$ are called the poles of the system. These are the same poles we discussed in Chapter 2. We called them then the eigenvalues of the system. $D(s)$ is actually the characteristic equation of the system or, as we called it before, the auxiliary equation of the system.

5.12.2 The Zeros of the System

The roots of the numerator $N(s)$ are called the zeros of the system.

5.12.3 The Stability of the System

The poles of the system determine the stability of the system under investigation. If the real parts of the poles are **all** negative, then the system at hand is stable and the transients will die as time gets larger. The stability of the system is determined by the poles not the zeros of the system. If one of the real parts of the poles is positive, then the system is not stable. You may have positive and negative zeros but the sign of their values have no effect on the stability of the system.

Given $H(s)$, the roots of the denominator will determine the general shape of the output, $y(t)$, which in this case is $h(t)$ because the input $x(t)$ is the impulse $\delta(t)$. If $D(s)$ has two roots (second order system) called α_1 and α_2, then the output will have the general form

$$y(t) = h(t) = c_1 e^{\alpha 1t} + c_2 e^{\alpha 2t}$$

where the constant c's are to be determined. The exponential terms will determine the shape of the transients. If one of the α's is positive, the output will grow without bounds. If the two α's are negative, the output will die as time progresses. The α's are the eigenvalues or the poles of the system.

The transfer function $Y(s)/X(s)$ is called $H(s)$ if the input $X(s)$ is 1 ($x(t) = \delta(t)$). The transfer function $Y(s)/X(s)$ is very important as we will see later in the design of linear time-invariant systems.

5.13 End-of-Chapter Examples

EOCE 5.1

What is the Laplace transform of the derivative of the signal $x(t)$?

Solution

We can use the defining integral

$$X(s) = \int_{0^-}^{\infty} x(t)e^{-st}dt$$

and write

$$L\left(\frac{d}{dt}x(t)\right) = \int_{0^-}^{\infty} \frac{d}{dt}x(t)e^{-st}dt$$

where "L" represents "the Laplace transform of."

Let $u = e^{-st}$ and $dv = \frac{d}{dt}x(t)dt$, then $du = -se^{-st}dt$ and $v = x(t)$. We now can integrate by parts and write

$$L\left(\frac{d}{dt}x(t)\right) = e^{-st}x(t)\Big|_{0^-}^{\infty} + s\int_{0^-}^{\infty} x(t)e^{-st}dt$$

$$L\left(\frac{d}{dt}x(t)\right) = -x(0^-) + sX(s)$$

Therefore, we can finally write

$$L\left(\frac{d}{dt}x(t)\right) \Leftrightarrow sX(s) - x(0^-)$$

EOCE 5.2

What is the Laplace transform of the second derivative of the signal $x(t)$?

Solution

We can use EOCE 5.1 to solve EOCE 5.2 as in the following.

The transform of the second derivative is

$$L\left(\frac{d^2}{dt}x(t)\right) = sL\left(\frac{d}{dt}x(t)\right) - \frac{d}{dt}x(0^-)$$

Substituting for $L\left(\frac{d}{dt}x(t)\right)$ from EOCE 5.1 in the above equation to get

$$L\left(\frac{d^2}{dt}x(t)\right) = s[sX(s) - x(0^-)] - \frac{d}{dt}x(0^-)$$

By simplification we can finally write

$$L\left(\frac{d^2}{dt}x(t)\right) = s^2X(s) - sx(0^-) - \frac{d}{dt}x(0^-)$$

EOCE 5.3

What is the Laplace transform of $\frac{d^3}{dt}x(t)$?

Solution

The transform of the third derivative is

$$L\left(\frac{d^3}{dt}x(t)\right) = sL\left(\frac{d^2}{dt}x(t)\right) - \frac{d^2}{dt}x(0^-)$$

Substituting for $L\left(\frac{d^2}{dt}x(t)\right)$ from EOCE 5.2 we get

$$L\left(\frac{d^3}{dt}x(t)\right) = s\left[s^2X(s) - sx(0^-) - \frac{d}{dt}x(0^-)\right] - \frac{d^2}{dt}x(0^-)$$

And with some manipulations we arrive at

$$L\left(\frac{d^3}{dt}x(t)\right) = s^3X(s) - s^2x(0^-) - s\frac{d}{dt}x(0^-) - \frac{d^2}{dt}x(0^-)$$

EOCE 5.4

Now that we know how to transform a derivative of $x(t)$, what about the integral of $x(t)$? Or what is the Laplace transform of

$$L\left(\int_{0^-}^{t} x(u)du\right)$$

Solution

Let $X(t)$ be the integral of $x(t)$. Then in this case $\frac{d}{dt}X(t) = x(t)$ and

$$L\left(\int_{0^-}^{t} x(u)du\right) = L(X(t))$$

But

$$L(x(t)) = L\left(\frac{d}{dt}X(t)\right) = sL(X(t)) - X(0^-)$$

$$L(x(t)) = sL\left(\int_{0^-}^{t} x(u)du - \int_{0^-}^{0^-} x(u)du\right)$$

$$L(x(t)) = sL\left(\int_{0^-}^{t} x(u)du\right)$$

Therefore

$$L\left(\int_{0^-}^{t} x(u)du\right) = \frac{1}{s}L(x(t)) = \frac{1}{s}X(s)$$

EOCE 5.5

If $x(0^-) = 0$, we claim that the inverse Laplace transform of $sX(s) = \frac{d}{dt}x(t)$. Is this claim true?

Solution

We can use EOCE 5.1 to write

$$L\left(\frac{d}{dt}x(t)\right) = sL(x(t)) - x(0^-) = sX(s)$$

This implies

$$L^{-1}(sX(s)) = \frac{d}{dt}x(t)$$

EOCE 5.6

Show that the inverse Laplace transform of $X(s)/s$ is the integral of $x(t)$.

Solution

We know that

$$L\left(\int_{0^-}^{t} x(u)du\right) = \frac{1}{s}X(s)$$

This means that

$$L^{-1}\left(\frac{1}{s}X(s)\right) = \int_{0^-}^{t} x(u)du$$

EOCE 5.7

Find the inverse Laplace transform of $\frac{1}{s}\frac{1}{s+1}$.

Solution

We can use EOCE 5.6 and ignore the term $1/s$ and find the inverse transform of $1/(s + 1)$.

$$\frac{1}{s+1} \Leftrightarrow e^{-t}u(t)$$

We can integrate $x(t)$ and get

$$\int_{0^-}^{t} e^{-u}du = -(e^{-t} - 1) = e^{-t} + 1$$

But

$$(e^{-t} + 1)u(t) \Leftrightarrow \frac{1}{s(s+1)}$$

and therefore

$$L^{-1}\left(\frac{1}{s(s+1)}\right) = (e^{-t}+1)u(t)$$

We can use partial fraction expansion to check this answer.

EOCE 5.8

Find the inverse Laplace transform of $\frac{s}{s+1}$ using EOCE 5.5.

Solution

Let us ignore s in the numerator and inverse transform $\frac{1}{s+1}$. By doing that we can write

$$\frac{1}{s+1} \Leftrightarrow e^{-t}u(t)$$

But at $t = 0,\ e^{-t}u(t)$ is not 0. Therefore, we cannot use the EOCE 5.5 result.

EOCE 5.9

Consider the two coupled first order differential equations with zero initial conditions

$$\frac{d}{dt}y(t) + 2y(t) + z(t) = 0$$

$$\frac{d}{dt}z(t) + z(t) = -u(t)$$

Solve for $y(t)$ using the Laplace transform approach.

Solution

In the Laplace domain the two coupled equations are

$$sY(s) + 2Y(s) + Z(s) = 0 \tag{5.7}$$

$$sZ(s) + Z(s) = -1/s \tag{5.8}$$

From (Equation 5.8) we get $Z(s)$ as

$$Z(s) = -\frac{1}{s(s+1)}$$

From (Equation 5.7) we have

$$Y(s)[s+2] = -Z(s)$$

Substitute for $Z(s)$ to get $Y(s)$ as

$$Y(s) = \frac{1}{s(s+1)(s+2)}$$

Using partial fraction expansion we get

$$Y(s) = \frac{\frac{1}{2}}{s} + \frac{-1}{s+1} + \frac{\frac{1}{2}}{s+2}$$

and then $y(t)$ is obtained by inverse transforming $Y(s)$ to get

$$y(t) = \frac{1}{2}u(t) - e^{-t}u(t) + \frac{1}{2}e^{-2t}u(t)$$

EOCE 5.10

What is $L(e^{-at}x(t))$?

Solution

Using the defining integral we get

$$L(e^{-at}x(t)) = \int_{0^-}^{\infty} e^{-at}x(t)e^{-st}dt$$

$$L(e^{-at}x(t)) = \int_{0^-}^{\infty} x(t)e^{-(s+a)t}dt = X(s)\Big|_{s+a}$$

Therefore, the pairs are

$$e^{-at}x(t) \Leftrightarrow X(s)|_{s+a}$$

EOCE 5.11

What is the Laplace transform of $e^{-2t}e^{-t}u(t)$? Use EOCE 5.10.

Solution

We can use EOCE 5.10 and write

$$e^{-t}u(t) \Leftrightarrow \frac{1}{s+1}$$

and

$$L(e^{-2t}e^{-t}u(t)) = \left.\frac{1}{s+1}\right|_{s+2}$$

$$L(e^{-2t}e^{-t}u(t)) = \frac{1}{(s+2)+1} = \frac{1}{s+3}$$

or

$$e^{-2t}e^{-t}u(t) \Leftrightarrow \frac{1}{s+3}$$

EOCE 5.12

Consider a shifted signal $x(t-a)$. What is the Laplace transform of $x(t-a)u(t-a)$?

Solution

Notice that the multiplication $x(t-a)u(t-a)$ starts at $t = a$.
Using the defining integral we can write

$$L(x(t-a)u(t-a)) = \int_{0^-}^{\infty} x(t-a)u(t-a)e^{-st}dt$$

$$L(x(t-a)u(t-a)) = \int_{a}^{\infty} x(t-a)e^{-st}dt$$

If we let $t - a = m$, then

$$L(x(t-a)u(t-a)) = \int_{0}^{\infty} x(m)e^{-s(m+a)}dm$$

or

$$L(x(t-a)u(t-a)) = e^{-sa}\int_{0}^{\infty} x(m)e^{-sm}dm = e^{-sa}X(s)$$

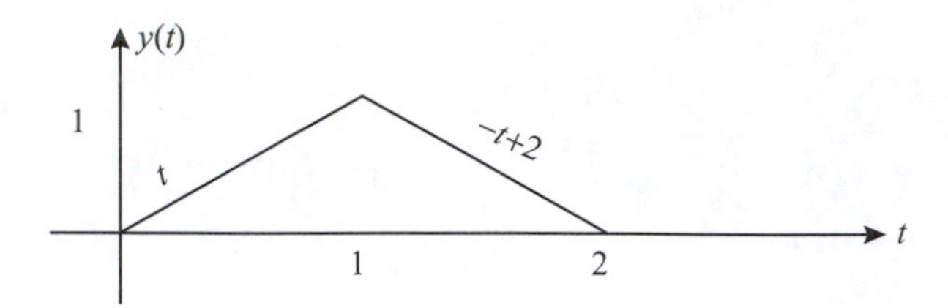

FIGURE 5.14
Signal for EOCE 5.13.

EOCE 5.13

Find the Laplace transform of the signal in Figure 5.14.

Solution

Mathematically, the signal can be written as

$$y(t) = (t)[u(t) - u(t-1)] + (-t+2)[u(t-1) - u(t-2)]$$

Multiplying the terms out we get

$$y(t) = tu(t) - tu(t-1) + (t-2)u(t-2) + (-t+2)u(t-1)$$

Expanding the terms we will have

$$\begin{aligned} y(t) = {} & tu(t) - (t+1-1)u(t-1) + (t-2)u(t-2) \\ & -(t-1+1)u(t-1) + 2u(t-1) \end{aligned}$$

$$\begin{aligned} y(t) = {} & tu(t) - (t-1)u(t-1) + u(t-1) + (t-2)u(t-2) \\ & -(t-1)\,u(t-1) - u(t-1) + 2u(t-1) \end{aligned}$$

We can easily show by using the defining integral that

$$u(t-a) \Leftrightarrow \frac{e^{-sa}}{s}$$

We can now find $Y(s)$ as

$$Y(s) = \frac{1}{s^2} - \frac{e^{-s}}{s^2} + \frac{e^{-s}}{s} + \frac{e^{-2s}}{s^2} - \frac{e^{-s}}{s^2} - \frac{e^{-s}}{s} + 2\frac{e^{-s}}{s}$$

$$Y(s) = \frac{1}{s^2} - 2\frac{e^{-s}}{s^2} + \frac{e^{-2s}}{s^2} + 2\frac{e^{-s}}{s}$$

EOCE 5.14

What is $L(tx(t))$?

Solution

Let us differentiate the defining integral

$$X(s) = \int_{0^-}^{\infty} x(t)e^{-st}dt$$

with respect to s on both sides to get

$$\frac{d}{ds}X(s) = \frac{d}{ds}\int_{0^-}^{\infty} x(t)e^{-st}dt$$

$$\frac{d}{ds}X(s) = \int_{0^-}^{\infty} \frac{d}{ds}x(t)e^{-st}dt = -\int_{0^-}^{\infty} tx(t)e^{-st}dt$$

Therefore

$$-\frac{d}{ds}X(s) = \int_{0^-}^{\infty} tx(t)e^{-st}dt$$

The last equation simply says that

$$(tx(t)) \Leftrightarrow -\frac{d}{ds}X(s)$$

EOCE 5.15

What is $L(x(t)/t)$ if $\lim_{t\to 0}[x(t)/t]$ exists?

Solution

By definition

$$X(s) = \int_{0^-}^{\infty} x(t)e^{-st}dt$$

If we integrate both sides from s to ∞ we get

$$\int_{s}^{\infty} X(s)ds = \int_{s}^{\infty}\left[\int_{0^-}^{\infty} x(t)e^{-st}dt\right]ds$$

By interchanging the integrals we get

$$\int_s^{\infty} X(s)ds = \int_{0^-}^{\infty}\int_s^{\infty} [x(t)e^{-st}ds]dt = \int_{0^-}^{\infty}\left[x(t)\frac{e^{-st}}{-t}\right]_s^{\infty} dt$$

or

$$\int_s^{\infty} X(s)ds = \int_{0^-}^{\infty}\left[\frac{x(t)}{t}e^{-st}\right]dt = L\left(\frac{x(t)}{t}\right)$$

and finally

$$L\left(\frac{x(t)}{t}\right) \Leftrightarrow \int_s^{\infty} X(s)ds$$

EOCE 5.16

What is $L\left(\frac{\sin(t)u(t)}{t}\right)$?

Solution

Using EOCE 5.15 we can write

$$L(\sin(t)u(t)/t) = \int_s^{\infty} L(\sin(t)u(t))ds$$

which simplifies to

$$L(\sin(t)u(t)/t) = \int_s^{\infty} \frac{1}{s^2+1}ds = \tan^{-1}(s)\Big|_s^{\infty}$$

and finally

$$L(\sin(t)u(t)/t) = \frac{\pi}{2} - \tan^{-1}(s) = \cot^{-1}(s)$$

EOCE 5.17

Consider the following output in the *s*-domain

$$Y(s) = \frac{1}{s^3 + 4s^2 + 2s + 1}$$

What are the initial and the final values of $y(t)$? Is the system stable?

Solution

We will give the solution in time-domain first. We will use MATLAB to do partial fraction expansion and write the script

```
num = [1];
den = [14 -2 -2];
[R, P, K] = residue(num, den)
```

The output is

```
R =
0.0499
0.1364
-0.1864
P =
-4.3539
0.8774
-0.5235
K =
[ ]
```

Therefore

$$Y(s) = \frac{0.049}{s + 4.3539} + \frac{0.1364}{s - 0.8774} - \frac{0.1864}{s + 0.5235}$$

K has no values. It will have values when the order of the numerator is greater than or equal to the order of the denominator.

We can use the Laplace transform table to inverse transform $Y(s)$ to get $y(t)$ as

$$y(t) = 0.049e^{-4.3539t}u(t) + 0.1364e^{0.8774t}u(t) - 0.1864e^{0.5235t}u(t)$$

We notice that this output is unstable because one of the poles of the system is positive. As t approaches $\propto$, $y(t)$ approaches $\propto$ and this indicates that we have a positive pole. As t approaches zero, $y(0) = 0.049 + 0.1364 - 0.1864 = -0.001$.

In the s-domain, we can see that one of the poles is positive and that indicates instability. We cannot use the final value theorem to find $y(\propto)$ because one of the poles is positive.

EOCE 5.18

What is the Laplace transform of the signal $x(t)$ that is periodic with period T on the interval [0 ∝]?

Solution

Using the defining integral again we can write

$$L(x(t)) = \int_{0^-}^{\infty} x(t)e^{-st}dt = \int_0^T x(t)e^{-st}dt + \int_T^{2T} x(t)e^{-st}dt + \int_{2T}^{3T} x(t)e^{-st}dt + \cdots$$

Let $t = P$ in the first integral, $t = P + T$ in the second integral, $t = P+2T$ in the third integral, and so on, to get

$$L(x(t)) = \int_{0^-}^{\infty} x(t)e^{-st}dt = \int_0^T x(p)e^{-sp}dp + \int_0^T x(P+T)e^{-s(P+T)}dP + \int_0^T x(P+2T)e^{-s(P+2T)}dP + \cdots$$

Using the fact that $x(P + T) = x(P)$, we get

$$L(x(t)) = \int_0^T x(P)e^{-sP}dP + e^{-sT}\int_0^T x(P+T)e^{-s(P)}dP + e^{-2sT}\int_0^T x(P+2T)e^{-s(P)}dP + \cdots$$

$$L(x(t)) = \int_0^T x(P)e^{-sP}dP + e^{-sT}\int_0^T x(P)e^{-s(P)}dP + e^{-2sT}\int_0^T x(P)e^{-s(P)}dP + \cdots$$

And by taking the common terms we arrive at

$$L(x(t)) = [1 + e^{-sT} + e^{-2sT} + \cdots]\int_0^T x(P)e^{-s(P)}dP$$

Also

$$1 + e^{-sT} + e^{-2sT} + e^{-3sT} + \cdots$$

is a geometric series and can be written as the summation

$$\frac{1}{1 - e^{-sT}}$$

Then

$$L(x(t)) = \left[\frac{1}{1+e^{-sT}}\right]\int_0^T x(P)e^{-s(P)}dP$$

EOCE 5.19

Consider the following system represented by the differential equation

$$\frac{d^2}{dt}y(t) + \frac{d}{dt}y(t) + y(t) = x(t)$$

Assume zero initial conditions and plot the output $y(t)$ vs. time for a step and an impulse input $x(t)$. Is the system stable? What are the final and the initial values of $y(t)$ when the input is step, impulsive, and ramp?

Solution

We can use MATLAB to find $y(t)$. First, we need to find the transfer function that relates the input to the output. To do that we take the Laplace transform of the given differential equation with a general input $x(t)$ to get

$$s^2Y(s) + sY(s) + Y(s) = X(s)$$

Taking $Y(s)$ as common factor gives

$$Y(s)[s^2 + s + 1] = X(s)$$

and the transfer function is

$$\frac{Y(s)}{X(s)} = H(s) = \frac{1}{s^2 + s + 1}$$

At this point, we can find the output $y(t)$ for any input $x(t)$ by inverse-transforming

$$Y(s) = H(s)X(s)$$

into its time domain. Notice that $H(s)$ is a ratio of two polynomials. To find the response (output) due to a step input, we type the following

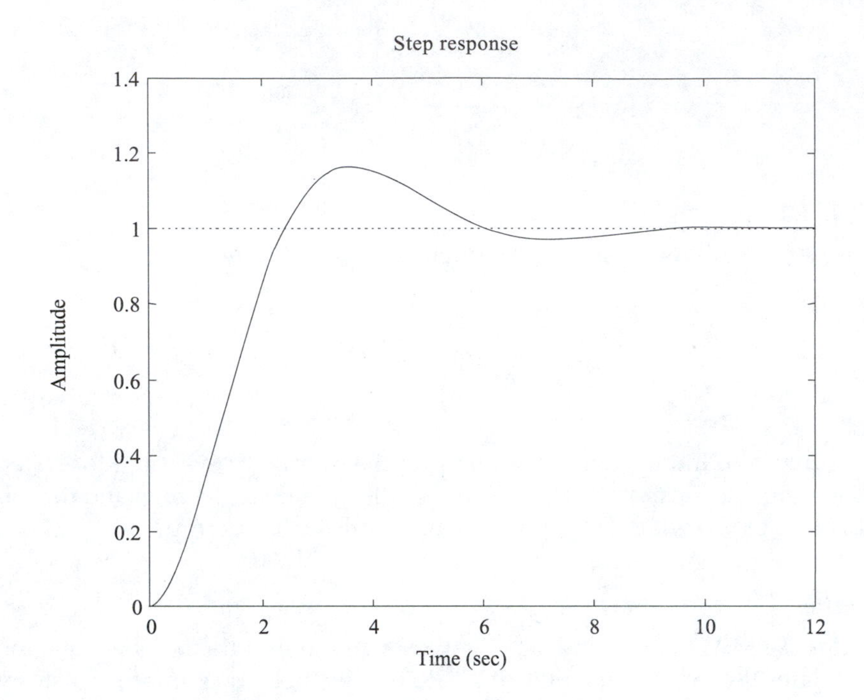

FIGURE 5.15
Plot for EOCE 5.19.

MATLAB script.

```
num = [1];
den = [1 1 1];
step(num, den)
```

The output plot is shown in Figure 5.15.

To find the output due to an impulsive input $x(t)$, we type the following MATLAB script.

```
num = [1];
den = [1 1 1];
impulse(num, den)
```

The output is shown in Figure 5.16.

To find the output if the input is $tu(t) = x(t)$, we need to find $Y(s)$ if $x(t) = tu(t)$ because MATLAB does not have a function to calculate a ramp response as we saw for the step and the impulse inputs. In this case, for

$$x(t) = tu(t)$$

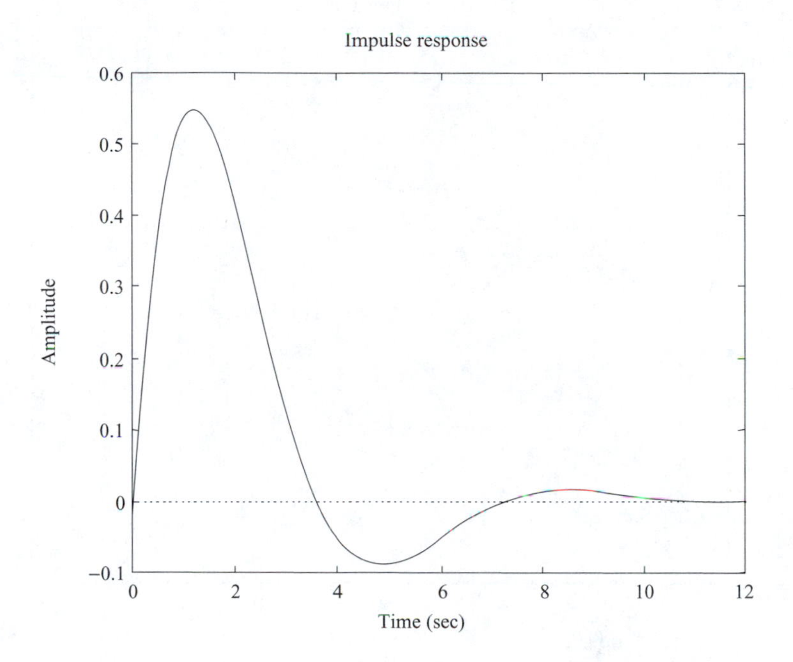

FIGURE 5.16
Plot for EOCE 5.19.

we have

$$X(s) = \frac{1}{s^2}$$

The output $Y(s)$ becomes

$$Y(s) = H(s)X(s) = \frac{1}{s^2}\frac{1}{s^2+s+1}$$

If we write $Y(s)$ as

$$Y(s) = \frac{1}{s}\frac{1}{s^2+s+1} = \frac{1}{s^3+s^2+s}$$

we can use the step response function to calculate the ramp response by writing this script

```
num = [1];
den = [1 1 1 0];
[y, t] = step(num, den);
```

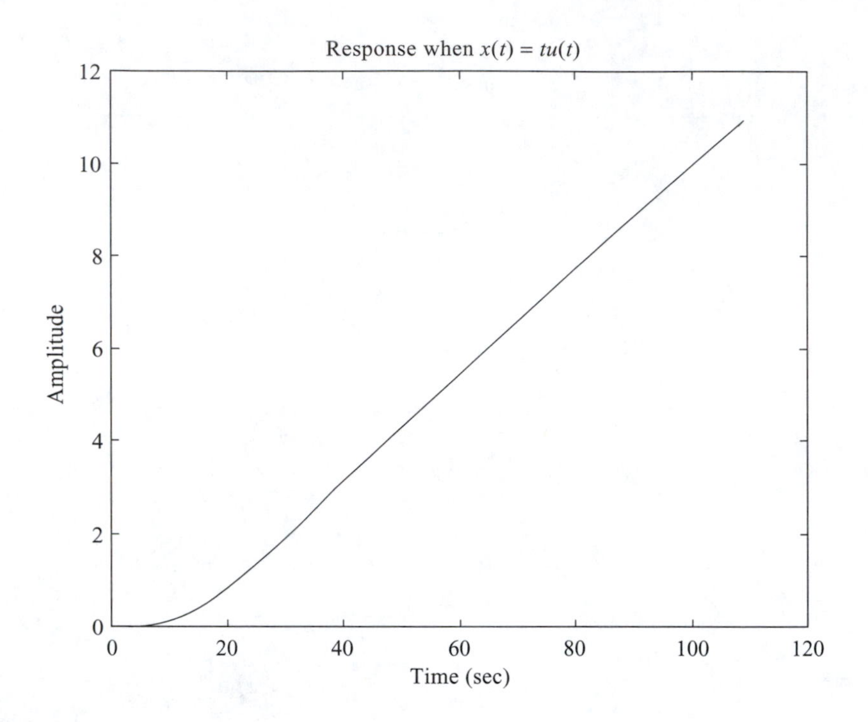

FIGURE 5.17
Plot for EOCE 5.19.

```
plot(y); xlabel('time in seconds'), ylabel('amplitude')
title('the response when x(t) = tu(t)');
```

with the plot shown in Figure 5.17.

It can be seen from the graphs that for the step response the initial value is 0 and the final value will approach 1.

For the impulsive input the initial value is 0 and the final value approaches 0 as well. For the ramp input the initial value is 0 and the final value approaches $\propto$.

We could use the final and the initial value theorems to confirm the above statements.

For the step input the output is

$$Y(s) = \frac{1}{s}\frac{1}{s^2 + s + 1}$$

And using the initial value theorem we can find the initial value of the output as

$$\text{limit}_{s\to\infty}\left[s\frac{1}{s}\frac{1}{s^2 + s + 1}\right] = 0 = y(0)$$

We can use the final value theorem to calculate the final value of the output as

$$\text{limit}_{s \to 0}\left[s\frac{1}{s}\frac{1}{s^2 + s + 1}\right] = 1 = y(\infty)$$

For the impulsive input, the output is

$$Y(s) = \frac{1}{s^2 + s + 1}$$

The initial value of the output is

$$\text{limit}_{s \to \infty}\left[s\frac{1}{s^2 + s + 1}\right] = 0 = y(0)$$

and the final value is

$$\text{limit}_{s \to 0}\left[s\frac{1}{s^2 + s + 1}\right] = 0 = y(\infty)$$

For the ramp input, the output is

$$Y(s) = \frac{1}{s^2}\frac{1}{s^2 + s + 1}$$

The initial value of $y(t)$ is

$$\text{limit}_{s \to \infty}\left[s\frac{1}{s^2}\frac{1}{s^2 + s + 1}\right] = 0 = y(0)$$

and its final value is

$$\text{limit}_{s \to 0}\left[s\frac{1}{s^2}\frac{1}{s^2 + s + 1}\right] = \infty = y(\infty)$$

To discuss stability of the system, we look at the roots of the denominator of $H(s)$ using MATLAB and type the following command at the MATLAB prompt.

```
poles = roots([1 1 1])
```

The poles are

```
poles =
-0.5000 + 0.8660i
-0.5000 - 0.8660i
```

We observe that the roots are complex but the real parts of the roots are negative and this indicates stability of the system represented by $H(s)$.

EOCE 5.20

Consider the following systems:

$$H(s) = \frac{s^3 + s}{s^3 + s^2 + 1}$$

and

$$H(s) = \frac{s^2 + s + 1}{s^3 + 4s^2 + 3s}$$

1. Find the impulse response of each system.
2. Discuss stability for each system.
3. Draw the block diagram for each system.
4. Find the final value and the initial value for $y(t)$ when the input is an impulse.

Solution

For the first system we write the following script

```
num = [1 0 1 0];
den = [1 1 0 1];
poles_sys1 = roots([1 1 0 1])
impulse(num, den)
title('of system 1')
```

with the output

```
poles_sys1 =
-1.4656
0.2328 + 0.7926i
0.2328 - 0.7926i
```

and the output showing the impulse response is shown in Figure 5.18.

We can see from MATLAB that the real parts of the poles are positive and this indicates instability. It is also seen that the final value of $y(t)$ is $\propto$. The initial value of $y(t)$ can be calculated from the expression for $Y(s)$

$$Y(s) = \frac{s^3 + s}{s^3 + s^2 + 1}$$

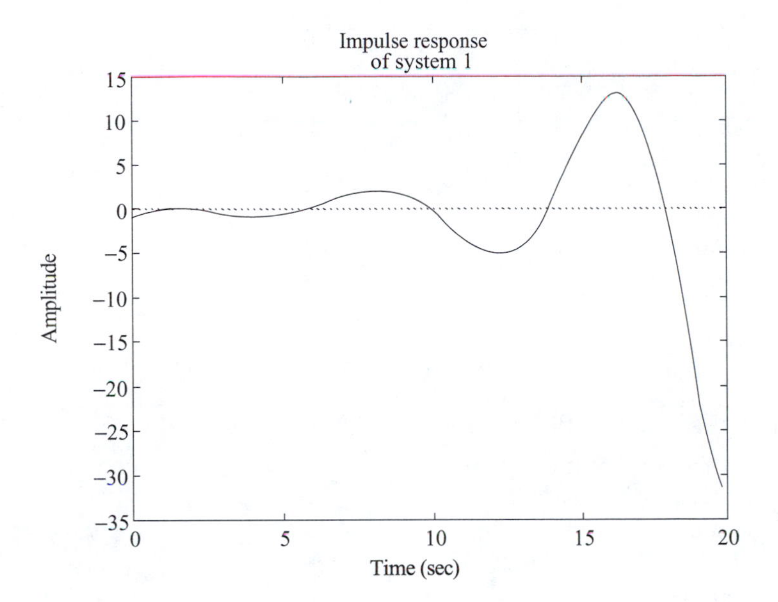

FIGURE 5.18
Plot for EOCE 5.20.

The initial value is then

$$\text{limit}_{s \to \infty}\left[s\frac{s^3 + s}{s^3 + s^2 + 1}\right] = \infty = y(\infty)$$

and the final value is

$$\text{limit}_{s \to 0}\left[s\frac{s^3 + s}{s^3 + s^2 + 1}\right] = 0 = y(\infty)$$

which contradict the results from Figure 5.18. The reason is that one of the poles of $sY(s)$ is positive.

The exact values can be calculated by solving for $y(t)$ in time domain and then taking the limits to find the initial and the final values for $y(t)$.

The block diagram representing the first system is sketched in Figure 5.19.

The block diagram in Figure 5.19 was drawn with the help of the example presented in Chapter 5. For the second system we type the script

```
num = [1 1 1];
den = [1 4 3 0];
poles_sys2 = roots([1 4 3 0])
```

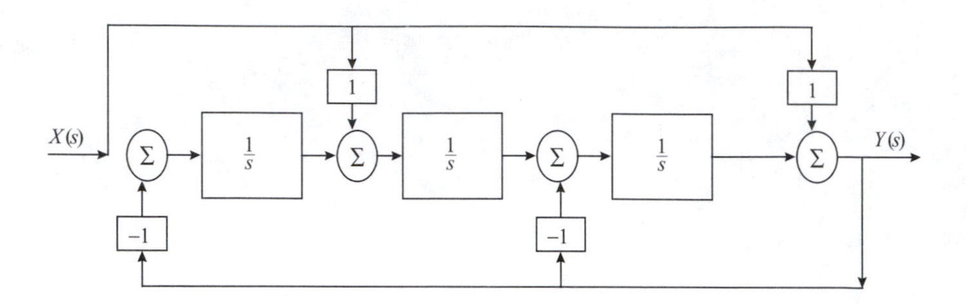

FIGURE 5.19
Block diagram for EOCE 5.20.

```
impulse(num, den)
title('of system 2')
```

with the output

```
poles_sys2 =
 0
-3
-1
```

and with the impulse response as shown in Figure 5.20.

We can see from MATLAB that the poles are negative and this indicates stability. The initial and final values of $y(t)$ can be calculated by first looking at the expression

$$Y(s) = \frac{s^2 + s + 1}{s^3 + 4s^2 + 3s}$$

The initial value for $y(t)$ is

$$\text{limit}_{s \to \infty}\left[s\frac{s^2 + s + 1}{s^3 + 4s^2 + 3s}\right] = 1 = y(0)$$

and the final value is

$$\text{limit}_{s \to 0}\left[s\frac{s^2 + s + 1}{s^3 + 4s^2 + 3s}\right] = 1/3 = y(\infty)$$

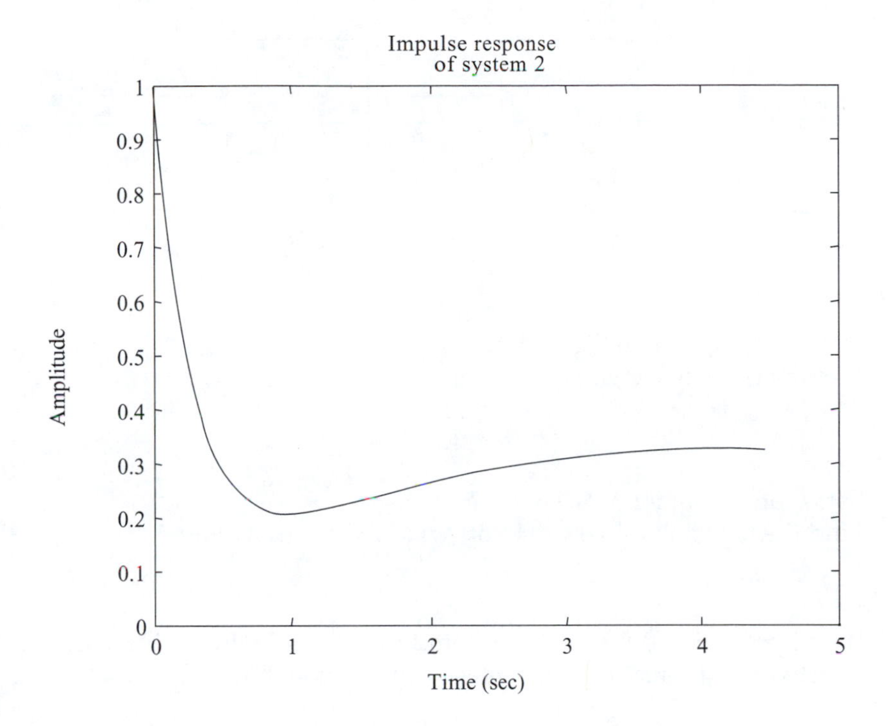

FIGURE 5.20
Plot for EOCE 5.20.

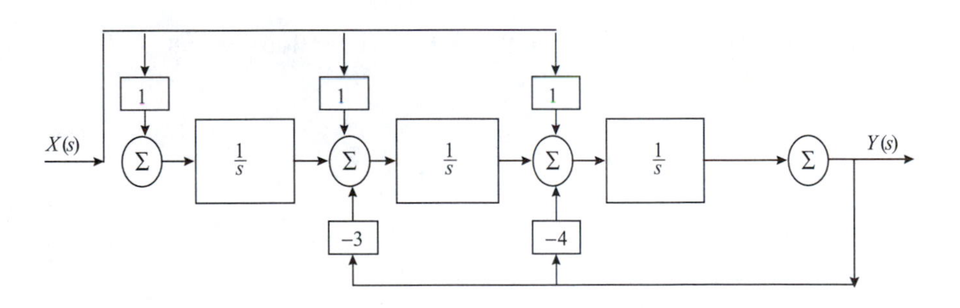

FIGURE 5.21
Block diagram for EOCE 5.20.

You can see that the results from the graphs agree with the results using the initial and the final value theorem since all the eigenvalues of $sY(s)$ are negative.

The block diagram representing the second system is sketched in Figure 5.21.

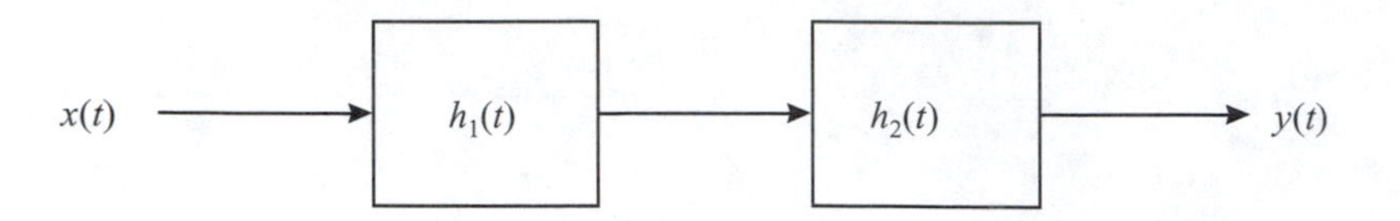

FIGURE 5.22
System for EOCE 5.21.

EOCE 5.21

Consider the system in Figure 5.22.

If the impulse response is

$$h_1(t) = e^{-t}u(t) = h_2(t)$$

what is the output $y(t)$ if $x(t) = u(t)$?

Find the final and initial values of $y(t)$ and discuss stability.

Solution

$x(t)$ will propagate through the system in Figure 5.22. In the block diagram and between $h_1(t)$ and $h_2(t)$ we can use convolution to write

$$z(t) = x(t)*h_1(t)$$

At the output we write

$$y(t) = z(t)*h_2(t)$$

Taking the Laplace transform of the last two equations we get

$$Z(s) = X(s)H_1(s)$$

and

$$Y(s) = Z(s)H_2(s)$$

Substituting $Z(s)$ in $Y(s)$ we get

$$Y(s) = X(s)H_1(s)H_2(s)$$

and

$$H(s) = H_1(s)H_2(s)$$

Therefore, we have

$$Y(s) = \frac{1}{s}\frac{1}{s+1}\frac{1}{s+1} = \frac{1}{s^3 + 2s^2 + s}$$

We can use MATLAB to solve for $y(t)$ with

$$H(s) = \frac{1}{s^2 + 2s + 1}$$

and by typing the script

```
num = [1];
den = [1 2 1];
poles_of_H = roots([1 2 1])
step(num, den);
```

with the output as

```
poles_of_H =
-1
-1
```

and with the plot shown in Figure 5.23.

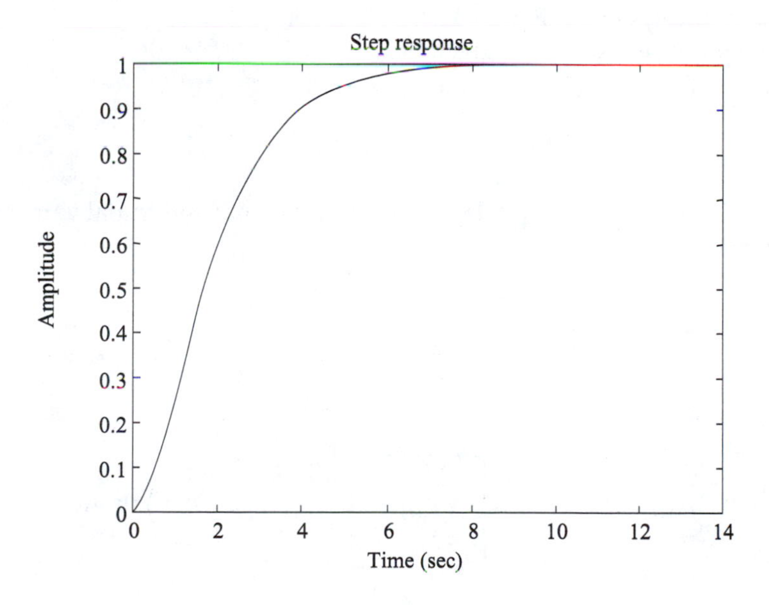

FIGURE 5.23
Plot for EOCE 5.21.

From the plot you can see that the initial value of $y(t)$ is 0 and the final value of $y(t)$ is 1. We also can see that the system is stable (it settles at a final magnitude of 1).

These observations can be confirmed mathematically by looking at the poles of $H(s)$ that are all negative and an indication of stability. Also the initial and final values for $y(t)$ can be calculated by using the expression for the output

$$Y(s) = \frac{1}{s}\frac{1}{s^3 + 2s + 1}$$

The initial value is

$$\text{limit}_{s\to\infty}\left[s\frac{1}{s}\frac{1}{s^3 + 2s + 1}\right] = 0 = y(0)$$

and the final value is

$$\text{limit}_{s\to 0}\left[s\frac{1}{s}\frac{1}{s^3 + 2s + 1}\right] = 1 = y(\infty)$$

EOCE 5.22

Consider the system in Figure 5.24.

If the impulse response is given by

$$h_1(t) = e^{-t}u(t) = h_2(t) = h_3(t).$$

What is the output $y(t)$ if $x(t) = u(t)$? Find the final and initial values of $y(t)$ and discuss stability.

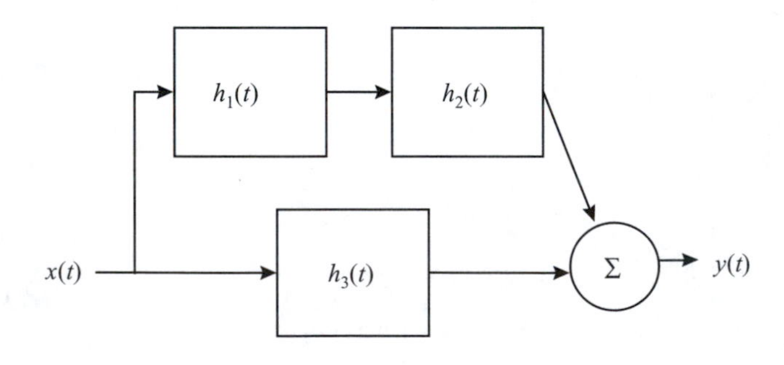

FIGURE 5.24
System for EOCE 5.22.

Solution

$x(t)$ will propagate through the system in Figure 5.24. In the block diagram, and between $h_1(t)$ and $h_2(t)$ we can use convolution to write

$$z_1(t) = x(t)*h_1(t)$$

Taking the Laplace transform we have

$$Z_1(s) = X(s)H_1(s)$$

Before the summer and at the lower part of the block in Figure 5.24 we also have

$$z_2(t) = x(t)*h_3(t)$$

and its transform

$$Z_2(s) = X(s)H_3(s)$$

Before the summer and at the upper part of the block we have

$$z_3(t) = z_1(t)*h_2(t)$$

with its transform

$$Z_3(s) = Z_1(s)H_2(s)$$

At the output we write

$$y(t) = z_2(t) + z_3(t)$$

or

$$Y(s) = Z_2(s) + Z_3(s) = X(s)H_3(s) + X(s)H_1(s)H_2(s)$$

and finally, the transfer function of the system is

$$H(s) = H_3(s) + H_1(s)H_2(s)$$

where

$$H(s) = \frac{1}{s+1} + \frac{1}{s+1}\frac{1}{s+1} = \frac{s+2}{s^2+2s+1}$$

and the output

$$Y(s) = X(s)H(s)$$

or

$$Y(s) = \frac{1}{s}\frac{s+2}{s^2+2s+1}$$

Now we are ready to use MATLAB and type the script

```
num = [1 2];
den = [1 2 1];
poles_of_H = roots([1 2 1])
step(num, den)
```

with the output as

```
poles_of_H =
-1
-1
```

The step response is shown in Figure 5.25.

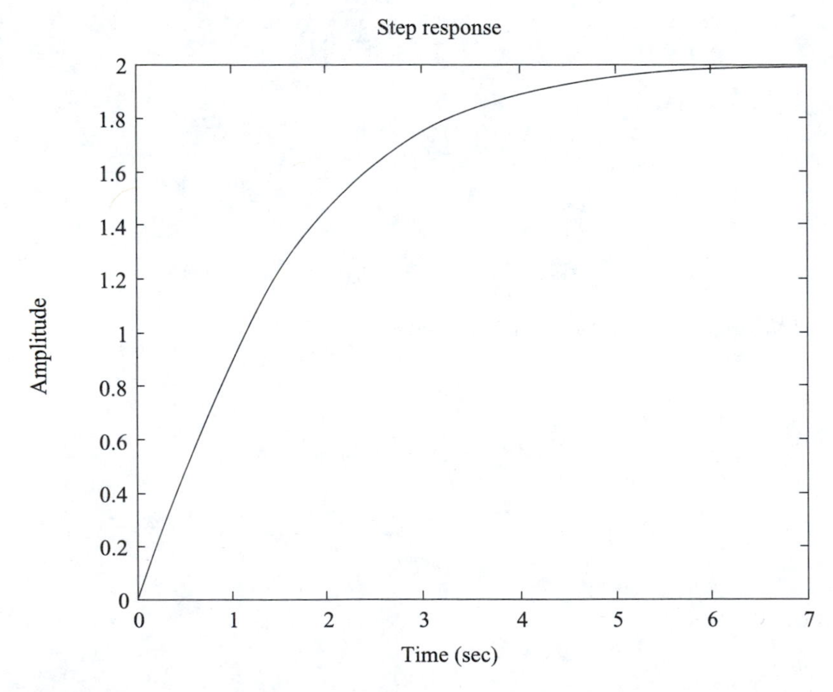

FIGURE 5.25
Plot for EOCE 5.22.

From the plot you can see that the initial value of $y(t)$ is 0 and the final value of $y(t)$ is 2. We also can see that the system is stable (it settles at a final magnitude of 2).

These observations can be confirmed mathematically by looking at the poles of $H(s)$ which are all negative and that is an indication of stability. Also, the initial and final values for $y(t)$ can be calculated with the output written in the Laplace domain as

$$Y(s) = \frac{1}{s}\frac{s+2}{s^3+2s+1}$$

The initial value of $y(t)$ is

$$\text{limit}_{s\to\infty}\left[s\frac{1}{s}\frac{s+2}{s^3+2s+1}\right] = 0 = y(0)$$

and the final value is

$$\text{limit}_{s\to 0}\left[s\frac{1}{s}\frac{s+2}{s^3+2s+1}\right] = 2 = y(\infty)$$

5.14 End-of-Chapter Problems

EOCP 5.1

Find the Laplace transform for each of the following signals.

1. $2u(t-10)+10\delta(t-1)$
2. $\sin(2(t-1))u(t-1)$
3. $\cos(3(t-5))u(t-5)$
4. $e^{-a(t-b)}u(t-b)+u(t-2)$
5. $(t-1)e^{-a(t-1)}u(t-1)$
6. $10e^{-2(t-2)^2}u(t-2)$
7. $2\delta(t-2)+3\delta(t-4)$
8. $e^{-3t}u(t)+\frac{t^5}{5!}e^{-10t}u(t)+u(t-5)$

EOCP 5.2

Find the Laplace transform of the derivative for each of the following signals.

1. $tu(t)$
2. $t\sin(t)u(t)$
3. $(e^{-3t} + \sin(t))u(t)$
4. $(\cos wt + \sin wt)u(t) + 10u(t)$

EOCP 5.3

Find the Laplace transform for each of the following signals, where * indicates convolution.

1. $e^{-t}u(t)^*e^{-t}u(t)$
2. $[\sin(t)u(t)^*u(t)]^*u(t)$
3. $[e^{-t}u(t) + e^{-2t}u(t)]^*\sin(t)u(t)$
4. $[e^{-3t}u(t)^*u(t)] + u(t)^*u(t)$
5. $\delta(t)^*\sin(10t)u(t) - \delta(t-1)^*u(t)$

EOCP 5.4

Consider the following systems represented by the differential equations. $x(t)$ is the input, and $y(t)$ is the output. Find the total solution.

1. $\frac{d}{dt}y + 2y = x(t) = \sin(t)u(t), \quad y(0) = 1$
2. $\frac{d^2}{dt}y + 3\frac{d}{dt}y + 2y = x(t) = u(t), \quad y(0) = 2, \quad \frac{d}{dt}y(0) = 0$
3. $\frac{d^2}{dt}y + 3\frac{d}{dt}y = x(t) = e^{-t}u(t), \quad y(0) = -1, \quad \frac{d}{dt}y(0) = 1$
4. $\frac{d^2}{dt}y + 5y = x(t) = e^{-t}u(t), \quad y(0) = 0, \quad \frac{d}{dt}y(0) = -3$
5. $\frac{d^2}{dt}y + \frac{d}{dt}y + y = x(t) = \delta(t), \quad y(0) = 2, \quad \frac{d}{dt}y(0) = 1$

EOCP 5.5

Given the following Laplace transforms for some signals, what are the time-domain signals?

1. $\dfrac{1}{s+2} + \dfrac{1}{s+3}$

2. $\dfrac{1}{s^2 + 7s + 10}$

3. $\dfrac{1}{s^2 + 5s + 6}$

4. $\dfrac{1}{s^2 + 4s + 4}$

5. $\dfrac{e^{-s}s}{s^2 + 4s + 4}$

6. $\dfrac{e^{-s}s + s}{s^2 + 4s + 4}$

7. $\dfrac{s}{s^2 + 7s + 10}$

8. $\dfrac{s^2}{s^2 + 7s + 10}$

9. $\dfrac{s^3 + 1}{s^2 + 7s + 10}$

10. $\dfrac{s^3 + s^2 + 1}{s^2 + 7s + 10}$

EOCP 5.6

Use the Laplace transform to find $y(t)$, a particular solution, in each of the systems shown in Figures 5.26, 5.27, 5.28, 5.29, and 5.30.

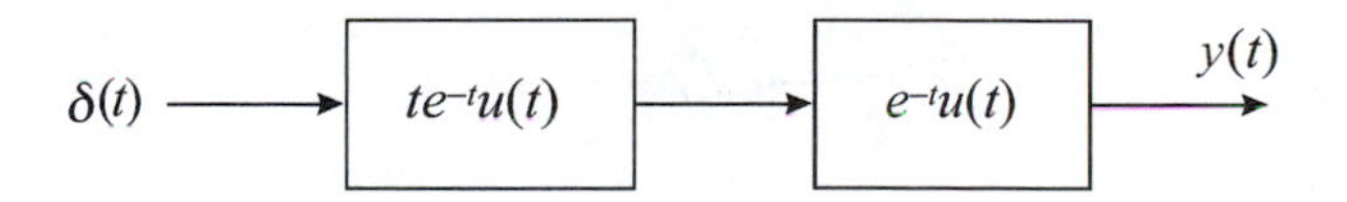

FIGURE 5.26
System for EOCP 5.6.

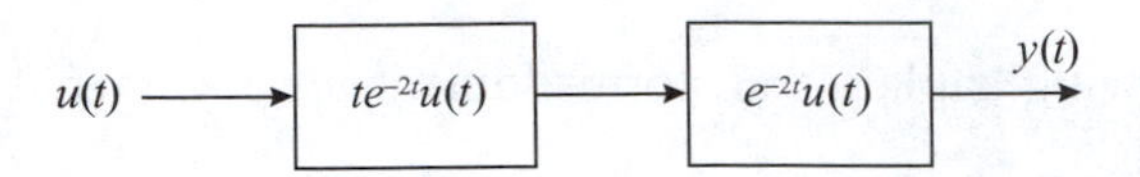

FIGURE 5.27
System for EOCP 5.6.

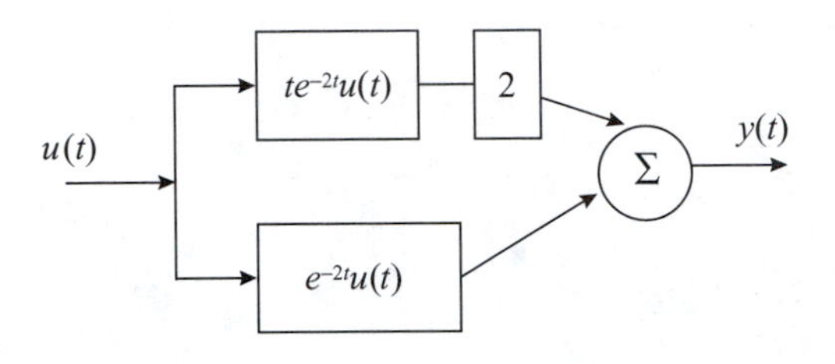

FIGURE 5.28
System for EOCP 5.6.

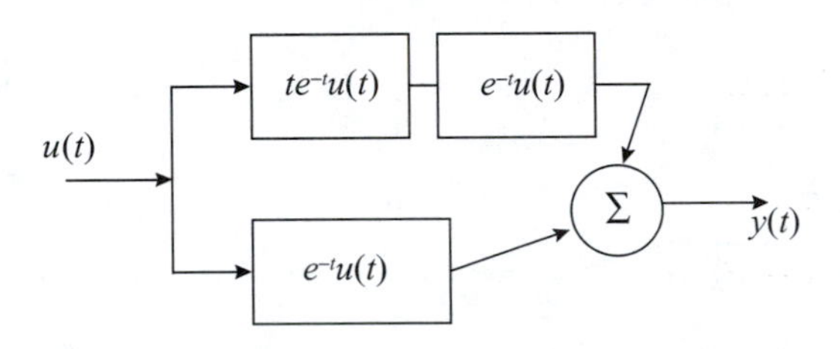

FIGURE 5.29
System for EOCP 5.6.

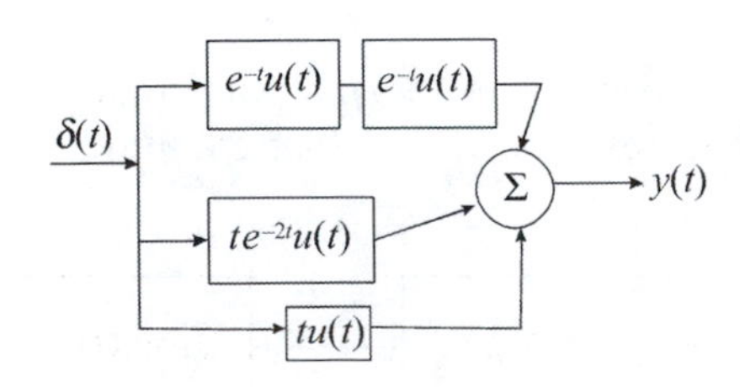

FIGURE 5.30
System for EOCP 5.6.

EOCP 5.7

What is the Laplace transform for each of the signals shown in Figures 5.31, 5.32, 5.33, and 5.34?

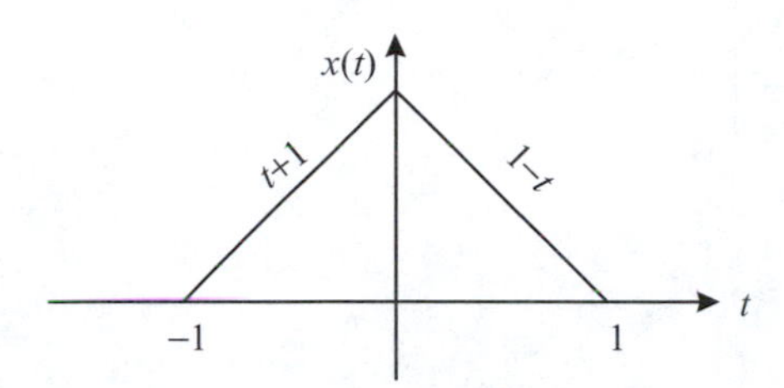

FIGURE 5.31
Signal for EOCP 5.7.

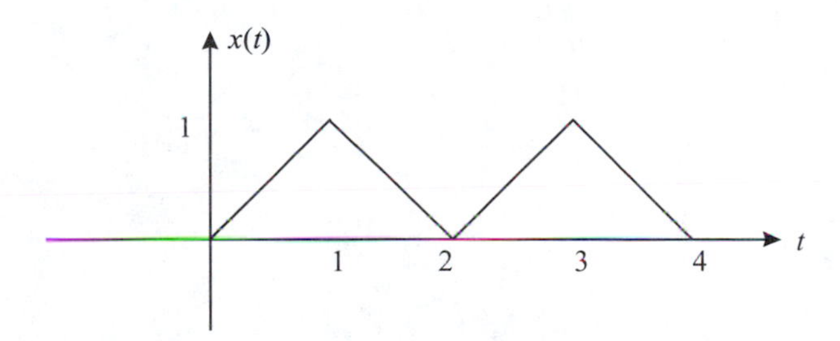

FIGURE 5.32
Signal for EOCP 5.7.

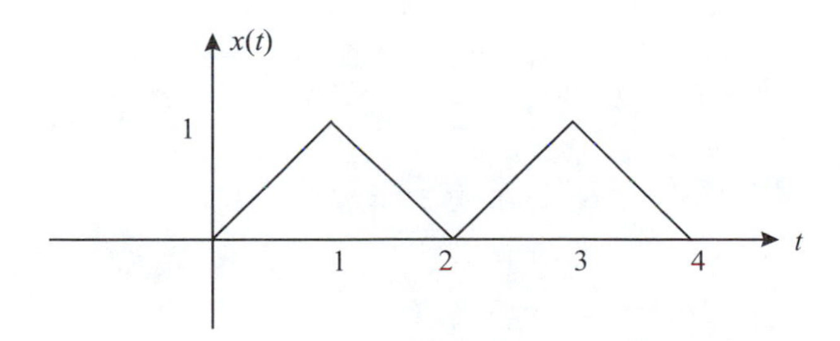

FIGURE 5.33
Signal for EOCP 5.7.

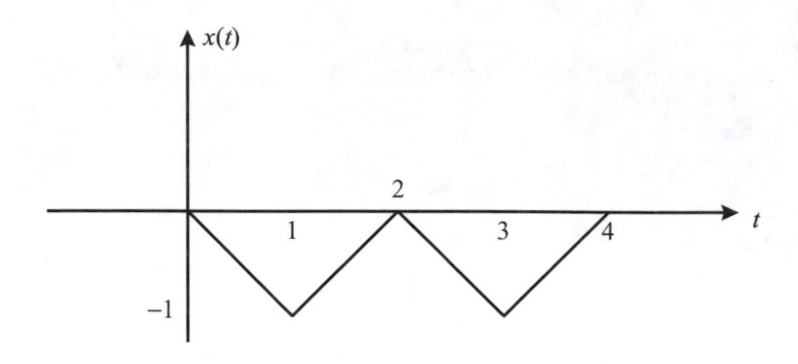

FIGURE 5.34
Signal for EOCP 5.7.

EOCP 5.8

Use the series and parallel models to draw the block diagram for each transfer function.

1. $\dfrac{10}{s^2 - 4}$

2. $\dfrac{10}{2s^2 - 32}$

3. $\dfrac{1}{s^3 + s^2 - 9s - 9}$

4. $\dfrac{s + 1}{s^2 + s - 20}$

5. $\dfrac{s^2 + s + 1}{s^3 - 3s^2 - s + 3}$

EOCP 5.9

Draw the block diagram using integrators, summers, and gain blocks for each of the following systems.

1. $\dfrac{s}{s + 1}$

2. $\dfrac{2s + 1}{s - 1}$

3. $\dfrac{2s^2 + s + 1}{s^2 - 1}$

4. $\dfrac{1}{s^2 + 2s + 10}$

5. $\dfrac{s}{s(s+1)}$

6. $\dfrac{100}{2s^2 - 4}$

7. $\dfrac{10}{3s^2 - 16}$

8. $\dfrac{10}{3s^3 + 2s^2 - 9s - 9}$

9. $\dfrac{s^2 + s + 1}{s^2 + s - 20}$

10. $\dfrac{s^2 + s + 1}{5s^2}$

EOCP 5.10

Find the initial and final values for each system using the final and the initial value theorem when applicable.

1. $\frac{d^2}{dt}y(t) + 3\frac{d}{dt}y(t) + 4y(t) = e^{-t}u(t)$

2. $\frac{d^2}{dt}y(t) + 3\frac{d}{dt}y(t) - 4y(t) = \sin(t)u(t)$

3. $\frac{d^3}{dt}y(t) + 3\frac{d}{dt}y(t) + 4y(t) = e^{-2t}u(t)$

4. $\frac{d^3}{dt}y(t) - \frac{d}{dt}y(t) = \cos(t)u(t)$

5. $\frac{d^3}{dt}y(t) + \frac{d^2}{dt}y(t) + \frac{d}{dt}y(t) + y(t) = u(t)$

EOCP 5.11

Consider the circuit in Figure 5.35.

1. Find the output voltage $y(t)$, for $t > 0$, using the Laplace transform-method. Assume the 5 volts are applied at $t = 0$ seconds.
2. Find the current in the resistor R using the Laplace method.

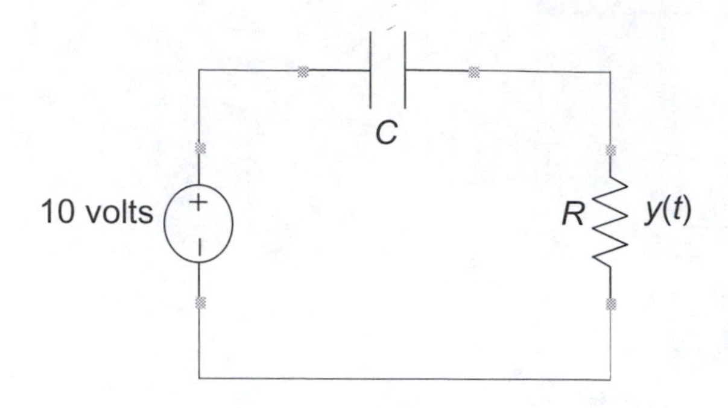

FIGURE 5.35
Circuit for EOCP 5.11.

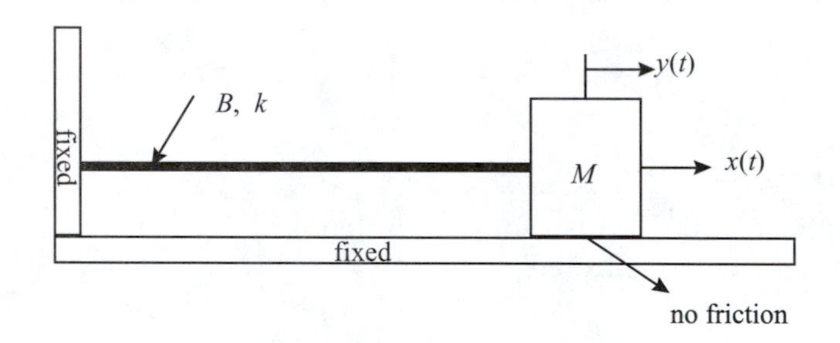

FIGURE 5.36
System for EOCP 5.12.

EOCP 5.12

Consider the mechanical system described by the differential equation and shown in Figure 5.36.

$$M\frac{d^2}{dt}y(t) + B\frac{d}{dt}y(t) + ky(t) = x(t)$$

The rod to the left of the mass, M, can be modeled by a translational spring with spring constant k, in parallel with a translational damper of constant B.

1. If $x(t)$ is an impulse of strength 10, use the Laplace transform method to find the displacement $y(t)$. Assume no friction and no stored energy.
2. Repeat if $x(t)$ is a sinusoidal shaking of magnitude 10 and period 1 second.

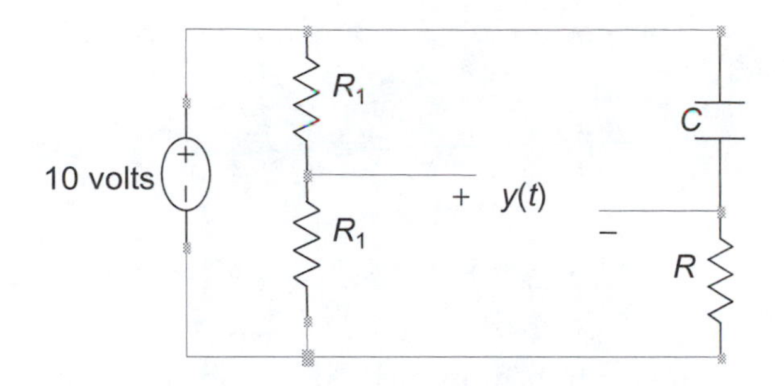

FIGURE 5.37
Circuit for EOCP 5.13.

EOCP 5.13

Consider the circuit in Figure 5.37.

1. If the 10 volts are applied at $t = 0$ seconds, find $y(t)$ using the Laplace transform method.
2. Repeat part 1 but calculate the current in the capacitor.
3. Draw a block diagram of the system using the basic blocks.
4. What are the final and the initial values of the output $y(t)$ and the current in the capacitor?

EOCP 5.14

Consider the mechanical system described in Figure 5.38. The masses were initially displaced. The rod to the left of the mass M_1 can be modeled by a translational spring with spring constant k_1. The rod to the right of the mass M_1 can be modeled by a translational spring with spring constant k_2. The rod to the right of the mass M_2 can be modeled by a translational spring with spring constant k_3. Both masses M_1 and M_2 move smoothly on the fixed surfaces.

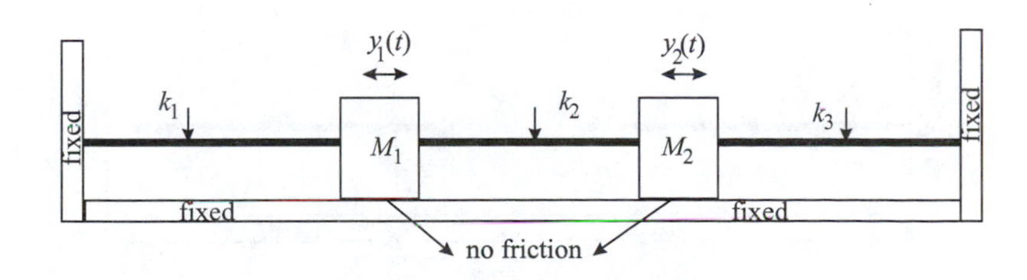

FIGURE 5.38
System for EOCP 5.14.

The two differential equations describing the system are

$$M_1\frac{d^2}{dt}y_1 + k_1y_1 + k_2y_1 - k_2y_2 = 0$$

$$M_2\frac{d^2}{dt}y_2 - k_2y_1 + k_2y_2 + k_3y_2 = 0$$

1. Use the Laplace transform method to find the outputs, $y_1(t)$ and $y_2(t)$.
2. Draw the block diagram for the entire system.

EOCP 5.15

Consider the dynamic system shown in Figure 5.39 and represented by the differential equation

$$M\frac{d^2}{dt}y(t) + (B_1 - B_2)\frac{d}{dt}y(t) + (k_1 - k_2)y(t) = x(t)$$

The rod to the left of the mass M can be modeled by a translational spring with spring constant k_1, in parallel with a translational damper of constant B_1. The rod to the right of the mass M can be modeled by a translational spring with spring constant k_2, in parallel with a translational damper of constant B_2. The mass M moves freely on the fixed surface.

1. Draw the block diagram for the system.
2. Find the displacement $y(t)$ if
   a) $x(t)$ is an impulse of unity magnitude
   b) $x(t)$ is a unit step force
3. Find the initial and the final values for $y(t)$.
4. Find the initial and the final values for the velocity of the mass M.

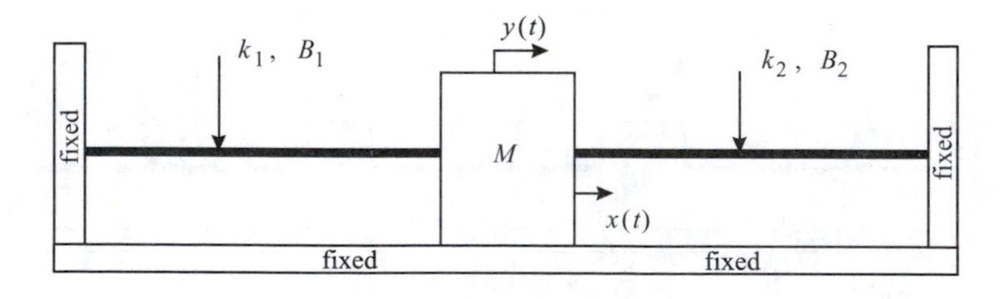

FIGURE 5.39
System for EOCP 5.15.

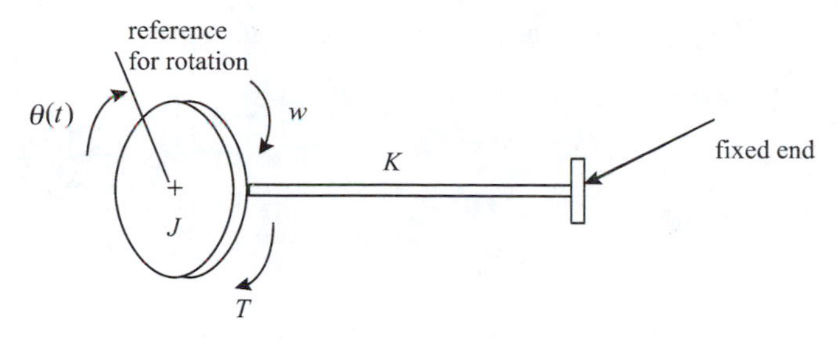

FIGURE 5.40
System for EOCP 5.16.

EOCP 5.16

Consider the dynamic system as represented by the differential equation below and shown in Figure 5.40.

$$J\frac{d^2}{dt}\theta(t) + k\theta(t) = T(t)$$

This is a rod attached to a fixed end. Assume that the rod can be represented as a rotational spring with spring constant k, and the mass of the rod is concentrated and modeled by the moment of inertia J.

1. Draw the block diagram representing the system.
2. If the torque is represented as a unit step signal, what is the angular rotation in radians?
3. Find the final and the initial values for $\theta(t)$.

EOCP 5.17

A dynamic system is given in Figure 5.41.

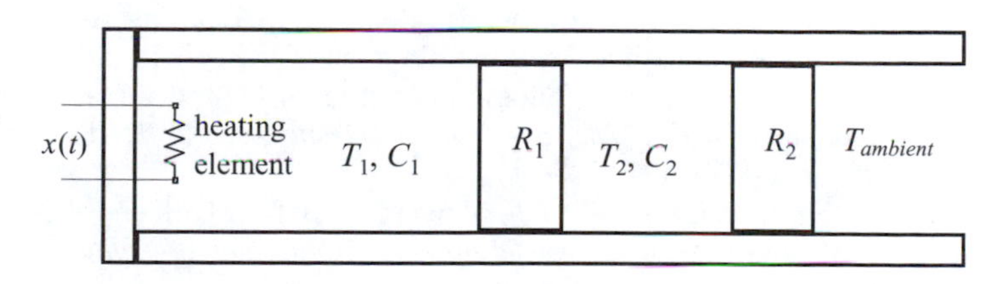

FIGURE 5.41
System for EOCP 5.17.

The differential equations representing the system are

$$\frac{d}{dt}T_1 = \frac{1}{C_1}\left[x(t) - \frac{1}{R_1}(T_1 - T_2)\right]$$

$$\frac{d}{dt}T_2 = \frac{1}{C_2}\left[\frac{1}{R_1}(T_1 - T_2) - \frac{1}{R_2}(T_2 - T)\right]$$

where T_1 and T_2 are the temperatures in the two containers as indicated in Figure 5.41. C_1 and C_2 represent the capacitance in each container. R_1 and R_2 are the resistances of the two separators.

1. Draw the block diagram for the system.
2. If $x(t)$ is a unit step flow of unity magnitude, what are the temperatures $T_1(t)$ and $T_2(t)$?

References

Bequette, B.W. *Process Dynamics*, Englewood Cliffs, NJ: Prentice-Hall, 1998.
Brogan, W.L. *Modern Control Theory*, 3rd ed., Englewood Cliffs, NJ: Prentice-Hall, 1991.
Close, M. and Frederick, K. *Modeling and Analysis of Dynamic Systems*, 2nd ed., New York: Wiley, 1995.
Cogdell, J.R. *Foundations of Electrical Engineering*, 2nd ed., Englewood Cliffs, NJ: Prentice-Hall, 1996.
Denbigh, P. *System Analysis and Signal Processing*, Reading, MA: Addison-Wesley, 1998.
Driels, M. *Linear Control System Engineering*, New York: McGraw-Hill, 1996.
Golubitsky, M. and Dellnitz, M. *Linear Algebra and Differential Equations Using MATLAB*, Stamford, CT: Brooks/Cole, 1999.
Harman, T.L., Dabney, J. and Richert, N. *Advanced Engineering Mathematics with MATLAB*, Stanford, CT: Brooks/Cole, 2000.
Kuo, B.C. *Automatic Control System*, 7th ed., Englewood Cliffs, NJ: Prentice-Hall, 1995.
Lewis, P.H. and Yang, C. *Basic Control Systems Engineering*, Englewood Cliffs, NJ: Prentice-Hall, 1997.
The MathWorks. *The Student Edition of MATLAB*, Englewood Cliffs, NJ: Prentice-Hall, 1997.
Nilson, W.J. and Riedel, S.A. *Electrical Circuits*, 6th ed., Englewood Cliffs, NJ: Prentice-Hall, 2000.
Nise, N.S. *Control Systems Engineering*, 2nd ed., Reading, MA: Addison-Wesley, 1995.
Ogata, K. *Modern Control Engineering*, 3rd ed., Englewood Cliffs, NJ: Prentice-Hall, 1997.
Ogata, K. *System Dynamics*, 3rd ed., Englewood Cliffs, NJ: Prentice-Hall, 1998.
Phillips, C.L. and Parr, J.M. *Signals, Systems, and Transforms*, 2nd ed., Englewood Cliffs, NJ: Prentice-Hall, 1999.
Pratap, R. *Getting Started with MATLAB 5*, New York: Oxford University Press, 1999.
Woods, R.L. and Lawrence, K.L. *Modeling and Simulation of Dynamic Systems*, Englewood Cliffs, NJ: Prentice-Hall, 1997.
Wylie, R.C. and Barrett, C.L. *Advanced Engineering Mathematics*, 6th ed., New York: McGraw-Hill, 1995.
Ziemer, R.E., Tranter, W.H., and Fannin, D.R. *Signals Systems Continuous and Discrete*, 4th ed., Englewood Cliffs, NJ: Prentice-Hall, 1998.

6

State-Space and Linear Systems

CONTENTS

6.1 Introduction

As the interest in many scientific fields increased, modeling systems using linear time-invariant differential equations and tools like transfer functions were not adequate. The state-space approach is superior to other classical methods of modeling. This modern approach can handle systems with non-zero initial conditions (modeling using transfer functions requires that initial conditions be set to zero) as well as time-variant systems. It can also handle linear and non-linear systems. We also have been considering systems with single-input single-output. State-space approach can handle multiple-input multiple-output systems.

The state-space approach can also be used to represent or model linear time-invariant single-input single-output systems. These are the systems we considered in the previous chapters. Systems can have many variables. An example is an electrical circuit where the variables are the inductor current, the capacitor voltage, and the resistor voltage, among others. Using state-space approach, we will use differential equations to solve for a selected set of these variables. The other variables in the circuit system can be found using the solution for the selected variables. In using state-space approach we will follow the procedure detailed below.

We will select specific variables in the system and call them state variables. No state variable selected can be written as a linear combination of the other state variables. Linear combination means that if

$$z_1(t) = 3z_3(t) + 2z_2(t)$$

where $z_1(t)$, $z_2(t)$, and $z_3(t)$ are state variables, we say that $z_1(t)$ is a linear combination of $z_2(t)$ and $z_3(t)$. If we have a first order differential equation, we will have only one state variable. If the differential equation is second order, we will have only two state variables. Similarly, if we have an nth order differential equation, we will have only n state variables. Once we select or decide on the state variables in the system under consideration, we will write a set of first order simultaneous differential equations where the right side of these equations is a function only of the state variables (no derivatives) and the inputs to the system, and the number of these equations is determined by the number of state variables selected. We will call this set the state equations set. These state equations will be solved for the selected state variables. All other variables in the system under consideration can be solved using the solutions of these selected state variables and the input to the system. We can use any approach we desire to solve for these selected states. The equations we write to find the other variables in the system are called output equations.

6.2 A Review of Matrix Algebra

What follows is a brief review of some of the concepts and definitions we need in this chapter. We will discuss second order systems when we deal with hand solutions. For matrices of higher dimensions you can consult any linear algebra book.

6.2.1 Definition, General Terms, and Notations

A matrix is a collection of elements arranged in a rectangular or square array. The size of the matrix is determined by the number of rows and the number of columns in the matrix. A matrix A of m rows and n column is represented as $A_{m\times n}$. If $m = 1$ then A is a row vector and written as $A_{1\times n}$. If $n = 1$ then A is a column vector and written as $A_{m\times 1}$. If $n = m$ then A is a square matrix and we write it as $A_{n\times n}$ or $A_{m\times m}$. If all elements in the matrix are zeros we say A is a null matrix or a zero matrix.

6.2.2 The Identity Matrix

The identity matrix is the square matrix where elements along the main diagonal are ones and elements off the main diagonal are zeros. A 2×2 identity matrix is

$$I_{2\times 2} = \begin{bmatrix} 1 & 0 \\ 0 & 1 \end{bmatrix} \tag{6.1}$$

6.2.3 Adding Two Matrices

If

$$A = \begin{bmatrix} a & b \\ c & d \end{bmatrix}$$

and

$$B = \begin{bmatrix} e & f \\ g & h \end{bmatrix}$$

then

$$A + B = \begin{bmatrix} a+e & b+f \\ c+g & d+h \end{bmatrix} \tag{6.2}$$

To add two matrices they must be of the same size. If the matrices are of higher order, the procedure is the same; we add the corresponding entries.

6.2.4 Subtracting Two Matrices

If

$$A = \begin{bmatrix} a & b \\ c & d \end{bmatrix}$$

and

$$B = \begin{bmatrix} e & f \\ g & h \end{bmatrix}$$

then

$$A - B = \begin{bmatrix} a-e & b-f \\ c-g & d-h \end{bmatrix} \tag{6.3}$$

To subtract two matrices they must be of the same size. If the matrices are of higher order, the procedure is the same; we subtract the corresponding entries.

6.2.5 Multiplying a Matrix by a Constant

If

$$A = \begin{bmatrix} a & b \\ c & d \end{bmatrix}$$

and k is any given constant, then

$$k\begin{bmatrix} a & b \\ c & d \end{bmatrix} = \begin{bmatrix} ka & kb \\ kc & kd \end{bmatrix} \tag{6.4}$$

If the matrix A is of higher order, then k is multiplied by each entry in A.

6.2.6 Determinant of a 2 × 2 Matrix

Consider the $A_{2\times2}$ matrix

$$A_{2\times2} = \begin{bmatrix} a & b \\ c & d \end{bmatrix}$$

The determinant of A is

$$\det(A) = ad - bc \qquad (6.5)$$

6.2.7 Transpose of a Matrix

If

$$A = \begin{bmatrix} a & b \\ c & d \end{bmatrix}$$

then the transpose of A is given by

$$A^T = \begin{bmatrix} a & c \\ b & d \end{bmatrix} \qquad (6.6)$$

This works for higher order matrices as well, where the first column in A becomes the first row in A^T and so on.

6.2.8 Inverse of a Matrix

If

$$A = \begin{bmatrix} a & b \\ c & d \end{bmatrix}$$

then the inverse of A is

$$A^{-1} = \frac{1}{ad - bc}\begin{bmatrix} d & -b \\ -c & a \end{bmatrix}$$

and since $\frac{1}{ad - bc}$ is a constant, we can write the inverse as

$$A^{-1} = \begin{bmatrix} \dfrac{d}{ad - bc} & \dfrac{-b}{ad - bc} \\ \dfrac{-c}{ad - bc} & \dfrac{a}{ad - bc} \end{bmatrix} \qquad (6.7)$$

The inverse of a square matrix exists if the determinant of the matrix is not zero. Also, to find an inverse of a certain matrix, that matrix has to be square. The procedure above for finding the inverse is only for a 2×2 matrix. For higher order matrices, the procedure is different and is found in any linear algebra book.

6.2.9 Matrix Multiplication

We can multiply two matrices A and B if the number of columns in A is equal to the number of rows in B.

If $A_{m \times n}$ is to be multiplied by $B_{r \times p}$ then n must be equal to r and the resulting matrix should have m rows and p columns.

If

$$A = \begin{bmatrix} a & b \\ c & d \end{bmatrix}$$

and

$$B = \begin{bmatrix} e & f \\ g & h \end{bmatrix}$$

then if we multiply A by B and let matrix C hold the resulting product, the size of C is 2×2. We could multiply A by B because the number of columns in A, which is two, is equal to the number of rows in B, which is also two. The multiplication of A by B is C and it is

$$C = AB = \begin{bmatrix} a & b \\ c & d \end{bmatrix} \begin{bmatrix} e & f \\ g & h \end{bmatrix}$$

$$C = AB = \begin{bmatrix} ae + bg & af + bh \\ ce + dg & cf + dh \end{bmatrix} \tag{6.8}$$

We multiply the first row of A, element by element, by all the columns of B. Similarly we take the second row of A and multiply it by all the columns of B.

Note that in general AB is not the same as BA. The rules for multiplication have to be observed.

6.2.10 Diagonal Form of a Matrix

A matrix A is in diagonal form if all elements in the matrix that are off the diagonal are zeros. If A is not diagonal, we can make it diagonal by finding the matrix P that contains the eigenvectors of A. So if A is not a diagonal matrix then $(P^{-1}AP)$ will transform A into a diagonal matrix.

6.2.11 Exponent of a Matrix

If A is a square matrix and t is a time variable, then

$$e^{At} = I + At + \frac{(At)^2}{2!} + \frac{(At)^3}{3!} + \frac{(At)^4}{4!} + \cdots \tag{6.9}$$

where I is the identity matrix with the same size as A.

If

$$A = \begin{bmatrix} a & 0 \\ 0 & b \end{bmatrix}$$

a diagonal matrix, then

$$e^{At} = \begin{bmatrix} e^{at} & 0 \\ 0 & e^{bt} \end{bmatrix}$$

This is true for matrices of bigger sizes provided that A is a diagonal matrix.

If

$$A = \begin{bmatrix} 0 & 0 \\ 0 & 2 \end{bmatrix}$$

a diagonal matrix too, then

$$e^{At} = \begin{bmatrix} e^{0t} = 1 & 0 \\ 0 & e^{2t} \end{bmatrix}$$

6.2.12 A Special Matrix

Let

$$A = \begin{bmatrix} 0 & a \\ 0 & 0 \end{bmatrix}$$

then

$$A^2 = \begin{bmatrix} 0 & a \\ 0 & 0 \end{bmatrix} \begin{bmatrix} 0 & a \\ 0 & 0 \end{bmatrix} = \begin{bmatrix} 0 & 0 \\ 0 & 0 \end{bmatrix}$$

and

$$A^3 = \begin{bmatrix} 0 & 0 \\ 0 & 0 \end{bmatrix}$$

A^4, A^5, and so on will have the same value as A^2.

6.2.13 Observation

Let

$$A = \begin{bmatrix} a & b \\ 0 & a \end{bmatrix}$$

A can be written as

$$A = \begin{bmatrix} a & 0 \\ 0 & a \end{bmatrix} + \begin{bmatrix} 0 & b \\ 0 & 0 \end{bmatrix}$$

and

$$e^{(At)} = e^{\begin{bmatrix} a & b \\ 0 & a \end{bmatrix} t} = e^{\left\{ \begin{bmatrix} a & 0 \\ 0 & a \end{bmatrix} + \begin{bmatrix} 0 & b \\ 0 & 0 \end{bmatrix} \right\} t}$$

$$e^{(At)} = e^{\begin{bmatrix} a & 0 \\ 0 & a \end{bmatrix} t} e^{\begin{bmatrix} 0 & b \\ 0 & 0 \end{bmatrix} t}$$

But as we saw earlier, if A is a matrix with all zeros except for the upper right element, then $e^{(At)} = 1 + (At)$. Knowing this we can finish evaluating the above expression as

$$e^{(At)} = \begin{bmatrix} e^{at} & 0 \\ 0 & e^{at} \end{bmatrix} \left\{ \begin{bmatrix} 1 & 0 \\ 0 & 1 \end{bmatrix} + \begin{bmatrix} 0 & bt \\ 0 & 0 \end{bmatrix} \right\}$$

$$e^{(At)} = \begin{bmatrix} e^{at} & 0 \\ 0 & e^{at} \end{bmatrix} \begin{bmatrix} 1 & bt \\ 0 & 1 \end{bmatrix}$$

and finally

$$e^{(At)} = \begin{bmatrix} e^{at} & bte^{at} \\ 0 & e^{at} \end{bmatrix}$$

The above result is very important when we solve state-space first order equations in real time.

6.2.14 Eigenvalues of a Matrix

The eigenvalues of a matrix A, are the roots of the determinant of $(\lambda I - A)$, where I is the identity matrix and λ is a variable.

6.2.15 Eigenvectors of a Matrix

The eigenvectors of a matrix A are the roots of the homogeneous matrix equation

$$(\lambda I - A)\underline{p} = \underline{0}$$

where $\underline{p}$ is a column vector that represents the eigenvector for a certain eigenvalue.

Example 6.1

Consider the first order differential equation

$$\frac{d}{dt}y(t) + 2y(t) = x(t)$$

where $y(0^-) = 0$, the initial condition for the output $y(t)$.

Let the input to this system be $x(t) = u(t)$, a unit step input. Find the output $y(t)$ for $t \geq 0$.

Solution

This is a first order differential equation and therefore we will have one state variable. Let us call that state variable $z_1(t)$.

Next we let

$$z_1(t) = y(t)$$

and then

$$\frac{d}{dt}z_1(t) = \frac{d}{dt}y(t)$$

Rewriting the given differential equation as a function of the state variables we get

$$\frac{d}{dt}z_1 = 2z_1 + x$$

where again when we write x we mean $x(t)$, the input to the system.

In state-space representation we can write

$$\frac{d}{dt}z_1 = 2z_1 + x$$

as

$$\frac{d}{dt}\underset{\sim}{z} = A\underset{\sim}{z} + B\underset{\sim}{x}$$

where $\underset{\sim}{z}$, $\frac{d}{dt}\underset{\sim}{z}$, and $\underset{\sim}{x}$ are vectors such that

$$\underset{\sim}{z} = [z_1], \frac{d}{dt}\underset{\sim}{z} = \left[\frac{d}{dt}z_1\right] \quad \text{and} \quad \underset{\sim}{x} = [x]$$

A and B are matrices where $A = [2]$ and $B = [1]$.

1. The Laplace method

   We can use the Laplace transform to solve this last matrix differential equation by taking the Laplace transform of

$$\frac{d}{dt}\underset{\sim}{z} = A\underset{\sim}{z} + B\underset{\sim}{x}$$

   and putting it in a simple algebraic equation in the variable s and write

$$[s\underline{Z}(s) - \underset{\sim}{z}(0^-)] = A\underline{Z}(s) + B\underline{X}(s)$$

   with $\underset{\sim}{z}(0^-) = z_1(0^-) = y(0^-) = 0$ and $\underline{X}(s) = 1/s$ since $x(t)$ is a unit step input. Rearranging terms we get

$$[s\underline{Z}(s)] = A\underline{Z}(s) + B\underline{X}(s) + \underset{\sim}{z}(0^-)$$

which is a Laplace representation of a first order single-input single-output system and therefore is one-dimensional. To take $\underline{Z}(s)$ as a common factor in the last equation, $s\underline{Z}(s)$ must be written as $sI\underline{Z}(s)$ where I is the 1×1 identity matrix. Now we write the last equation as

$$[sI - A]\underline{Z}(s) = B\underline{X}(s) + \underline{z}(0^-)$$

By multiplying the above equation by $[sI - A]^{-1}$ to solve for $\underline{Z}(s)$ and substituting for the initial conditions we get

$$\underline{Z}(s) = [sI - A]^{-1}B\underline{X}(s) + [sI - A]^{-1}\underline{z}(0^-)$$
$$\underline{Z}(s) = [s - 2]^{-1}\underline{X}(s) + [s - 2]^{-1}[0]$$

Since $[s - 2]$ is a 1×1 square matrix, its inverse is just its reciprocal. Therefore,

$$\underline{Z}(s) = \{1/[s - 2]\}\underline{X}(s)$$

With $\underline{X}(s) = U(s)$, the unit step in Laplace domain, $U(s) = 1/s$. Then

$$\underline{Z}(s) = \{1/[s - 2]\}\{1/s\}$$

Using partial fraction expansion we have

$$\underline{Z}(s) = [1]/[s(s - 2)] = [-1/2]/[s] + [1/2]/[s - 2]$$

Taking the inverse Laplace transform and solving for $\underline{z}(t)$ we get

$$\underline{z}(t) = [-1/2]u(t) + [1/2]e^{2t}u(t)$$

or

$$\underline{z}(t) = 1/2[e^{2t} - 1]u(t)$$

$\underline{z}(t)$, which is a 1×1 column vector, is the same as $y(t)$ as defined previously.

2. Real-time method

If we have a differential equation of the form

$$\frac{d}{dt}y(t) + ay(t) = x(t)$$

with initial condition $y(0) = y_0$, then the solution to this equation will have two parts: a solution due to the initial conditions with $x(t) = 0$, the homogeneous part, and a solution due to the input $x(t)$, the particular solution. The solution due to the initial conditions will have the form

$$y_{ic}(t) = y_h(t) = e^{-at}y_0$$

and the solution due to the input $x(t)$ involves the convolution integral

$$y_p(t) = \int_0^t e^{-a(t-\tau)}x(\tau)d\tau$$

The state-space equation in this example is

$$\frac{d}{dt}\underline{z} = A\underline{z} + B\underline{x}$$

where again, $\underline{z}$, $\frac{d}{dt}\underline{z}$, and $\underline{x}$ are vectors such that $\underline{z} = [z_1]$, $\frac{d}{dt}\underline{z} = [\frac{d}{dt}z_1]$ and $\underline{x} = [x]$. A and B are matrices where $A = [2]$ and $B = [1]$.

The total solution for

$$\frac{d}{dt}\underline{z} = A\underline{z} + B\underline{x}$$

with the initial condition vector $\underline{z}(0^-)$ is

$$\underline{z}(t) = \exp[At]\underline{z}(0) + \int_{t_0}^t \exp[A(t-\tau)]Bu(\tau)d\tau$$

where t_0 in our example is 0 and $u(t)$ is unity in the range from zero to infinity. Notice that A is a 1×1 matrix and therefore is already in a diagonal form. Therefore, the solution with $\underline{z}(0^-) = \underline{0}$ becomes

$$\underline{z}(t) = \int_0^t \exp[2(t-\tau)]d\tau = \exp(2t)\int_0^t \exp[-2\tau]d\tau$$

$$\underline{z}(t) = \exp(2t)\frac{\exp[-2\tau]}{-2}\bigg|_0^t = \frac{1}{-2}\exp(2t)[\exp(-2t)-1]$$

and finally

$$\underline{z}(t) = \frac{1}{2}[\exp(2t) - 1]u(t)$$

which is the same solution we get using the Laplace transform technique.

We really do not need to use state-space approach to solve first order single-input single-output systems. This example is just to introduce you to the steps needed to deal with problems using the state-space approach.

You will start to feel the power of state-space approach with the following example.

Example 6.2

Consider the following single-input single-output linear second order differential equation

$$\frac{d^2}{dt}y(t) + 3\frac{d}{dt}y(t) + 2y(t) = x(t)$$

with $x(t) = u(t)$ and initial conditions

$$\frac{d}{dt}y(0^-) = y(0^-) = 0$$

Find the output $y(t)$ for $t \geq 0$.

Solution

In transforming this second order differential equation into state-space form we need to define two state variables since the differential equation is given in second order.

Let

$$z_1 = y(t)$$

and

$$z_2 = \frac{d}{dt}y(t)$$

Next, we take derivatives of the two states and write

$$\frac{d}{dt}z_1 = \frac{d}{dt}y(t)$$

and

$$\frac{d}{dt}z_2 = \frac{d^2}{dt}y(t)$$

Substituting for $y(t)$ and $\frac{d}{dt}y(t)$ in the above state equations we get the state equations

$$\frac{d}{dt}z_1 = z_2$$
$$\frac{d}{dt}z_2 = -2z_1 - 3z_2 + x$$

These are the two simultaneous state equations. Putting them in state-space matrix form with $x(t) = u(t)$ we get

$$\frac{d}{dt}\underline{z} = A\underline{z} + B\underline{x}$$

where

$$A = \begin{bmatrix} 0 & 1 \\ -2 & -3 \end{bmatrix}, \quad B = \begin{bmatrix} 0 \\ 1 \end{bmatrix}, \quad \frac{d}{dt}\underline{z} = \begin{bmatrix} \frac{d}{dt}z_1 \\ \frac{d}{dt}z_2 \end{bmatrix}, \quad \underline{z} = \begin{bmatrix} z_1 \\ z_2 \end{bmatrix}, \quad \text{and} \quad \underline{x} = [x]$$

The first row in A is obtained by reading the coefficients of z_1 and z_2 in the first state equation. The second row is obtained in a similar way. The first row in B is the coefficient of the input $x(t)$ in the first state equation and the second row is obtained similarly.

1. The Laplace method

   We now try to solve this matrix state equation using the Laplace transform approach. Taking the Laplace transform of the matrix state equation we get

$$[s\underline{Z}(s) - \underline{z}(0^-)] = A\underline{Z}(s) + B\underline{X}(s)$$

   $\underline{z}(0^-) = \underline{0}$, where $\underline{0}$ is a vector of zeros (initial conditions vector).
   To take $\underline{Z}(s)$ as a common factor we need to write $s\underline{Z}(s)$ as $sI\underline{Z}(s)$, where I is the 2×2 identity matrix to get

$$[sI - A]\underline{Z}(s) = B\underline{X}(s) + \underline{z}(0^-)$$

   $[sI - A]$ is a 2×2 matrix. To solve for $\underline{Z}(s)$ we need to multiply the last equation by the inverse of $[sI - A]$ and get

$$\underline{Z}(s) = [sI - A]^{-1}B\underline{X}(s) + [sI - A]^{-1}\underline{z}(0^-)$$

Substituting for A, B, and $\underset{\sim}{z}(0^-)$ we get

$$\underline{Z}(s) = \left\{ s\begin{bmatrix} 1 & 0 \\ 0 & 1 \end{bmatrix} - \begin{bmatrix} 0 & 1 \\ -2 & -3 \end{bmatrix} \right\}^{-1} \begin{bmatrix} 0 \\ 1 \end{bmatrix} \begin{bmatrix} \frac{1}{s} \end{bmatrix}$$

This last equation simplifies to

$$\underline{Z}(s) = \begin{bmatrix} s & -1 \\ 2 & s+3 \end{bmatrix}^{-1} \begin{bmatrix} 0 \\ 1/s \end{bmatrix}$$

Finding the inverse and rewriting we get

$$\underline{Z}(s) = \frac{1}{s^2 + 3s + 2} \begin{bmatrix} s+3 & 1 \\ -2 & s \end{bmatrix} \begin{bmatrix} 0 \\ 1/s \end{bmatrix}$$

or

$$\underline{Z}(s) = \begin{bmatrix} \dfrac{s+3}{s^2+3s+2} & \dfrac{1}{s^2+3s+2} \\ \dfrac{-2}{s^2+3s+2} & \dfrac{s}{s^2+3s+2} \end{bmatrix} \begin{bmatrix} 0 \\ 1/s \end{bmatrix}$$

Finally

$$\underline{Z}(s) = \begin{bmatrix} \dfrac{1}{s(s^2+3s+2)} \\ \dfrac{1}{(s^2+3s+2)} \end{bmatrix}$$

Using partial fraction expansion on the transfer function elements in the above matrix, we have

$$\underline{Z}(s) = \begin{bmatrix} \dfrac{a}{s} + \dfrac{b}{s+1} + \dfrac{c}{s+2} \\ \dfrac{e}{s+1} + \dfrac{f}{s+2} \end{bmatrix}$$

which after calculating the constants gives

$$\underline{Z}(s) = \begin{bmatrix} \dfrac{1/2}{s} + \dfrac{-1}{s+1} + \dfrac{1/2}{s+2} \\ \dfrac{1}{s+1} + \dfrac{-1}{s+2} \end{bmatrix}$$

Taking the inverse Laplace transform of the above vector we get

$$\underline{z}(t) = \begin{bmatrix} z_1(t) \\ z_2(t) \end{bmatrix} = \begin{bmatrix} \frac{1}{2}u(t) - e^{-t}u(t) + \frac{1}{2}e^{-2t}u(t) \\ e^{-t}u(t) - e^{-2t}u(t) \end{bmatrix}$$

Our goal is to solve for $y(t)$, the output of the system. In this example, $y(t)$ was assigned to the state variable $z_1(t)$. $\frac{d}{dt}y(t)$ was assigned to the other state variable $z_2(t)$.
Therefore,

$$y(t) = \frac{1}{2}u(t) - e^{-t}u(t) + \frac{1}{2}e^{-2t}u(t)$$

or

$$y(t) = \left(\frac{1}{2} - e^{-t} + \frac{1}{2}e^{-2t}\right)u(t)$$

2. Real-time method

The state-space equation is

$$\frac{d}{dt}\underline{z} = A\underline{z} + B\underline{x}$$

where

$$A = \begin{bmatrix} 0 & 1 \\ -2 & -3 \end{bmatrix}, \quad B = \begin{bmatrix} 0 \\ 1 \end{bmatrix}, \quad \frac{d}{dt}\underline{z} = \begin{bmatrix} \frac{d}{dt}z_1 \\ \frac{d}{dt}z_2 \end{bmatrix}, \quad \underline{z} = \begin{bmatrix} z_1 \\ z_2 \end{bmatrix}, \quad \text{and} \quad \underline{x} = [x]$$

with $\underline{z}(t_0) = \underline{z}(0) = \underline{0}$, and $x(t) = u(t)$.

The first thing we do here is to notice that A is not in a diagonal form. Hence, we need to diagonalize it. Let us transform the whole system into a new system where A is transformed into a diagonal matrix $P^{-1}AP$.

Let $\underline{z} = P\underline{w}$, where $\underline{w}$ is a new state vector. Take the derivative to get

$$\frac{d}{dt}\underline{z} = P\frac{d}{dt}\underline{w}$$

By substituting in the state equation

$$\frac{d}{dt}\underline{z} = A\underline{z} + B\underline{x}$$

we get

$$P\frac{d}{dt}\underline{w} = AP\underline{w} + B\underline{x}$$

or

$$\frac{d}{dt}\underline{w} = P^{-1}AP\underline{w} + P^{-1}B\underline{x}$$

Now this is a new state-space equation. The last equation is obtained by multiplying the equation that precedes it by P^{-1}. The initial condition vector can be transformed into the new space by manipulating the equation

$$\frac{d}{dt}\underline{z} = P\frac{d}{dt}\underline{w}$$

Evaluating this equation at $t = 0$ and multiplying it by P^{-1} we arrive at

$$\underline{w}(0) = P^{-1}\underline{z}(0)$$

The total solution now can be written as

$$\underline{w}(t) = \exp[P^{-1}APt]\underline{w}(0) + \int_0^t \exp[P^{-1}AP(t-\tau)]P^{-1}Bu(\tau)d\tau$$

We need to find $P^{-1}AP$ to substitute in the total solution equation. We start first by finding the eigenvalues of A. We look at the roots of $\det(\lambda I - A)$.

$$\det(\lambda I - A) = \det\left\{\lambda\begin{bmatrix}1 & 0\\0 & 1\end{bmatrix} - \begin{bmatrix}0 & 1\\-2 & -3\end{bmatrix}\right\}$$

$$\det(\lambda I - A) = \det\left\{\begin{bmatrix}\lambda & -1\\2 & \lambda+3\end{bmatrix}\right\}$$

which simplifies to

$$\det(\lambda I - A) = \lambda(\lambda + 3) - 2(-1)$$
$$\det(\lambda I - A) = \lambda^2 + 3\lambda + 2$$

The roots of $\det(\lambda I - A)$ are $\lambda_1 = -1$ and $\lambda_2 = -2$.

Now it is time to find the two eigenvectors that correspond to each of the eigenvalues: -1 and -2. To find these vectors we need to find a solution for the equation

$$(\lambda I - A)\underline{p} = \underline{0}$$

for each value of the eigenvalues.

For $\lambda_1 = -1$

$$(\lambda I - A)\underline{p} = \begin{bmatrix} -1 & -1 \\ 2 & 2 \end{bmatrix}\underline{p} = \begin{bmatrix} -1 & -1 \\ 2 & 2 \end{bmatrix}\begin{bmatrix} p_1 \\ p_2 \end{bmatrix} = \underline{0}$$

The two equations to solve are

$$(-1)p_1 + (-1)p_2 = 0$$

and

$$2p_1 + 2p_2 = 0$$

These two equations are the same (divide the second by -2 to get the first). To solve them we let $p_1 = c$, then $p_2 = -c$. Therefore the first eigenvector $\underline{p}_1$ will be

$$\underline{p}_1 = c\begin{bmatrix} 1 \\ -1 \end{bmatrix}$$

Choosing c to have a value of 1 will give

$$\underline{p}_1 = \begin{bmatrix} 1 \\ -1 \end{bmatrix}$$

which is the first eigenvector of the matrix P.

For $\lambda_2 = -2$

$$(\lambda I - A)\underline{p} = \begin{bmatrix} -2 & -1 \\ 2 & 1 \end{bmatrix}\underline{p} = \begin{bmatrix} -2 & -1 \\ 2 & 1 \end{bmatrix}\begin{bmatrix} p_1 \\ p_2 \end{bmatrix} = \underline{0}$$

and the two equations to solve are

$$(-2)p_1 + (-1)p_2 = 0$$

and

$$2p_1 + 1p_2 = 0$$

These two equations are the same (multiply the second by -1 to get the first). To solve them we let $p_1 = c$, then $p_2 = -2c$. Therefore the second eigenvector $\underline{p}_2$ will be

$$\underline{p}_2 = c\begin{bmatrix} 1 \\ -2 \end{bmatrix}$$

Choosing c to have a value of 1 will give

$$\underline{p}_2 = \begin{bmatrix} 1 \\ -2 \end{bmatrix}$$

And the matrix P of the two eigenvectors is

$$P = \begin{bmatrix} \underline{p}_1 & \underline{p}_2 \end{bmatrix} = \begin{bmatrix} 1 & 1 \\ -1 & -2 \end{bmatrix}$$

and

$$P^{-1} = \frac{1}{\det(P)}\begin{bmatrix} -2 & -1 \\ 1 & 1 \end{bmatrix} = -1\begin{bmatrix} -2 & -1 \\ 1 & 1 \end{bmatrix} = \begin{bmatrix} 2 & 1 \\ -1 & -1 \end{bmatrix}$$

Now let us look at $P^{-1}AP$ and see if it is really a diagonal matrix.

$$P^{-1}AP = \begin{bmatrix} 2 & 1 \\ -1 & -1 \end{bmatrix}\begin{bmatrix} 0 & 1 \\ -2 & -3 \end{bmatrix}\begin{bmatrix} 1 & 1 \\ -1 & -2 \end{bmatrix}$$

By carrying out the matrix multiplication we arrive at

$$P^{-1}AP = \begin{bmatrix} -2 & -1 \\ 2 & 2 \end{bmatrix}\begin{bmatrix} 1 & 1 \\ -1 & -2 \end{bmatrix} = \begin{bmatrix} -1 & 0 \\ 0 & -2 \end{bmatrix}$$

Notice that the eigenvalues of A are the same as the eigenvalues of $P^{-1}AP$ (check that out). This means that the shape of the transients is not affected by the transformations we made when we went from the z to the w space.

Now go back to the equation that should give us a solution for $\underline{w}(t)$ and notice that $u(t)$ is the step signal that is unity from time $t = 0$ until infinity.

The solution now can be written as

$$\underline{w}(t) = \exp[P^{-1}APt]\underline{w}(0) + \int_0^t \exp[P^{-1}AP(t-\tau)]P^{-1}Bu(\tau)d\tau$$

$$\underline{w}(t) = \int_0^t \begin{bmatrix} e^{-(t-\tau)} & 0 \\ 0 & e^{-2(t-\tau)} \end{bmatrix} \begin{bmatrix} 1 \\ -1 \end{bmatrix} d\tau = \int_0^t \begin{bmatrix} e^{-(t-\tau)} \\ e^{-2(t-\tau)} \end{bmatrix} d\tau$$

We can integrate element by element to get

$$\underline{w}(t) = \int_0^t \begin{bmatrix} e^{-(t-\tau)} \\ e^{-2(t-\tau)} \end{bmatrix} d\tau = \begin{bmatrix} e^{-t}[e^{\tau}]_0^t \\ -e^{-2t}[e^{2\tau}]_0^t \end{bmatrix}$$

This simplifies to

$$\underline{w}(t) = \begin{bmatrix} e^{-t}[e^{t}-1] \\ \frac{1}{-2}e^{-2t}[e^{2t}-1] \end{bmatrix} = \begin{bmatrix} 1-e^{-t} \\ \frac{1}{2}[e^{-2t}-1] \end{bmatrix}$$

This is the solution for $\underline{w}(t)$. We need a solution for $\underline{z}(t)$ and therefore we write

$$\underline{z}(t) = P\underline{w}(t)$$

$$\underline{z}(t) = P\underline{w}(t) = \begin{bmatrix} 1 & 1 \\ -1 & -2 \end{bmatrix} \begin{bmatrix} 1-e^{-t} \\ \frac{1}{2}[e^{-2t}-1] \end{bmatrix}$$

$$\underline{z}(t) = P\underline{w}(t) = \begin{bmatrix} \frac{1}{2} - e^{-t} + \frac{1}{2}e^{-2t} \\ (e^{-t} - e^{-2t})u(t) \end{bmatrix}$$

from which $y(t)$ is

$$y(t) = \frac{1}{2} - e^{-t} + \frac{1}{2}e^{-2t}$$

These solutions do agree with the solutions obtained using the Laplace transform approach taken earlier in the example.

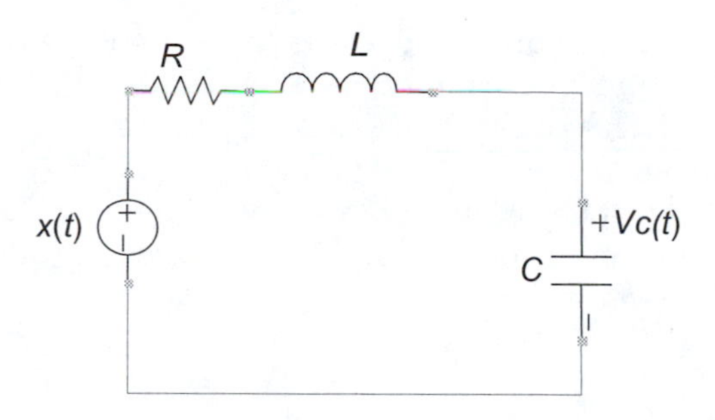

FIGURE 6.1
Circuit for Example 6.3.

Example 6.3

Consider the series RLC circuit shown in Figure 6.1, where $x(t)$ is the input.

We are interested in finding the output $v_R(t)$, the voltage across the resistor with $R = 1$ ohm, $C = 1$ farad, and $L = 1$ henry. Let $x(t)$ be an impulse input with unity strength.

Solution

This case is different from the case of a linear constant-coefficient differential equation where the number of states was set to the order of the differential equation at hand. Now we have a real physical system, the RLC circuit. How do we select the number of states? The system in this example has two energy storage elements (the resistor does not store energy). This means that we will need a second order differential equation to solve for all variables in the system. Thus, we will have two states. We will select the voltage across the capacitor as one state and the current in the inductor as the second state because they relate to the storage elements. We also selected the voltage in the capacitor because we know from circuit transient analysis that the voltage in the capacitor does not change instantaneously. We also selected the current in the inductor because it does not change instantaneously. Notice that other variables in the circuit were not selected as states. The reason is that the only two variables in the circuit that are independent are the voltage in the capacitor and the current in the inductor. Take, for example, the voltage in the resistor

$$v_R(t) = Ri_L(t)$$

where $i_L(t)$ is the current in the inductor as seen in the circuit. $v_R(t)$ is a linear combination of the current in the inductor. Thus, it is not selected and its value can be calculated easily after we find the current in the inductor. For a circuit that has many resistors, n capacitors that cannot be combined, and m inductors that also cannot be combined, we will have $n + m$ states, where the n states are the n voltages across the capacitors, and the m states are the m currents in the inductors.

Therefore, our two states in this example are the voltage in the capacitor and the current in the inductor, $v_C(t)$ and $i_L(t)$.

In the storage elements

$$L\frac{d}{dt}i_L(t) = v_L(t)$$

and

$$C\frac{d}{dt}v_C(t) = i_C(t)$$

Remember that the right side of the above state equations must be in terms of the state variables and the input $x(t)$ only.

A loop equation in the circuit in Figure 6.1 will yield the following equations

$$v_L(t) = -Ri_L(t) - v_C(t) + x(t)$$

and

$$i_C(t) = i_L(t)$$

which can be written as

$$L\frac{d}{dt}i_L(t) = -Ri_L(t) - v_C(t) + x(t)$$

and

$$C\frac{d}{dt}v_C(t) = i_L(t)$$

We will let $z_1(t) = i_L(t)$ and $z_2(t) = v_C(t)$. Rewriting the above state equations with $L = 1H$, $C = 1F$, and $R = 1$ ohm we get

$$\frac{d}{dt}z_1(t) = -z_1(t) - z_2(t) + x(t)$$

and

$$\frac{d}{dt}z_2(t) = z_1(t)$$

Assuming zero initial conditions, we can put the two simultaneous state equations in state-space matrix form as

$$\frac{d}{dt}\underset{\sim}{z} = A\underset{\sim}{z} + B\underset{\sim}{x}$$

where

$$A = \begin{bmatrix} -1 & -1 \\ 1 & 0 \end{bmatrix}, \; B = \begin{bmatrix} 0 \\ 1 \end{bmatrix}, \; \frac{d}{dt}\underset{\sim}{z} = \begin{bmatrix} \frac{d}{dt}z_1 \\ \frac{d}{dt}z_2 \end{bmatrix}, \text{ and } \underset{\sim}{z} = \begin{bmatrix} z_1 \\ z_2 \end{bmatrix}$$

Following the procedure in the second example and after taking the Laplace transform of the time domain state matrix equations we get

$$\underset{\sim}{Z}(s) = [sI - A]^{-1}B\underset{\sim}{X}(s) + [sI - A]^{-1}\underset{\sim}{z}(0^-)$$

Substituting we get

$$\underset{\sim}{Z}(s) = \left\{ s\begin{bmatrix} 1 & 0 \\ 0 & 1 \end{bmatrix} - \begin{bmatrix} -1 & -1 \\ 1 & 0 \end{bmatrix} \right\}^{-1} \begin{bmatrix} 1 \\ 0 \end{bmatrix} [1]$$

which simplifies to

$$\underset{\sim}{Z}(s) = \begin{bmatrix} s+1 & 1 \\ -1 & s \end{bmatrix}^{-1} \begin{bmatrix} 1 \\ 0 \end{bmatrix}$$

$$\underset{\sim}{Z}(s) = \frac{1}{s^2 + s + 1} \begin{bmatrix} s & -1 \\ 1 & s+1 \end{bmatrix} \begin{bmatrix} 1 \\ 0 \end{bmatrix}$$

$$\underset{\sim}{Z}(s) = \begin{bmatrix} \frac{s}{s^2 + s + 1} & \frac{-1}{s^2 + s + 1} \\ \frac{1}{s^2 + s + 1} & \frac{s+1}{s^2 + s + 1} \end{bmatrix} \begin{bmatrix} 1 \\ 0 \end{bmatrix}$$

Finally

$$\underline{Z}(s) = \begin{bmatrix} \dfrac{s}{s^2+s+1} \\ \dfrac{1}{s^2+s+1} \end{bmatrix}$$

Taking the inverse Laplace transform of the above state-space Laplace matrix representation we get

$$\underline{z}(t) = \begin{bmatrix} \dfrac{-1}{2}\sqrt{\dfrac{4}{3}}e^{-\frac{1}{2}t}\sin\left(\sqrt{\dfrac{3}{4}}t\right)u(t) + e^{-\frac{1}{2}t}\cos\left(\sqrt{\dfrac{3}{4}}t\right)u(t) \\ \sqrt{\dfrac{4}{3}}e^{-\frac{1}{2}t}\sin\left(\sqrt{\dfrac{3}{4}}t\right)u(t) \end{bmatrix}$$

The states are

$$z_1(t) = -\frac{1}{2}\sqrt{\frac{4}{3}}e^{-\frac{1}{2}t}\sin\left(\sqrt{\frac{3}{4}}t\right)u(t) + e^{-\frac{1}{2}t}\cos\left(\sqrt{\frac{3}{4}}t\right)u(t)$$

and

$$z_2(t) = \sqrt{\frac{4}{3}}e^{-\frac{1}{2}t}\sin\left(\sqrt{\frac{3}{4}}t\right)u(t)$$

$z_1(t) = i_L(t)$ and $z_2(t) = v_C(t)$. But we were asked to find $v_R(t)$, which is

$$v_R(t) = Ri_L(t)$$

Therefore

$$v_R(t) = \frac{-1}{2}\sqrt{\frac{4}{3}}e^{-\frac{1}{2}t}\sin\left(\sqrt{\frac{3}{4}}t\right)u(t) + e^{-\frac{1}{2}t}\cos\left(\sqrt{\frac{3}{4}}t\right)u(t)$$

So the input to the system was $x(t) = \delta(t)$ and the output to calculate was $y(t) = v_R(t)$. This example demonstrates how a variable in a system like $v_R(t)$ can be computed from the selected independent states $z_1(t) = i_L(t)$ and $z_2(t) = v_C(t)$. Knowing the states $z_1(t) = i_L(t)$ and $z_2(t) = v_C(t)$ we can find any variable in the circuit. Actually, we can find the following variables:

1. The current in the resistor R: $i_R(t) = i_L(t)$
2. The voltage in the inductor L: $v_L(t) = L\frac{d}{dt}i_L(t)$
3. The current in the capacitor C: $i_C(t) = C\frac{d}{dt}v_C(t)$

Example 6.4

Consider the following mechanical system with the input $x(t)$ as an impulsive force of unity strength, $\delta(t)$, and initial conditions

$$\frac{d}{dt}y(0^-) = y(0^-) = 0$$

The system is shown in Figure 6.2.

The rod to the left of the mass M can be modeled by a translational spring with spring constant K, in parallel with a translational damper of constant B. $x(t)$ is the input force, and $y(t)$ is the output translation measured from a specific reference as indicated in the graph in Figure 6.2. For $M = 1$ kg, $B = 1$ newton/meter/second (N.s/m), $K = 1$ newton/meter, and with $x(t) = \delta(t)$, find the output displacement $y(t)$.

Solution

A free-body diagram is constructed to help us write the equations of motion. This diagram is shown in Figure 6.3.

Since we are interested in the output $y(t)$ and if we choose not to use state-space techniques we can Laplace-transform the free-body diagram in Figure 6.3 and obtain the diagram in Figure 6.4.

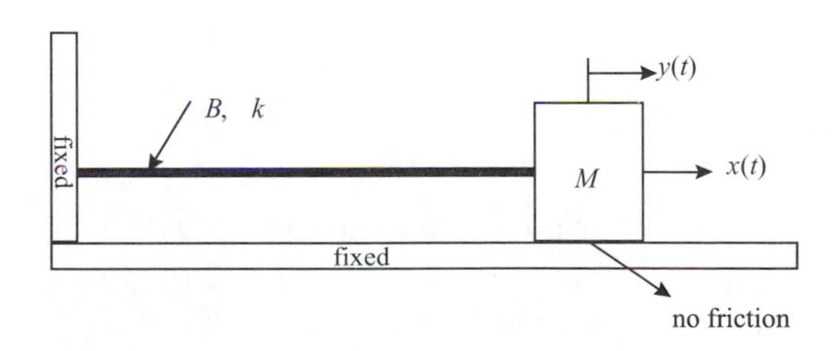

FIGURE 6.2
System for Example 6.4.

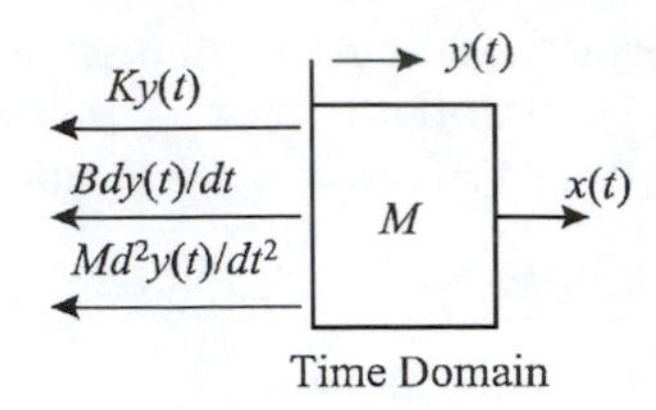

FIGURE 6.3
Free-body diagram for Example 6.4.

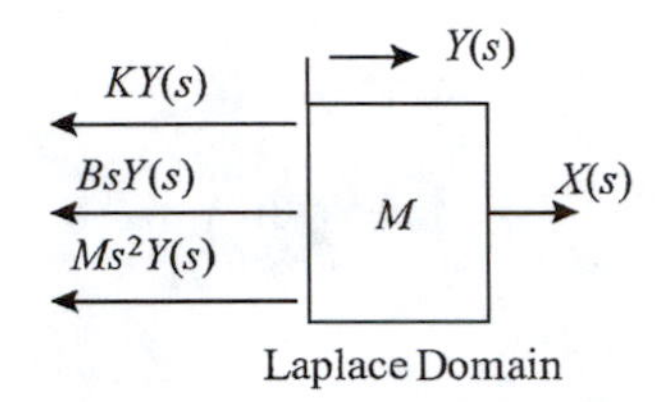

FIGURE 6.4
Free-body diagram for Example 6.4.

Summing the forces using the Laplace-transformed free-body diagram we get

$$KY(s) + BsY(s) + Ms^2Y(s) = X(s)$$

For illustrative purposes only, assuming $M = 1$ kg, $B = 1$ newton.second/meter (N.s/m), $K = 1$ newton/meter, and with $x(t) = \delta(t)$ we can write

$$Y(s) + sY(s) + s^2Y(s) = 1$$

$$Y(s)[s^2 + s + 1] = 1$$

Finally the output is

$$Y(s) = 1/[s^2 + s + 1]$$

Taking the inverse Laplace transform of $Y(s)$ we get

$$y(t) = \sqrt{\frac{4}{3}}e^{-\frac{1}{2}t}\sin\left(\sqrt{\frac{3}{4}}t\right)u(t)$$

But we are interested in using the state-space approach in solving for the output $y(t)$. We will consider the free-body diagram in time domain and sum the forces acting on the object of mass M to get

$$M\frac{d^2}{dt}y(t) + B\frac{d}{dt}y(t) + Ky(t) = x(t)$$

In transforming this second order differential equation into state-space form we need to define two state variables since the differential equation given is second order.

Let

$$z_1 = y(t)$$

and

$$z_2 = \frac{d}{dt}y(t)$$

Next, we take derivatives of the two states to get

$$\frac{d}{dt}z_1 = \frac{d}{dt}y(t)$$

and

$$\frac{d}{dt}z_2 = \frac{d^2}{dt}y(t)$$

In terms of state variables and inputs to the system we have

$$\frac{d}{dt}z_1 = z_2$$

and

$$\frac{d}{dt}z_2 = -z_1 - z_2 + x$$

These are the two simultaneous state equations. Putting them in state-space matrix form with $x(t) = \delta(t)$ we get

$$\frac{d}{dt}\underset{\sim}{z} = A\underset{\sim}{z} + B\underset{\sim}{x}$$

where

$$A = \begin{bmatrix} 0 & 1 \\ -1 & -1 \end{bmatrix}, \quad B = \begin{bmatrix} 0 \\ 1 \end{bmatrix}, \quad \frac{d}{dt}\underset{\sim}{z} = \begin{bmatrix} \frac{d}{dt}z_1 \\ \frac{d}{dt}z_2 \end{bmatrix}, \quad \underset{\sim}{z} = \begin{bmatrix} z_1 \\ z_2 \end{bmatrix}, \quad \text{and} \quad \underset{\sim}{x} = [x]$$

Up to this point we are still in time domain. We now try to solve this matrix state equation using the Laplace transform approach. Notice that this system is similar to the system in Example 6.3. We can use the same procedure to get

$$\underline{Z}(s) = \begin{bmatrix} \dfrac{1}{(s^2 + 3s + 2)} \\ \dfrac{s}{(s^2 + 3s + 2)} \end{bmatrix}$$

Taking the inverse Laplace transform on the above state-space Laplace matrix representation, we get

$$\underline{z}(t) = \begin{bmatrix} \sqrt{\frac{4}{3}} e^{-\frac{1}{2}t} \sin\left(\sqrt{\frac{3}{4}}t\right) u(t) \\ \frac{-1}{2}\sqrt{\frac{4}{3}} e^{-\frac{1}{2}t} \sin\left(\sqrt{\frac{3}{4}}t\right) u(t) + e^{-\frac{1}{2}t} \cos\left(\sqrt{\frac{3}{4}}t\right) u(t) \end{bmatrix}$$

Therefore, the states are

$$z_1(t) = \sqrt{\frac{4}{3}} e^{-\frac{1}{2}t} \sin\left(\sqrt{\frac{3}{4}}t\right) u(t)$$

and

$$z_2(t) = \frac{-1}{2}\sqrt{\frac{4}{3}} e^{-\frac{1}{2}t} \sin\left(\sqrt{\frac{3}{4}}t\right) u(t) + e^{-\frac{1}{2}t} \cos\left(\sqrt{\frac{3}{4}}t\right) u(t)$$

$z_1(t) = y(t)$ and $z_2(t) = \frac{d}{dt}y(t)$ which is the velocity of the displacement $y(t)$.

Notice that Example 6.3 and Example 6.4 are two different systems in the sense that one is electrical and one is mechanical and yet the solutions are similar. The reason is that both systems were modeled mathematically as differential equations.

6.3 General Representation of Systems in State-Space

We will consider systems not only with single-input single-output, but our study will be general and we will generalize to systems of multiple-inputs multiple-outputs. We also have indicated that the states we select among the many variables in the system under consideration must be linearly independent. This means that no selected state can be written as a linear combination of the others. A linear combination of n states, z_k, where $k = 1$

to n can be represented as

$$\sum_{k=1}^{k=n} \alpha_k z_k = \alpha_1 z_1 + \alpha_2 z_2 + \alpha_3 z_3 + \cdots + \alpha_n z_n \tag{6.10}$$

where the α's are constants. Further, the selected states are linearly independent if the sum

$$\sum_{k=1}^{k=n} \alpha_k z_k = \alpha_1 z_1 + \alpha_2 z_2 + \alpha_3 z_3 + \cdots + \alpha_n z_n$$

is equal to zero only if all the α's are zero and no selected state is zero.

System variables are the variables in the system where when an input is applied to the system they will have a response whether this input is an actual external input or an internal input (initial condition).

The state variables of the system are the minimum linearly independent set of variables in the same system.

When we express the output variables in the system as a linear combination of the state variables, these output variables are called output equations for the system. In a similar way we say that the set of the first order differential equations that describe the system completely are called state equations. When we group our state variables in a vector form we will call them state vectors. The n-dimensional space whose variables are the state variables is called the state-space. Multiple-input multiple-output systems can be represented in state space as the two state-space matrix equations

$$\begin{aligned} \frac{d}{dt}\underline{z} &= A\underline{z} + B\underline{x} \\ \underline{y} &= C\underline{z} + D\underline{x} \end{aligned} \tag{6.11}$$

In the above $\underline{z}$ is the state vector, A is the system coefficient matrix, B is the input matrix, C is the output matrix, and D is the feed forward matrix.

The first of the equations above is called the state equation, and the second is called the output equation.

6.4 General Solution of State-Space Equations Using the Laplace Transform

Given the state-space matrix equations in time domain as

$$\begin{aligned} \frac{d}{dt}\underline{z} &= A\underline{z} + B\underline{x} \\ \underline{y} &= C\underline{z} + D\underline{x} \end{aligned}$$

the solution for these equations in the Laplace domain is

$$\begin{aligned} \underline{Z}(s) &= (sI - A)^{-1}\underline{z}(0) + (sI - A)^{-1}B\underline{X}(s) \\ \underline{Y}(s) &= C\underline{Z}(s) + D\underline{X}(s) \end{aligned} \tag{6.12}$$

6.5 General Solution of the State-Space Equations in Real Time

Given the state-space matrix equations in time domain as

$$\begin{aligned} \frac{d}{dt}\underline{z} &= A\underline{z} + B\underline{x} \\ \underline{y} &= C\underline{z} + D\underline{x} \end{aligned}$$

we will examine the two cases where the matrix A can have two different forms.

If A is a diagonal matrix, then $\exp(At)$ is straightforward and explained in the examples above. The solution for the states in this case is

$$\underline{z}(t) = \exp[At]\underline{z}(0) + \int_0^t \exp[A(t-\tau)]Bx(\tau)d\tau \tag{6.13}$$

and for the outputs it is

$$\underline{y}(t) = C\underline{z}(t) + D\underline{x}(t) \tag{6.14}$$

If A is not a diagonal matrix, it can be diagonalized using the techniques explained in the previous examples and the representation in state-space can be transformed to a new space, $\underline{w}(t)$, as a result. The solution for the states is then

$$\underline{z}(t) = P\underline{w}(t) \tag{6.15}$$

and for the outputs it is

$$\underline{y}(t) = CP\underline{w}(t) + D\underline{x}(t) \tag{6.16}$$

where

$$\underline{w}(t) = \exp[P^{-1}APt]\underline{w}(0) + \int_0^t \exp[P^{-1}AP(t-\tau)]P^{-1}Bx(\tau)d\tau \tag{6.17}$$

and P is the transformation matrix that transforms A into $P^{-1}AP$, the diagonalized matrix.

6.6 Ways of Evaluating e^{At}

There are many ways of evaluating the exponent of the matrix A. We will consider six of them in this chapter.

6.6.1 First Method: A is a Diagonal Matrix

If A is diagonal square matrix, then e^{At} is a square matrix, too. It is calculated by leaving alone all elements in A which are zeros and using the other elements multiplied by t as exponents. As an example:

If

$$A = \begin{bmatrix} a & 0 \\ 0 & b \end{bmatrix}$$

is a diagonal matrix, then

$$e^{At} = \begin{bmatrix} e^{at} & 0 \\ 0 & e^{bt} \end{bmatrix} \tag{6.18}$$

6.6.2 Second Method: A is of the Form $\begin{bmatrix} a & b \\ 0 & a \end{bmatrix}$

This was shown previously and is presented here for completeness as

$$\exp(At) = \begin{bmatrix} e^{at} & bte^{at} \\ 0 & e^{at} \end{bmatrix} \tag{6.19}$$

6.6.3 Third Method: Numerical Evaluation, A of Any Form

From Taylor's series, we can write the exponent of the matrix A as

$$e^{At} = I + At + \frac{(At^2)}{2!} + \frac{(At)^3}{3!} + \frac{(At)^4}{4!} + \cdots \tag{6.20}$$

This method is not desirable if a complete closed form solution is sought.

6.6.4 Fourth Method: The Cayley–Hamilton Approach

Using the Cayley–Hamilton theorem, we can write

$$e^{At} = a_0 I + a_1 A + a_2 A^2 + a_3 A^3 + \cdots + a_{n-1} A^{n-1} \tag{6.21}$$

where the α's are to be calculated and n is the size of the matrix A. Let us denote the eigenvalues for A as λ_1 through λ_n and substitute A by each of the eigenvalues to get the following set of equations

$$\begin{aligned} e^{\lambda_1 t} &= a_0 + a_1\lambda_1 + a_2\lambda_1^2 + \cdots + a_{n-1}\lambda_1^{n-1} \\ e^{\lambda_2 t} &= a_0 + a_1\lambda_2 + a_2\lambda_2^2 + \cdots + a_{n-1}\lambda_2^{n-1} \\ e^{\lambda_3 t} &= a_0 + a_1\lambda_3 + a_2\lambda_3^2 + \cdots + a_{n-1}\lambda_3^{n-1} \\ &\vdots \\ e^{\lambda_n t} &= a_0 + a_1\lambda_n + a_2\lambda_n^2 + \cdots + a_{n-1}\lambda_n^{n-1} \end{aligned} \tag{6.22}$$

If the eigenvalues of A are distinct, we will have n simultaneous equations to solve for the coefficients a_0 through a_{n-1}. If the eigenvalues of A are not distinct, then the above set of equations will be dependent. If an eigenvalue has, for example, a multiplicity of two then we will need an extra equation involving the derivative of the equation corresponding to that particular eigenvalue with respect to the eigenvalue itself. As an example, consider

$$A = \begin{bmatrix} -2 & 0 \\ 0 & -2 \end{bmatrix}$$

The characteristic equation ($\det(\lambda I - A) = 0$) is

$$\lambda^2 + 4\lambda + 4 = 0$$

The roots are at $\lambda = -2$ and $\lambda = -2$.
Using the general equations presented before we get

$$e^{\lambda t} = a_0 + \lambda a_1$$

and its derivative

$$te^{\lambda t} = a_1$$

With the eigenvalues both at -2, the above equations become

$$\begin{aligned} e^{-2t} &= a_0 - 2a_1 \\ te^{-2t} &= a_1 \end{aligned}$$

Therefore,

$$
\begin{aligned}
a_1 &= te^{-2t} \\
a_0 &= 2te^{-2t} + e^{-2t}
\end{aligned}
$$

Since A is a 2×2 matrix, we have

$$
\begin{aligned}
e^{At} &= a_0 I + a_1 A \\
e^{At} &= a_0 I + a_1 A = \begin{bmatrix} a_0 - 2a_1 & 0 \\ 0 & a_0 - 2a_1 \end{bmatrix} = \begin{bmatrix} e^{-2t} & 0 \\ 0 & e^{-2t} \end{bmatrix}
\end{aligned}
$$

As another example, let

$$
A = \begin{bmatrix} 0 & 1 \\ -6 & -5 \end{bmatrix}
$$

We will first find the eigenvalues for A. The characteristic equation is

$$
\lambda^2 + 5\lambda + 6 = 0
$$

The roots are at $\lambda = -2$ and $\lambda = -3$.
Using the general equations presented before we write

$$
\begin{aligned}
e^{-2t} &= a_0 - 2a_1 \\
e^{-3t} &= a_0 - 3a_1
\end{aligned}
$$

Solving these simultaneous equations leads to

$$
\begin{aligned}
a_0 &= -2e^{-3t} + 3e^{-2t} \\
a_1 &= e^{-2t} - e^{-3t}
\end{aligned}
$$

Since A is a 2×2 matrix we have

$$
\begin{aligned}
e^{At} &= a_0 I + a_1 A \\
e^{At} &= a_0 I + a_1 A = \begin{bmatrix} a_0 & a_1 \\ -6a_1 & a_0 - 5a_1 \end{bmatrix} = \begin{bmatrix} 3e^{-2t} - 2e^{-2t} & e^{-2t} - e^{-3t} \\ -6e^{-2t} + 6e^{-3t} & -2e^{-2t} + 3e^{-3t} \end{bmatrix}
\end{aligned}
$$

6.6.5 Fifth Method: The Inverse-Laplace Method

e^{At} is known as the state-transition matrix and usually denoted as $\Phi(t)$. Let us look at the Laplace transform solution of the state equations

$$
\underline{Z}(s) = (sI - A)^{-1}\underline{z}(0) + (sI - A)^{-1}B\underline{X}(s)
$$

and for comparison, let us give the time domain solution for the same equations

$$\underset{\sim}{z}(t) = \exp[At]\underset{\sim}{z}(0) + \int_0^t \exp[A(t-\tau)]Bx(\tau)d\tau$$

A close look at these equations tells us that e^{At} is the inverse Laplace transform of $(sI - A)^{-1}$. Or mathematically, we write

$$e^{At} = L^{-1}[(sI - A)^{-1}] \tag{6.23}$$

As an example, let

$$A = \begin{bmatrix} -2 & 0 \\ 0 & -2 \end{bmatrix}$$

Then

$$(sI - A)^{-1} = \begin{bmatrix} s+2 & 0 \\ 0 & s+2 \end{bmatrix}^{-1} = \frac{1}{(s+2)^2}\begin{bmatrix} s+2 & 0 \\ 0 & s+2 \end{bmatrix}$$

$$(sI - A)^{-1} = \begin{bmatrix} \frac{1}{s+2} & 0 \\ 0 & \frac{1}{s+2} \end{bmatrix}$$

Taking the inverse transform we get

$$\begin{bmatrix} e^{-2t} & 0 \\ 0 & e^{-2t} \end{bmatrix} = e^{At} = \Phi(t)$$

Before we go into the sixth method of finding e^{At}, we will look at some important properties of the state-transition matrix.

We know that for systems with zero input

$$\underset{\sim}{z}(t) = \Phi(t)\underset{\sim}{z}(0)$$

Let $t = 0$. Then

$$\underset{\sim}{z}(0) = \Phi(0)\underset{\sim}{z}(0)$$

says that

$$\Phi(0) = I$$

which is the identity matrix. Let us differentiate

$$\underline{z}(t) = \Phi(t)\underline{z}(0)$$

with respect to time. We will get

$$\frac{d}{dt}\underline{z}(t) = \frac{d}{dt}\Phi(t)\underline{z}(0) = A\underline{z}(t)$$

Let us set $t = 0$ in the last equation to get

$$\frac{d}{dt}\Phi(0)\underline{z}(0) = A\underline{z}(0)$$

This last equation tells us that

$$\frac{d}{dt}\Phi(0) = A$$

We also noted that each term in the transition matrix is generated by the system poles. So, in general, for a 2×2 system with poles λ_1 and λ_2, the form of the transition matrix is

$$\Phi(t) = \begin{bmatrix} a_1 e^{\lambda_1 t} + a_2 e^{\lambda_2 t} & a_3 e^{\lambda_1 t} + a_4 e^{\lambda_2 t} \\ a_5 e^{\lambda_1 t} + a_6 e^{\lambda_2 t} & a_7 e^{\lambda_1 t} + a_8 e^{\lambda_2 t} \end{bmatrix}$$

where the constants a_1 through a_8 are to be determined.

6.6.6 Sixth Method: Using the General Form of $\Phi(t) = e^{At}$ and Its Properties

Let us consider an example to illustrate the procedure.

Let

$$A = \begin{bmatrix} 0 & 1 \\ -8 & -6 \end{bmatrix}$$

Since A is of dimension 2×2,

$$\Phi(t) = \begin{bmatrix} a_1 e^{\lambda_1 t} + a_2 e^{\lambda_2 t} & a_3 e^{\lambda_1 t} + a_4 e^{\lambda_2 t} \\ a_5 e^{\lambda_1 t} + a_6 e^{\lambda_2 t} & a_7 e^{\lambda_1 t} + a_8 e^{\lambda_2 t} \end{bmatrix} \tag{6.24}$$

We will use the properties of the transition matrix to find the constants a_1 to a_8.

$$\Phi(0) = \begin{bmatrix} a_1 + a_2 & a_3 + a_4 \\ a_5 + a_6 & a_7 + a_8 \end{bmatrix} = \begin{bmatrix} 1 & 0 \\ 0 & 1 \end{bmatrix} \tag{6.25}$$

$$\frac{d}{dt}\Phi(t) = \begin{bmatrix} \lambda_1 a_1 e^{\lambda_1 t} + \lambda_2 a_2 e^{\lambda_2 t} & \lambda_1 a_3 e^{\lambda_1 t} + \lambda_2 a_4 e^{\lambda_2 t} \\ \lambda_1 a_5 e^{\lambda_1 t} + \lambda_2 a_6 e^{\lambda_2 t} & \lambda_1 a_7 e^{\lambda_1 t} + \lambda_2 a_8 e^{\lambda_2 t} \end{bmatrix} \tag{6.26}$$

Evaluating at $t = 0$, we get

$$\frac{d}{dt}\Phi(0) = \begin{bmatrix} \lambda_1 a_1 + \lambda_2 a_2 & \lambda_1 a_3 + \lambda_2 a_4 \\ \lambda_1 a_5 + \lambda_2 a_6 & \lambda_1 a_7 + \lambda_2 a_8 \end{bmatrix} = \begin{bmatrix} 0 & 1 \\ -8 & -6 \end{bmatrix}$$

We will have eight algebraic equations to solve. These equations are

$$\begin{aligned} a_1 + a_2 &= 1 \\ a_3 + a_4 &= 0 \\ a_5 + a_6 &= 0 \\ a_7 + a_8 &= 1 \end{aligned} \tag{6.27}$$

and

$$\begin{aligned} \lambda_1 a_1 + \lambda_2 a_2 &= 0 \\ \lambda_1 a_3 + \lambda_2 a_4 &= 1 \\ \lambda_1 a_5 + \lambda_2 a_6 &= -8 \\ \lambda_1 a_7 + \lambda_2 a_8 &= -6 \end{aligned} \tag{6.28}$$

With λ_1 and λ_2 both at -2 and -4, we can solve Equation 1 of set (Equation 6.27) and Equation 1 of set (Equation 6.28) and so forth to get all the constant terms. The constants are $a_1 = 2$, $a_2 = -1$, $a_3 = 0.5$, $a_4 = -0.5$, $a_5 = -4$, $a_6 = 4$, $a_7 = -1$, and $a_8 = 2$. Finally the transition matrix $\Phi(t)$ is

$$\Phi(t) = \begin{bmatrix} 2e^{\lambda_1 t} - e^{\lambda_2 t} & 0.5e^{\lambda_1 t} - 0.5e^{\lambda_2 t} \\ -4e^{\lambda_1 t} + 4e^{\lambda_2 t} & -e^{\lambda_1 t} + 2e^{\lambda_2 t} \end{bmatrix}$$

Example 6.4
Consider the state-space system represented as

$$\frac{d}{dt}\underset{\sim}{z} = \begin{bmatrix} 0 & 1 \\ -6 & -5 \end{bmatrix}\underset{\sim}{z}$$

for the states and

$$\underset{\sim}{y} = \begin{bmatrix} 0 & 1 \\ 1 & 0 \end{bmatrix}\underset{\sim}{z}$$

for the outputs.
With the initial conditions

$$\underset{\sim}{z}(0) = \begin{bmatrix} 0 \\ 1 \end{bmatrix},$$

what are the outputs, y_1 and y_2? Calculate the state-transition matrix using the methods presented neglecting the numerical method.

Solution

1. Using the Cayley–Hamilton method
   Note that this A matrix is the same matrix we used when we presented the Cayley–Hamilton method. We will present the transition matrix here for completeness.

$$e^{At} = a_0 I + a_1 A$$

which is finally

$$e^{At} = \begin{bmatrix} 3e^{-2t} - 2e^{-3t} & e^{-2t} - e^{-3t} \\ -6e^{-2t} + 6e^{-3t} & -2e^{-2t} + 3e^{-3t} \end{bmatrix}$$

2. Using the inverse transform method
   First we find

$$(sI - A)^{-1} = \begin{bmatrix} s & -1 \\ 6 & s+5 \end{bmatrix}^{-1} = \frac{1}{(s^2 + 5s + 6)}\begin{bmatrix} s+5 & 1 \\ -6 & s \end{bmatrix}$$

which simplifies to

$$(sI - A)^{-1} = \begin{bmatrix} \frac{s+5}{(s+2)(s+3)} & \frac{1}{(s+2)(s+3)} \\ \frac{-6}{(s+2)(s+3)} & \frac{s}{(s+2)(s+3)} \end{bmatrix}$$

Using partial fraction expansion we can write

$$(sI - A)^{-1} = \begin{bmatrix} \frac{3}{(s+2)} - \frac{2}{(s+3)} & \frac{1}{(s+2)} - \frac{1}{(s+3)} \\ \frac{-6}{(s+2)} + \frac{6}{(s+3)} & \frac{-2}{(s+2)} + \frac{3}{(s+3)} \end{bmatrix}$$

Taking the inverse transform using partial fraction expansion we will get $\Phi(t)$ as

$$\Phi(t) = \begin{bmatrix} 3e^{-2t} - 2e^{-3t} & e^{-2t} - e^{-3t} \\ -6e^{-2t} + 6e^{-3t} & -2e^{-2t} + 3e^{-3t} \end{bmatrix}$$

3. Using the general form of the transition matrix and its properties.

Since A is of dimension 2×2,

$$\Phi(t) = \begin{bmatrix} a_1 e^{\lambda_1 t} + a_2 e^{\lambda_2 t} & a_3 e^{\lambda_1 t} + a_4 e^{\lambda_2 t} \\ a_5 e^{\lambda_1 t} + a_6 e^{\lambda_2 t} & a_7 e^{\lambda_1 t} + a_8 e^{\lambda_2 t} \end{bmatrix}$$

By evaluating at $t = 0$ we get

$$\Phi(0) = \begin{bmatrix} a_1 + a_2 & a_3 + a_4 \\ a_5 + a_6 & a_7 + a_8 \end{bmatrix} = \begin{bmatrix} 1 & 0 \\ 0 & 1 \end{bmatrix}$$

and

$$\frac{d}{dt}\Phi(0) = \begin{bmatrix} \lambda_1 a_1 + \lambda_2 a_2 & \lambda_1 a_3 + \lambda_2 a_4 \\ \lambda_1 a_5 + \lambda_2 a_6 & \lambda_1 a_7 + \lambda_2 a_8 \end{bmatrix} = \begin{bmatrix} 0 & 1 \\ -6 & -5 \end{bmatrix}$$

Again, we will have eight algebraic equations to solve. These equations are

$$\begin{aligned} a_1 + a_2 &= 1 \\ a_3 + a_4 &= 0 \\ a_5 + a_6 &= 0 \\ a_7 + a_8 &= 1 \end{aligned} \tag{6.29}$$

and

$$\begin{aligned}\lambda_1 a_1 + \lambda_2 a_2 &= 0 \\ \lambda_1 a_3 + \lambda_2 a_4 &= 1 \\ \lambda_1 a_5 + \lambda_2 a_6 &= -6 \\ \lambda_1 a_7 + \lambda_2 a_8 &= -5\end{aligned} \tag{6.30}$$

With λ_1 and λ_2 both at -2 and -3, we can solve Equation 1 of set (Equaton 6.29) and Equation 1 of set (Equation 6.30) and so forth to get all the constant terms. In doing that we find $a_1 = 3$, $a_2 = -2$, $a_3 = 1$, $a_4 = -1$, $a_5 = -6$, $a_6 = 6$, $a_7 = -2$, and $a_8 = 3$. With these values the transition matrix $\Phi(t)$ is obtained as

$$\Phi(t) = \begin{bmatrix} a_1 e^{\lambda_1 t} + a_2 e^{\lambda_2 t} & a_3 e^{\lambda_1 t} + a_4 e^{\lambda_2 t} \\ a_5 e^{\lambda_1 t} + a_6 e^{\lambda_2 t} & a_7 e^{\lambda_1 t} + a_8 e^{\lambda_2 t} \end{bmatrix} = \begin{bmatrix} 3e^{\lambda_1 t} - 2e^{\lambda_2 t} & e^{\lambda_1 t} - e^{\lambda_2 t} \\ -6e^{\lambda_1 t} + 6e^{\lambda_2 t} & -2e^{\lambda_1 t} + 3e^{\lambda_2 t} \end{bmatrix}$$

The solution for the states is

$$\underset{\sim}{z} = \begin{bmatrix} 3e^{-2t} - 2e^{-3t} & e^{-2t} - e^{-3t} \\ -6e^{-2t} + 6e^{-3t} & -2e^{-2t} + 3e^{-3t} \end{bmatrix} \begin{bmatrix} 0 \\ 1 \end{bmatrix} = \begin{bmatrix} e^{-2t} - e^{-3t} \\ -2e^{-2t} + 3e^{-3t} \end{bmatrix}$$

and for the outputs

$$\underset{\sim}{y} = \begin{bmatrix} 0 & 1 \\ 1 & 0 \end{bmatrix} \begin{bmatrix} e^{-2t} - e^{-3t} \\ -2e^{-2t} + 3e^{-3t} \end{bmatrix} = \begin{bmatrix} -2e^{-2t} + 3e^{-3t} \\ e^{-2t} - e^{-3t} \end{bmatrix}$$

For hand examples, the inverse Laplace method is the fastest for second order systems.

6.7 Some Insights: Poles and Stability

The objective of this chapter was to represent linear systems in state-space form and to look for ways of solving for the states. The process was to represent an nth order system (nth order differential equation) as n first order differential equations and arrange these equations in what we call state-space representation as

$$\begin{aligned}\frac{d}{dt}\underset{\sim}{z} &= A\underset{\sim}{z} + B\underset{\sim}{x} \\ \underset{\sim}{y} &= C\underset{\sim}{z} + D\underset{\sim}{x}\end{aligned}$$

where $\underset{\sim}{x}$ is the input vector (assuming multiple inputs) and $\underset{\sim}{y}$ is the output vector (assuming multiple outputs). The $\underset{\sim}{z}$ vector is the vector that contains the states of the system. The A matrix is the matrix that contains the parameters that control the dynamics of the system. As we saw in previous chapters, in every system representation there was a way to find the eigenvalues of the system. In state-space representation, the roots of the determinant of the matrix, $(sI - A)$, where I is the identity matrix, are the eigenvalues of the system, the poles. And as we mentioned before, these poles determine the shape of the transients of the system under investigation.

Consider the following single-input single-output linear second order differential equation that we discussed in Example 6.2 earlier in this chapter

$$\frac{d^2}{dt}y(t) + 3\frac{d}{dt}y(t) + 2y(t) = x(t)$$

with $x(t) = u(t)$ and initial conditions

$$\frac{d}{dt}y(0^-) = y(0^-) = 0$$

Putting this differential equation in state-space matrix form with $x(t) = u(t)$ we get

$$\frac{d}{dt}\underset{\sim}{z} = A\underset{\sim}{z} + B\underset{\sim}{x}$$

where

$$A = \begin{bmatrix} 0 & 1 \\ -2 & -3 \end{bmatrix}, \quad B = \begin{bmatrix} 0 \\ 1 \end{bmatrix}, \quad \frac{d}{dt}\underset{\sim}{z} = \begin{bmatrix} \frac{d}{dt}z_1 \\ \frac{d}{dt}z_2 \end{bmatrix}, \quad \underset{\sim}{z} = \begin{bmatrix} z_1 \\ z_1 \end{bmatrix}, \quad \text{and} \quad \underset{\sim}{x} = [x]$$

The roots of the determinant of $(sI - A)$ are calculated as

$$sI - A = s\begin{bmatrix} 1 & 0 \\ 0 & 1 \end{bmatrix} - \begin{bmatrix} 0 & 1 \\ -2 & -3 \end{bmatrix} = \begin{bmatrix} s & -1 \\ 2 & s+3 \end{bmatrix}$$

The determinant of the above matrix is

$$s^2 + 3s + 2 = (s+1)(s+2)$$

The roots of the above equation are at $s = -1$ and at $s = -2$. These are the eigenvalues or the poles of the system. Therefore, as we mentioned earlier, we expect a solution that contains the terms $c_1e^{-1t} + c_2e^{-2t}$. As you can see, this solution is stable (eigenvalues are all negative). Let us look at the solution we arrived at earlier

$$y(t) = \left(1 + \frac{1}{2}e^{-t} - e^{-2t}\right)u(t)$$

You can see that $c_1 = 1/2$ and $c_2 = -1$. These coefficients have no effect on the stability of the system. The extra term in the output $y(t)$ is the term $u(t)$. This term is due to the input $x(t)$. So if the input is bounded, and the eigenvalues are all negative, the output has to be stable; it dies as time progresses or settles at a certain bounded value, in this case $u(t)$.

To summarize, if the system is given in state-space form, the stability of the system is determined by finding the roots of $\det[sI - A]$. If the roots are all negative (in case of complex roots we look at the real part of the roots), the system is stable. If one of the roots is positive, the system is unstable. And again, the roots will determine the shape of the transients.

6.8 End-of-Chapter Examples

EOCE 6.1

Consider the following two matrices:

$$A = \begin{bmatrix} -14 & 4 \\ 1 & -18 \end{bmatrix}$$

and

$$B = \begin{bmatrix} 2 & 1 \\ 3 & 0 \end{bmatrix}$$

We will utilize MATLAB to find each of the following:

$A + B$, $B - A$, $23A$, $2(A + B)$, AB, BA, $2AB$, transpose(A), and inverse(A).

Solution

Let us write the following script

```
A = [-14  4;1 -18];
B = [2 1;3 0];
sum_A_and_B=A+B
```

```
Diff_B_A=B-A
A_times_23=23*A
A_plus_B_times_2=2*(A+B)
A_times_B=A*B
A_times_A=B*A
Two_times_A_times_B=2*(A*B)
transpose_A=A'
Inverse_A=inv(A)
```

The output is then

```
sum_A_and_B=
-12   5
4  -18
Diff_B_A=
16   -3
2   18
A_times_23=
-322   92
23 -414
A_plus_B_times_2=
-24   10
8  -36
A_times_B=
-16  -14
-52    1
B_times_A=
-27  -10
-42   12
Two_times_A_times_B=
-32  -28
-104  2
transpose_A=
-14    1
4  -18
Inverse_A=
-0.0726  -0.0161
-0.0040  -0.0565
```

In the above example, notice that AB is not equal to BA. This is to say that multiplication may not commute.

EOCE 6.2

Consider the two matrices given in EOCE 6.1 again.

$$A = \begin{bmatrix} -14 & 4 \\ 1 & -18 \end{bmatrix} \quad \text{and} \quad B = \begin{bmatrix} 2 & 1 \\ 3 & 0 \end{bmatrix}$$

Find the eigenvalues and eigenvectors of the above two matrices using MATLAB.

Solution

To do that we write the script

```
A = [-14 4;1 -18];
B = [2 1;3 0];
[VA,DA]=eig(A)%VA vector will contain eigenvectors
% DA will hold the eigenvalues
[VB,DB]=eig(B)
```

The output is

```
VA=
0.9792     -0.6380
0.2028     0.7701
DA=
-13.1716     0
0   -18.8284
VB=
0.7071     -0.3162
0.7071     0.9487
DB=
3   0
0   -1
```

As we can see, the eigenvalues for A are -13.17 and -18.82 and the eigenvectors are

$$\begin{bmatrix} .97 \\ .2 \end{bmatrix} \text{ and } \begin{bmatrix} -.63 \\ .77 \end{bmatrix}$$

EOCE 6.3

Consider the matrix B as given in EOCE 6.2.

Find e^{Bt}.

Solution

B is not diagonal. As discussed in this chapter, we need to find the matrix P that contains the eigenvectors of B, then we will have

$$e^{Bt} = Pe^{inv(P)BPt}P^{-1}$$

We can use MATLAB to find $P^{-1}BP$ as in the following script

```
B = [4 1;3 2];
[VB,DB] = eig(B)
P = VB;
Inverse_of_P = inv(P)
Diagonal_form_of_B = inv(P)*B*P
```

with the output

```
VB=
0.7071     -0.3162
0.7071     0.9487
DB=
5  0
0  1
Inverse_of_P=
1.0607    0.3536
-0.7906     0.7906
Diagonal_form_of_B=
5   0
0   1
```

We can see that the diagonal form of B is

$$P^{-1}BP = \begin{bmatrix} 5 & 0 \\ 0 & 1 \end{bmatrix}$$

and

$$e^{P^{-1}BPt} = \begin{bmatrix} e^{5t} & 0 \\ 0 & e^{t} \end{bmatrix}$$

$$e^{Bt} = Pe^{p^{-1}BPt}P^{-1} = \begin{bmatrix} .7071 & -.3162 \\ .7071 & .9487 \end{bmatrix} \begin{bmatrix} e^{5t} & 0 \\ 0 & e^{t} \end{bmatrix} \begin{bmatrix} 1.0607 & .3536 \\ -.7906 & .7906 \end{bmatrix}$$

$$e^{Bt} = \begin{bmatrix} .7071e^{5t} & -.3162e^{t} \\ 7071e^{5t} & .9487e^{t} \end{bmatrix} \begin{bmatrix} 1.0607 & .3536 \\ -.7906 & .7906 \end{bmatrix}$$

And finally

$$e^{Bt} = \begin{bmatrix} .75e^{5t} + .25e^{t} & .25e^{5t} - .25e^{t} \\ .75e^{5t} - .75e^{t} & .25e^{5t} + .75e^{t} \end{bmatrix}$$

EOCE 6.4

Consider the following state equations.

$$\frac{d}{dt}z_1 = z_2$$

$$\frac{d}{dt}z_2 = 6z_2 - 5z_1$$

We can see that these state equations are a result of a second order system with zero input, since no inputs appear to the right side of the two equations.

Assume that the initial conditions are $z_1(0) = 1$ and $z_2(0) = -1$. What are the states of this system?

Solution

1. Using the Laplace domain

   The general state-space solution in Laplace domain is

   $$\underline{Z}(s) = (sI - A)^{-1}\underline{z}(0) + (sI - A)^{-1}B\underline{X}(s)$$

   where A is the dynamic matrix of the system, B is the input matrix ($B = [0]$, no inputs), $\underline{X}$ is the input vector, and $\underline{Z}$ is the state vector.

   In this example

   $$A = \begin{bmatrix} 0 & 1 \\ -5 & 6 \end{bmatrix} \text{ and } B = \begin{bmatrix} 0 \\ 0 \end{bmatrix}$$

with the initial conditions

$$\underline{z}(0) = \begin{bmatrix} 1 \\ -1 \end{bmatrix}$$

Substituting these values in the state equation we get

$$\underline{Z}(s) = \frac{1}{s^2 - 6s + 5}\begin{bmatrix} s-6 & 1 \\ -5 & s \end{bmatrix}\begin{bmatrix} 1 \\ -1 \end{bmatrix} = \begin{bmatrix} \dfrac{s-7}{s^2 - 6s + 5} \\ \dfrac{-5-s}{s^2 - 6s + 5} \end{bmatrix}$$

Finally, the states are

$$\underline{Z}(s) = \begin{bmatrix} \dfrac{-1/2}{s-5} + \dfrac{3/2}{s-1} \\ \dfrac{-5/2}{s-5} + \dfrac{3/2}{s-1} \end{bmatrix}$$

If we take the inverse transform we get

$$\underline{z}(t) = \begin{bmatrix} z_1(t) \\ z_2(t) \end{bmatrix} = \begin{bmatrix} -0.5e^{5t} & +1.5e^{t} \\ -2.5e^{5t} & +1.5e^{t} \end{bmatrix} t \geq 0$$

2. Using the time domain

Since A is not in the diagonal form we need to use the following form of the solution

$$\underline{w}(t) = \exp[P^{-1}APt]\underline{w}(0) + \int_0^t \exp[P^{-1}AP(t-\tau)]P^{-1}Bx(\tau)d\tau$$

where $\underline{z}(t) = P\underline{w}(t)$ and $\underline{w}(0) = P^{-1}\underline{z}(0)$. Since B is a zero matrix

$$\underline{z}(t) = P\exp[P^{-1}APt]P^{-1}\underline{z}(0)$$

To solve for $\underline{z}(t)$ we need to find P, P^{-1}, and $P^{-1}AP$. We do that using the MATLAB script

```
A=[0 1;-5 6];
[VA,DA]=eig(A);
P=VA
Pinverse=inv(P)
PinverseAP=inv(P)*A*P
```

The output is

```
P =
-0.7071 -0.1961
-0.7071 -0.9806
Pinverse =
-1.7678      0.3536
1.2748 -1.2748
PinverseAP =
1.0000  0.0000
0   5.0000
```

Therefore, the solution for the states is

$$\underline{Z}(t) = \begin{bmatrix} -0.7071 & -0.1961 \\ -0.7071 & -0.9806 \end{bmatrix} \begin{bmatrix} e^{t} & 0 \\ 0 & e^{5t} \end{bmatrix} \begin{bmatrix} -2.1213 \\ 2.5495 \end{bmatrix}$$

$$\underline{Z}(t) = \begin{bmatrix} -0.7071 & -0.1961 \\ -0.7071 & -0.9806 \end{bmatrix} \begin{bmatrix} -2.1213e^{t} \\ 2.5495e^{5t} \end{bmatrix} t \geq 0$$

Finally

$$\begin{bmatrix} z_1(t) \\ z_2(t) \end{bmatrix} = \begin{bmatrix} 1.5e^{t} - 0.5e^{5t} \\ 1.5e^{t} - 2.5e^{5t} \end{bmatrix} t \geq 0$$

EOCE 6.5

Consider the mechanical system in Figure 6.5.

The rod to the left of the mass M can be modeled by a translational spring with spring constant K, in parallel with a translational damper of constant B. Assume the mass moves freely. As seen before, the differential equation representing the system is

$$M\frac{d^2}{dt}y(t) + B\frac{d}{dt}y(t) + Ky(t) = x(t)$$

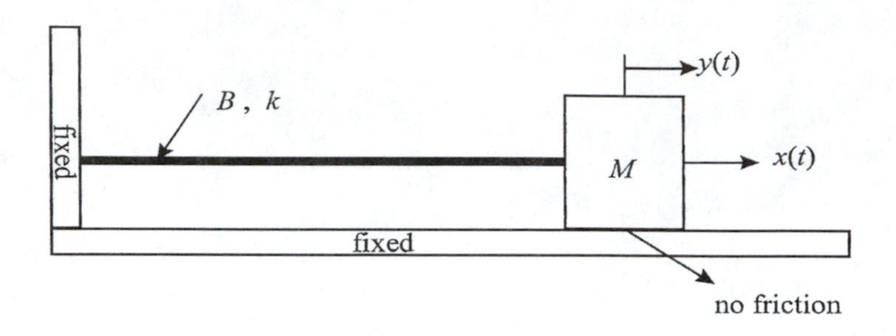

FIGURE 6.5
System for EOCE 6.5.

In a general form we can write the above equation as

$$\frac{d^2}{dt}y(t) + 2\xi w_n \frac{d}{dt} y(t) + w_n^2 y(t) = \frac{x(t)}{M}$$

The characteristic equation is

$$m^2 + 2\xi w_n m + w_n^2 = 0$$

with roots at

$$m_{1,2} = -\xi w_n \pm w_n\sqrt{\xi^2 - 1}$$

We call ξ the damping ratio where

$$\xi = \frac{B}{2\sqrt{KM}}$$

In our example

$$w_n = \sqrt{\frac{K}{M}}$$

We have critical damping if $\xi = 1$, underdamping if ξ is less than one and greater than zero, and overdamping if $\xi > 1$.
Let $M = 1$ kg and $K = 40$ newtons/meter.
For critical damping, $\xi = 1$, and $w_n = \sqrt{\frac{K}{M}} = 6.3246$.
For an underdamped case, let $\xi = 0.5$ while keeping the same value for w_n.
For the overdamped case let $\xi = 2$, while keeping the same value for w_n as well.

Assume zero initial conditions and plot the displacement $y(t)$ vs. time for all cases above if $x(t)$ is a constant force of 1 newton.

Solution

We will use state equations to solve this problem using MATLAB.

Let $z_1(t) = y(t)$ and $z_2(t) = \frac{d}{dt}y(t)$ to get the state equations as

$$\frac{d}{dt}\underset{\sim}{z}(t) = \begin{bmatrix} 0 & 1 \\ -w_n^2 & -2\xi w_n \end{bmatrix}\underset{\sim}{z}(t) + \begin{bmatrix} 0 \\ \frac{1}{M} \end{bmatrix}\underset{\sim}{x}(t)$$

with the matrices

$$A = \begin{bmatrix} 0 & 1 \\ -w_n^2 & -2\xi w_n \end{bmatrix} \text{ and } B = \begin{bmatrix} 0 \\ \frac{1}{M} \end{bmatrix}$$

and the initial conditions

$$\underset{\sim}{z}(0) = \begin{bmatrix} 0 \\ 0 \end{bmatrix}$$

A MATLAB function will be written to define the state equation as

```
function zdot=eoce5(t,z)
global A
global B
zdot=A*z+B;
```

Then a script is written that we will call the "ode23" function, which is a MATLAB function with initial conditions and initial and final values for the time needed for the simulation. The script is

```
clf % to clear the graph area
global A % declaring the dynamic matrix A as global
global B
k=40;
m=1;
t0=0;
```

```
tf=10;
z0=[0;0]; % vector of initial conditions
wn=sqrt(k/m);
zeta=[0.5 2 1]; % 3 values for the damping zeta
for i=1:3
A=[0 1;-wn^2 -2*zeta(i)*wn];
B=[0;1/m];
[t,z]=ode23('eoce5',t0,tf,z0); % a call to the ode23
y1=z(:,1);
plot(t,y1);
hold on % to hold for plotting the second graph
end
xlabel('Time in sec)')
ylabel('The output y(t)')
title('The output y(t) for different damping values')
gtext('zeta=0.5: underdamped case') % to label the plot
gtext('zeta=2: overdamped case')
gtext('zeta=1: critically damped case')
```

The plots are given in Figure 6.6.

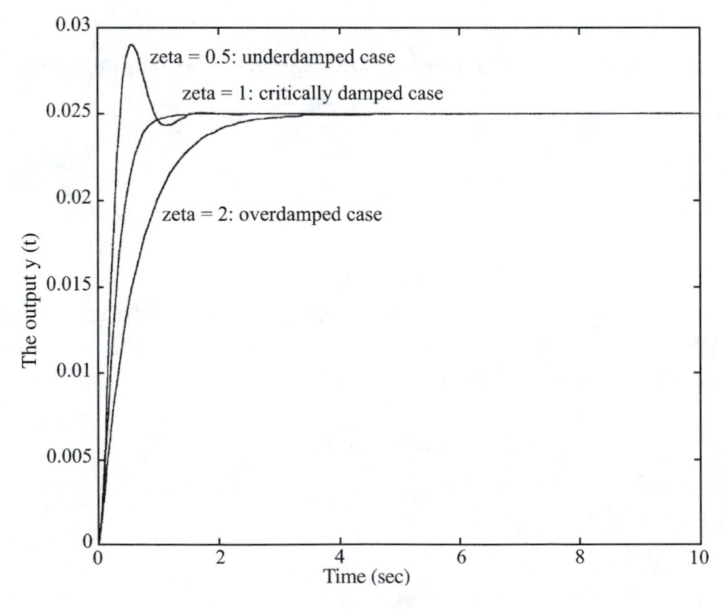

FIGURE 6.6
Plots for EOCE 6.5.

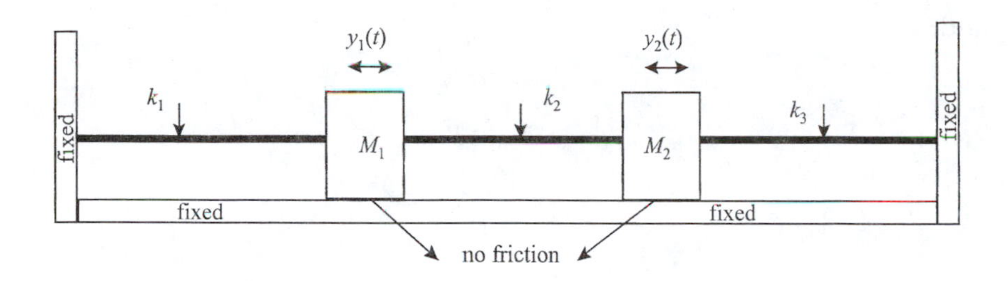

FIGURE 6.7
System for EOCE 6.6.

EOCE 6.6

Consider the system in Figure 6.7.

The rod to the left of the mass M_1 can be modeled by a translational spring with spring constant k_1. The rod to the right of the mass M_1 can be modeled by a translational spring with spring constant k_2. The rod to the right of the mass M_2 can be modeled by a translational spring with spring constant k_3.

Find the displacements $y_1(t)$ and $y_2(t)$ for different initial conditions.

Solution

The system has two degrees of freedom (movement of M_1 and M_2) and it will have two coupled second order differential equations. Each differential equation can be represented as two first order differential equations. Hence the system given is a system with four state variables. We can use Newton's law to write the two coupled differential equations as

$$M_1\frac{d^2}{dt}y_1 + k_1y_1 + k_2y_1 - k_2y_2 = 0$$

$$M_2\frac{d^2}{dt}y_2 - k_2y_1 + k_2y_2 + k_3y_2 = 0$$

If we let all the spring constants be 1 and the values of the masses be 1, we will have

$$\frac{d^2}{dt}y_1 + y_1 + y_1 - y_2 = 0$$

$$\frac{d^2}{dt}y_2 - y_1 + y_2 + y_2 = 0$$

If we now let

$$z_1 = y_1$$

$$z_2 = y_2$$

and

$$z_3 = \frac{d}{dt}y_1$$
$$z_4 = \frac{d}{dt}y_2$$

then the following state equation is obtained

$$\frac{d}{dt}\underline{z} = \begin{bmatrix} \frac{d}{dt}z_1 \\ \frac{d}{dt}z_2 \\ \frac{d}{dt}z_3 \\ \frac{d}{dt}z_4 \end{bmatrix} = \begin{bmatrix} 0 & 0 & 1 & 0 \\ 0 & 0 & 0 & 1 \\ -2 & 1 & 0 & 0 \\ 1 & -2 & 0 & 0 \end{bmatrix} \underline{z}$$

We will let the system vibrate at its natural frequencies with specified initial conditions. The natural frequencies can be calculated by using the MATLAB command to find the eigenvalues of the system as

```
A=[0 0 1 0;0 0 0 1;-2 1 0 0;1 -2 0 0];
poles=eig(A)
```

with the output

```
poles =
0.0000+ 1.7321i
0.0000- 1.7321i
0.0000+ 1.0000i
0.0000- 1.0000i
```

Thus, the natural frequencies are at 1 and 1.7321 rad/sec. We will use MATLAB to study the response of the system for different initial conditions. We will start writing the function that defines the system as

```
function zdot=eoce6(t,z)
global A
zdot=A*z;
```

Next, we write a script that we will call the MATLAB "ode23" function and the function that defines the system. Since we will have different initial conditions, we will write a very general script that will communicate with the user.

```
global A
A=[0 0 1 0;0 0 0 1;-2 1 0 0;1 -2 0 0];
t_initial=input('Initial time for simulation = ')
t_final=input('Final time for simulation = ')
z0=input('[z1(t_initial) z2(t_initial)
dz1(t_initial)  dz2(t_initial)]= ')
z0=z0';
[t,z]=ode23('eoce6' t_initial,t_final,z0);
z1=z(:,1);
z2=z(:,2);
plot_title=input('Title= ','s')
subplot(2,1,1),plot(t,z1)
ylabel('Displacementz1')
subplot(2,1,2),plot(t,z2)
ylabel('Displacement  z2')
title(eval('plot_title'))
```

We will first run the script above and let the masses M_1 and M_2 move two units to the right. In this case, the motions of M_1 and M_2 are in-phase. The following is a sample dialogue obtained by typing "vibration" at the MATLAB prompt, where "vibration" is the name of the script just written.

When we type "vibration" and hit the Enter key, the script will display the following dialogue on the screen

```
Initial time for simulation = 0
t_initial =
0
Final time for simulation = 10
t_final =
10
[z1(t_initial) z2(t_initial) dz1(t_initial)
dz2(t_initial)] = [2 2 0 0]
z0 =
2 2 0 0
Title = The two movements are in-phase: M1 and M2
are moved each 2 units to the right
```

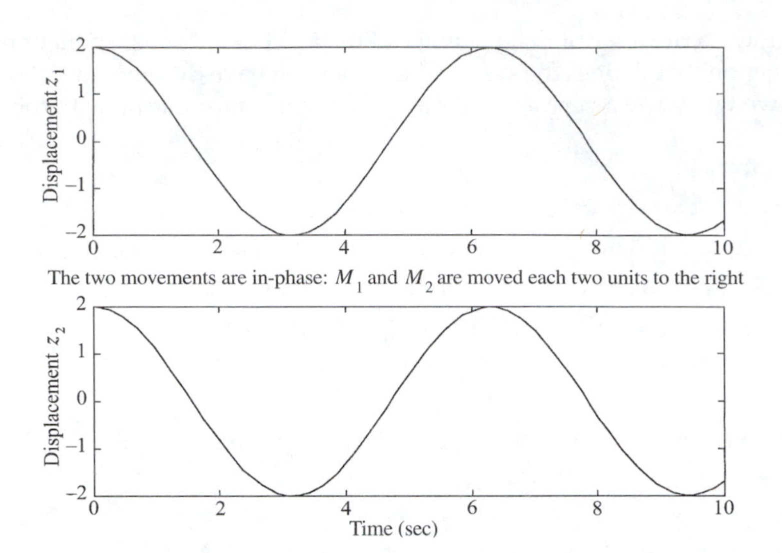

FIGURE 6.8
Plots for EOCE 6.6.

```
plot_title =
The two movements are in-phase: M1 and M2 are moved
each 2 units to the right
```

The plots are shown in Figure 6.8.

In this first case, the central spring with the spring constant k_2 is not stretched or compressed. Since the initial displacements are equal, the output displacements are in-phase and the fourth order system behaves as a second order system.

Now let us try moving M_1 two units to the right and moving M_2 two units to the left at the same time and then release the masses. This is done by setting $z_1(0) = 2$ and $z_2(0) = -2$.

This is another dialogue with MATLAB and we initiate it by typing "vibration" at the MATLAB prompt and then pressing the Enter key to see the following on the screen.

```
Initial time for simulation = 0
t_Initial =
0
Final time for simulation = 10
t_final =
10
```

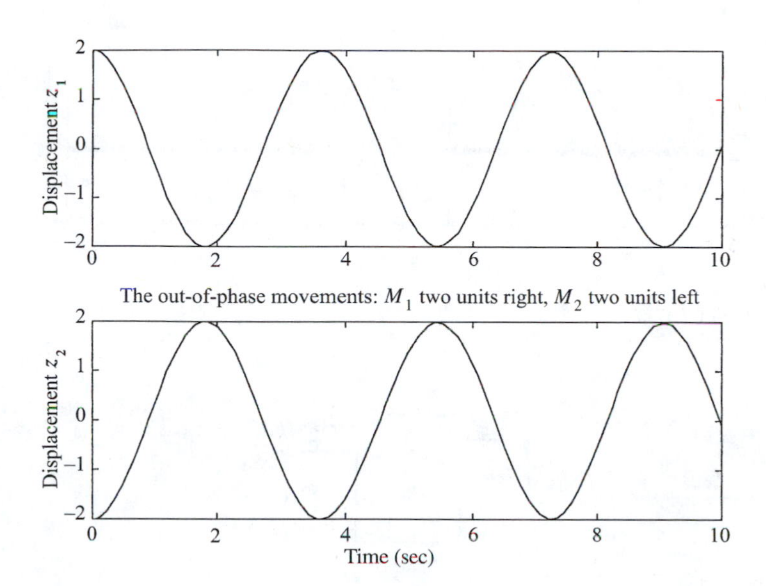

FIGURE 6.9
Plots for EOCE 6.6.

```
z1(t_initial) z2(t_initial) dz1(t_initial)
dz2(t_initial)] = [2 -2 0 0]

z0 =
2 -2 0 0
Title = The out-of-phase movements: M1 two units
right, M2 two units left
plot_title =
The out-of-phase movements: M1 two units right, M2
two units left
```

The plots are shown in Figure 6.9.

In this second case, the central spring is compressed and all spring constants are involved. This example is constructed for equal k and M values. You may experiment with the functions presented using various values for the masses, the spring constants, and the initial conditions.

EOCE 6.7

Consider the system in Figure 6.10.

The rod to the left of the mass M can be modeled by a translational spring with spring constant k_1, in parallel with a translational damper of constant B_1.

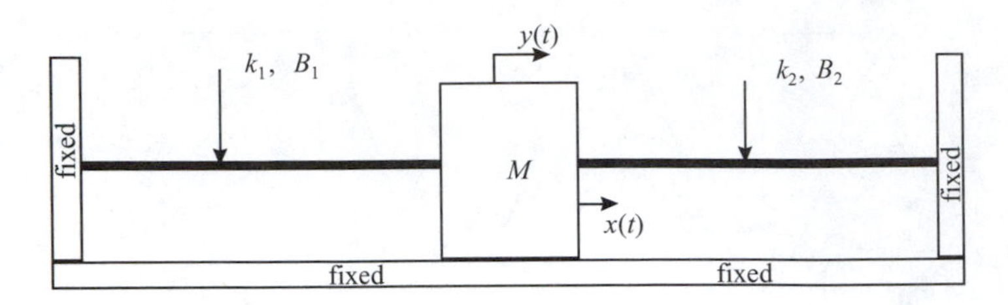

FIGURE 6.10
System for EOCE 6.7.

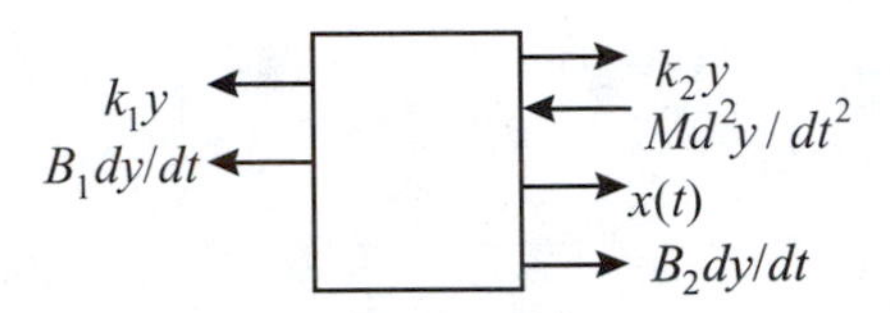

FIGURE 6.11
Free-body diagram for EOCE 6.7.

The rod to the right of the mass M can be modeled by a translational spring with spring constant k_2, in parallel with a translational damper of constant B_2.

Find the displacement $y(t)$ due to an input $x(t)$.

Solution

The free-body diagram that describes the system is shown in Figure 6.11 where $x(t)$ is the input displacement and $y(t)$ is the output displacement. k_1 and k_2 are the stiffness elements and B_1 and B_2 are the friction elements. The differential equation describing the system is

$$M\frac{d^2}{dt}y + (B1 + B2)\frac{d}{dt}y + (k1 + k2)y = x$$

If we let $k = k_1 + k_2$ and $B = B_1 + B_2$, the differential equation becomes

$$M\frac{d^2}{dt}x + B\frac{d}{dt}x + kx = x$$

This system is similar to the system given previously in EOCE 6.5. Let $M = 1$ and $k = 40$ and $B = 12$. Let $z_1(t) = y(t)$ and $z_2(t) = \frac{d}{dt}y(t)$. Therefore the state equation is

$$\frac{d}{dt}\underline{z}(t) = \begin{bmatrix} 0 & 1 \\ -40 & -12 \end{bmatrix}\underline{z}(t) + \begin{bmatrix} 0 \\ 1 \end{bmatrix}\underline{x}(t)$$

and the output equation is

$$y = \begin{bmatrix} 1 & 0 \end{bmatrix} \begin{bmatrix} z_1 \\ z_2 \end{bmatrix} + \begin{bmatrix} 0 \end{bmatrix} x$$

with the matrices

$$A = \begin{bmatrix} 0 & 1 \\ -40 & -12 \end{bmatrix}, \; B = \begin{bmatrix} 0 \\ 1 \end{bmatrix}, \; C = \begin{bmatrix} 1 & 0 \end{bmatrix}, \text{ and } D = \begin{bmatrix} 0 \end{bmatrix}$$

and the initial condition vector

$$\underset{\sim}{z}(0) = \begin{bmatrix} 0 \\ 0 \end{bmatrix}$$

where we assumed that the system is relaxed (zero initial conditions).

The solution for state equation is

$$\underset{\sim}{z}(t) = \int_0^t \exp[A(t-\tau)]Bx(\tau)d\tau$$

where e^{At} is the inverse Laplace transform of $(sI - A)^{-1}$.

$$(sI - A)^{-1} = \begin{bmatrix} s & -1 \\ 40 & s+12 \end{bmatrix} = \frac{1}{(s^2 + 12s + 40)} \begin{bmatrix} s+12 & 1 \\ -40 & s \end{bmatrix}$$

$$(sI - A)^{-1} = \begin{bmatrix} \dfrac{s+12}{(s^2 + 12s + 40)} & \dfrac{1}{(s^2 + 12s + 40)} \\ \dfrac{-40}{(s^2 + 12s + 40)} & \dfrac{s}{(s^2 + 12s + 40)} \end{bmatrix}$$

Finally

$$(sI - A)^{-1} = \begin{bmatrix} \dfrac{s+6+6}{(s+6)^2 + 2^2} & \dfrac{1}{(s+6)^2 + 2^2} \\ \dfrac{-40}{(s+6)^2 + 2^2} & \dfrac{s+6-6}{(s+6)^2 + 2^2} \end{bmatrix}$$

The inverse transform of the last matrix is

$$\begin{bmatrix} e^{-6t}\cos(2t) + 3e^{-6t}\sin(2t) & 1/2(e^{-6t}\sin(2t)) \\ -20e^{-6t}\sin(2t) & e^{-6t}\cos(2t) - 3e^{-6t}\sin(2t) \end{bmatrix} = e^{At} = \Phi(t)$$

Therefore, using the transition matrix we can write the solution for the state vector as

$$\underset{\sim}{z}(t) = \int_0^t \exp[A(t-\tau)]Bx(\tau)d\tau$$

$$\underset{\sim}{z}(t) = \int_0^t \begin{bmatrix} e^{-6(t-\tau)}\cos(2(t-\tau)) + 3e^{-6(t-\tau)}\sin(2(t-\tau)) \\ -20e^{-6(t-\tau)}\sin(2(t-\tau)) \end{bmatrix} \delta(\tau)d\tau$$

and finally

$$\underset{\sim}{z}(t) = \begin{bmatrix} e^{-6t}\cos(2t) + 3e^{-6t}\sin(2t) \\ -20e^{-6t}\sin(2t) \end{bmatrix}$$

The last result was obtained using the sifting property of the impulse function. The output equation $y(t)$ is

$$y = \begin{bmatrix} 1 & 0 \end{bmatrix} \begin{bmatrix} z_1 \\ z_2 \end{bmatrix} = [e^{-6t}\cos(2t) + 3e^{-6t}\sin(2t)]u(t)$$

We can see that the system is stable because the real part of the poles is negative and at -6. We can use MATLAB to find the impulse response of the system and comparing it with the analytical solution that we just found. To do that we write the script

```
clf
A=[0 1;-40 -12];
B=[1;0];
C=[1 0];
D=[0];
[z y t]=impulse(A,B,C,D);
hold on
y_analytical=exp(-6*t).*cos(2*t)+3*exp
(-6*t).*sin(2*t);
plot(t,z(:,1),'*',t,y_analytical,'o')% z(:,1)means z1
gtext('* analytical, o simulation');
title('analytical and MATLAB simulation for eoce7')
xlabel('Time (sec)')
ylabel('Amplitudes')
```

The output is shown in Figure 6.12.

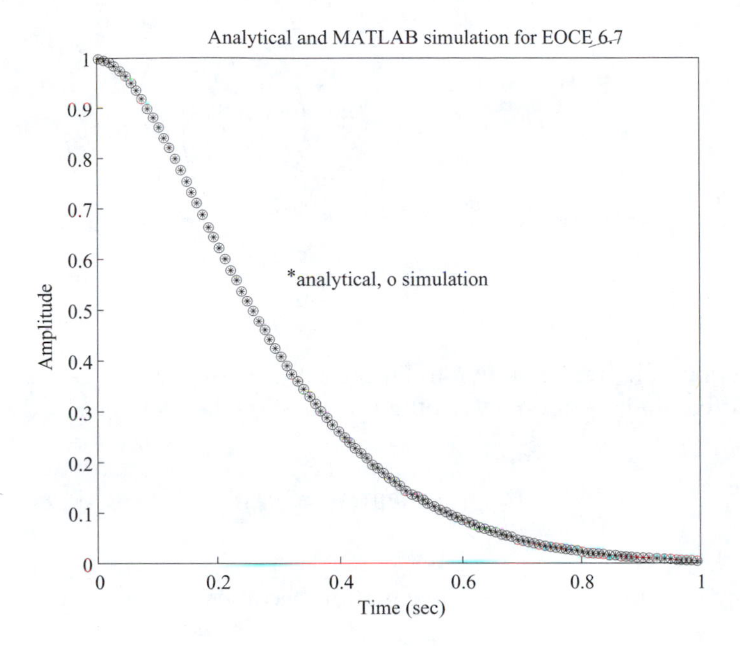

FIGURE 6.12
Plots for EOCE 6.7.

EOCE 6.8

Consider the system shown in Figure 6.13.

Where $\theta(t)$ is the rotational angle, J represents the moment of inertia where the mass of the rod is concentrated, k is the stiffness coefficient, w is the angular velocity, and τ is the torque applied. Assume the system was relaxed before applying the torque. Find the rotational angle resulting from the applied torque.

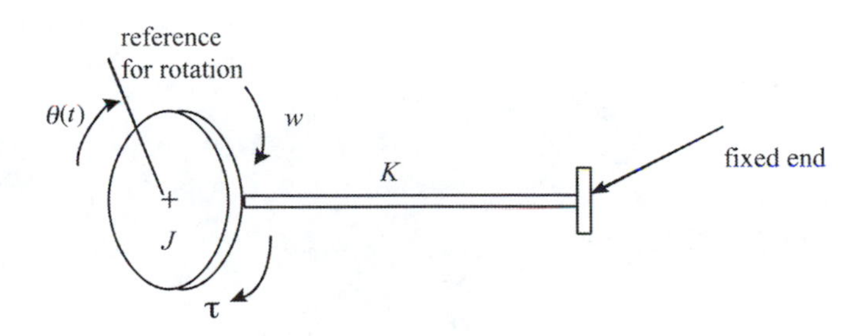

FIGURE 6.13
System for EOCE 6.8.

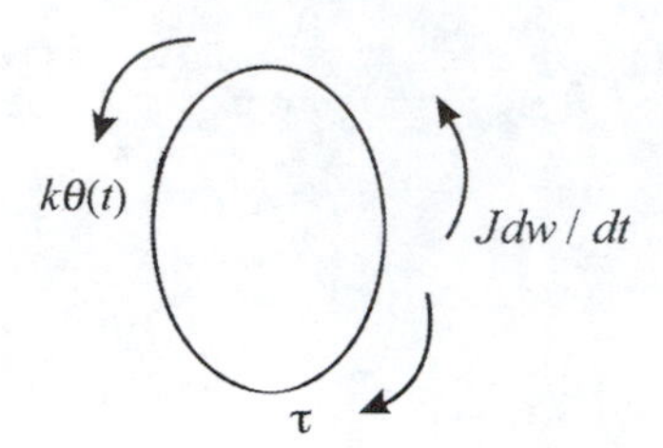

FIGURE 6.14
Free-body diagram for EOCE 6.8.

Solution

The free-body diagram is shown in Figure 6.14.

The differential equation describing this rotational system is

$$J\frac{d}{dt}w(t) + k\theta(t) = \tau(t)$$

Rewriting as a function of $\theta(t)$, the equation becomes

$$J\frac{d^2}{dt}\theta(t) + k\theta(t) = \tau(t)$$

This is a second order differential equation in $\theta(t)$, where $\tau(t)$ is the input torque.

If we let $z_1 = \theta(t)$, and $z_2 = \frac{d}{dt}\theta(t)$, then we can write the system in state-space as

$$\frac{d}{dt}\underline{z}(t) = \begin{bmatrix} 0 & 1 \\ -k/J & 0 \end{bmatrix}\underline{z}(t) + \begin{bmatrix} 0 \\ 1/J \end{bmatrix}\underline{x}(t)$$

and the output equation as

$$y = \begin{bmatrix} 1 & 0 \end{bmatrix}\begin{bmatrix} z_1 \\ z_2 \end{bmatrix} + \begin{bmatrix} 0 \end{bmatrix}x$$

The initial condition vector is

$$\underline{z}(0) = \begin{bmatrix} 0 \\ 0 \end{bmatrix}$$

If the input torque is an impulse of unity strength, $J = 1$ kg.m^2, and $k = 1$ newton.meter/rad, we can find the displacement angle $\theta(t)$.

For the given values the system state equation becomes

$$\frac{d}{dt}\underset{\sim}{z}(t) = \begin{bmatrix} 0 & 1 \\ -1 & 0 \end{bmatrix}\underset{\sim}{z}(t) + \begin{bmatrix} 0 \\ 1 \end{bmatrix}\underset{\sim}{x}(t)$$

and the output equation becomes

$$y = \begin{bmatrix} 1 & 0 \end{bmatrix}\begin{bmatrix} z_1 \\ z_2 \end{bmatrix} + \begin{bmatrix} 0 \end{bmatrix} x$$

with the initial conditions

$$\underset{\sim}{z}(0) = \begin{bmatrix} 0 \\ 0 \end{bmatrix}$$

e^{At} is the inverse Laplace transform of $(sI - A)^{-1}$ and is calculated next.

$$(sI - A)^{-1} = \begin{bmatrix} s & -1 \\ 1 & s \end{bmatrix}^{-1} = \frac{1}{(s^2 + 1)}\begin{bmatrix} s & 1 \\ -1 & s \end{bmatrix}$$

$$(sI - A)^{-1} = \begin{bmatrix} \frac{s}{(s^2 + 1)} & \frac{1}{(s^2 + 1)} \\ \frac{-1}{(s^2 + 1)} & \frac{s}{(s^2 + 1)} \end{bmatrix}$$

$$e^{At} = \Phi(t) = \begin{bmatrix} \cos(t) & \sin(t) \\ -\sin(t) & \cos(t) \end{bmatrix}$$

The solution for the states now is calculated as

$$\underset{\sim}{z}(t) = \int_0^t \exp[A(t - \tau)]Bx(\tau)d\tau$$

With $\Phi(t)$ as calculated above

$$\underline{z}(t) = \int_0^t \begin{bmatrix} \cos(t-\tau) & \sin(t-\tau) \\ -\sin(t-\tau) & \cos(t-\tau) \end{bmatrix} \begin{bmatrix} 1 \\ 0 \end{bmatrix} \delta(\tau) d\tau$$

$$\underline{z}(t) = \int_0^t \begin{bmatrix} \cos(t-\tau) \\ -\sin(t-\tau) \end{bmatrix} \delta(\tau) d\tau$$

Using the sifting property we have the displacement and the velocity angles as

$$\underline{z}(t) = \begin{bmatrix} \theta(t) \\ \frac{d}{dt}\theta(t) \end{bmatrix} = \begin{bmatrix} \cos(t) \\ -\sin(t) \end{bmatrix}$$

We can use MATLAB to display the angular displacement and the angular velocity by writing the following script.

```
clf
A=[0 1;-1 0];
B=[1;0];
C=[1 0];
D=[0];
[y z t]=impulse(A,B,C,D);%return the states and the
outputs
theta_analytical=cos(t);
omega_analytical=-sin(t);
plot(t,z(:,1),'*',t,theta_analytical,'o',t,z(:,2),
t,omega_analytical,'o')
gtext('*displacement');
gtext('o velocity');
title('analytical and MATLAB simulation for EOCE
6.8')
xlabel('Time in sec)')
ylabel('Amplitudes')
```

The output plot is shown in Figure 6.15.

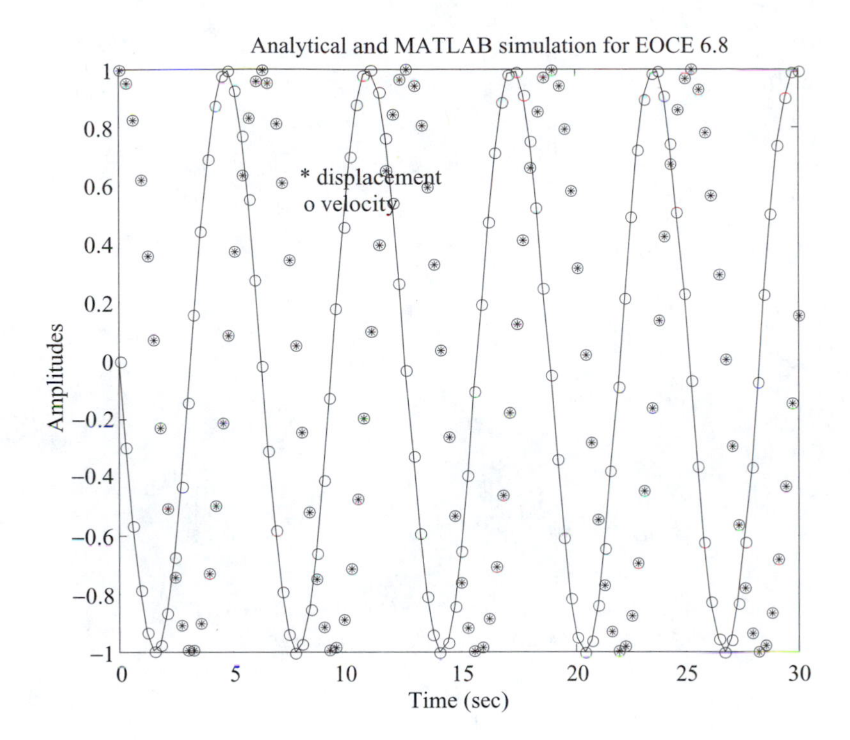

FIGURE 6.15
Plots for EOCE 6.8.

EOCE 6.9

Consider the thermal system shown in Figure 6.16 where $x(t)$ is the rate at which heat is supplied to the left capacitor. R_1 and R_2 are the thermal resistors, T_1 and T_2 are the temperatures in the two sections as shown, and T is the outside temperature. Assume, excluding R_1 and R_2, that the system is insulated.

Let T_1 and T_2 be the two state variables. Find T_1 and T_2.

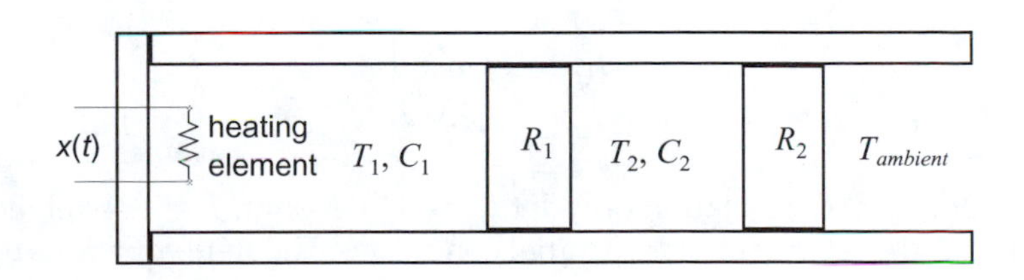

FIGURE 6.16
System for EOCE 6.9.

Solution

Using the general relation

$$\frac{d}{dt}T = \frac{1}{C}(rate\_heat\_in - rate\_heat\_out)$$

we can write

$$\frac{d}{dt}T_1 = \frac{1}{C_1}\left[x(t) - \frac{1}{R_1}(T_1 - T_2)\right]$$

and

$$\frac{d}{dt}T_2 = \frac{1}{C_2}\left[\frac{1}{R_1}(T_1 - T_2) - \frac{1}{R_2}(T_2 - T)\right]$$

As you may notice, this system has two inputs, T and $x(t)$.

Let $z_1 = T_1$ and $z_2 = T_2$. Then the state equation is

$$\frac{d}{dt}\underline{z}(t) = \begin{bmatrix} \dfrac{-1}{C_1R_1} & \dfrac{1}{C_1R_1} \\ \dfrac{1}{C_2R_1} & \dfrac{-(R_2 + R_1)}{C_2R_2R_1} \end{bmatrix}\underline{z}(t) + \begin{bmatrix} 1/C_1 & 0 \\ 0 & 1/R_2C_2 \end{bmatrix}\begin{bmatrix} x(t) \\ T \end{bmatrix}$$

and the output equation is

$$y = \begin{bmatrix} 1 & 0 \end{bmatrix}\begin{bmatrix} z_1 \\ z_2 \end{bmatrix} + \begin{bmatrix} 0 \end{bmatrix} x$$

with the initial condition

$$\underline{z}(0) = \begin{bmatrix} z_1(0) \\ z_2(0) \end{bmatrix}$$

For the purpose of simulation only, let $R_1 = R_2 = 1$ ohm, $T = 104$ kelvin, $C_1 = C_2 = 1$, and $x(t) = 0$. Assume zero initial conditions. The state equations become

$$\frac{d}{dt}\underline{z}(t) = \begin{bmatrix} -1 & 1 \\ 1 & -2 \end{bmatrix}\underline{z}(t) + \begin{bmatrix} 0 \\ 104 \end{bmatrix}$$

and the output becomes

$$y = \begin{bmatrix} 1 & 0 \end{bmatrix} \begin{bmatrix} z_1 \\ z_2 \end{bmatrix} + \begin{bmatrix} 0 \end{bmatrix} x$$

with the initial conditions

$$\underline{z}(0) = \begin{bmatrix} 0 \\ 0 \end{bmatrix}$$

The eigenvalues of this system are found by setting $(\det(sI - \mathrm{A})) = 0)$, or by using MATLAB as in the following.

```
A=[-1 1;1 -2];
eig(A)
```

The output is then

```
-0.3820
-2.6180
```

e^{At} is the inverse Laplace transform of $(sI - A)^{-1}$ and is needed to find the solution of the state equation.

$$(sI - A)^{-1} = \begin{bmatrix} s+1 & -1 \\ -1 & s+2 \end{bmatrix}^{-1} = \frac{1}{(s^2 + 3s + 1)} \begin{bmatrix} s+2 & 1 \\ 1 & s+1 \end{bmatrix}$$

$$(sI - A)^{-1} = \begin{bmatrix} \dfrac{s+2}{(s^2 + 3s + 1)} & \dfrac{a}{(s^2 + 3s + 1)} \\ \dfrac{1}{(s^2 + 3s + 1)} & \dfrac{s+1}{(s^2 + 4s + 1)} \end{bmatrix}$$

Using partial fraction expansion we get

$$(sI - A)^{-1} = \begin{bmatrix} \dfrac{0.2764}{s+2.618} + \dfrac{0.7236}{s+0.382} & \dfrac{-0.4472}{s+2.618} + \dfrac{0.4472}{s+0.382} \\ \dfrac{-0.4472}{s+2.618} + \dfrac{0.4472}{s+0.382} & \dfrac{0.7236}{s+2.618} + \dfrac{0.2764}{s+0.382} \end{bmatrix}$$

The inverse transform of the last matrix is the state transition matrix.

$$\begin{bmatrix} 0.2764e^{-2.618t} + 0.7236e^{-0.382t} & -0.4472e^{-2.618t} + 0.4472e^{-0.382t} \\ -0.4472e^{-2.618t} + 0.4472e^{-0.382t} & 0.7236e^{-2.618t} + 0.2764e^{-0.382t} \end{bmatrix} = e^{At} = \Phi(t)$$

The state solution is

$$\underline{z}(t) = \int_0^t \exp[A(t-\tau)]Bx(\tau)d\tau$$

With $\Phi(t)$ substituted in the above convolution integral we have

$$\underline{z}(t) = \int_0^t \begin{bmatrix} 104(-0.4472e^{-2.618(t-\tau)} + 0.4472e^{-0.382(t-\tau)}) \\ 104(0.7236e^{-2.618(t-\tau)} + 0.2764e^{-0.382(t-\tau)}) \end{bmatrix} d\tau$$

$$\underline{z}(t) = \begin{bmatrix} 104[0.1708(e^{-2.618t} - 1) - 1.1707(e^{-0.382t} - 1)] \\ 104[-0.2764(e^{-2.618t} - 1) - 0.7236(e^{-0.382t} - 1)] \end{bmatrix}$$

where

$$z_1(t) = T_1(t) = 104(0.1708(e^{-2.618t} - 1) - 1.1707(e^{-0.382t} - 1)).$$

We can plot the temperature in the first capacitor using MATLAB by typing the script

```
clf
A = [-1 1;1 -2];
B = [0;104];
C = [1 0];
D = [0];
[y z t]=step(A,B,C,D);
T1 = 104*0.1708*(exp(-2.618*t)-1)-104*1.1707*
(exp(-0.382*t) -1);
plot(t,y(:,1),'*',t,T1,'-');
title('the step response for EOCE 6.9')
ylabel('Temperature within capacitance C1')
xlabel('Time (sec)')
gtext('* MATLAB simulation, - analytical result')
```

The output is shown in Figure 6.17.

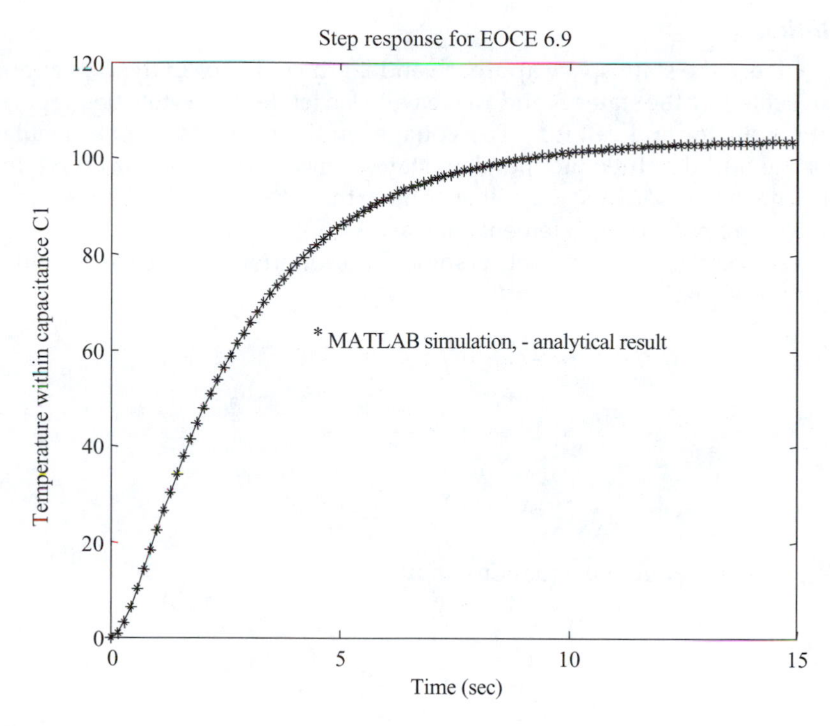

FIGURE 6.17
Plots for EOCE 6.9.

EOCE 6.10

Consider the circuit in Figure 6.18.

Find the voltages in the capacitors and the resistor as well as the current in the inductor with $x(t)$ as a step input of unity magnitude. Use MATLAB to verify the results. What are the poles of the system? What can you say about its stability?

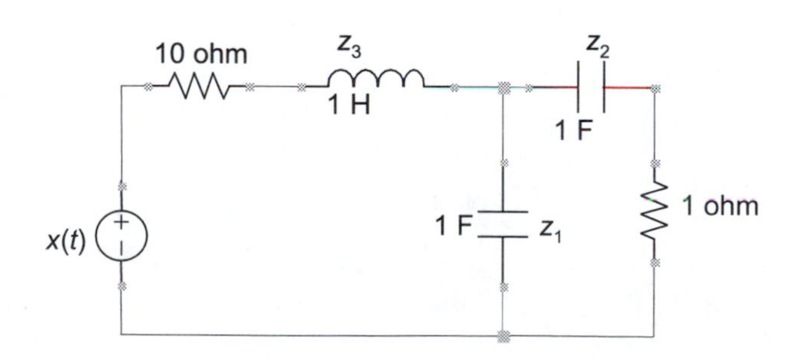

FIGURE 6.18
Circuit for EOCE 6.10.

Solution

We will use the state-space approach and let the voltages in the capacitors be represented by the states z_2 and z_3. We will also let the third state be the current in the inductor and call it z_1. The voltages in the resistors will be calculated after we find the three independent states. Since our system has three independent energy sources, we will only have these three states. Remember that resistors are not storage elements to start with.

Using nodal analysis as well as some loop equations, we come up with the following equations:

$$\frac{d}{dt}z_1 + \frac{d}{dt}z_2 - z_3 = 0$$

$$z_1 = z_2 + \frac{d}{dt}z_2$$

$$x = 10z_3 + \frac{d}{dt}z_3 + z_1$$

We can rearrange the equations above to get

$$\frac{d}{dt}z_1 = -z_1 + z_2 + z_3$$

$$\frac{d}{dt}z_2 = z_1 - z_2$$

$$\frac{d}{dt}z_3 = -z_1 - 10z_3 + x$$

In the state space form we write the above system as

$$\begin{bmatrix} \frac{d}{dt}z_1 \\ \frac{d}{dt}z_2 \\ \frac{d}{dt}z_3 \end{bmatrix} = \begin{bmatrix} -1 & 1 & 1 \\ 1 & -1 & 0 \\ -1 & 0 & -10 \end{bmatrix} \begin{bmatrix} z_1 \\ z_2 \\ z_3 \end{bmatrix} + \begin{bmatrix} 0 \\ 0 \\ 1 \end{bmatrix} x$$

where

$$A = \begin{bmatrix} -1 & 1 & 1 \\ 1 & -1 & 0 \\ -1 & 0 & -10 \end{bmatrix}$$

and

$$B = \begin{bmatrix} 0 \\ 0 \\ 1 \end{bmatrix}$$

Since A is not in the diagonal form, we need to diagonalize it in order to arrive at the analytical solution for the states.

The real-time solution in this case is

$$\underline{w}(t) = \exp[P^{-1}APt]\underline{w}(0) + \int_0^t \exp[P^{-1}AP(t-\tau)]P^{-1}Bx(\tau)d\tau$$

with $\underline{z}(t) = P\underline{w}(t)$, where P is the matrix that diagonalize A. Since $\underline{w}(0) = P^{-1}\underline{z}(0)$, and because $\underline{z}(0) = \underline{0}$, and $\underline{w}(0) = \underline{0}$, we get

$$\underline{w}(t) = \int_0^t \exp[P^{-1}AP(t-\tau)]P^{-1}Bx(\tau)d\tau$$

We will start by finding the matrix P that diagonalizes A. P is the matrix that contains the eigenvectors of A. We will use MATLAB to find the eigenvalues and eigenvectors of A. Knowing the eigenvalues of A, we can determine the stability of the system. If the real part of the eigenvalues are all negative, then the system is stable. We will write the following script in MATLAB to do that.

```
A = [-1  1  1;1  -1  0;-1  0  -10];
[VA,DA] = eig(A);
P = VA
Pinverse = inv(P)
PinverseAP = inv(P)*A*P
```

The output is

```
P  =
-0.6875     -0.7259     -0.1132
-0.7229      0.6816      0.0127
 0.0691      0.0915      0.9935
Pinverse  =
-0.6941     -0.7299     -0.0698
-0.7383      0.6932     -0.0930
 0.1163     -0.0131      1.0200
PinverseAP  =
-0.0490      0.0000      0.0000
 0.0000     -2.0650      0.0000
 0.0000      0.0000     -9.8860
```

Since the eigenvalues are all negative, the system is stable. To find $\underline{w}(t)$, we need the following matrices first.

$$P^{-1}B = \begin{bmatrix} -0.0698 \\ -0.0930 \\ 1.0200 \end{bmatrix}$$

$$e^{P^{-1}APt} = \begin{bmatrix} e^{-.049t} & 0 & 0 \\ 0 & e^{-2.065t} & 0 \\ 0 & 0 & e^{-9.886t} \end{bmatrix}$$

$$e^{P^{-1}AP(t-\tau)}P^{-1}B = \begin{bmatrix} e^{-.049(t-\tau)} & 0 & 0 \\ 0 & e^{-2.065(t-\tau)} & 0 \\ 0 & 0 & e^{-9.886(t-\tau)} \end{bmatrix} \begin{bmatrix} -0.069 \\ -0.093 \\ 1.02 \end{bmatrix}$$

$$e^{P^{-1}AP(t-\tau)}P^{-1}B = \begin{bmatrix} -0.069e^{-.049(t-\tau)} \\ -0.093e^{-2.065(t-\tau)} \\ 1.02e^{-9.886(t-\tau)} \end{bmatrix}$$

The new state solution is

$$\underline{w}(t) = \int_0^t \begin{bmatrix} -0.069e^{-.049(t-\tau)} \\ -0.093e^{-2.065(t-\tau)} \\ 1.02e^{-9.886(t-\tau)} \end{bmatrix} d\tau = \begin{bmatrix} 1.408(e^{-0.049t} - 1) \\ 0.045(e^{-2.065t} - 1) \\ -0.1032(e^{-9.886t} - 1) \end{bmatrix}$$

With $\underline{z}(t) = P\underline{w}(t)$

$$\underline{z}(t) =$$
$$\begin{bmatrix} -0.6875[1.408(e^{-0.049t} - 1)] - 0.7259[0.045(e^{-2.065t} - 1)] - 0.1132[-0.1032(e^{-9.886t} - 1)] \\ -0.7229[1.408(e^{-0.049t} - 1)] + 0.6816[0.045(e^{-2.065t} - 1)] + 0.0127[-0.1032(e^{-9.886t} - 1)] \\ 0.0691[1.408(e^{-0.049t} - 1)] + 0.0915[0.045(e^{-2.065t} - 1)] + 0.9935[-0.1032(e^{-9.886t} - 1)] \end{bmatrix}$$

or

$$\begin{bmatrix} z_1(t) \\ z_2(t) \\ z_3(t) \end{bmatrix} = \begin{bmatrix} -0.968e^{-0.049t} - 0.0327e^{-2.065t} + 0.0117e^{-9.886t} + 1 \\ -1.0294e^{-0.049t} + 0.0307e^{-2.065t} - 0.0013e^{-9.886t} + 1 \\ 0.0973e^{-0.049t} + 0.0041e^{-2.065t} - 0.1025e^{-9.886t} \end{bmatrix}$$

Notice that as time becomes large, the two capacitors will have the same voltage. We will demonstrate that using plots. Now let us write the following MATLAB function that we will use to compare the output from MATLAB and the output using the results just obtained. We start with the function that defines the system.

```
function zdot = EOCE 6.10(t,z); % return the state
derivatives
A = [-1 1 1;1 -1 0;-1 0 -10];
B = [0; 0; 1];
zdot = A*z1B;
```

Next we write the script that calls the functions "eoce6" and "ode23":

```
clf
A = [-1 1 1;1 -1 0;-1 0 -10];
B = [0; 0; 1];
tspan = [0 10]; % the time span for the simulation
z0 = [0 0 0];
[t,z] = ode23('eoce10',tspan, z0);
plot(t,z(:,1),'*');% for z1
hold on
plot(t,z(:,2),'+'); % for z2
plot(t,z(:,3),'-'); % for z3
title('Matlab simulation for eoce10');
gtext('z1(t)');
gtext('z2(t)');
gtext('z3(t)');
xlabel('Time (sec)')
```

The plots are shown in Figure 6.19.

If we increase the simulation time we see that the voltages in the two capacitors approach the input step signal of unity magnitude as shown in Figure 6.20.

Now we write the MATLAB script that simulates the hand calculations we obtained.

```
clf
t=0:.05:10;
z1=-0.968*exp(-.049*t)-.0327*exp(-2.065*t)+···
0.011*exp(-9.886*t)+1;
z2=-1.0294*exp(-.049*t)+0.0307*exp(-2.065*t)-···
```

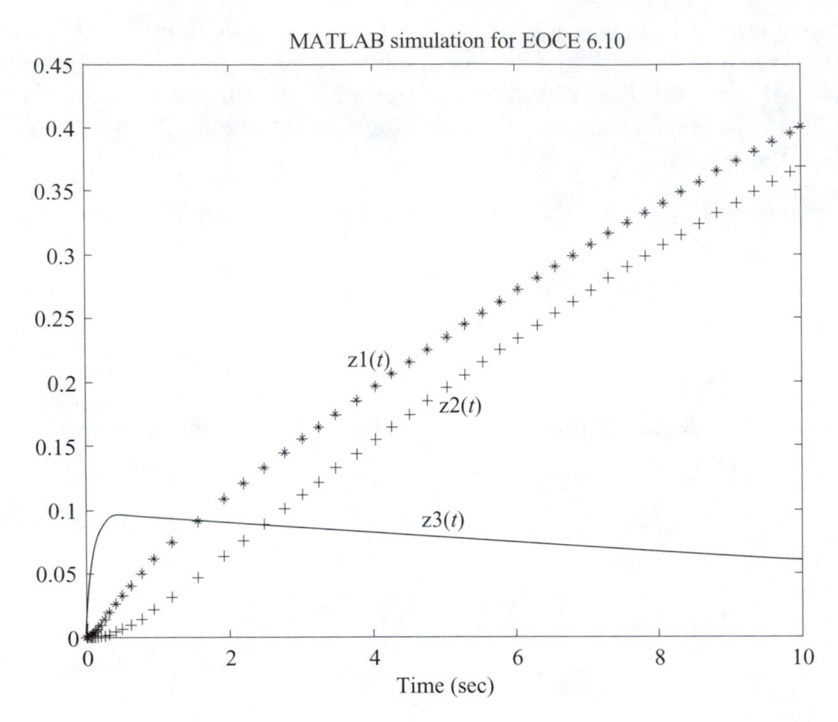

FIGURE 6.19
Plots for EOCE 6.10.

```
0.0013*exp(-9.886*t)+1;
z3=0.097*exp(-.049*t)+0.0041*exp(-2.065*t)-...
0.1025*exp(-9.886*t);
plot(t,z1,'*');% for z1
hold on
plot(t,z2,'+'); % for z2
plot(t,z3,'-'); % for z3
title('Hand simulation for EOCE 6.10');
gtext('z1(t)');
gtext('z2(t)');
gtext('z3(t)');
xlabel('Time (sec)')
```

The plots are shown in Figure 6.21.

If we increase the simulation time to 100 seconds we obtain the plots in Figure 6.22.

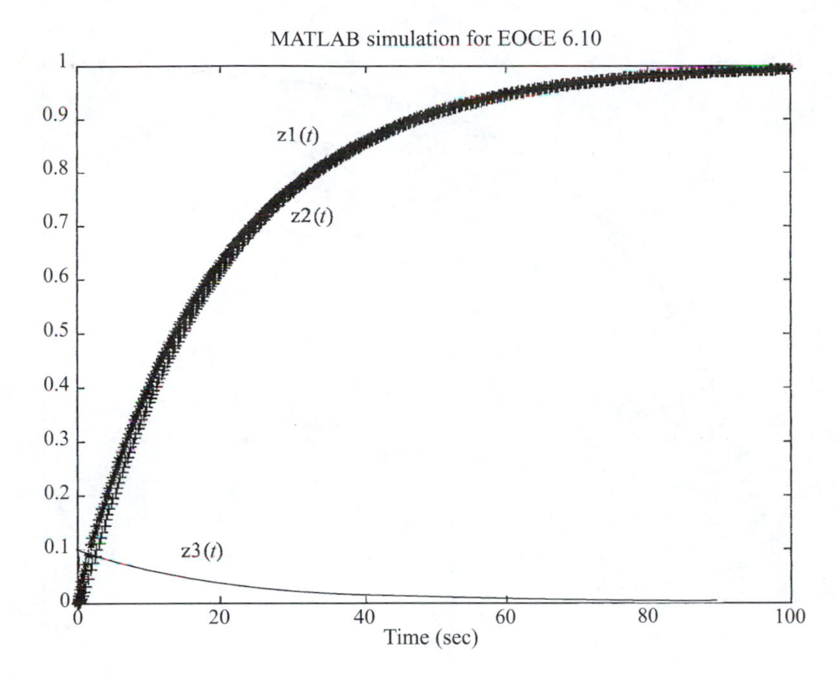

FIGURE 6.20
Plots for EOCE 6.10.

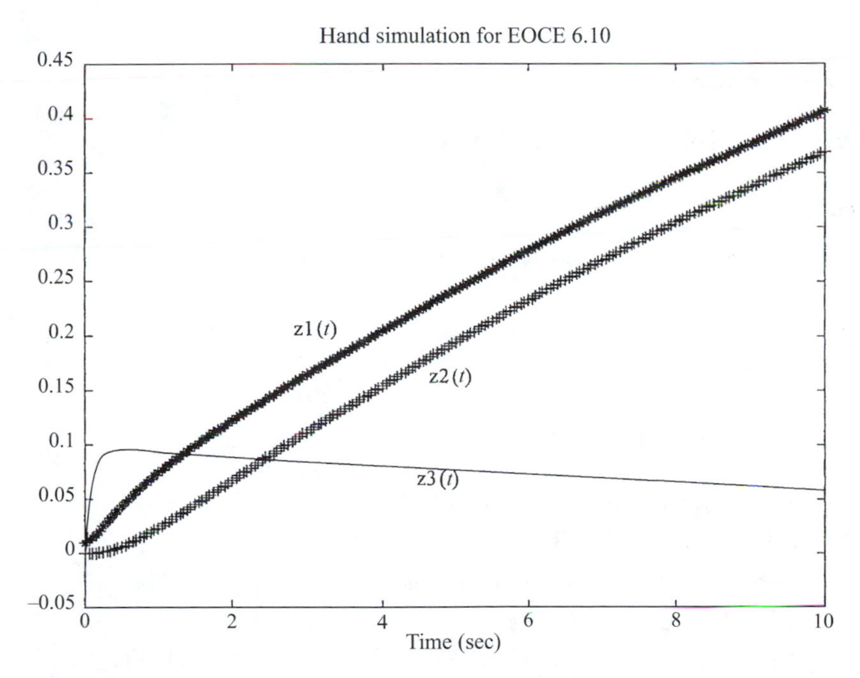

FIGURE 6.21
Plots for EOCE 6.10.

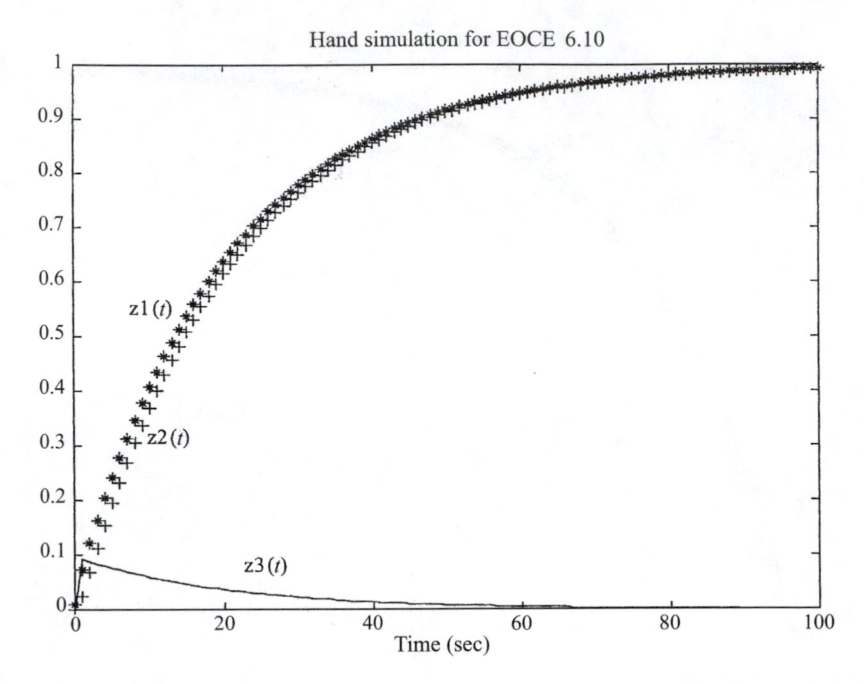

FIGURE 6.22
Plots for EOCE 6.10.

6.9 End-of-Chapter Problems

EOCP 6.1

Consider the following matrices.

$$A = \begin{bmatrix} 1 & 0 \\ 0 & 2 \end{bmatrix}, \ B = \begin{bmatrix} 3 & 0 \\ 0 & 4 \end{bmatrix}, \ C = \begin{bmatrix} 1 & 1 \\ 1 & 2 \end{bmatrix}, \quad \text{and} \quad D = \begin{bmatrix} 1 & 1 \\ 0 & 2 \end{bmatrix}.$$

Find the following.

1. $(A + B)$, $(A - B)$, $(A - C)$, and $(A + C)$
2. AB, BA, AC, CA, ABC, and CBA
3. $\det(A)$, $\det(B)$, $\det(C)$, $\det(AB)$, and $\det(AC)$
4. $A^{-1}, B^{-1}, C^{-1}, (AB)^{-1}, (BA)^{-1}, A^{-1}B^{-1}$, and $(ABC)^{-1}$
5. The transpose: A', B', C', $(AB)'$, $(ABC)'$, $A'B'$, and $(AB)'C'$
6. Eigenvalues for A and B
7. Eigenvectors for A and B

8. e^{At} and e^{Bt}
9. Diagonalize C and D as P_1 and P_2, respectively
10. $P_1^{-1}CP_1$ and $P_2^{-1}DP_2$
11. Eigenvalues for C, $P_1^{-1}CP_1$, D, and $P_2^{-1}DP_2$

EOCP 6.2

Let

$$A = \begin{bmatrix} 11 & 2 \\ 0 & 1 \end{bmatrix}$$

and

$$B = \begin{bmatrix} 2 & 5 \\ 1 & 6 \end{bmatrix}$$

Use the six methods described in this chapter to find

1. e^{At} and e^{Bt}
2. e^{ABt} and e^{BAt}

EOCP 6.3

Consider the following systems.

1. $\frac{d}{dt}y(t) + ay(t) = x(t)$
2. $\frac{d^2}{dt}y(t) + a\frac{d}{dt}y(t) + by(t) = x(t)$
3. $\frac{d^2}{dt}y(t) + a\frac{d}{dt}y(t) = x(t)$
4. $\frac{d^2}{dt}y(t) + by(t) = x(t)$
5. $\frac{d^3}{dt}y(t) + a\frac{d^2}{dt}y(t) + b\frac{d}{dt}y(t) + cy(t) = x(t)$
6. $\frac{d^3}{dt}y(t) + a\frac{d^2}{dt}y(t) + b\frac{d}{dt}y(t) = x(t)$
7. $\frac{d^3}{dt}y(t) + a\frac{d^2}{dt}y(t) = x(t)$
8. $\frac{d^3}{dt}y(t) + b\frac{d}{dt}y(t) = x(t)$
9. $\frac{d^3}{dt}y(t) + cy(t) = x(t)$

Write the state equations for each system but do not solve for the states.

Assume $y(0) = y_{00}$, $\frac{d}{dt}y(0) = y_{01}$, and $\frac{d^2}{dt}y(0) = y_{02}$. Under what conditions are the above systems stable?

EOCP 6.4

Consider the system shown in Figure 6.23.

Let $R_1 = R_2 = 1$ ohm, $L_1 = 0$ henry, $L_2 = 1$ henry, and $C = 1$ farad. Solve for the output $y(t)$ using state variables in the time domain and the Laplace domain if the input is an impulse of unity strength.

Is the system stable?

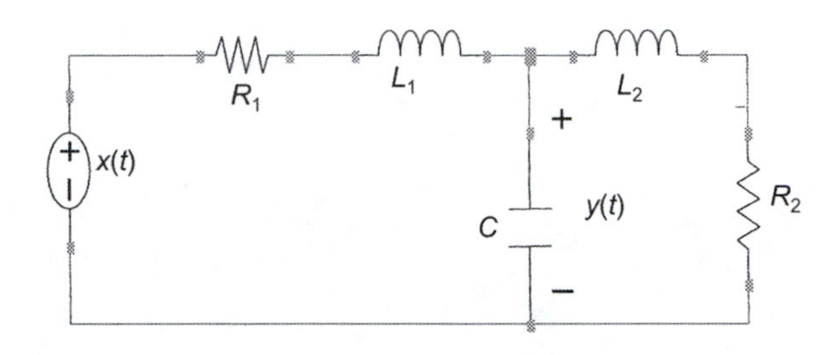

FIGURE 6.23
System for EOCP 6.4.

EOCP 6.5

Consider the system shown in Figure 6.24.

The rod to the left of the mass M can be modeled by a translational spring with spring constant k. The rod to the right of the mass M can be modeled by a translational damper of constant B. Let $M = 1$ kg, $k = 10$ N/m and $B = 5$ N/m/sec. The equation describing the system is

$$\frac{d^2}{dt}y(t) = -10\frac{d}{dt}y(t) + x(t)$$

Solve for the output $y(t)$ using state variables in the time domain and the Laplace domain if the input is an impulse of unity strength.

Is the system stable?

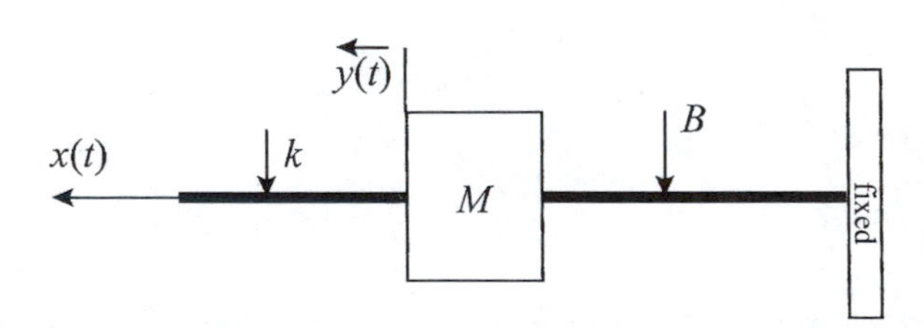

FIGURE 6.24
System for EOCP 6.5.

EOCP 6.6

Consider the system in Figure 6.25.

Let $R_1 = R_2 = 1$, and $C = 1$ farad. Solve for the output $y(t)$ using state variables in the time domain and the Laplace domain if the input is a unit step signal.

Is the system stable?

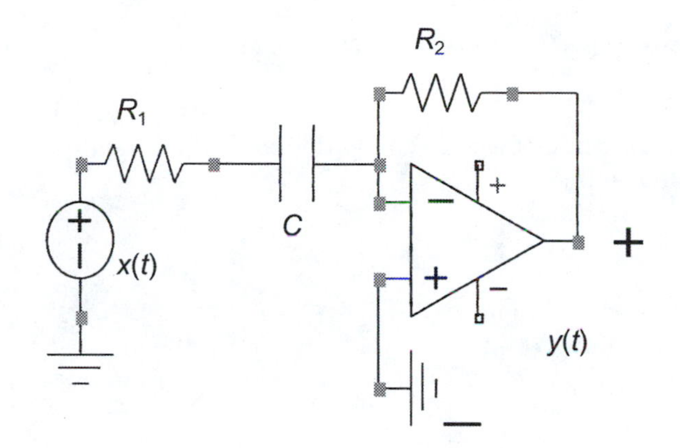

FIGURE 6.25
System for EOCP 6.5.

EOCP 6.7

Consider the system in Figure 6.26.

The rod above the mass M_1 can be modeled by a translational spring with spring constant k_1, in parallel with a translational damper of constant B_1.

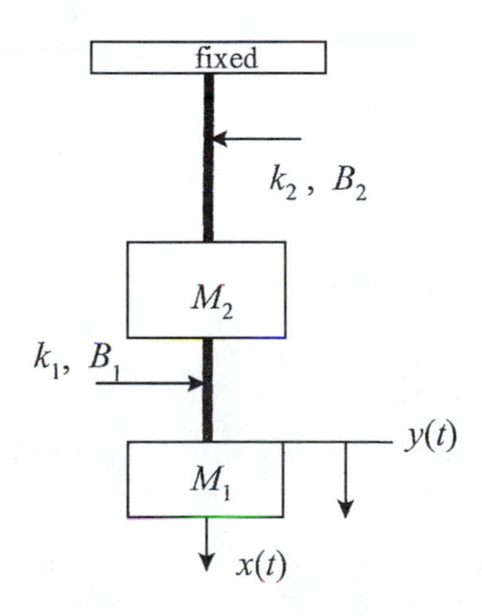

FIGURE 6.26
System for EOCP 6.7.

The rod above the mass M_2 can be modeled by a translational spring with spring constant k_2, in parallel with a translational damper of constant B_2.

Ignore the effect of gravity and let $k_1 = k_2 = 1$ N/m, $B_1 = B_2 = 2$ N/m/sec, $x(t) = \sin(t)$, and $M_1 = M_2 = 1$ kg.

The equation describing the system is

$$\frac{d^4}{dt}y(t) + 6\frac{d^3}{dt}y(t) + 7\frac{d^2}{dt}y(t) + 4\frac{d}{dt}y(t) + y(t) = \frac{d^2}{dt}x(t) + 4\frac{d}{dt}x(t) + 2x(t)$$

Solve for the output $y(t)$ using state variables in the time domain.

Is the system stable?

EOCP 6.8

Consider the system in Figure 6.27.

If $R_1 = R_2 = 1$ ohm, $C_1 = C_2 = 1$ farad, and $x(t)$ is the impulse function, solve for the output $y(t)$ using state variables in the time domain and the Laplace domain.

Is the system stable?

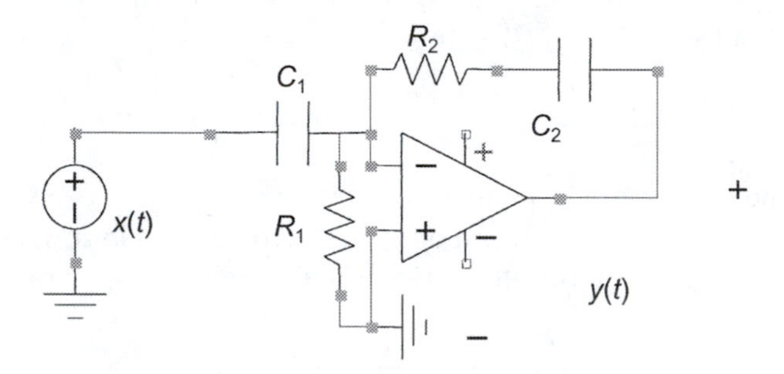

FIGURE 6.27
System for EOCP 6.8.

EOCP 6.9

Consider the system in Figure 6.28.

The differential equation describing the system is

$$\frac{d^3}{dt}\theta(t) = -(B/J)\frac{d}{dt}\theta(t) + (1/J)T(t)$$

The rod is attached to a fixed end and can be represented as a rotational damper only. The moment of inertia, J, represents the concentration of the mass, M. A torque T is applied and it produces the angular rotation.

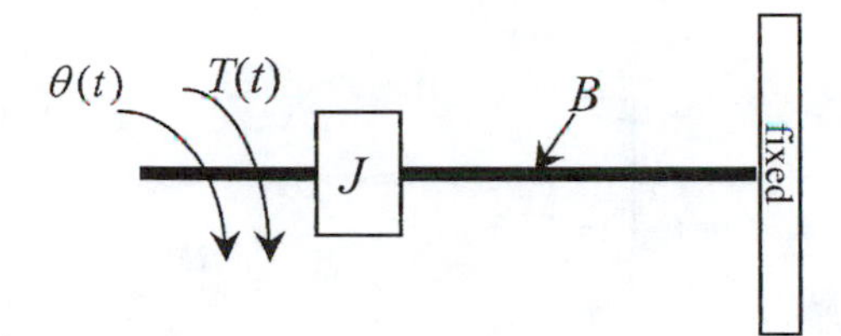

FIGURE 6.28
System for EOCP 6.9.

With $T(t) = 10$ N–m, solve for the output $\theta(t)$ using state variables in the time domain and the Laplace domain.

Is the system stable?

EOCP 6.10

Consider the system in Figure 6.29.

If $R = 1$ ohm, $C = 1$ farad, $L = 1$ henry, and $x(t) = 10\cos(2t)$, solve for the output $y(t)$ using state variables in the time domain and the Laplace domain.

Is the system stable?

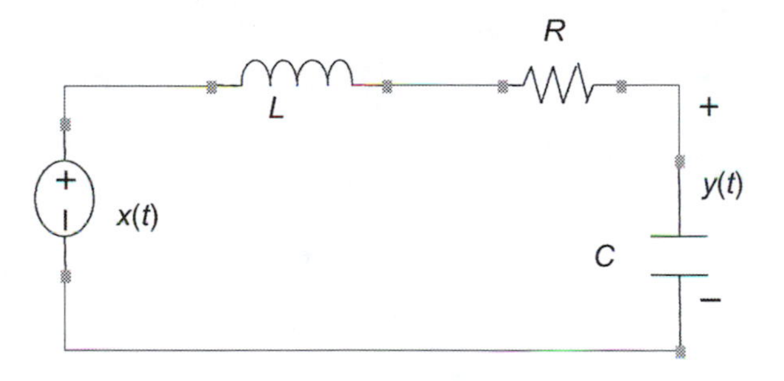

FIGURE 6.29
System for EOCP 6.10.

EOCP 6.11

Consider the system in Figure 6.30.

Two rods with different masses will give rise to the two moments of inertia J_1 and J_2. If $J_1 = J_2 = J = 10$ kg–m^2, $k = 5$ N.m/rad, and the torque is applied at a very short time that can be represented as an impulse, the relation between $\theta(t)$ and $T(t)$ can be described by the differential equation

$$J\frac{d^4}{dt}\theta(t) + 2kJ\frac{d^2}{dt}\theta(t) = k\delta(t)$$

Find $\theta(t)$ using the real-time state solution.

Is the system stable?

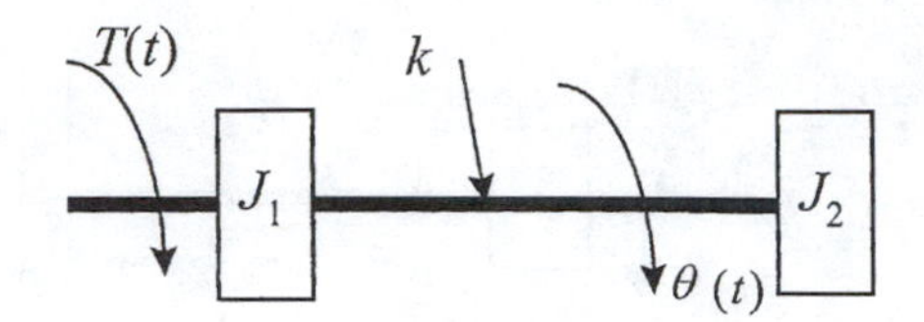

FIGURE 6.30
System for EOCP 6.11.

EOCP 6.12

Consider the system in Figure 6.31.

If $R_1 = R_2 = 1$ ohm, $C = 1$ farad, $L = 1$ henry, and $x(t)$ is $u(t)$, solve for the output $y(t)$ using state variables in the time and the Laplace domains.

Is the system stable?

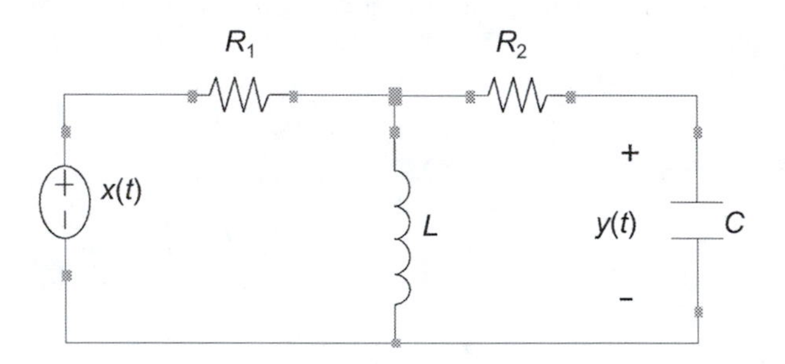

FIGURE 6.31
System for EOCP 6.12.

EOCP 6.13

Consider the following rotational system.

$$2\frac{d^3}{dt}\theta(t) + \frac{d}{dt}\theta(t) + 2\theta(t) = \delta(t)$$

Find $\theta(t)$ using the real-time state equation solution.

Is the system stable?

EOCP 6.14

Consider the system shown in Figure 6.32.

If $R_1 = R_2 = 1$ ohm, $C_1 = C_2 = 1$ farad, and $x(t)$ is $u(t)$, solve for the output $y(t)$ using state variables in the time domain and the Laplace domain.

Is the system stable?

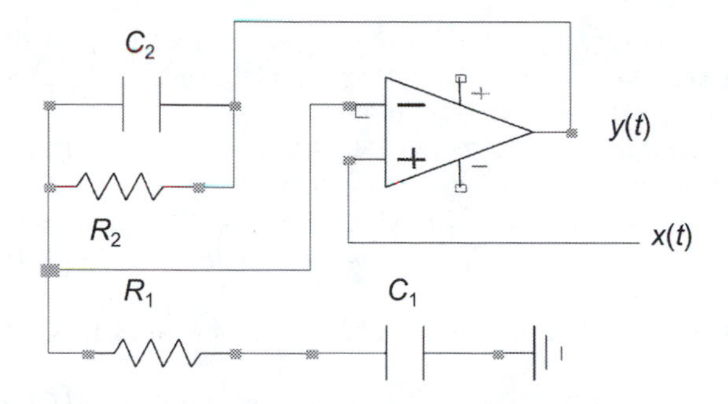

FIGURE 6.32
System for EOCP 6.14.

EOCP 6.15

Consider the system in Figure 6.33.

With $J_{load} = J_{motor} = J = 2$ kg–m^2, $k = 10$ N/m, and $T(t) = u(t)$, the dynamics that relate the angular rotation at the load side and the torque are given as

$$2\frac{d^5}{dt}\theta_{load} + 20\frac{d^2}{dt}\theta_{load} = 10u(t)$$

Find the angular rotation at the load using the state equations in real-time.

Is the system stable?

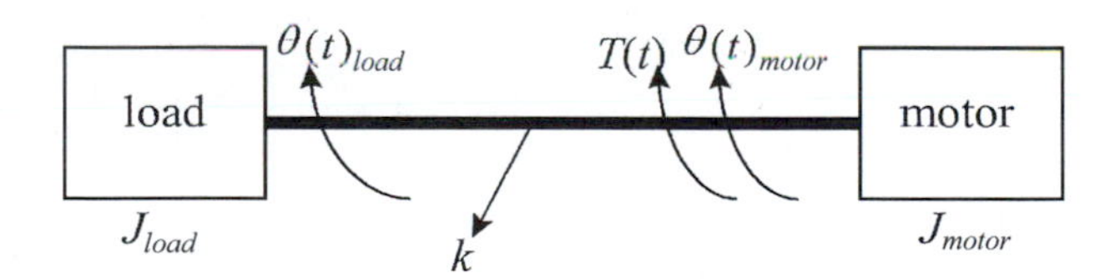

FIGURE 6.33
System for EOCP 6.15.

References

Bequette, B.W. *Process Dynamics,* Englewood Cliffs, NJ: Prentice-Hall, 1998.

Brogan, W.L. *Modern Control Theory,* 3rd ed., Englewood Cliffs, NJ: Prentice-Hall, 1991.

Close, M. and Frederick, K. *Modeling and Analysis of Dynamic Systems,* 2nd ed., New York: Wiley, 1995.

Cogdell, J.R. *Foundations of Electrical Engineering,* 2nd ed., Englewood Cliffs, NJ: Prentice-Hall, 1996.

Denbigh, P. *System Analysis and Signal Processing,* Reading, MA: Addison-Wesley, 1998.

Driels, M. *Linear Control System Engineering,* New York: McGraw-Hill, 1996.

Golubitsky, M. and Dellnitz, M. *Linear Algebra and Differential Equations Using MATLAB,* Stamford, CT: Brooks/Cole, 1999.

Harman, T.L., Dabney, J., and Richert, N. *Advanced Engineering Mathematics with MATLAB,* Stamford, CT: Brooks/Cole, 2000.

Kuo, B.C. *Automatic Control System,* 7th ed., Englewood Cliffs, NJ: Prentice-Hall, 1995.

Lewis, P.H. and Yang, C. *Basic Control Systems Engineering,* Englewood Cliffs, NJ: Prentice-Hall, 1997.

The MathWorks. *The Student Edition of MATLAB,* Englewood Cliffs, NJ: Prentice-Hall, 1997.

Nilson, W.J. and Riedel, S.A. *Electrical Circuits,* 6th ed., Englewood Cliffs, NJ: Prentice-Hall, 2000.

Nise, N.S. *Control Systems Engineering,* 2nd ed., Reading, MA: Addison-Wesley, 1995.

Ogata, K. *Modern Control Engineering,* 3rd ed., Englewood Cliffs, NJ: Prentice-Hall, 1997.

Ogata, K. *System Dynamics,* 3rd ed., Englewood Cliffs, NJ: Prentice-Hall, 1998.

Phillips, C.L. and Parr, J.M. *Signals, Systems, and Transforms,* 2nd ed., Englewood Cliffs, NJ: Prentice-Hall, 1999.

Pratap, R. *Getting Started with MATLAB 5,* New York: Oxford University Press, 1999.

Woods, R.L. and Lawrence, K.L. *Modeling and Simulation of Dynamics Systems,* Englewood Cliffs, NJ: Prentice-Hall, 1997.

Wylie, R.C. and Barrett, C.L. *Advanced Engineering Mathematics,* 6th ed., New York: McGraw-Hill, 1995.

Ziemer, R.E., Tranter, W.H., and Fannin, D.R. *Signals Systems Continuous and Discrete,* 4th ed., Englewood Cliffs, NJ: Prentice-Hall, 1998.

7

Modeling and Representation of Linear Systems

CONTENTS

7.1 Introduction

The transfer function, $H(s)$, for linear time-invariant systems, can be inverse-transformed to produce $h(t)$, the impulse response of the system. Thus $h(t)$, the signal, can be modeled using the Laplace transform approach to represent the transfer function of the system. In communications, we broadcast signals of various types, and in most situations we sample the signal to be transmitted and then transmit it via a communication channel using a huge number of data points. Due to many factors, such as interference, these data values get distorted. However, we can represent the signal to be transmitted using the Laplace transform approach and transmit only the coefficients of the numerator and the denominator of the transfer function $H(s)$.

When we studied the general form of the sinusoidal signal in previous chapters, what we did was actually model the signal as

$$x(t) = A\cos(wt + \varphi)$$

To avoid losing data points during transmission, we can send only the amplitude of the signal, its frequency, and its phase. That would be easier and more efficient.

The transfer function, $H(s)$, the signal $x(t) = A\cos(wt + \varphi)$, among many other representations, are models that we use for certain purposes. In this book we use modeling primarily for analysis and design of linear time-invariant systems.

We can model mechanical systems, for example, using the laws of Newton to write a set of differential equations. We can also model electrical systems using similar laws to derive differential equations that relate different components in the system. We do similar things in chemical systems, mechanics, and dynamics. Basically, we derive differential equations from existing systems. If various parameters in these systems are unknown, we can approximate these systems by finding their impulse responses and then deriving the differential equations represented by these responses.

7.2 Five Ways of Representing Linear Systems

In this chapter, we will consider representing linear time-invariant systems using all the techniques that we have studied thus far in this book. We can represent linear time-invariant systems in many ways. The following five ways are considered.

1. Representing a linear time-invariant system using ordinary linear differential equations with constant coefficients
2. Representing a linear time-invariant system using the impulse response function, $h(t)$
3. Representing a linear time-invariant system using the transfer function approach
4. Representing a linear time-invariant system using block diagrams
5. Representing a linear time-invariant system using state-space approach

To do so we will learn by example. We will consider many examples in this chapter and see how to go from one representation to the other. Given one representation from among the five representations that are listed above, we should be able to deduce the other representations. We will also find the output, $y(t)$, given the input, $x(t)$, along with the necessary initial conditions and demonstrate that the output $y(t)$ will be the same for the same input $x(t)$ and the same given initial conditions for the same system.

Example 7.1

Consider the circuit in Figure 7.1 with $R_1 = R_2 = 1$ ohm and $L = 1$H.

The input is the voltage, $x(t)$, and the output $y(t)$ is the voltage in resistor R_2. We want to find the output $y(t)$ due to an impulsive input $\delta(t)$. Assume zero initial conditions.

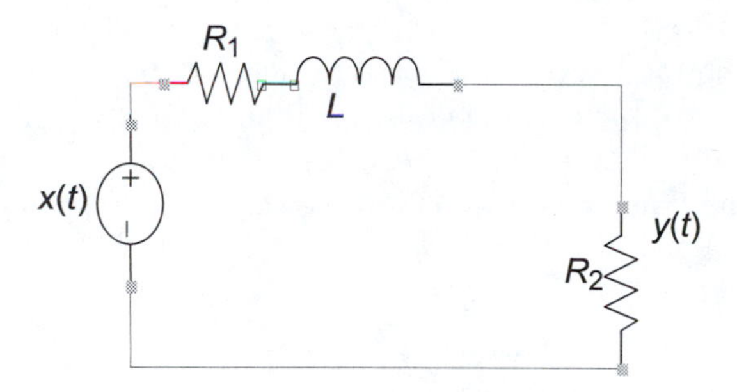

FIGURE 7.1
Circuit for Example 7.1.

Solution

1. Linear differential equation representation
   Since the circuit has only one storage element, the inductor, we expect that we will end up with a first order differential equation. Writing a loop equation we get

$$x(t) = i_L(t)R_1 + v_L(t) + y(t)$$

We need to put this equation in terms of $y(t)$ and its derivatives and $x(t)$ and its derivatives only. But the voltage in the inductor as a function of $y(t)$ can be found as

$$v_L(t) = L\frac{d}{dt}i_L(t)$$

with

$$i_L(t) = y(t)/R_2$$

Therefore,

$$x(t) = \frac{R_1}{R_2}y(t) + \frac{L}{R_2}\frac{d}{dt}y(t) + y(t)$$

Finally, the loop equation, with values for variables substituted, becomes

$$\frac{d}{dt}y(t) + 2y(t) = x(t)$$

The solution of this differential equation is $y(t)$, the desired output.

With $x(t) = \delta(t)$, we can solve this differential equation in real time. The total solution for $y(t)$ which is $h(t)$ for this particular input, as we have seen before, will have two parts; the homogenous part and the particular part.

For the homogenous solution we set $x(t)$ to zero and this kills the $\delta(t)$ signal to get

$$\frac{d}{dt}y(t) + 2y(t) = 0$$

The auxiliary equation is

$$m + 2 = 0$$

which gives m the value -2. The homogenous solution is then

$$y_h(t) = c_1 e^{-2t} \qquad t > 0$$

To find c_1 we need the total solution for $y(t)$ first.

For the particular solution

$$y_p(t) = 0$$

Therefore, the total solution is

$$y(t) = y_h(t) + y_p(t)$$

or

$$y(t) = c_1 e^{-2t} \qquad t > 0$$

which is the same as writing

$$y(t) = c_1 e^{-2t} u(t)$$

where $u(t)$ is the unit step signal that triggers signals at $t = 0$.

Finally, we have to find c_1. To do so we substitute the total solution for $y(t)$ in the original differential equation noticing that the derivative of $u(t)$ is $\delta(t)$ and

$$\frac{d}{dt}y(t) = c_1(-2)e^{-2t}u(t) + c_1 e^{-2t}\delta(t)$$

In the last equation, we used the product rule for derivatives. We next substitute the last equation in the original differential equation to get

$$[c_1(-2)e^{-2t}u(t) + c_1e^{-2t}\delta(t)] + 2[c_1e^{-2t}u(t)] = \delta(t)$$

Remember that the $\delta(t)$ signal is only valid at $t = 0$, the time of application of the signal, and in this case any function multiplied by $\delta(t)$ is evaluated at $t = 0$. Therefore,

$$[c_1(-2)e^{-2t}u(t) + c_1\delta(t)] + 2[c_1e^{-2t}u(t)] = \delta(t)$$

This last equation reduces to

$$c_1\delta(t) = \delta(t)$$

Equating coefficients we find that $c_1 = 1$, and the final solution is

$$y(t) = e^{-2t}u(t)$$

2. The impulse response representation, $h(t)$

   The impulse response, $h(t)$, is the output of the system, $y(t)$, due to an input, $x(t)$, where $x(t) = \delta(t)$, the impulse signal. The differential equation

$$\frac{d}{dt}y(t) + 2y(t) = x(t)$$

can then be rewritten as

$$\frac{d}{dt}h(t) + 2h(t) = \delta(t)$$

This is a first order differential equation with a forcing input signal $\delta(t)$. It can be solved in-time or in the Laplace domain. Taking the Laplace transform of this differential equation will yield the following algebraic equation in the variable s.

$$[sH(s) - h(0^-)] + 2H(s) = 1$$

But $h(0^-) = 0$ (system has zero initial conditions; it is relaxed). Therefore,

$$H(s)[s + 2] = 1$$

and

$$H(s) = 1/[s+2]$$

Taking the inverse Laplace transform of $H(s)$ will yield $h(t)$, the impulse response as

$$h(t) = e^{-2t}u(t)$$

Now we need to find $y(t)$ using $h(t)$ and the input $\delta(t)$. We will use the convolution integral and write

$$y(t) = \int_{-\infty}^{+\infty} x(\tau)h(t-\tau)d\tau$$

or by substituting we get

$$y(t) = \int_{-\infty}^{+\infty} \delta(\tau)e^{-2(t-\tau)}u(t-\tau)d\tau$$

Using the sifting property of the impulse signal

$$\int_{-\infty}^{\infty} x(t)\delta(t-a)dt = x(a)$$

we get

$$y(t) = e^{-2t}u(t)$$

3. The transfer function representation

   We learned before that the transfer function relates the input $X(s)$ to the output $Y(s)$.

   Let us Laplace-transform the differential equation

$$\frac{d}{dt}y(t) + 2y(t) = x(t)$$

   with zero initial conditions to get

$$[sY(s) - y(0^-)] + 2Y(s) = X(s)$$

   or

$$Y(s)[s+2] = X(s)$$

and finally, the transfer function is

$$\frac{Y(s)}{X(s)} = \frac{1}{s+2}$$

Given the transfer function in the Laplace domain with $X(s) = 1$ ($x(t) = \delta(t)$) we have

$$\frac{Y(s)}{1} = \frac{1}{s+2}$$

or

$$Y(s) = \frac{1}{s+2}$$

Taking the Laplace transform of $Y(s)$ we get

$$y(t) = e^{-2t}u(t)$$

4. The block diagram representation

   There are two approaches that we studied to represent systems in block diagrams. We need to mention again that block diagram representation is not unique. The two methods previously studied are first, the D operator method that works directly with the differential equation, and second, the method explained in Chapter 5 that deals with the transfer function of the system. We will use the second method here. Remember again that $1/s$ represents integration in the Laplace domain. Our transfer function was calculated as

$$\frac{Y(s)}{X(s)} = \frac{1}{s+2}$$

   We will follow the steps explained in Chapter 5.

   1. Our system is first order and hence we need one integrator. The initial block is shown in Figure 7.2.

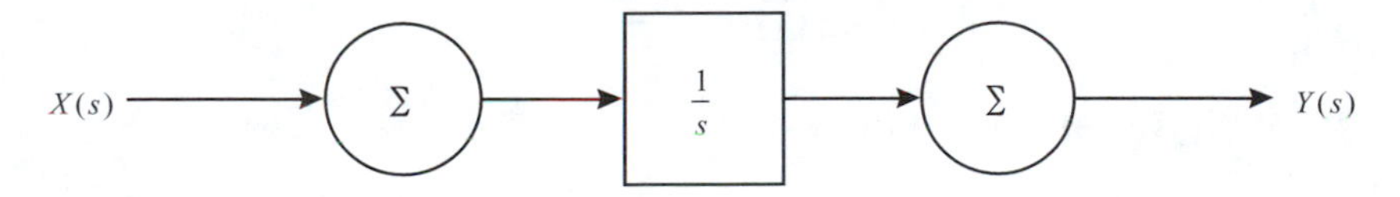

FIGURE 7.2
Block for Example 7.1.

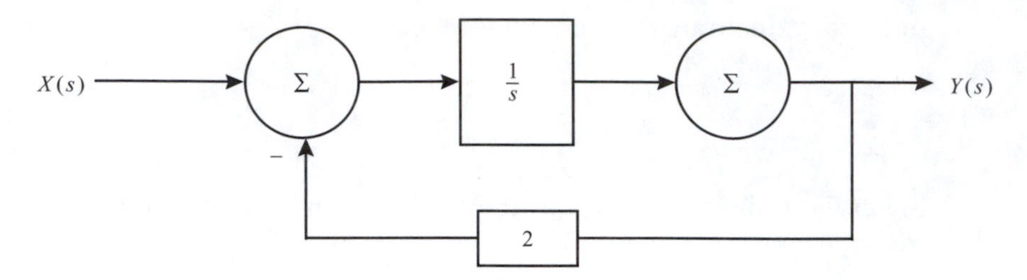

FIGURE 7.3
Block for Example 7.1.

2. We will feedback $Y(s)$ multiplied by 2 with a negative sign associated with it to the summer before the integrator in Figure 7.2. The result is shown in Figure 7.3.
3. There is nothing to feed forward (you may need to go back to Chapter 5 to see the process in details) and the final block diagram is as given in step two as Figure 7.3.

From the block diagram we can find the transfer function representation. At the output of the first summer from the left we have the signal

$$X(s) - 2Y(s)$$

This signal is then multiplied by $1/s$. Thus, we will have the signal

$$[1/s][X(s) - 2Y(s)]$$

as an input to the second summer. $Y(s)$ is the output of the second summer and since nothing other than

$$[1/s][X(s) - 2Y(s)]$$

is presented as an input to this second summer, we have

$$Y(s) = [X(s)/s - 2Y(s)/s]$$

or finally

$$\frac{Y(s)}{X(s)} = \frac{1}{s+2}$$

With $X(s) = 1$, the output is

$$Y(s) = \frac{1}{s+2}$$

By taking the inverse transform we get

$$y(t) = e^{-2t}u(t)$$

5. The state-space representation

We will start with the differential equation

$$\frac{d}{dt}y(t) + 2y(t) = x(t)$$

This is a first order differential equation and therefore we will have only one state. Let us call it $z_1(t)$.
Let $z_1(t) = y(t)$. Then

$$\frac{d}{dt}z_1 = \frac{d}{dt}y$$

Therefore, the state equation is

$$\frac{d}{dt}z_1 = -2z_1 + x$$

and the output equation is

$$y = z_1$$

In state-space matrix form we have

$$\frac{d}{dt}\underline{z} = A\underline{z} + B\underline{x}$$

$$\underline{y} = C\underline{z} + D\underline{x}$$

where

$$A = [-2],\ B = [1],\ C = [1],\ \text{and}\ D = [0]$$

The solution of these state-space matrix equations can be found in real time and using the Laplace approach. We will use the Laplace approach here to write the state solution as

$$\underline{Z}(s) = (sI - A)^{-1}\underline{z}(0) + (sI - A)^{-1}B\underline{X}(s)$$

and the output solution as

$$\underline{Y}(s) = C\underline{Z}(s) + D\underline{X}(s)$$

The identity matrix here is a 1×1 or $I = [1]$ and the state equation is nothing but a scalar equation; $\underline{Z}(s)$ is $Z_1(s)$ and $\underline{X}(s) = X(s)$. Therefore,

$$Z_1(s) = (sI - A)^{-1}z(0) + (sI - A)^{-1}BX(s)$$

Or by simplification we arrive at

$$Z_1(s) = \frac{1}{s+2}$$

and by inverse transforming we get

$$z_1(t) = e^{-2t}u(t)$$

Since $D = [0]$ and $C = [0]$, the output equation $\underline{Y}(s)$ reduces to

$$\underline{Y}(s) = \underline{Z}(s)$$

But $\underline{Y}(s)$ is a vector that contains only $Y(s)$ and $\underline{Z}(s)$ is also a vector that contains $Z_1(s)$. Therefore,

$$Y(s) = Z(s)$$

and

$$y(t) = e^{-2t}u(t)$$

Example 7.2

Consider the series RLC circuit which was presented in Chapter 6 and which is shown here in Figure 7.4, where $x(t) = u(t)$ is the input and initial conditions are all set to zero.

Now we are interested in finding the output $v_C(t)$, the voltage across the capacitor and the current in the inductor.

Solution

1. Linear differential equation representation

   Let $y_1(t)$ be the voltage across the capacitor and $y_2(t)$ be the current in the inductor. We could easily write one second order differential equation relating either $x(t)$ to $y_1(t)$ or $x(t)$ to $y_2(t)$ and then solve for either $y_1(t)$ or $y_2(t)$.

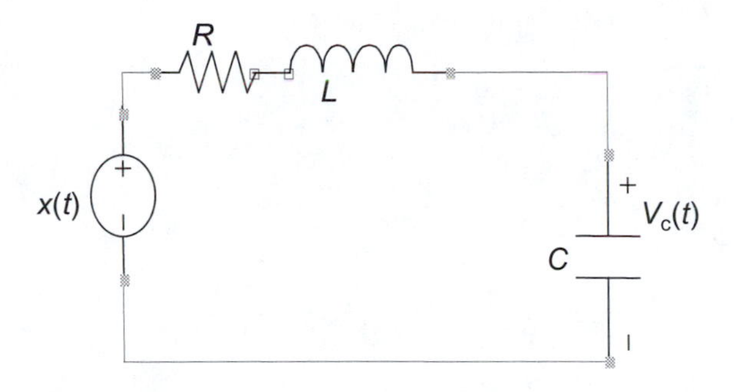

FIGURE 7.4
Circuit for Example 7.2.

The loop equation for the circuit in Figure 7.4 with $R = 3$ ohm, $C = 0.5$ F, and $L = 1$H is

$$x(t) = y_1(t) + v_L(t) + 3y_2(t)$$

But

$$y_1(t) = 2\int_0^t y_2(\tau)d\tau$$

and

$$v_L(t) = \frac{d}{dt}y_2(t)$$

Therefore, we can write

$$x(t) = 2\int_0^t y_2(\tau)d\tau + \frac{d}{dt}y_2(t) + 3y_2(t).$$

Differentiating the above equation and arranging terms we get

$$\frac{d^2}{dt}y_2(t) + 3\frac{d}{dt}y_2(t) + 2y_2(t) = \frac{d}{dt}x(t)$$

We will present the solution to this equation using the Laplace transform and write the differential equation in the Laplace domain as

$$s^2Y_2(s) + 3sY_2(s) + 2Y_2(s) = 1$$

or

$$Y_2(s)[s^2 + 3s + 2] = 1$$

Finally

$$Y_2(s) = \frac{1}{[s^2 + 3s + 2]} = \frac{A}{s+2} + \frac{B}{s+1} = \frac{-1}{s+2} + \frac{1}{s+1}$$

Taking the inverse Laplace transform we have

$$y_2(t) = (-e^{-2t} + e^{-t})u(t)$$

But

$$y_1(t) = 2\int_0^t y_2(\tau)d\tau$$

Therefore,

$$y_1(t) = (e^{-2t} - 2e^{-t} + 1)u(t)$$

2. The impulse response representation
   We will start with the differential equation representation

$$\frac{d^2}{dt}y_2(t) + 3\frac{d}{dt}y_2(t) + 2y_2(t) = \frac{d}{dt}x(t)$$

Since $x(t)$ now is $\delta(t)$ and $y_2(t)$ becomes $h_2(t)$ we can write

$$\frac{d^2}{dt}h_2(t) + 3\frac{d}{dt}h_2(t) + 2h_2(t) = \frac{d}{dt}x(t)$$

Taking the Laplace transform we get

$$s^2H_2(s) + 3sH_2(s) + 2H_2(s) = sX(s) - x(0^-)$$

or

$$H_2(s)[s^2 + 3s + 2] = s(1) - 0$$

and

$$H_2(s) = \frac{S}{[s^2 + 3s + 2]} = \frac{A}{s+2} + \frac{B}{s+1} = \frac{2}{s+2} + \frac{-1}{s+1}$$

Taking the inverse Laplace transform we have

$$h_2(t) = (2e^{-2t} - e^{-t})u(t)$$

Now we use convolution to solve for $y_2(t)$ as

$$y_2(t) = \int_{-\infty}^{+\infty} u(\tau)[2e^{-2(t-\tau)} - e^{-(t-\tau)}]u(t-\tau)d\tau$$

Simplifying we get

$$y_2(t) = (-e^{-2t} + e^{-t})u(t)$$

and

$$y_1(t) = (e^{-2t} - 2e^{-t} + 1)u(t)$$

3. The transfer function representation

   Now we will take circuit analysis approach in finding the transfer function that relates $x(t)$ to $y_1(t)$, the voltage in the capacitor, and $y_2(t)$, the current in the inductor.

   Laplace-transform the circuit itself to obtain the circuit in Figure 7.5.

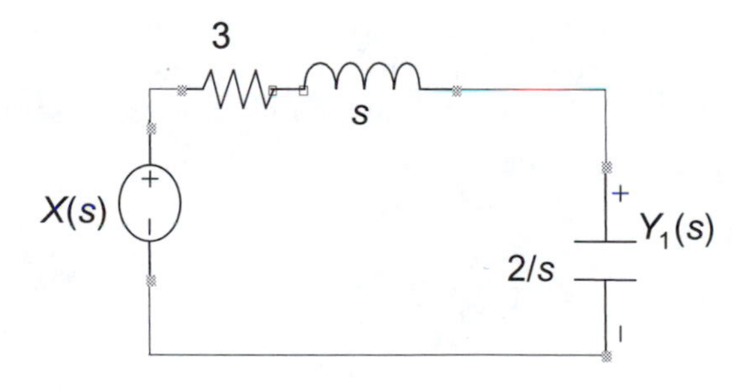

FIGURE 7.5
Circuit for Example 7.2.

Using voltage divider approach we get

$$Y_1(s) = \frac{\frac{2}{s}}{s + \frac{2}{s} + 3} X(s)$$

and the transfer function is

$$\frac{Y_1(s)}{X(s)} = \frac{2}{s^2 + 3s + 2}$$

From which

$$Y_1(s) = \frac{2}{s(s^2 + 3s + 2)} = \frac{A}{s} + \frac{B}{s+1} + \frac{C}{s+2} = \frac{1}{s} + \frac{-2}{s+1} + \frac{1}{s+2}$$

Taking the inverse Laplace transform we get

$$y_1(t) = (e^{-2t} - 2e^{-t} + 1)u(t)$$

Also, we can use Ohm's law in the circuit above to write

$$Y_2(s) = \frac{X(s)}{s + \frac{2}{s} + 3} = \frac{1}{s^2 + 3s + 2} = \frac{1}{s+1} + \frac{-1}{s+2}$$

where finally

$$y_2(t) = (-e^{-2t} + e^{-t})u(t)$$

4. The block diagram representation

   Our goal was to find the current in the inductor and the voltage in the capacitor. We will consider two transfer functions: one relating $y_1(t)$, the voltage in the capacitor, to the input $x(t) = u(t)$, and the other relating $y_2(t)$, the current in the inductor, to the same input $x(t)$. We had previously that

$$\frac{Y_1(s)}{X(s)} = \frac{2}{s^2 + 3s + 2}$$

   and

$$\frac{Y_2(s)}{X(s)} = \frac{S}{s^2 + 3s + 2}$$

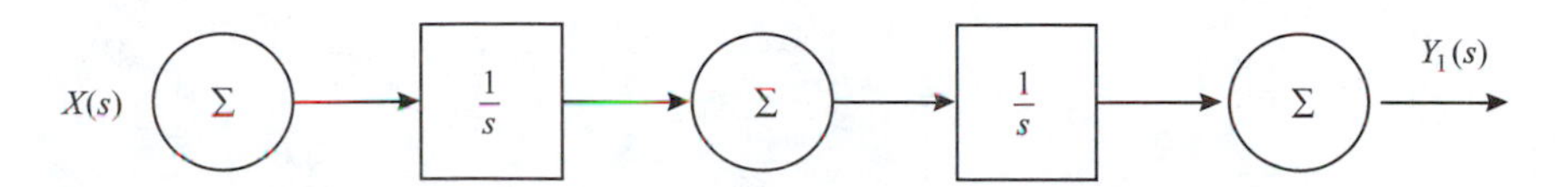

FIGURE 7.6
Block for Example 7.2.

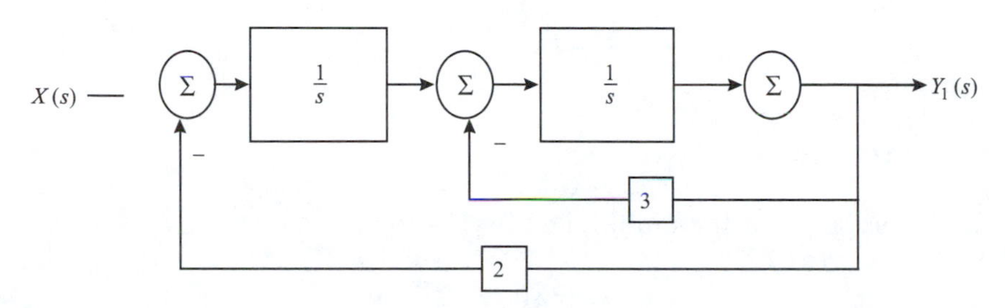

FIGURE 7.7
Block for Example 7.2.

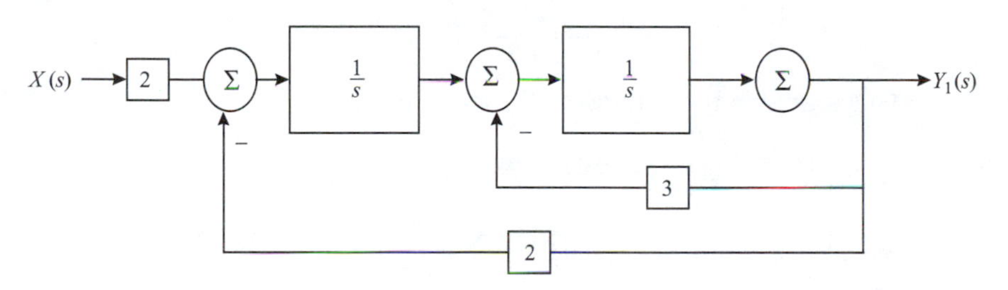

FIGURE 7.8
Block for Example 7.2.

We will start with the first transfer function. There are two integrators (the order of the denominator is 2) and so we will have the initial block in Figure 7.6.

We will feed backward $-3Y(s)$ to the summer before the second integrator and $-2Y(s)$ to the summer before the first integrator. The second block is shown in Figure 7.7.

Next, we feed forward $2X(s)$ to the first summer before the first integrator as seen in Figure 7.8.

The block diagram of the second transfer function can be drawn in a similar way as shown in Figure 7.9 where $1X(s)$ is fed forward to the summer after the first integrator.

Next, we will obtain the solutions $y_1(t)$ and $y_2(t)$ with $x(t) = u(t)$. At the output of the first summer in Figure 7.8, we have the signal

$$2X(s) - 2Y_1(s)$$

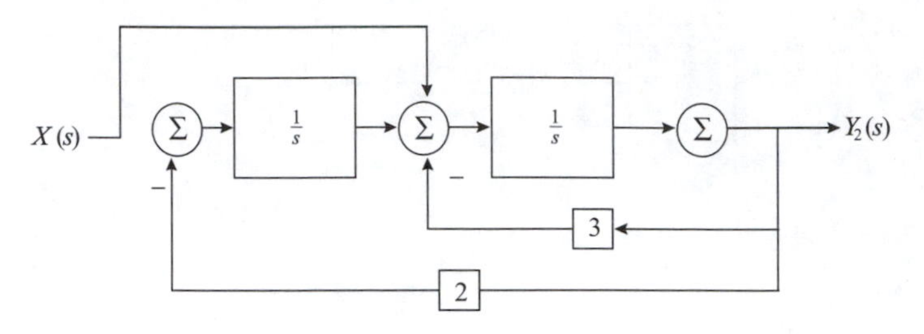

FIGURE 7.9
Block for Example 7.2

This signal is then multiplied by $1/s$ to get

$$(1/s)[2X(s) - 2Y_1(s)]$$

At the output of the second summer we have the signal

$$(1/s)[2X(s) - 2Y_1(s)] - 3Y_1(s)$$

This signal is then multiplied by $1/s$ to get

$$[(1/s)[2X(s) - 2Y_1(s)] - 3Y_1(s)](1/s)$$

which is the output signal $Y_1(s)$ since the output of the third summer is $Y_1(s)$. Finally, we can write

$$Y_1(s) = [(1/s)[2X(s) - 2Y_1(s)] - 3Y_1(s)](1/s)$$

Rearranging terms, we have the transfer function

$$\frac{Y_1(s)}{X(s)} = \frac{2}{s^2 + 3s + 2}$$

With $x(t) = u(t)$, $X(s)$ is $1/s$ and

$$Y_1(s) = \frac{2}{s(s^2 + 3s + 2)} = \frac{A}{s} + \frac{B}{s+1} + \frac{C}{s+2} = \frac{1}{s} + \frac{-2}{s+1} + \frac{1}{s+2}$$

with

$$y_1(t) = (e^{-2t} - 2e^{-t} + 1)u(t)$$

For the second block diagram in Figure 7.9, the output of the first summer is the signal $-2Y_2(s)$ which is then multiplied by $1/s$ to give the signal

$$(1/s)(-2Y_2(s))$$

The output of the second summer is the signal

$$[(1/s)(-2Y_2(s))] - 3Y_2(s) + X(s)$$

Then this last signal is multiplied by $1/s$ to give the signal

$$[[(1/s)(-2Y_2(s))] - 3Y_2(s) + X(s)]1/s$$

The output of the last summer is the signal $Y_2(s)$ which must be equal to the input of the summer. Therefore,

$$Y_2(s) = [[(1/s)(-2Y_2(s))] - 3Y_2(s) + X(s)]1/s$$

By rearranging terms we get

$$\frac{Y_2(s)}{X(s)} = \frac{s}{s^2 + 3s + 2}$$

With $x(t) = u(t)$ and $X(s) = 1/s$ we have

$$Y_2(s) = \frac{s}{s(s^2 + 3s + 2)} = \frac{1}{(s^2 + 3s + 2)}$$

As before

$$y_2(t) = (-e^{-2t} + e^{-t})u(t)$$

5. The state-space representation

   The system in this example has two energy storage elements; thus, we will have two states. We will select the voltage across the capacitor as one state and the current in the inductor as the second state because they relate to the storage elements. We selected the voltage in the capacitor because as we know from circuit transient analysis that the voltage in the capacitor does not change instantaneously. We also selected the current in the inductor because it does not change instantaneously. Notice that other variables in the circuits were not selected as states. Take, for example, the voltage in the resistor

$$v_R(t) = R\, i_L(t)$$

where $i_L(t)$ is the current in the inductor as seen in the circuit. As you can see, $v_R(t)$ is a linear combination of the current in the inductor. Thus, it is not selected and it can be calculated after we find the current in the inductor easily.

Therefore, our two states are the voltage in the capacitor and the current in the inductor: $y_1(t)$ and $y_2(t)$.

In the storage elements

$$L\frac{d}{dt}y_2(t) = v_L(t) = x(t) - Ry_2(t) - y_1(t)$$

and

$$C\frac{d}{dt}y_1(t) = y_2(t)$$

We will let $z_1(t) = y_2(t)$ and $z_2(t) = y_1(t)$. Rewriting the above state equations with $L = 1\text{H}$, $C = 0.5\text{F}$, and $R = 3$ ohms we get

$$\frac{d}{dt}z_1(t) = -3z_1(t) - z_2(t) + x(t)$$

$$\frac{d}{dt}z_2(t) = 2z_1(t)$$

Assuming zero initial conditions, we can put the two simultaneous state equations in state-space matrix form as

$$\frac{d}{dt}\underline{z} = A\underline{z} + B\underline{x}$$

where

$$A = \begin{bmatrix} -3 & -1 \\ 2 & 0 \end{bmatrix}, \quad B = \begin{bmatrix} 1 \\ 0 \end{bmatrix}, \quad \frac{d}{dt}\underline{z} = \begin{bmatrix} \frac{d}{dt}z_1 \\ \frac{d}{dt}z_2 \end{bmatrix}, \quad \text{and} \quad \underline{z} = \begin{bmatrix} z_1 \\ z_2 \end{bmatrix}$$

The state solution is

$$\underline{Z}(s) = [sI - A]^{-1}B\underline{X}(s) + [sI - A]^{-1}\underline{z}(0^-)$$

Expand to get

$$\underline{Z}(s) = \left\{ s\begin{bmatrix} 1 & 0 \\ 0 & 1 \end{bmatrix} - \begin{bmatrix} -3 & -1 \\ 2 & 0 \end{bmatrix} \right\}^{-1} \begin{bmatrix} 1 \\ 0 \end{bmatrix} [1/s]$$

By simplification we arrive at

$$\underline{Z}(s) = \begin{bmatrix} s+3 & 1 \\ -2 & s \end{bmatrix}^{-1} \begin{bmatrix} 1/s \\ 0 \end{bmatrix}$$

or

$$\underline{Z}(s) = \frac{1}{s^2+3s+2} \begin{bmatrix} s & -1 \\ 2 & s+3 \end{bmatrix} \begin{bmatrix} 1/s \\ 0 \end{bmatrix}$$

$$\underline{Z}(s) = \begin{bmatrix} \dfrac{s}{s^2+3s+2} & \dfrac{-1}{s^2+3s+2} \\ \dfrac{2}{s^2+3s+2} & \dfrac{s+3}{s^2+3s+2} \end{bmatrix} \begin{bmatrix} 1/s \\ 0 \end{bmatrix}$$

Finally

$$\underline{Z}(s) = \begin{bmatrix} \dfrac{1}{(s^2+3s+2)} \\ \dfrac{2}{s(s^2+3s+2)} \end{bmatrix}$$

Taking the inverse Laplace transform of the above equation leads to

$$y_1(t) = (e^{-2t} - 2e^{-t} + 1)u(t)$$

and

$$y_2(t) = (-e^{-2t} + e^{-t})u(t)$$

The desired outputs were the states themselves where we did not need to solve the output equations in this case. Knowing the states $z_1(t) = i_L(t)$ and $z_2(t) = v_C(t)$, we can find any variable in the circuit. Actually, we can find the following variables:

1. The current in the resistor R

$$i_R(t) = i_L(t) = z_1(t)$$

2. The voltage in the inductor L

$$v_L(t) = L\frac{d}{dt}i_L(t) = x(t) - 3z_1(t) - z_2(t)$$

3. The current in the capacitor C

$$i_C(t) = C\frac{d}{dt}v_C(t) = 0.5z_2(t)$$

Considering these variables as outputs we can write the output equation as

$$\underline{y}(t) = C\underline{z}(t) + D\underline{x}(t)$$

where

$$\underline{y}(t) = [i_R(t)\ v_L(t)\ i_C(t)]^T$$

and T indicates transpose.
In this case we have single-input multiple-output (3 outputs) where before we had 2 outputs, the states themselves. In this example

$$D = \begin{bmatrix} 0 \\ 1 \\ 0 \end{bmatrix}$$

and

$$C = \begin{bmatrix} 1 & 0 \\ -3 & -1 \\ 0 & 0.5 \end{bmatrix}$$

The output equation is written as

$$\underline{y}(t) = \begin{bmatrix} 1 & 0 \\ -3 & -1 \\ 0 & 0.5 \end{bmatrix} \begin{bmatrix} z_1(t) \\ z_2(t) \end{bmatrix} + \begin{bmatrix} 0 \\ 1 \\ 0 \end{bmatrix} u(t)$$

$$\underline{y}(t) = \begin{bmatrix} z_1(t) \\ -3z_1(t) - z_2(t) \\ 0.5z_2(t) \end{bmatrix} + \begin{bmatrix} 0 \\ u(t) \\ 0 \end{bmatrix}$$

which gives

$$\begin{aligned} i_R(t) &= (-e^{-2t} + e^{-t})u(t) \\ v_L(t) &= (2e^{-2t} - e^{-t})u(t) \\ i_C(t) &= 0.5(e^{-2t} - 2e^{-t} + 1)u(t) \end{aligned}$$

We can see that all solutions agree in using all five representations.

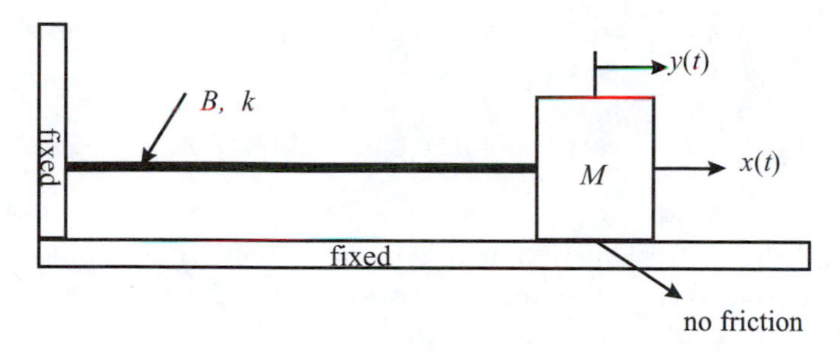

FIGURE 7.10
System for Example 7.3.

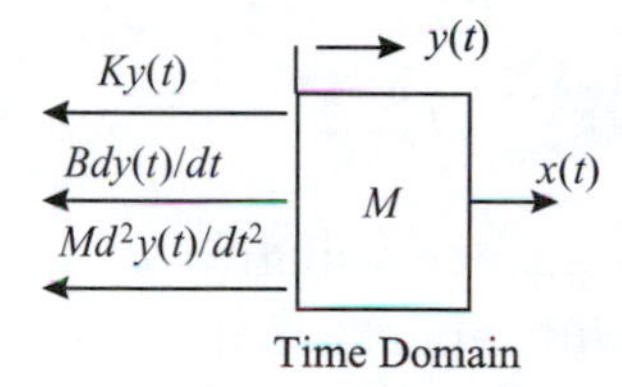

FIGURE 7.11
System for Example 7.3.

Example 7.3

Consider the mechanical system in Figure 7.10 with the input $x(t)$, an impulsive force of unity strength, $\delta(t)$, and initial conditions

$$\frac{d}{dt}y(0^-) = y(0^-) = 0$$

The rod to the left of the mass M can be modeled by a translational spring with spring constant K in parallel with a translational damper of constant B.

What is $y(t)$?

Solution

For illustrative purposes only assume $M = 1$kg, $B = 1$ newton/meter/second (N/m/s), $K = 1$ newton/meter, and let $x(t) = \delta(t)$.

1. The differential equation representation

   A free-body diagram is constructed to help us write the equations of motion as shown in Figure 7.11.

   We will consider the free-body diagram in time domain and sum the forces acting on the object of mass M to get

$$M\frac{d^2}{dt}y(t) + B\frac{d}{dt}y(t) + Ky(t) = x(t)$$

With the values for the constants as given above, the equation becomes

$$\frac{d^2}{dt}y(t) + \frac{d}{dt}y(t) + y(t) = \delta(t)$$

As we learned earlier, if the input to the system is the impulse signal, the output will be the impulse response, $h(t)$. Therefore, $y(t)$ can be determined using the Laplace transform techniques discussed previously to get

$$y(t) = \sqrt{\frac{4}{3}}e^{-\frac{1}{2}t}\sin\left(\sqrt{\frac{3}{4}}t\right)u(t)$$

2. The impulse response representation

   We start with the differential equation letting the input be $x(t) = \delta(t)$. Since the input is the $\delta(t)$ signal, the output will be called $h(t)$. The Laplace transform with initial conditions set to zero is

$$H(s) + sH(s) + s^2H(s) = 1$$

   or

$$H(s) = 1/[s^2 + s + 1]$$

   Next, we inverse transform $H(s)$ to get

$$h(t) = \sqrt{\frac{4}{3}}e^{-\frac{1}{2}t}\sin\left(\sqrt{\frac{3}{4}}t\right)u(t)$$

   We then use the convolution integral to find $y(t)$ as

$$y(t) = \int_{-\infty}^{\infty}\delta(\tau)\sqrt{\frac{4}{3}}e^{-\frac{1}{2}(t-\tau)}\sin\left(\sqrt{\frac{3}{4}}(t-\tau)\right)u(t-\tau)d\tau$$

   Here again, we can use the sifting property of the delta function to get

$$y(t) = \sqrt{\frac{4}{3}}e^{-\frac{1}{2}t}\sin\left(\sqrt{\frac{3}{4}}t\right)u(t)$$

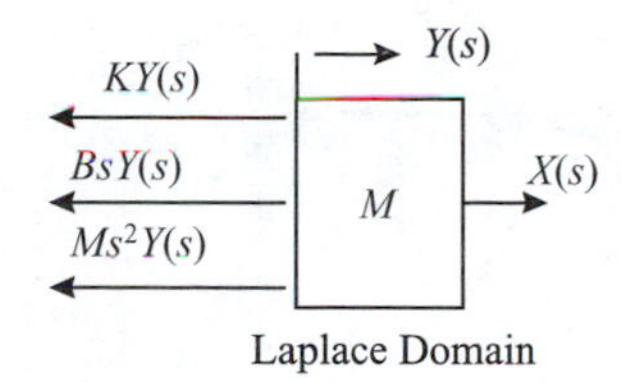

FIGURE 7.12
System for Example 7.3.

3. The transfer function representation

   We can Laplace-transform the free-body diagram as shown in Figure 7.11 to get the diagram in Figure 7.12.

   Summing the forces using the Laplace-transformed free-body diagram we get

$$KY(s) + BsY(s) + Ms^2Y(s) = X(s)$$

   With $M = 1$ kg, $B = 1$ N/m/sec, and $K = 1$ N/m, we can write

$$Y(s) + sY(s) + s^2Y(s) = X(s)$$

   or

$$Y(s)[s^2 + s + 1] = X(s)$$

   And finally, the transfer function is

$$Y(s)/X(s) = 1/[s^2 + s + 1]$$

   With $X(s) = 1$, we can take the inverse Laplace transform of $Y(s)$ to get

$$y(t) = \sqrt{\frac{4}{3}}e^{-\frac{1}{2}t}\sin\left(\sqrt{\frac{3}{4}}t\right)u(t)$$

4. The block diagram representation

   We start with the transfer function

$$Y(s)/X(s) = 1/[s^2 + s + 1]$$

   This is a second order system, and hence, we need two integrators as seen in the initial block in Figure 7.13.

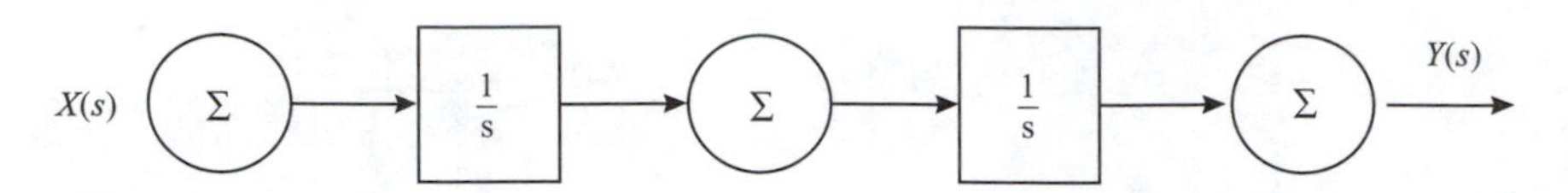

FIGURE 7.13
Block For Example 7.3.

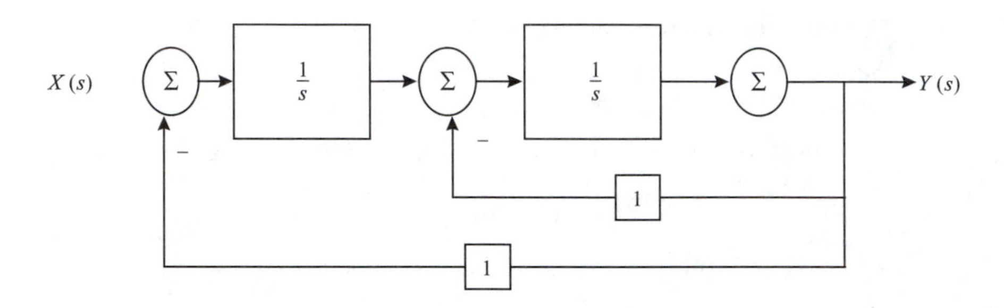

FIGURE 7.14
Block for Example 7.3.

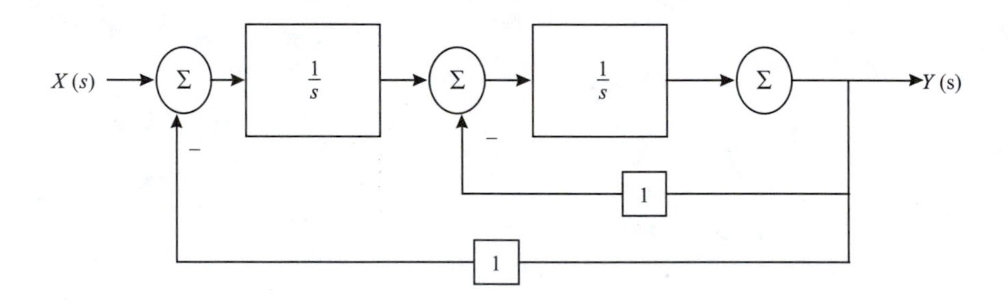

FIGURE 7.15
Block for Example 7.3.

Then we need to feed backward $(-1Y(s))$, the constant term in the denominator, to the summer preceding the first integrator and $(-1Y(s))$, the coefficient of s in the denominator, to the summer that precedes the second integrator and then obtain Figure 7.14.

Next, we feed forward $(1X(s))$ to the first summer as seen in Figure 7.15.

At the output of the first summer we have the signal $(X(s) - Y(s))$. This signal is then multiplied by $1/s$ to get

$$(1/s)[X(s) - Y(s)]$$

At the output of the second summer we have the signal

$$(1/s)[X(s) - Y(s)] - Y(s)$$

This signal is then multiplied by $1/s$ to get

$$[(1/s)[X(s) - Y(s)] - Y(s)](1/s)$$

which is the output signal $Y(s)$ since the output of the third summer is $Y(s)$. Finally, we can write

$$Y(s) = [(1/s)\,[X(s) - Y(s)] - Y(s)](1/s)$$

Rearranging terms we have the transfer function

$$Y(s)/X(s) = 1/[s^2 + s + 1]$$

With $X(s) = 1$, we can inverse transform $Y(s)$ to get

$$y(t) = \sqrt{\frac{4}{3}}e^{-\frac{1}{2}t}\sin\left(\sqrt{\frac{3}{4}}t\right)u(t)$$

5. The state-space representation

   This was discussed in the previous chapter and is repeated here for completeness. We are interested in using state-space approach in solving for the output $y(t)$. We will consider the free-body diagram in time domain and sum the forces acting on the object of mass M to get

$$M\frac{d^2}{dt}y(t) + B\frac{d}{dt}y(t) + Ky(t) = x(t)$$

   In transforming this second order differential equation into state-space form we need to define two state variables since the differential equation given is second order.

   Let

$$z_1 = y(t)$$

$$z_2 = \frac{d}{dt}y(t)$$

   Next, we take derivatives of the two states to get

$$\frac{d}{dt}z_1 = \frac{d}{dt}y(t)$$

$$\frac{d}{dt}z_2 = \frac{d^2}{dt}y(t)$$

In terms of state variables and inputs to the system we have

$$\frac{d}{dt}z_1 = z_2$$

$$\frac{d}{dt}z_2 = -z_1 - z_2 + x$$

These are the two simultaneous state equations. Putting them in state-space matrix form with $x(t) = \delta(t)$ we get

$$\frac{d}{dt}\underline{z} = A\underline{z} + B\underline{x}$$

where

$$A = \begin{bmatrix} 0 & 1 \\ -1 & -1 \end{bmatrix},\ B = \begin{bmatrix} 0 \\ 1 \end{bmatrix},\ \frac{d}{dt}\underline{z} = \begin{bmatrix} \frac{d}{dt}z_1 \\ \frac{d}{dt}z_2 \end{bmatrix},\ \underline{z} = \begin{bmatrix} z_1 \\ z_2 \end{bmatrix},\ \text{and } \underline{x} = [x]$$

Up to this point we are still in time domain. We now try to solve this matrix state equation using the Laplace transform approach. The state solution is

$$\underline{Z}(s) = [sI - A]^{-1}B\underline{X}(s) + [sI - A]^{-1}\underline{z}(0^-)$$

Substituting for A, B, and the initial conditions we get

$$\underline{Z}(s) = \left\{s\begin{bmatrix} 1 & 0 \\ 0 & 1 \end{bmatrix} - \begin{bmatrix} 0 & 1 \\ -1 & -1 \end{bmatrix}\right\}^{-1}\begin{bmatrix} 0 \\ 1 \end{bmatrix}[1]$$

$$\underline{Z}(s) = \begin{bmatrix} s & -1 \\ 1 & s+1 \end{bmatrix}^{-1}\begin{bmatrix} 0 \\ 1 \end{bmatrix}$$

$$\underline{Z}(s) = \frac{1}{s^2 + s + 1}\begin{bmatrix} s+1 & 1 \\ -1 & s \end{bmatrix}\begin{bmatrix} 0 \\ 1 \end{bmatrix}$$

$$\underline{Z}(s) = \begin{bmatrix} \dfrac{s+1}{s^2+s+1} & \dfrac{1}{s^2+s+1} \\ \dfrac{-1}{s^2+s+1} & \dfrac{s}{s^2+s+1} \end{bmatrix}\begin{bmatrix} 0 \\ 1 \end{bmatrix}$$

and finally,

$$\underline{Z}(s) = \begin{bmatrix} \dfrac{1}{(s^2+s+1)} \\ \dfrac{s}{(s^2+s+1)} \end{bmatrix}.$$

The first entry in the above vector is really $y(t)$ as defined above. So we need only to inverse transform this entry to get $y(t)$ as

$$y(t) = \sqrt{\frac{4}{3}} e^{-\frac{1}{2}t} \sin\left(\sqrt{\frac{3}{4}}t\right) u(t)$$

7.3 Some Insights: The Poles Considering Different Outputs within the Same System

Given a linear time-invariant system with multiple-input multiple-output, the poles of the system, regardless of what output you choose and what input you consider, will stay the same. This means that in such a system of many inputs and many outputs, the shape of the transients will be dominated by the fixed number poles of the system.

As we saw in Example 7.2 in this chapter, we considered extra outputs:

1. The current in the resistor R

$$i_R(t) = i_L(t) = z_1(t)$$

2. The voltage in the inductor L

$$v_L(t) = L\frac{d}{dt}i_L(t) = x(t) - 3z_1(t) - z_2(t)$$

3. The current in the capacitor C

$$i_C(t) = C\frac{d}{dt}v_C(t) = 0.5z_2(t)$$

The solution for these outputs was found to be

$$i_R(t) = (-e^{-2t} + e^{-t})u(t)$$
$$v_L(t) = (2e^{-2t} - e^{-t})u(t)$$
$$i_C(t) = 0.5(e^{-2t} - 2e^{-t} + 1)u(t)$$

In all of these outputs, what is shaping the transients? Clearly, the eigenvalues of -1 and -2 appear in all the solutions. All of these outputs are stable because their eigenvalues are the same and all are negative. Again, the eigenvalues here are the coefficients of the time variable t that appears in the exponential terms.

In summary, in a system of many inputs and many outputs, if you consider the transfer function relating any input to any output, the eigenvalues or the poles will be the same. You may have different zeros for different transfer functions. But different transfer functions within the same system will have the same exact poles.

7.4 End-of-Chapter Examples

EOCE 7.1

Consider the mechanical system shown in Figure 7.16 with initial conditions

$$y(0) = 1 \quad \text{and} \quad \frac{d}{dt}y(0) = 0$$

Find the output, $y(t)$, using all five representations with $x(t) = 0$ (response due to initial conditions). The rod to the left of the mass M can be modeled by a translational spring with spring constant K, in parallel with a translational damper of constant B.

Solution

1. The differential equation representation

   B is the damping and K is the spring constant. As seen in Chapter 6,

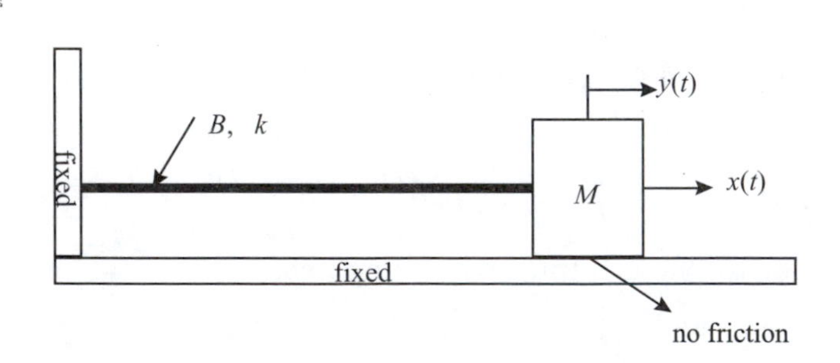

FIGURE 7.16
System for EOCE 7.1.

the differential equation representing the system is

$$M\frac{d^2}{dt}y(t) + B\frac{d}{dt}y(t) + Ky(t) = x(t)$$

In a general form, we can write the above equation as

$$\frac{d^2}{dt}y(t) + 2\xi w_n\frac{d}{dt}y(t) + w_n^2 y(t) = \frac{x(t)}{M}$$

If ξ is 0.5 and $w_n = 2$, and $M = 1$, the equation becomes

$$\frac{d^2}{dt}y(t) + 2\frac{d}{dt}y(t) + 4y(t) = 0$$

with the initial conditions

$$y(0) = 1$$
$$\frac{d}{dt}y(0) = 0$$

We can use MATLAB and write the following script.

```
y = dsolve('D2y + 2*Dy + 4* y = 0','y(0) = 1','Dy(0) = 0')
axis ([0 10 -1 1])
ezplot(y, [0 10])
ylabel ('The displacement y')
title ('y = cos(3^1/2)*t)+1/3*sin(3^(1/2)*t)*3^(1/2))...
/exp(t)')
xlabel('Time (sec)')
```

with the output

```
y = (cos(3^(1/2) * t) + 1/3 * sin(3^(1/2) * t) *
3^(1/2))/exp(t)
```

and the output plot is shown in Figure 7.17.

2. The impulse response representation

   When we calculate the impulse response of the system we let $x(t)$ be the impulse function $\delta(t)$ with zero initial conditions. To do that we use the following MATLAB script to find the step response and then take the derivative of the step response to get the impulse response $h(t)$ as in the following

```
step = dsolve('D2y + 2*Dy + 4*y = 1','y(0) =
0','Dy(0) = 0')
impulse=diff(step) % taking the derivative of the step
```

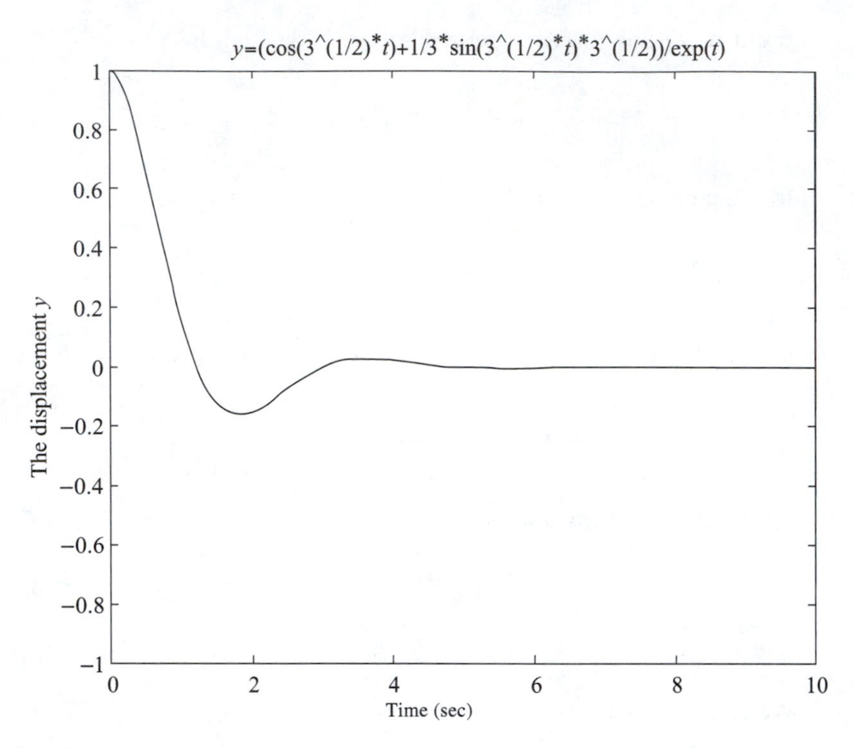

FIGURE 7.17
Plot for EOCE 7.1.

```
axis([0 10 -1 1]) % set the axis
impulse = simple(impulse) % to simplify
clf % to clear the screen
ezplot(impulse,[0 10])
ylabel('The impulse response')
title('h(t) = 1/3*sin(3^(1/2)*t)*3^(1/2)/exp(t)')
xlabel('Time (sec)')
```

The output is

```
impulse =

1/4*(exp(t) - cos(3^(1/2)*t) + sin(3^(1/2)
*t)*3^(1/2))/exp(t)-1/4*(exp(t)-1/3*sin(3^(1/2)
*t)*3^(1/2) - cos(3^(1/2)*t))/exp(t)

impulse =

1/3*sin(3^(1/2)*t)*3^(1/2)/exp(t)
```

with the plot shown in Figure 7.18.

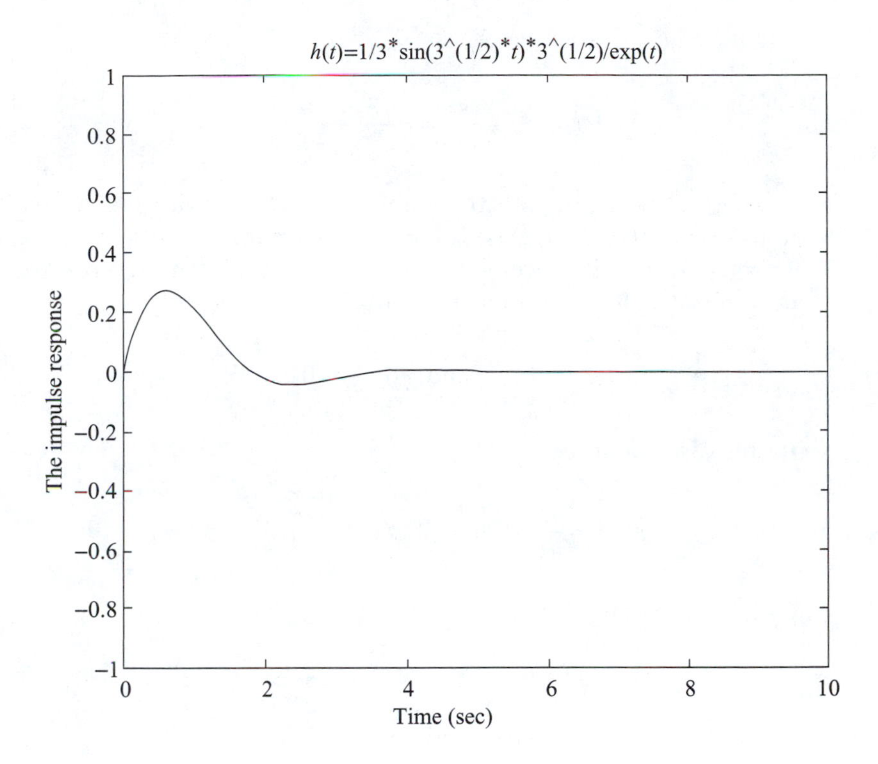

FIGURE 7.18
Plot for EOCE 7.1.

Knowing the impulse response we can find the output of the system to any input by using the convolution formula.

3. The transfer function representation

   We can Laplace-transform the differential equation

$$\frac{d^2}{dt}y(t) + 2\frac{d}{dt}y(t) + 4y(t) = 0$$

   by setting the initial conditions to zero.

   In Laplace domain we have

$$s^2Y(s) + 2sY(s) + 4Y(s) = X(s)$$

   Taking $Y(s)$ as a common factor we have

$$Y(s)[s^2 + 2s + 4] = X(s)$$

and the transfer function is

$$H(s) = \frac{Y(s)}{X(s)} = \frac{1}{s^2 + 2s + 4}$$

Knowing $H(s)$ we can find $y(t)$ for any input $x(t)$ by taking the inverse transform of $[H(s)X(s)]$. $X(s)$ is always known for any system.
To solve for the displacement output as our problem requires, we need to set $x(t) = 0$ and use the initial conditions

$$y(0) = 1 \quad \text{and} \quad \frac{d}{dt}y(0) = 0$$

In this case, we have

$$\left[s^2Y(s) - sy(0) - \frac{d}{dt}y(0)\right] + 2[sY(s) - y(0)] + 4Y(s) = 0$$

Or by simplifying, we get

$$Y(s)[s^2 + 2s + 4] = s + 2$$

The output in the Laplace domain is

$$Y(s) = \frac{s+2}{s^2 + 2s + 4} = \frac{s+1}{(s+1)^2 + (\sqrt{3})^2} + \frac{1}{\sqrt{3}}\frac{1\sqrt{3}}{(s+1)^2 + (\sqrt{3})^2}$$

We can use the Laplace transform tables to find $y(t)$ as

$$y(t) = \left[e^{-t}\cos(\sqrt{3}t) + \frac{1}{\sqrt{3}}e^{-t}\sin(\sqrt{3}t)\right]u(t)$$

Let us plot $y(t)$ to see if it matches the result we obtained in the differential equation representation. We do that by writing the MATLAB script

```
t = 0:0.05:10;
y = exp(-t).*(cos(sqrt(3)*t) + (1/sqrt(3))*sin(sqrt(3)*t));
plot(t,y);
title('Response due to initial conditions');
xlabel('Time (sec)');
ylabel('The output y(t)');
axis([0 10 -1 1]);
```

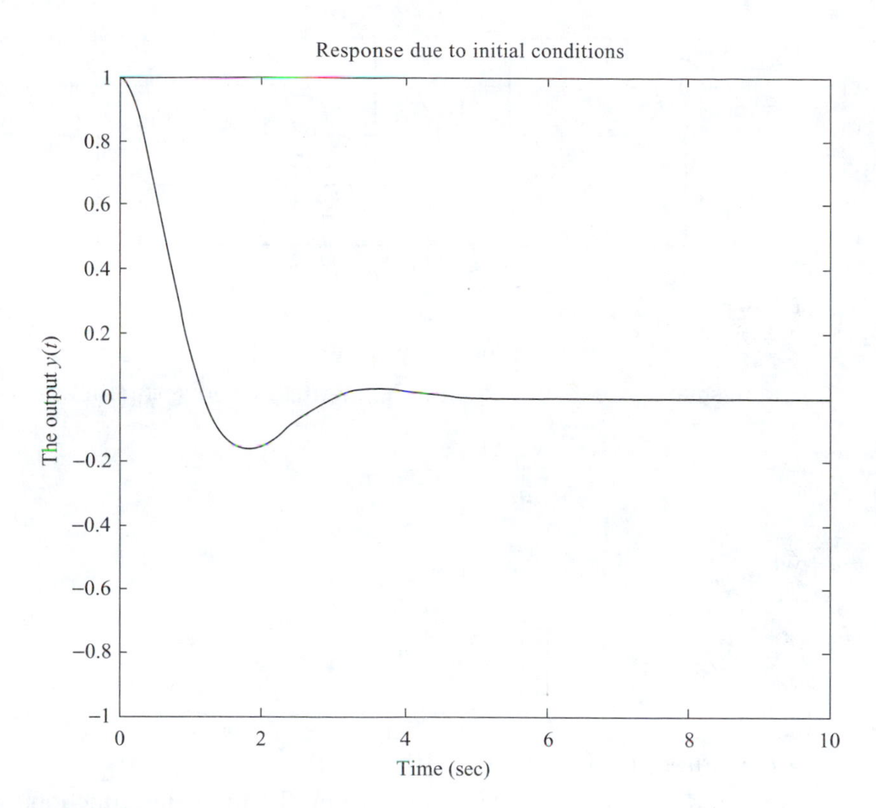

FIGURE 7.19
Plot for EOCE 7.1.

The plot is shown in Figure 7.19.

We can see that the result here matches the result we obtained previously in the first representation.

4. The block diagram representation

From the transfer function

$$H(s) = \frac{Y(s)}{X(s)} = \frac{1}{s^2 + 2s + 4}$$

we can draw the block diagram as in Figure 7.20.

5. The state-space representation

We will let $z_1(t) = y(t)$ and $z_2(t) = \frac{d}{dt}y(t)$. Therefore, we can write

$$\frac{d}{dt}z_1(t) = z_2(t)$$

$$\frac{d}{dt}z_2(t) = -4z_1(t) - 2z_2(t) + x(t)$$

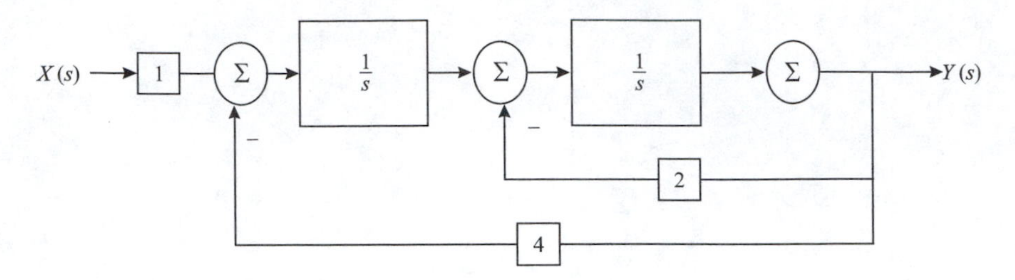

FIGURE 7.20
Block for EOCE 7.1.

and in state-space matrix equation we write the above equations as

$$\frac{d}{dt}\underset{\sim}{z} = A\underset{\sim}{z} + B\underset{\sim}{x}$$

where

$$A = \begin{bmatrix} 0 & 1 \\ -4 & -2 \end{bmatrix}, \quad B = \begin{bmatrix} 0 \\ 0 \end{bmatrix}, \quad \text{and} \quad \underset{\sim}{z}(0) = \begin{bmatrix} 1 \\ 0 \end{bmatrix}$$

We will use MATLAB to find $y(t) = z_1(t)$.
We will define the system first by writing the following function.

```
function zdot = eoce1(t, z)
global A
global B
zdot = A * z + B
```

Then we write the following script that will call the functions "eoce1" and "ode23."

```
clf % to clear the graph area
global A
global B
t0 = 0;
tf = 10;
z0 = [1; 0]; % vector of initial conditions
A = [0 1;-4 -2];
```

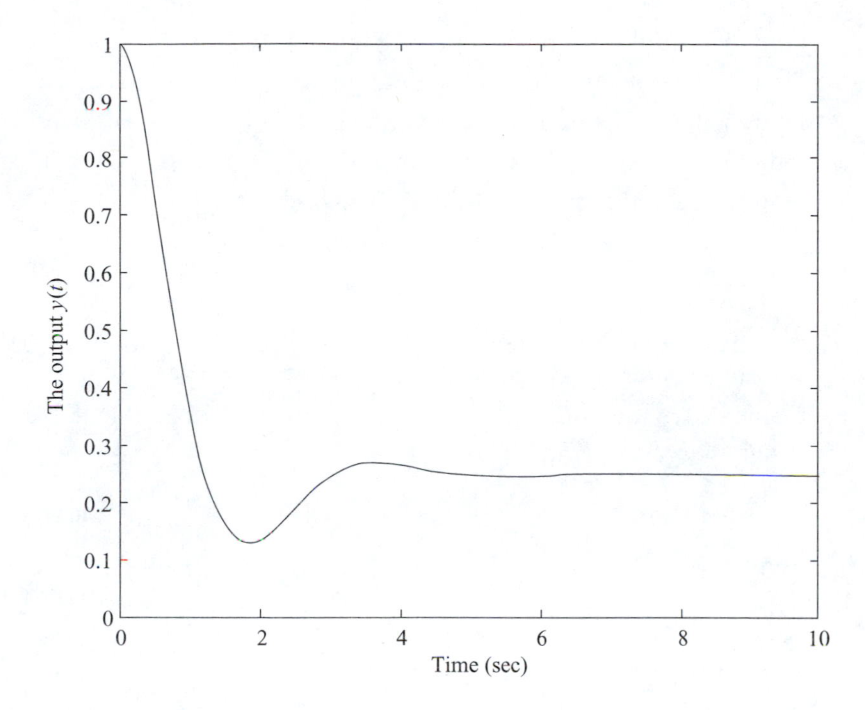

FIGURE 7.21
Plot for EOCE 7.1.

```
B = [0; 1];
[t, z]=ode23('eoce1', t0, tf, z0); % a call to the ode23
y1=z(:, 1);
plot (t, y1);
xlabel('Time (sec)')
ylabel('The output y(t)')
```

The output plot is shown in Figure 7.21.

This graph indeed agrees with all of the other plots for $y(t)$ in all representations.

EOCE 7.2

Consider the following differential equation.

$$\frac{d^2}{dt}y(t) + 7\frac{d}{dt}y(t) + 12y(t) = 10x(t)$$

1. Find the impulse response representation.
2. Find the transfer function representation.
3. Find the block diagram representation.
4. Find the state-space representation
5. Find the step and the impulse response of the system.

Solution

1. The impulse response representation

   We can calculate the impulse response in many ways.

   Graphically, the impulse response representation is calculated using the MATLAB script

```
num = [10];
den = [1  7  12];
impulse(num, den)
```

   The output plot is shown in Figure 7.22.

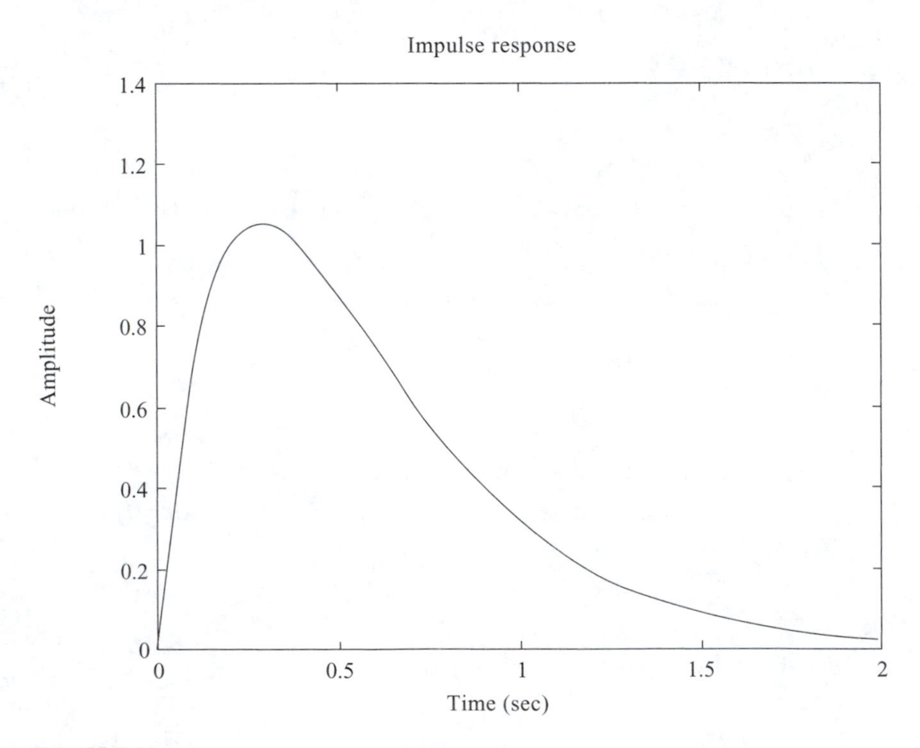

FIGURE 7.22
Plot for EOCE 7.2.

Using MATLAB symbolic toolbox function "dsolve" we can write the following script.

```
step = dsolve('D2y  + 7*Dy + 12*y = 10', 'y(0) =
0', 'Dy(0) = 0')

impulse = diff(step); % taking the derivative of the step

axis([0 10 0 1.5]) % set the axis

impulse = simple(impulse) % to simplify

clf % to clear the screen

ezplot(impulse, [0 10])

ylabel ('The impulse response')
```

The output is

```
impulse =

-10/exp(t)^4 + 10/exp(t)^3
```

and the plot is shown in Figure 7.23.
Using partial fraction expansion with MATLAB, we write the script

```
num = [10];

den = [1 7 12];

[R, P, K] = residue(num, den)
```

The output is

```
R =

-10

 10

P =

-4

-3

K =

            []
```

In this case

$$H(s) = \frac{-10}{s+4} + \frac{10}{s+3}$$

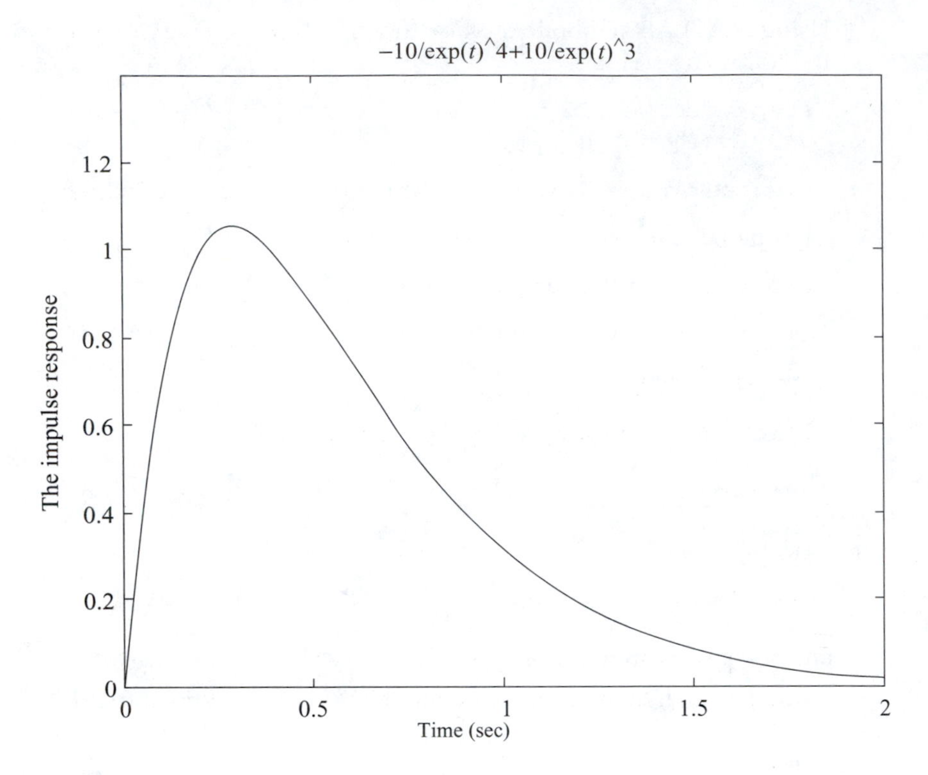

FIGURE 7.23
Plot for EOCE 7.2.

and

$$h(t) = (-10e^{-4t} + 10e^{-3t})u(t)$$

which if plotted against time will produce the same figure as Figure 7.23.

2. The transfer function representation

   We can Laplace-transform the differential equation with zero initial conditions to get

$$s^2Y(s) + 7sY(s) + 12Y(s) = 10X(s)$$

   with the transfer function

$$H(s) = \frac{Y(s)}{X(s)} = \frac{10}{s^2 + 7s + 12}$$

3. The block diagram representation

   The block diagram can be drawn directly from $H(s)$ as shown in Figure 7.24.

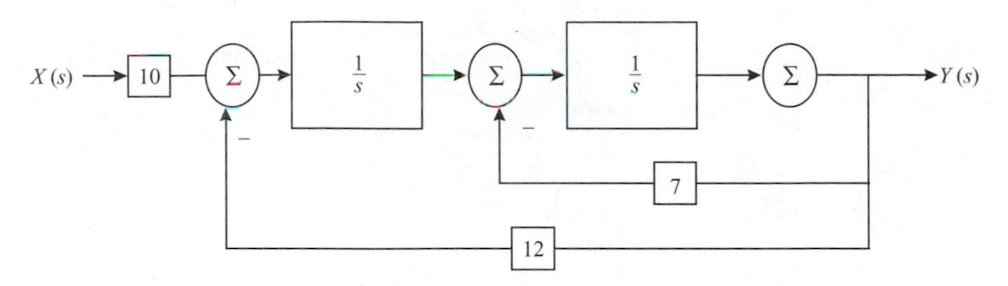

FIGURE 7.24
Block for EOCE 7.2.

4. The state-space representation
   We will let $z_1(t) = y(t)$ and $z_2(t) = \frac{d}{dt}y(t)$. Then

$$\frac{d}{dt}z_1(t) = z_2(t)$$

$$\frac{d}{dt}z_2(t) = -12z_1(t) - 7z_2(t) + 10x(t)$$

   and the state matrix equation is

$$\frac{d}{dt}\underline{z} = A\underline{z} + B\underline{x}$$

   with

$$A = \begin{bmatrix} 0 & 1 \\ -12 & -7 \end{bmatrix}, \quad B = \begin{bmatrix} 0 \\ 10 \end{bmatrix}, \quad \text{and} \quad \underline{z}(0) = \begin{bmatrix} 0 \\ 0 \end{bmatrix}$$

5. The step and the impulse responses

   To accomplish that we use the MATLAB script as follows.

```
num = [10]
den = [1  7  12];
step(num, den);
hold on
impulse(num, den);
ylabel('The step and the impulse response');
title('and the impulse response using transfer
functions');
```

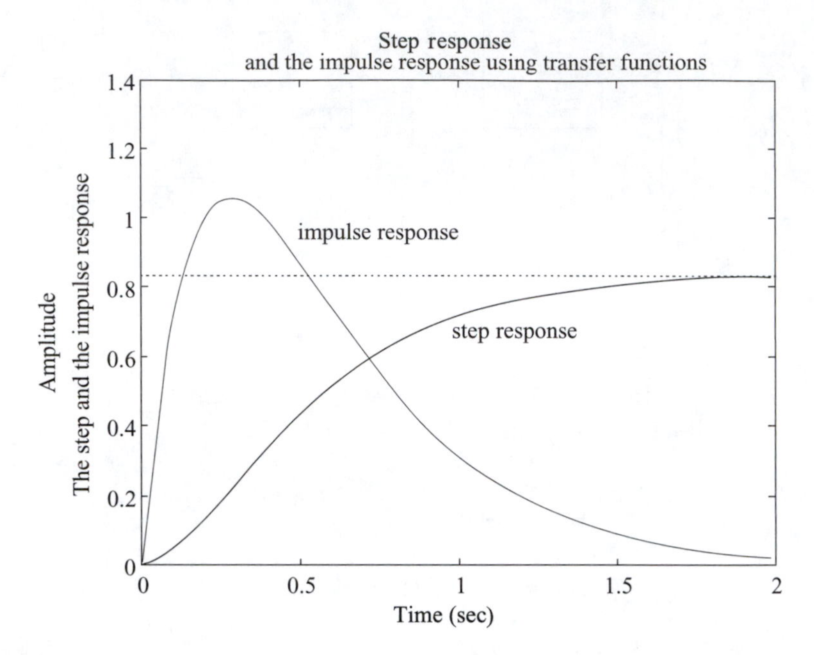

FIGURE 7.25
Plots for EOCE 7.2.

```
gtext('step response');
gtext('impulse response')
```

The output is shown in Figure 7.25.

EOCE 7.3

Consider the block diagram representation shown in Figure 7.26.

Represent the system using the other representations discussed previously.

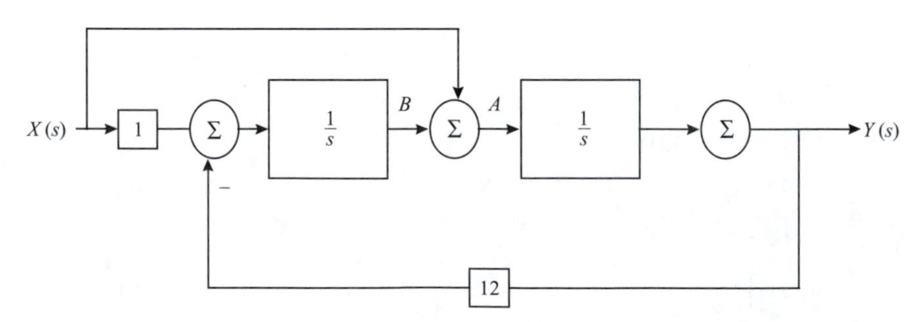

FIGURE 7.26
System for EOCE 7.3.

Solution

1. The differential equation representation

   The output of the second integrator is y. Let the input to the second integrator be A. Let the output of the first integrator be B. The input to the first integrator is $(x - 12y)$. But

$$(x - 12y) = \frac{d}{dt}B$$

   and

$$A = x + B = \frac{d}{dt}y$$

   Also

$$\frac{d}{dt}A = \frac{d}{dt}x + \frac{d}{dt}B = \frac{d}{dt}x + x - 12y = \frac{d^2}{dt}y$$

   The differential equation representing the system is then

$$\frac{d^2}{dt}y(t) + 12y(t) = x(t) + \frac{d}{dt}x(t)$$

2. The transfer function representation

   Let us Laplace-transform the differential equation with zero initial conditions to get

$$s^2Y(s) + 12Y(s) = X(s) + sX(s)$$

   with the transfer function

$$\frac{Y(s)}{X(s)} = \frac{s+1}{s^2+12} = H(s)$$

   Let us work with the block diagram directly. From the diagram

$$Y(s) = 1/s[A] = 1/s[X(s) + B]$$

   Substituting for B we get

$$Y(s) = 1/s[X(s) + 1/s[X(s) - 12Y(s)]]$$

or

$$Y(s)[1 + 12/s^2] = X(s)[1/s + 1/s^2]$$
$$Y(s)[s^2 + 12] = X(s)[s + 1]$$

The transfer function finally is

$$H(s) = \frac{Y(s)}{X(s)} = \frac{s + 1}{s^2 + 12}$$

3. The impulse response representation

   The impulse response representation can be calculated graphically and analytically. Graphically it can be found using the MATLAB command

```
impulse([1 1], [1 0 12]);
```

   The plot is shown in Figure 7.27.

   Analytically

$$H(s) = \frac{s + 1}{s^2 + 12} = \frac{s}{s^2 + 12} + \frac{1}{s^2 + 12}$$

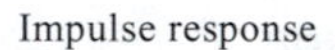

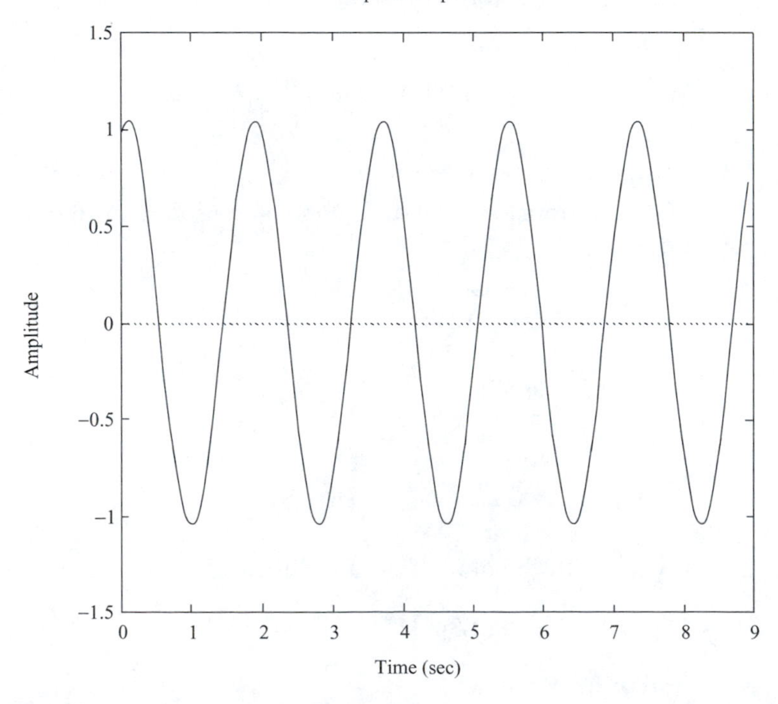

FIGURE 7.27
Plot for EOCE 7.3.

Using the table for the Laplace transform we can write $h(t)$ as

$$h(t) = \cos(\sqrt{12}t)u(t) + \frac{1}{\sqrt{12}}\sin(\sqrt{12}t)u(t)$$

This $h(t)$ agrees with Figure 7.27.

4. The state-space representation

   We will use MATLAB to find the state-space representation. Let us create the following script in MATLAB.

```
num = [1  1];
den = [1  0  12];
t = tf (num, den) % create the transfer function model
ssr  = ss (t) % create the state-space representation
```

The output is

```
Transfer  function:
s + 1
---------
s^2 + 12
a =
                       x1           x2
         x1             0     -3.00000
         x2       4.00000            0
b =
                       u1
         x1       1.00000
         x2             0
c =
                       x1           x2
         y1       1.00000      0.25000
d =
                       u1
         y1             0
```

This indicates that

$$A = \begin{bmatrix} 0 & -3 \\ 4 & 0 \end{bmatrix}, \quad B = \begin{bmatrix} 1 \\ 0 \end{bmatrix}, \quad C = \begin{bmatrix} 1 & 0.25 \end{bmatrix}, \quad \text{and} \quad D = \begin{bmatrix} 0 \end{bmatrix}$$

If we do not use MATLAB and try to write the state-space representation, we will let $z_1(t) = y(t)$ and $z_2(t) = \frac{d}{dt}y(t) - x(t)$.

From the differential equations

$$\frac{d}{dt}z_1(t) = \frac{d}{dt}y(t) = z_2(t) + x(t)$$

$$\frac{d}{dt}z_2(t) = \frac{d^2}{dt}y(t) - \frac{d}{dt}x(t) = x(t) - 12z_1(t)$$

the state and the output matrix equations are

$$\frac{d}{dt}\underset{\sim}{z} = A\underset{\sim}{z} + B\underset{\sim}{x}$$

$$\underset{\sim}{y} = C\underset{\sim}{z} + D\underset{\sim}{x}$$

with

$$A = \begin{bmatrix} 0 & 1 \\ -12 & 0 \end{bmatrix}, B = \begin{bmatrix} 1 \\ 1 \end{bmatrix}, \underset{\sim}{z}(0) = \begin{bmatrix} 0 \\ 0 \end{bmatrix}, C = \begin{bmatrix} 1 & 0 \end{bmatrix}, \text{ and } D = \begin{bmatrix} 0 \end{bmatrix}$$

With C as given, $y = z_1(t)$.

You also can see that the two state-space models or representations are different. Let us plot the output $y(t)$ in both cases and then we will make a conclusion.

For both models we first define the system in MATLAB by writing the following function.

```
function zdot = eoce3 (t, z)
global A
global B
zdot = A*z + B
```

For the model we derived we write the MATLAB script

```
clf % to clear the graph area
global A % declaring the dynamic matrix A as global
global B
t0 = 0;
tf = 10;
z0 = [0; 0]; % vector of initial conditions
A = [0 1;-12 0];
B = [1; 1];
[t, z]=ode23('eoce3', t0, tf, z0); % a call to the ode23
```

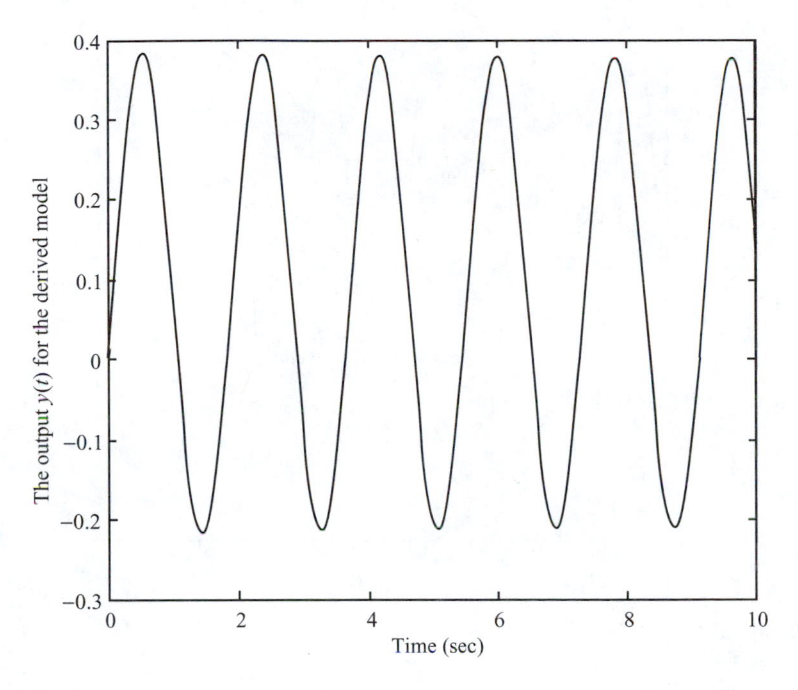

FIGURE 7.28
Plot for EOCE 7.3.

```
y1 = z(:,1);
plot(t, y1);
xlabel('Time (sec)')
ylabel('The output y(t) for the derived model')
```

The plot is shown in Figure 7.28.
For the MATLAB model the script is

```
clf % to clear the graph area
global A
global B
t0 = 0;
tf = 10;
z0 = [0;0]; % vector of initial conditions
A = [0 -3;4 0];
B = [1;0];
[t,z] = ode23('eoce3',t0,tf,z0); % a call to the ode23
```

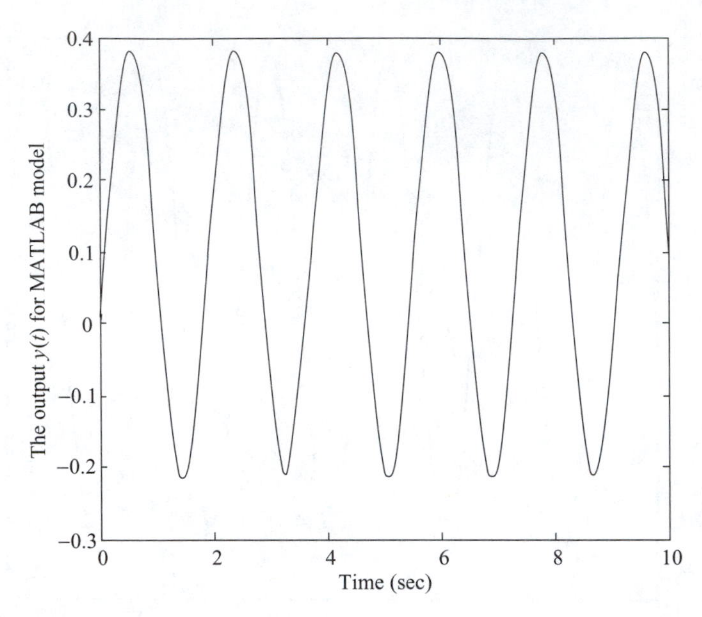

FIGURE 7.29
Plot for EOCE 7.3.

```
y1 = z(:,1) + 0.25*z(:,2);
plot(t, y1);
xlabel('Time (sec)')
ylabel('The output y(t) for MATLAB model')
```

The plot is shown in Figure 7.29.

It is clear that the two plots are identical and therefore, the state-space representations for the same system are not unique.

We can also obtain the state-space representation directly from the block diagram. Let the output of the second integrator be z_1 and the output of the first integrator be z_2.

In this case

$$y = z_1$$

By differentiating we get

$$\frac{d}{dt}z_1 = z_2 + x$$

Also

$$\frac{d}{dt}z_2 = x - 12z_1$$

The resulting state matrices are

$$A = \begin{bmatrix} 0 & 1 \\ -12 & 0 \end{bmatrix}, \quad B = \begin{bmatrix} 1 \\ 1 \end{bmatrix}, \quad C = \begin{bmatrix} 1 & 0 \end{bmatrix}, \quad \text{and} \quad D = \begin{bmatrix} 0 \end{bmatrix}$$

EOCE 7.4

Consider the impulse response of a certain system as

$$h(t) = 10e^{-2t}u(t) - 10e^{-4t}u(t)$$

Derive the other representations.

Solution

1. The transfer function representation

   The inverse Laplace transform of the impulse response gives the transfer function. Laplace-transform the given $h(t)$ to get

$$H(s) = \frac{10}{s+2} - \frac{10}{s+4}$$

   or

$$H(s) = \frac{10(s+4) - 10(s+2)}{s^2 + 6s + 8} = \frac{20}{s^2 + 6s + 8}$$

2. The differential equation representation

   Remember that the impulse response is obtained by setting the initial conditions to zero.

$$H(s) = \frac{Y(s)}{X(s)} = \frac{20}{s^2 + 6s + 8}$$

   or we can write

$$Y(s)[s^2 + 6s + 8] = X(s)[20]$$

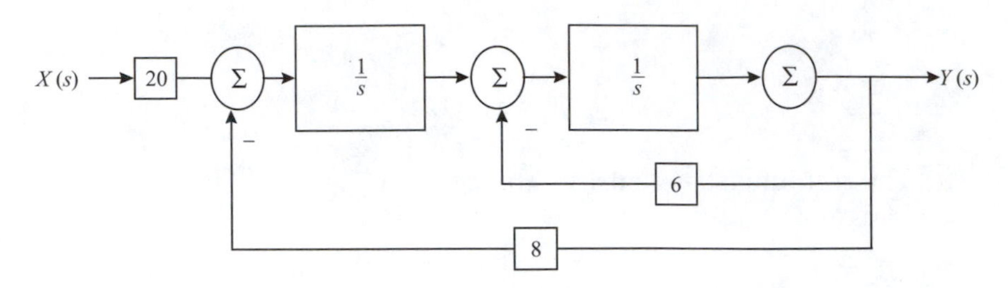

FIGURE 7.30
Block for EOCE 7.4.

Let us now inverse transform term by term to get

$$\frac{d^2}{dt}y(t) + 6\frac{d}{dt}y(t) + 8y(t) = 20x(t)$$

3. The block diagram representation
   The block diagram is shown in Figure 7.30.
4. The state-space representation
   First we will use MATLAB to do that. Here is the MATLAB script.

```
num = [20];
den = [1  6  8];
t = tf(num, den); % create the transfer function model
ssr = ss(t) % create the state-space representation
```

The output is

```
a =
                        x1          x2
        x1        -6.00000    -2.00000
        x2         4.00000           0
b =
                        u1
        x1         2.00000
        x2               0
c =
                        x1          x2
        y1               0     2.50000
d =
                        u1
        y1               0
```

The state space matrices are then

$$A = \begin{bmatrix} -6 & -2 \\ 4 & 0 \end{bmatrix}, \quad B = \begin{bmatrix} 2 \\ 0 \end{bmatrix}, \quad C = \begin{bmatrix} 0 & 2.5 \end{bmatrix}, \quad \text{and} \quad D = \begin{bmatrix} 0 \end{bmatrix}$$

The output is

$$y(t) = C\underset{\sim}{z} + D\underset{\sim}{x} = 2.5(z_2(t))$$

Second, we can find the state-space model in another way. Let us define the states as

$$\frac{d}{dt}z_1(t) = \frac{d}{dt}y(t) = z_2(t)$$

$$\frac{d}{dt}z_2(t) = \frac{d^2}{dt}y(t) = 20x(t) - 8z_1(t) - 6z_2(t)$$

and the state and output matrix equations are then

$$\frac{d}{dt}\underset{\sim}{z} = A\underset{\sim}{z} + B\underset{\sim}{x}$$

$$\underset{\sim}{y} = C\underset{\sim}{z} + D\underset{\sim}{x}$$

with

$$A = \begin{bmatrix} 0 & 1 \\ -8 & -6 \end{bmatrix}, \ B = \begin{bmatrix} 0 \\ 20 \end{bmatrix}, \ \underset{\sim}{z}(0) = \begin{bmatrix} 0 \\ 0 \end{bmatrix}, \ C = \begin{bmatrix} 1 & 0 \end{bmatrix}, \ \text{and } D = \begin{bmatrix} 0 \end{bmatrix}$$

The output is

$$y(t) = C\underset{\sim}{z} + D\underset{\sim}{x} = z_1(t)$$

You can see that the two models are not the same by comparing the matrices that we obtained in both cases. But the plots will show that the two models produce the same response. The script to create the plots is given below. The function that defines the system is the same as the function in the last example.

```
clf % to clear the graph area
global A
global B
t0 = 0;
tf = 10;
```

```
z0 = [0; 0]; % vector of initial conditions
A = [-6 -2; 4 0];
B = [2; 0];
[t, z] = ode23('eoce4',t0, tf, z0); % a call to the ode23
y1 = 2.5*z(:,2);
plot(t, y1,'go');
xlabel('Time (sec)')
A = [0 1;-8 -6];
B = [0; 20];
[t, z] = ode23('eoce3', t0, tf, z0); % a call to the ode23
y1 = z(:,1);
hold on
plot(t, y1,'b*');
title('The output y(t) for the MATLAB and the
derived model')
```

The plot is shown in Figure 7.31.

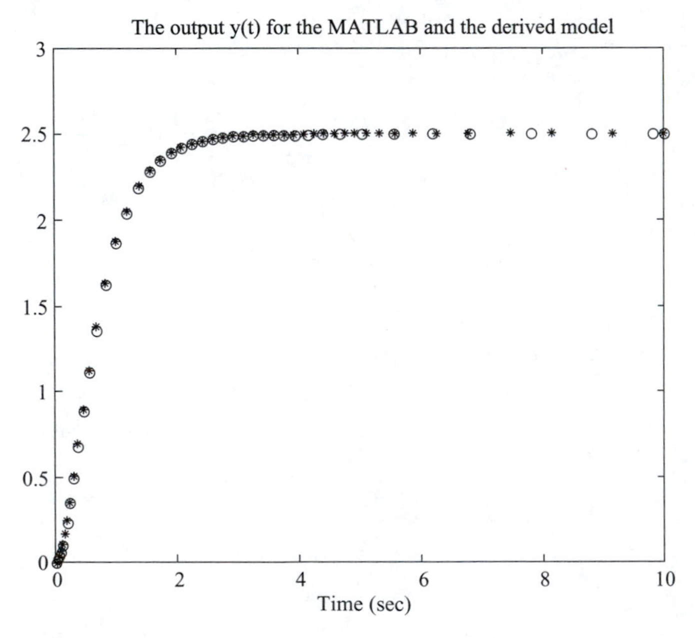

FIGURE 7.31
Plots for EOCE 7.4.

EOCE 7.5

Consider the single-input single-output system in state-space.

$$A = \begin{bmatrix} 0 & 1 \\ -8 & -6 \end{bmatrix}, \quad B = \begin{bmatrix} 0 \\ 20 \end{bmatrix}, \quad C = \begin{bmatrix} 1 & 0 \end{bmatrix}, \quad \text{and} \quad D = \begin{bmatrix} 0 \end{bmatrix}$$

Derive the other representations.

Solution

1. The transfer function representation

   We will do that in two ways: one is analytical and the other is by using MATLAB.

   Analytically, consider the solutions of the state-space model in the Laplace domain

$$\underline{Z}(s) = (sI - A)^{-1}BX(s)$$

and

$$\underline{Y}(s) = C\underline{Z}(s) + D\underline{X}(s)$$

Substituting for $\underline{Z}(s)$ in $\underline{Y}(s)$ to get

$$\underline{Y}(s) = C(sI - A)^{-1}BX(s) + D\underline{X}(s)$$

and the transfer function is

$$\frac{\underline{Y}(s)}{\underline{X}(s)} = C(sI - A)^{-1}B + D$$

We will substitute for the matrices and get

$$H(s) = \begin{bmatrix} 1 & 0 \end{bmatrix} \begin{bmatrix} \dfrac{s+6}{s^2 + 6s + 8} & \dfrac{1}{s^2 + 6s + 8} \\ \dfrac{-8}{s^2 + 6s + 8} & \dfrac{s}{s^2 + 6s + 8} \end{bmatrix} \begin{bmatrix} 0 \\ 20 \end{bmatrix}$$

And finally, the transfer function is

$$H(s) = \frac{20}{s^2 + 6s + 8}$$

Using MATLAB, we write the MATLAB script

```
A = [0 1; -8 -6];
B = [0; 20];
C = [1 0];
D = 0;
[num, den] = ss2tf(A,B,C,D)
```

The output is

```
num =
    0          0       20.0000
den =
    1          6       8
```

Therefore, the transfer function is

$$H(s) = \frac{0s^2 + 0s + 20}{1s^2 + 6s + 8} = \frac{20}{s^2 + 6s + 8}$$

2. The block diagram representation
   The system is second order and hence we need two integrators. The diagram is shown in Figure 7.32.
3. The differential equation representation
   The state equation is

$$\frac{d}{dt}z = \begin{bmatrix} 0 & 1 \\ -8 & -6 \end{bmatrix}\begin{bmatrix} z_1 \\ z_2 \end{bmatrix} + \begin{bmatrix} 0 \\ 20 \end{bmatrix} x(t)$$

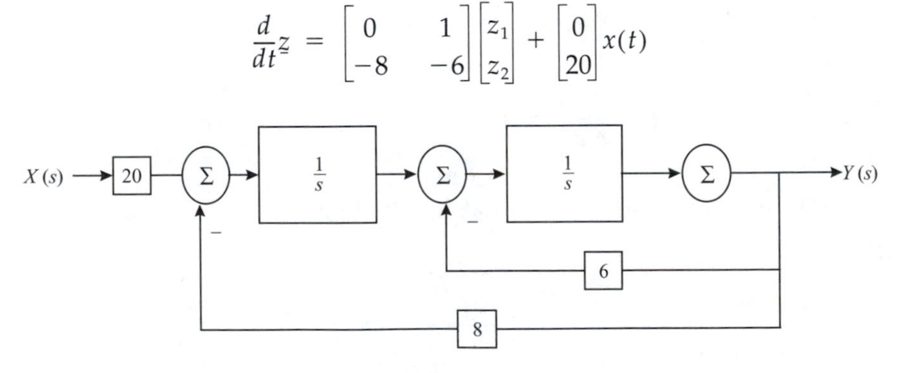

FIGURE 7.32
Block for EOCE 7.5.

where

$$\frac{d}{dt}z_1(t) = 0z_1 + 1z_2(t) + 0x(t)$$
$$\frac{d}{dt}z_2(t) = -8z_1(t) - 6z_2(t) + 20x(t)$$

If we let $z_1(t) = y(t)$ and $z_2(t) = \frac{d}{dt}y(t)$ then

$$\frac{d}{dt}z_1(t) = \frac{d}{dt}y(t)$$
$$\frac{d}{dt}z_2(t) = \frac{d^2}{dt}y(t)$$

Now substitute in for the second derivative of y to get

$$\frac{d}{dt}z_2(t) = -8z_1(t) - 6z_2(t) + 20x(t)$$

or finally

$$\frac{d^2}{dt}y(t) = -8y(t) - 6\frac{d}{dt}y(t) + 20x(t)$$

7.5 End-of-Chapter Problems

EOCP 7.1

Consider the following systems represented as differential equations.

1. $\frac{d}{dt}y + 2y = x(t)$

2. $\frac{d^2}{dt}y + 3\frac{d}{dt}y + 2y = x(t)$

3. $\frac{d^2}{dt}y + 3\frac{d}{dt}y = x(t)$

4. $\frac{d^2}{dt}y + 5y = x(t)$

5. $\frac{d^2}{dt}y + \frac{d}{dt}y + y = x(t)$

6. $\frac{d^2}{dt}y + 3\frac{d}{dt}y + 4y = x(t)$

7. $\frac{d^2}{dt}y + 3\frac{d}{dt}y - 4y = x(t)$

8. $\frac{d^3}{dt}y + 3\frac{d}{dt}y + 4y = x(t)$

9. $\frac{d^3}{dt}y - \frac{d}{dt}y = x(t)$

10. $\frac{d^3}{dt}y + \frac{d^2}{dt}y + \frac{d}{dt}y + y = x(t)$

Derive all other representations.

EOCP 7.2

Consider the following systems represented as transfer functions.

1. $\frac{1}{s^2 + 5s + 6}$

2. $\frac{1}{s^2 + 7s}$

3. $\frac{1}{s^2 + 6}$

4. $\frac{1}{4s + 4}$

5. $\frac{s}{s^3 + 4s + 4}$

6. $\frac{s^2 + s}{s^2 + 4s + 4}$

7. $\dfrac{s}{s^4 + 7s + 10}$

8. $\dfrac{s^2 + 1}{s^4 + 7s^2 + 10}$

9. $\dfrac{s^3 + 1}{s^2 + 7s + 10}$

10. $\dfrac{s^3 + s^2 + 1}{s^2 + 7s + 10}$

Derive all other representations.

EOCP 7.3

Consider the following systems represented by the impulse response signal.

1. $h(t) = 10e^{-2t}u(t)$
2. $h(t) = (10e^{-2t} + 2e^{-3t})u(t)$
3. $h(t) = 10e^{-2t}\sin(2t)u(t)$
4. $h(t) = 10e^{-2t}u(t) + \sin(2t)u(t)$
5. $h(t) = te^{-2t}u(t) + \cos(2t)u(t)$
6. $h(t) = \delta(t) + te^{-3t}u(t) + \cos(t)u(t)$
7. $h(t) = 10e^{-2t}u(t) + 11\delta(t) + u(t)$
8. $h(t) = 10t^2e^{-2t}u(t)$
9. $h(t) = e^{-t}u(t) + 5e^{-3t}u(t) + 3\cos(t)u(t)$
10. $h(t) = te^{-t}u(t) + 5te^{-3t}u(t) + 3\sin(t)u(t)$

Derive all other representations.

EOCP 7.4

Consider the following systems represented as block diagrams.

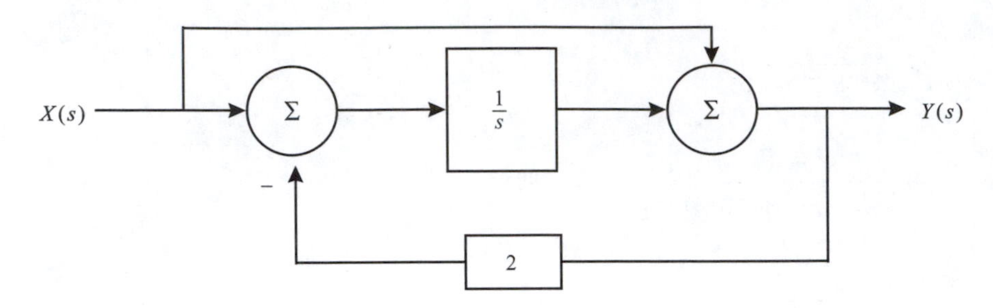

FIGURE 7.33
Block 1 for EOCP 7.4.

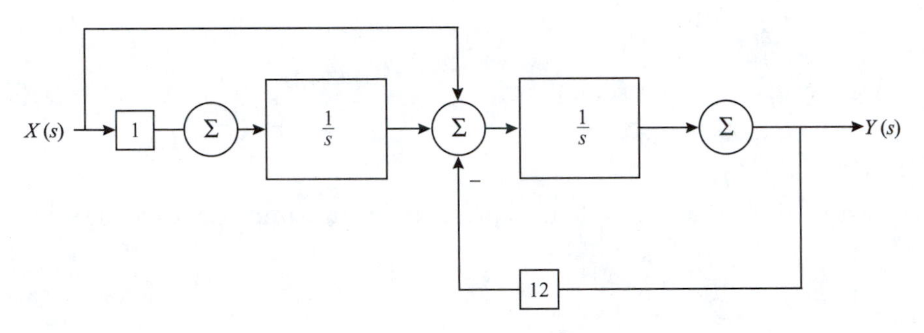

FIGURE 7.34
Block 2 for EOCP 7.4.

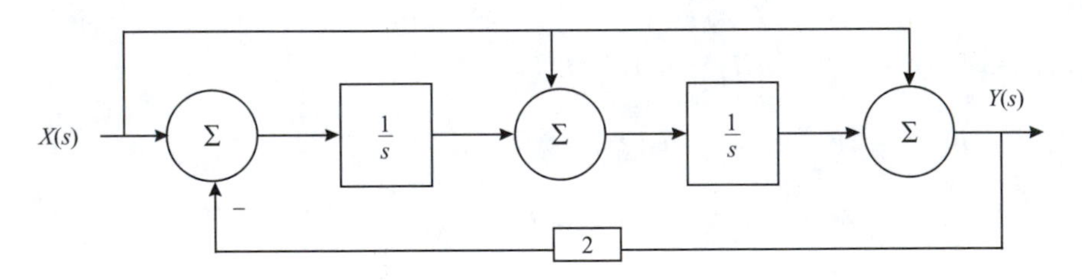

FIGURE 7.35
Block 3 for EOCP 7.4.

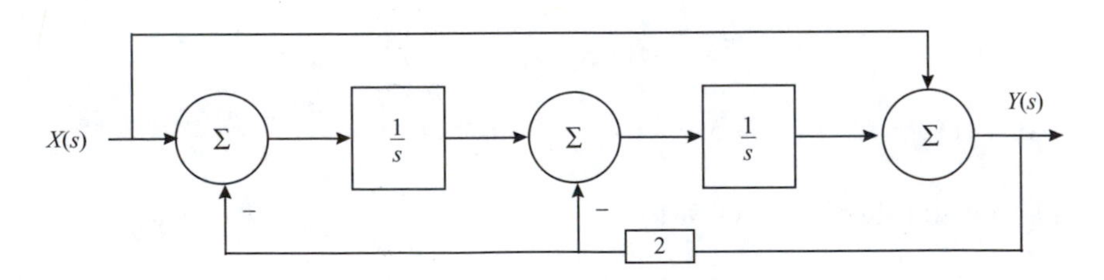

FIGURE 7.36
Block 4 for EOCP 7.4.

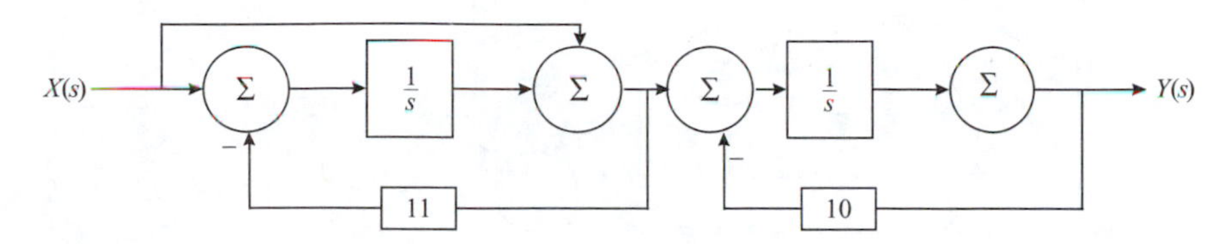

FIGURE 7.37
Block 5 for EOCP 7.4.

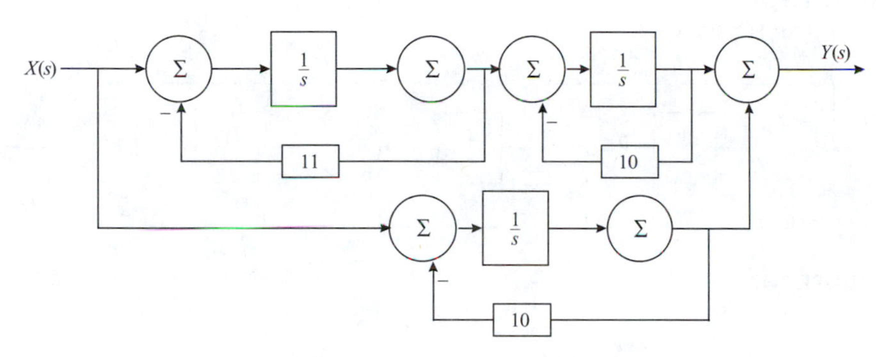

FIGURE 7.38
Block 6 for EOCP 7.4.

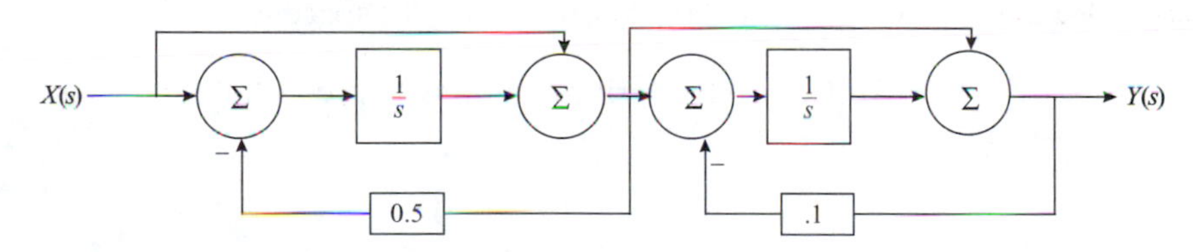

FIGURE 7.39
Block 7 for EOCP 7.4.

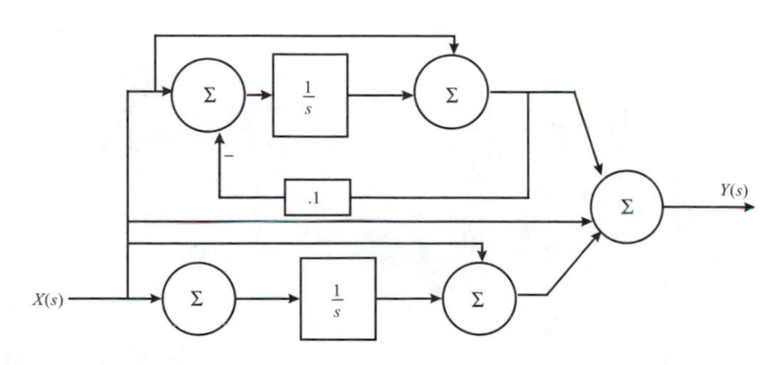

FIGURE 7.40
Block 8 for EOCP 7.4.

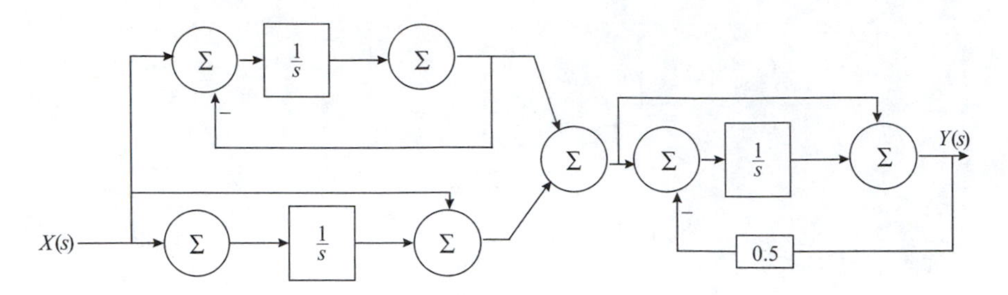

FIGURE 7.41
Block 9 for EOCP 7.4.

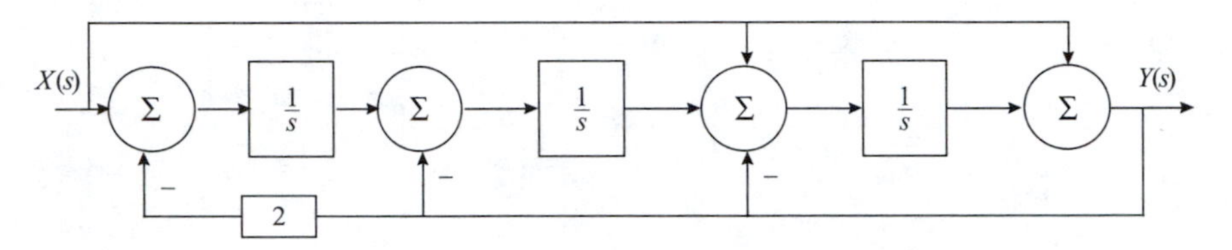

FIGURE 7.42
Block 10 for EOCP 7.4

Derive all other representations.

EOCP 7.5

Consider the following systems represented in state-space form.

1. For the following systems derive the other representations as discussed in this chapter.

   a. $A = \begin{bmatrix} 1 & 0 \\ 0 & 2 \end{bmatrix}, \quad B = \begin{bmatrix} 0 \\ 1 \end{bmatrix}, \quad C = \begin{bmatrix} 1 & 0 \end{bmatrix}, \quad D = \begin{bmatrix} 0 \end{bmatrix}$

   b. $A = \begin{bmatrix} 1 & 0 \\ -2 & -2 \end{bmatrix}, \quad B = \begin{bmatrix} 0 \\ 1 \end{bmatrix}, \quad C = \begin{bmatrix} 1 & 0 \end{bmatrix}, \quad D = \begin{bmatrix} 0 \end{bmatrix}$

   c. $A = \begin{bmatrix} 1 & 1 \\ 0 & 2 \end{bmatrix}, \quad B = \begin{bmatrix} 1 \\ 1 \end{bmatrix}, \quad C = \begin{bmatrix} 1 & 1 \end{bmatrix}, \quad D = \begin{bmatrix} 0 \end{bmatrix}$

   d. $A = \begin{bmatrix} 1 & 0 \\ 3 & 2 \end{bmatrix}, \quad B = \begin{bmatrix} 0 \\ 1 \end{bmatrix}, \quad C = \begin{bmatrix} 1 & 0 \end{bmatrix}, \quad D = \begin{bmatrix} 1 \end{bmatrix}$

   e. $A = \begin{bmatrix} 1 & 0 \\ 0 & 2 \end{bmatrix}, \quad B = \begin{bmatrix} 1 \\ 1 \end{bmatrix}, \quad C = \begin{bmatrix} 1 & 1 \end{bmatrix}, \quad D = \begin{bmatrix} 1 \end{bmatrix}$

2. For the following systems use MATLAB to derive the transfer function representation. Then from the transfer function, derive the differential equation representation. Use MATLAB to obtain the impulse response representation. Draw the block diagram for each.

a. $A = \begin{bmatrix} -1 & 0 & 0 \\ 0 & -2 & 0 \\ 0 & 0 & -2 \end{bmatrix}, \quad B = \begin{bmatrix} 1 \\ 0 \\ 0 \end{bmatrix}, \quad C = \begin{bmatrix} 1 & 0 & 0 \end{bmatrix}, \quad D = \begin{bmatrix} 0 \end{bmatrix}$

b. $A = \begin{bmatrix} -1 & 2 & 2 \\ 0 & -2 & 0 \\ 0 & 0 & -2 \end{bmatrix}, \quad B = \begin{bmatrix} 1 \\ 0 \\ 0 \end{bmatrix}, \quad C = \begin{bmatrix} 1 & 1 & 0 \end{bmatrix}, \quad D = \begin{bmatrix} 0 \end{bmatrix}$

c. $A = \begin{bmatrix} -1 & 0 & 0 \\ 0 & -2 & 0 \\ 0 & 0 & -2 \end{bmatrix}, \quad B = \begin{bmatrix} 1 \\ 1 \\ 1 \end{bmatrix}, \quad C = \begin{bmatrix} 1 & 0 & 0 \end{bmatrix}, \quad D = \begin{bmatrix} 0 \end{bmatrix}$

d. $A = \begin{bmatrix} 1 & 0 & 0 \\ 2 & 2 & 0 \\ 2 & 2 & 2 \end{bmatrix}, \quad B = \begin{bmatrix} 1 \\ 0 \\ 0 \end{bmatrix}, \quad C = \begin{bmatrix} 1 & 0 & 0 \end{bmatrix}, \quad D = \begin{bmatrix} 0 \end{bmatrix}$

e. $A = \begin{bmatrix} -1 & 0 & 0 \\ 0 & -2 & 0 \\ 0 & 0 & -2 \end{bmatrix}, \quad B = \begin{bmatrix} 1 \\ 0 \\ 0 \end{bmatrix}, \quad C = \begin{bmatrix} 1 & 0 & 0 \\ 0 & 0 & 1 \end{bmatrix}, \quad D = \begin{bmatrix} 0 \\ 1 \end{bmatrix}$

EOCP 7.6

Consider the system in Figure 7.43.

Let $L_1 = L_2 = 1$ henry, $R_1 = R_2 = 1$ ohm, and $C = 1$ farad.

1. Let the currents in the inductors and the voltage in the capacitor be the state variables. Write the state equations describing the system.
2. Obtain the transfer function relating the output $y(t)$ to the input $x(t)$.
3. Let $x(t)$ be the step signal. Find the output $y(t)$.
4. With $x(t)$ as the impulse input, obtain the output currents in the inductors.
5. Draw the block diagram relating $x(t)$ to $y(t)$.

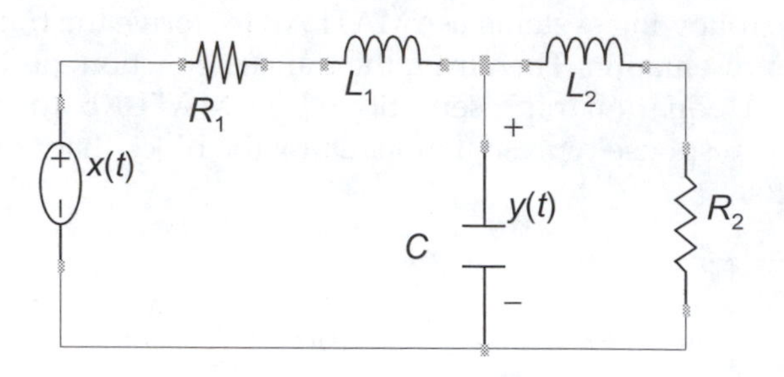

FIGURE 7.43
Circuit for EOCP 7.6.

EOCP 7.7

Consider the system in Figure 7.44.

The rod to the left of the mass, M, can be modeled by a translational spring with spring constant K, in parallel with a translational damper of constant B.

1. Show that the differential equation relating the input $x(t)$ to the output $y(t)$ is

$$M\frac{d^2}{dt}y(t) + B\frac{d}{dt}y(t) + Ky(t) = x(t)$$

2. What is the transfer function representation?
3. Draw the block diagram representing the system.
4. What is the impulse response representation?
5. Represent the system in state-space.
6. Find the output $y(t)$ using any representation due to a step input of unity magnitude.

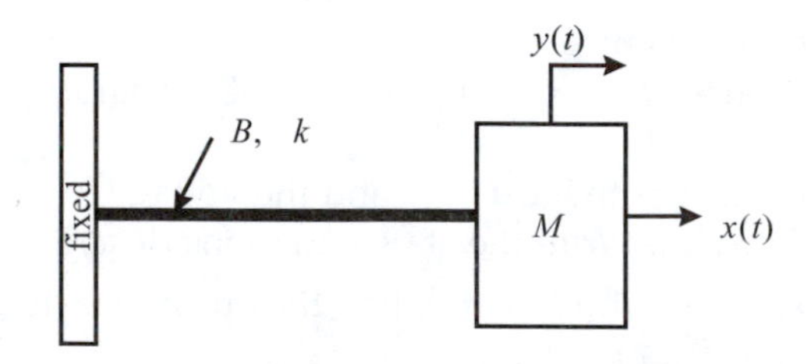

FIGURE 7.44
System for EOCP 7.7.

EOCP 7.8

Consider the following rotational system.

$$J^2\frac{d^4}{dt}\theta(t) + 2JB\frac{d^3}{dt}\theta(t) + (2kJ + B^2)\frac{d^2}{dt}\theta(t) + 2Bk\frac{d}{dt}\theta(t) = kT(t)$$

1. Obtain the state-space representation for this system.
2. Obtain the transfer function representation.
3. For the purpose of calculations only, let $J = B = k = 1$ and obtain the impulse response representation.
4. Obtain the step response representation as you did in 3.
5. Obtain the block diagram representation.

EOCP 7.9

Consider the system in Figure 7.45.

The transfer function relating the input voltage $x(t)$ to the output angular rotation is

$$\frac{\theta(s)}{x(s)} = \frac{k_1/(RJ)}{s(s + (1/J)(D + (k_1k_2/R)))}$$

where k_1 and k_2 are proportionality constants, and D and J are the viscosity and the inertia, respectively.

1. Find the differential equation representation.
2. Obtain the state-space representation.
3. Draw the block diagram.
4. Give values to the parameters shown in the transfer function. Obtain the impulse response representation.

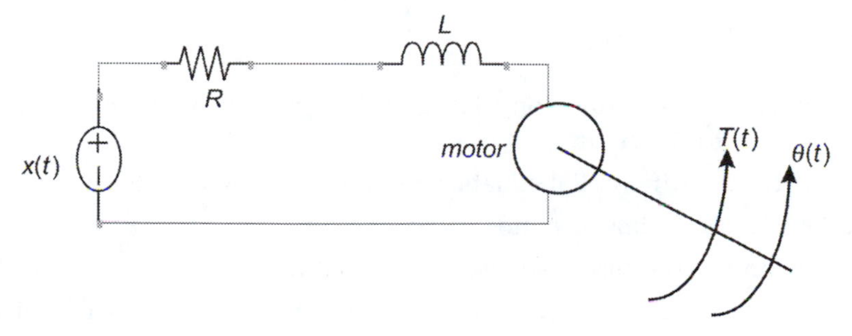

FIGURE 7.45
System for EOCP 7.9.

5. Is the system stable?
6. Repeat by choosing other values for the parameters and obtain the impulse response until the system is stabilized.

EOCP 7.10

Consider the system in Figure 7.46.

1. Find the transfer function relating $X(s)$ to $Y(s)$.
2. Derive the differential equation representing the transfer function.
3. Obtain the state-space model.
4. Obtain the impulse response representation.
5. Use any representation to find $y(t)$ due to a step input of unity magnitude.
6. Draw the block diagram.

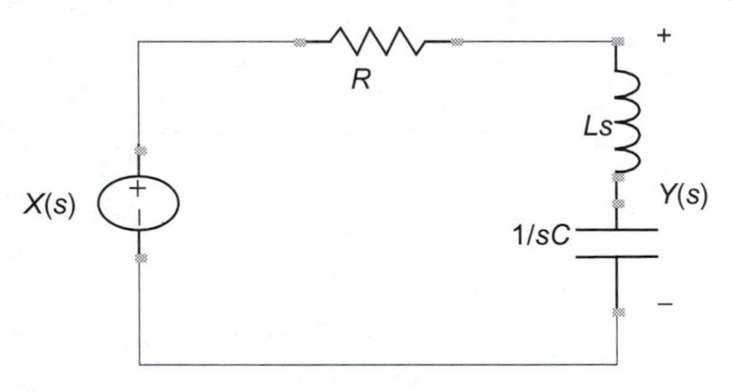

FIGURE 7.46
System for EOCP 7.10.

EOCP 7.11

Consider the circuit in Figure 7.47.

1. Find the transfer function relating $X(s)$ to the current in the inductor flowing to the right.
2. Derive the differential equation representing the transfer function.
3. Obtain the state-space model.
4. Obtain the impulse response representation.
5. Use any representation to find $y(t)$ due to a step input of unity magnitude.
6. Draw the block diagram.

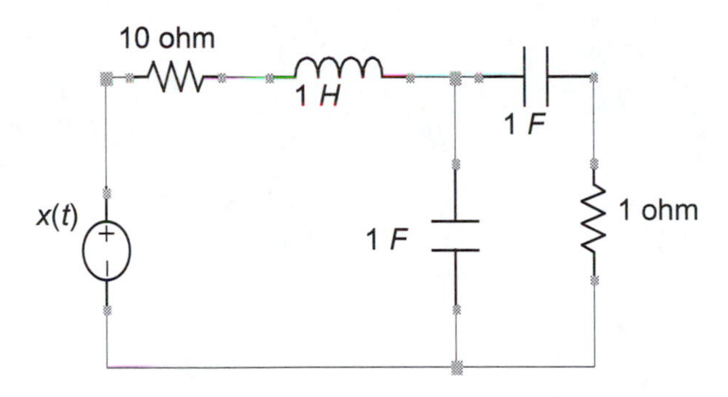

FIGURE 7.47
System for EOCP 7.11.

References

Bequette, B.W. *Process Dynamics*, Englewood Cliffs, NJ: Prentice-Hall, 1998.
Brogan, W.L. *Modern Control Theory*, 3rd ed., Englewood Cliffs, NJ: Prentice-Hall, 1991.
Close, M. and Frederick, K. *Modeling and Analysis of Dynamic Systems*, 2nd ed., New York: Wiley, 1995.
Driels, M. *Linear Control System Engineering*, New York: McGraw-Hill, 1996.
Golubitsky, M. and Dellnitz, M. *Linear Algebra and Differential Equations Using MATLAB*, Stamford, CT: Brooks/Cole, 1999.
Kuo, B.C. *Automatic Control System*, 7th ed., Englewood Cliffs, NJ: Prentice-Hall, 1995.
Lewis, P.H. and Yang, C. *Basic Control Systems Engineering*, Englewood Cliffs, NJ: Prentice-Hall, 1997.
The MathWorks. *The Student Edition of MATLAB*, Englewood Cliffs, NJ: Prentice-Hall, 1997.
Nise, N.S. *Control Systems Engineering*, 2nd ed., Reading, MA: Addison-Wesley, 1995.
Ogata, K. *Modern Control Engineering*, 3rd ed., Englewood Cliffs, NJ: Prentice-Hall, 1997.
Ogata, K. *System Dynamics*, 3rd ed., Englewood Cliffs, NJ: Prentice-Hall, 1998.
Pratap, R. *Getting Started with MATLAB 5*, New York: Oxford University Press, 1999.
Woods, R.L. and Lawrence, K.L. *Modeling and Simulation of Dynamic Systems*, Englewood Cliffs, NJ: Prentice-Hall, 1997.
Wylie, R.C. and Barrett, C.L. *Advanced Engineering Mathematics*, 6th ed., New York: McGraw-Hill, 1995.

8

Introduction to the Design of Systems

CONTENTS

8.1 Introduction

At this point in this book we should have a very good understanding of the concept of input, output, and a linear time-invariant system. We will stress again that if $x(t)$ is an input to a linear time-invariant system, the output will reshape the input due to the characteristics of the system itself. Let us look at the following system transfer function. $T(jw)$ is used here because most books that discuss the design of filters use $T(jw)$ instead of $H(jw)$.

$$T(jw) = \frac{Y(jw)}{X(jw)} = \frac{1}{jw + 1} \tag{8.1}$$

The magnitude of this transfer function is

$$\left|\frac{Y(jw)}{X(jw)}\right| = |T(jw)| = \frac{1}{\sqrt{1 + w^2}} \tag{8.2}$$

At low frequencies, frequencies close to zero, the magnitude of the transfer function is close to unity. As the frequency increases, the magnitude decreases to a value close to zero. It seems that this system will pass low frequencies and prevent high frequencies from passing through the system. This is a process similar to the process of filtering frequencies. Most input signals have frequencies. It seems to us that the impulse signal has no frequency, but if we take the Laplace transform of the impulse signal, we see that the transform is a constant that is spread over all frequency values; thus the impulse signal has a constant value at all frequencies. If this input (the impulse function) is passed as an input to the above system, only frequencies in the low range will pass.

From the discussion above we see that the words system and filter are the same. In reality, the word system, meaning filter, is used for electrical, mechanical, and any dynamical systems.

8.2 Kinds of Filters

In this chapter we will use the word filter instead of system. Filters can have different characteristics in terms of the frequencies they pass and reject. We have the following kinds of filters.

1. **Low-pass:** will pass low frequencies
2. **High-pass:** will pass high frequencies

3. **All-pass:** will pass all frequencies
4. **Band-pass:** will pass a band of frequencies
5. **Band-reject:** will reject a band of frequencies

We will do the analysis using passive circuit elements representation. The analysis using active circuit elements is left as an exercise. The use of active or passive elements depends on the kind of the application you are designing. In most communication systems we use active filters because inductors are heavy and very noisy. Also, if high gain is required, we use active filters.

8.3 The Ideal Low-Pass Filter

The ideal low-pass filter has the following characteristics:

$$\left|\frac{Y(jw)}{X(jw)}\right| = |T(jw)| = \begin{cases} A & 0 < w < w_0 \\ 0 & w_0 < w < \infty \end{cases} \tag{8.3}$$

where w_0 is the cut-off frequency. The cut-off frequency is the largest frequency that is allowed to pass.

Graphically the low-pass filter is represented as in Figure 8.1.

We call this filter an ideal filter because there is no way in real life that we can create sharp edges like the edges we have at the cut-off frequency w_0 in the characteristics shown in Figure 8.1.

In real life we can have something like Figure 8.2. You can see in this figure that the system will attenuate high frequencies and pass those between zero and $\frac{1}{2}$ radians without considerable attenuation.

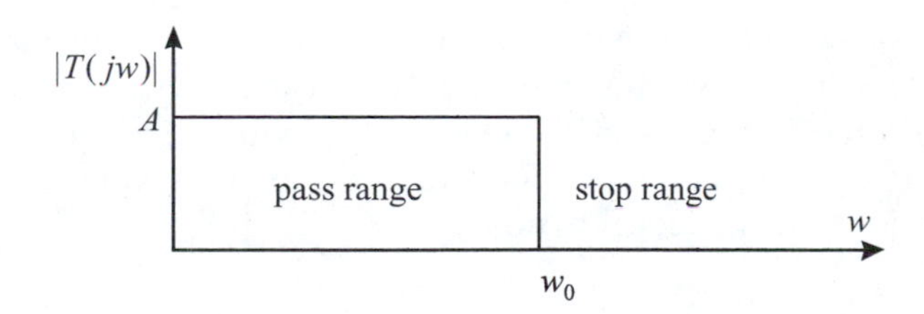

FIGURE 8.1
Ideal low-pass filter.

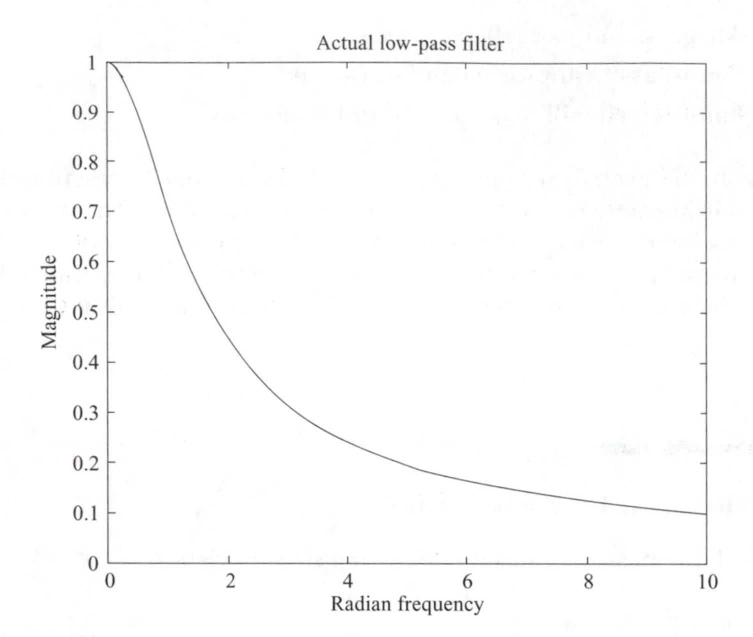

FIGURE 8.2
Actual low-pass filter.

8.3.1 How Do We Find the Cut-Off Frequency w_0?

Ideally, the cut-off frequency is the frequency where the magnitude of the transfer function of the filter falls close to zero. As we said before, the sharp falling edge in the magnitude plot in Figure 8.1 is ideal. In filter design and analysis we calculate the frequency w_0 as in the following.

Given the transfer function $T(jw)$, we set

$$\frac{\max(|T(jw)|)}{\sqrt{2}} = |T(jw)|_{w_0} \tag{8.4}$$

We can solve this equation by first finding the maximum value of the magnitude of $T(jw)$ and divide that value by $\sqrt{2}$, and then set it equal to $|T(jw)|$ evaluated at w_0. The maximum value for $T(jw)$ can either be calculated using calculus (find the derivative of $T(jw)$ and set it equal to zero) or by plotting the magnitude of $T(jw)$ vs. w and locating the maximum value.

If we consider the transfer function

$$\frac{Y(jw)}{X(jw)} = T(jw) = \frac{1}{jw+1}$$

we can calculate w_0 as in the following.

Using the plot of $|T(jw)|$ vs. w, we see that the maximum amplitude is 1.

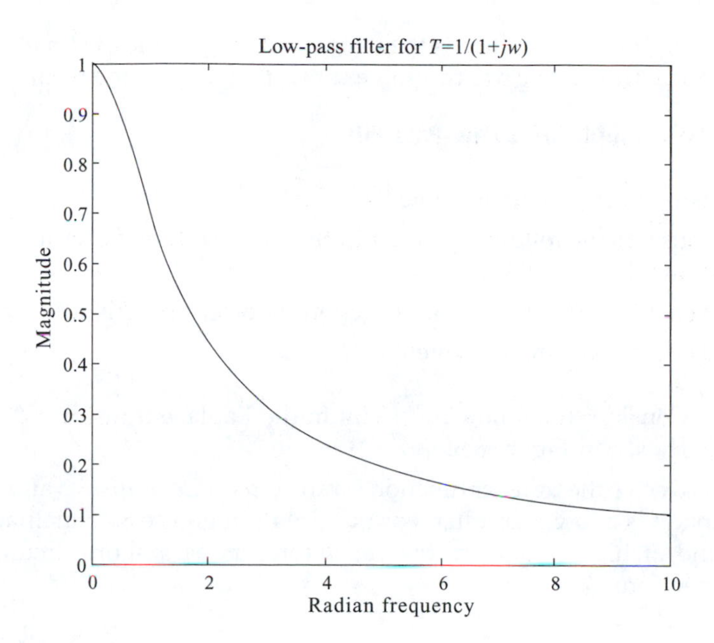

FIGURE 8.3
Low-pass filter example.

We now use Equation 8.4 and write

$$\frac{\max(|T(jw)|)}{\sqrt{2}} = |T(jw)|_{w_0}$$

or

$$\frac{1}{\sqrt{2}} = |T(jw)|_{w_0} = \frac{1}{\sqrt{1 + w_0^2}}$$

Squaring both sides gives

$$\frac{1}{2} = \frac{1}{1 + w_0^2}$$

Solving the above equation, by considering the positive side of the w-axis, gives $w_0 = 1$.

Using calculus, we can find the maximum magnitude of the transfer function by finding the derivative of $|T(jw)|$ with respect to w as

$$\frac{d}{dw}\left(\frac{1}{\sqrt{1 + w^2}}\right) = \frac{-2w}{2(1 + w^2)^{3/2}}$$

Setting the above equation equals to zero results in $w = 0$.

At $w = 0$, $|T(j0)| = 1$, which is the maximum magnitude of $|T(jw)|$. Following the procedure above, we can solve for w_0 to get the same result; $w_0 = 1$.

8.3.2 An Example of a Low-Pass Filter

1. Using passive circuit elements

   Consider the following circuit in the Laplace transform domain as shown in Figure 8.4.

   We will derive the transfer function representing this filter shortly.

2. Using active circuit elements

   Consider the following circuit in the Laplace transform domain as shown in Figure 8.5.

   Derive the transfer function for the circuit in Figure 8.5 and show that it is a low-pass filter. For the circuit in Figure 8.4, qualitatively and at high frequency, the capacitor acts as a short circuit, and therefore

$$Y(s) = 0$$

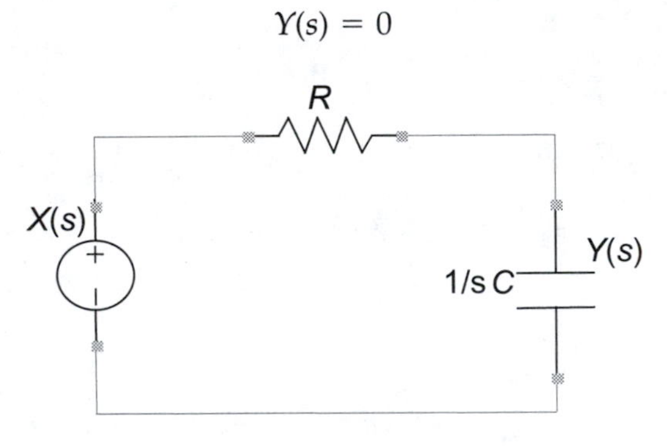

FIGURE 8.4
Passive low-pass filter.

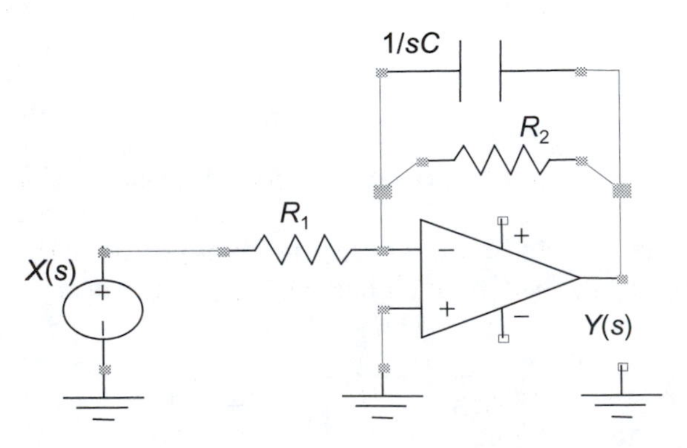

FIGURE 8.5
Active low-pass filter.

At low frequency, the capacitor is an open circuit, and

$$Y(s) = X(s)$$

Quantitatively, the transfer function can easily be derived using voltage divider techniques as

$$T(s) = \frac{Y(s)}{X(s)} = \frac{1}{RC}\frac{1}{s + 1/RC} \tag{8.5}$$

With $s = jw$ the transfer function becomes

$$T(jw) = \frac{Y(jw)}{X(jw)} = \frac{1}{RC}\frac{1}{jw + 1/RC}$$

with $w_0 = 1/RC$ (show that for yourself). In this case we can rewrite the transfer function as

$$T(jw) = \frac{Y(jw)}{X(jw)} = \frac{1}{1 + jw/w_0}$$

and

$$|T(jw)|^2 = \left|\frac{Y(jw)}{X(jw)}\right|^2 = \frac{1}{1 + (w/w_0)^2}$$

This filter has the following characteristics:

1. $|T(j0)| = 1$
2. $|T(jw_0)| = \frac{1}{\sqrt{2}} \approx .707$
3. $|T(j\infty)| = 0$

8.3.3 What Is the Phase?

The phase angle of any transfer function has the lead-lag information that relates the input to the output. If the phase angle is positive, we say the output leads the input, and if the phase angle is negative, it means that the output lags behind the input. If the phase angle is zero, then the input and the output appear at the same time and we say the input and the output are in-phase.

In the example above, the phase is calculated as

$$T(jw) = -\tan^{-1}(wRC) \tag{8.6}$$

For any value of w, the phase is always negative. Thus the output leads the input, or $y(t)$ leads $x(t)$.

8.3.4 What Is the DC Gain?

The gain of any filter (the DC gain) can be calculated at the zero frequency. If we are interested in the gain of the filter given in the above example, we need to put the transfer function describing the filter in the following form:

$$T(s) = \frac{Y(s)}{X(s)} = \frac{1/RC}{s + 1/RC} \tag{8.7}$$

Remember that we are dealing with steady-state analysis ($s = jw$) when we are dealing with the problem of designing filters. As the frequency approaches zero, w will approach zero, and thus s approaches zero. If we substitute $s = 0$, then the magnitude of the transfer function is

$$|T(jw)| = \{1/RC\}/\{1/RC\} = 1$$

So the gain of this filter is unity. To increase or decrease the gain of a filter we simply multiply the transfer function by the desired constant. If the constant is positive, it will have no effect on the phase angle of the filter. If the constant is negative, 180° will be added to the phase of the filter. The sign of the constant is important.

8.4 The Ideal High-Pass Filter

The ideal high-pass filter has the following characteristics:

$$\left|\frac{Y(jw)}{X(jw)}\right| = |T(jw)| = \begin{cases} 0 & 0 < w < w_0 \\ A & w_0 < w < \infty \end{cases} \tag{8.8}$$

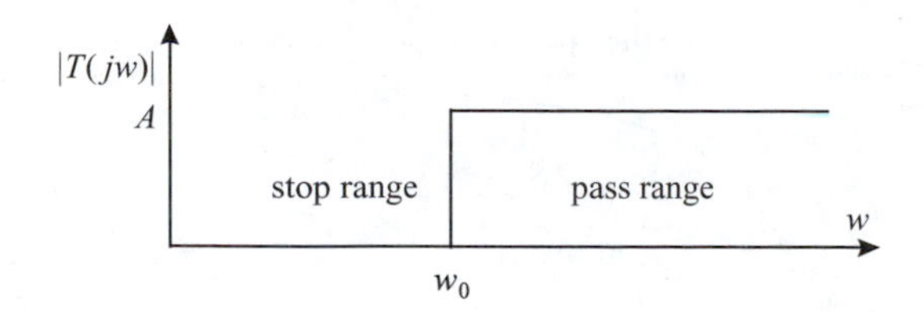

FIGURE 8.6
Ideal high-pass filter.

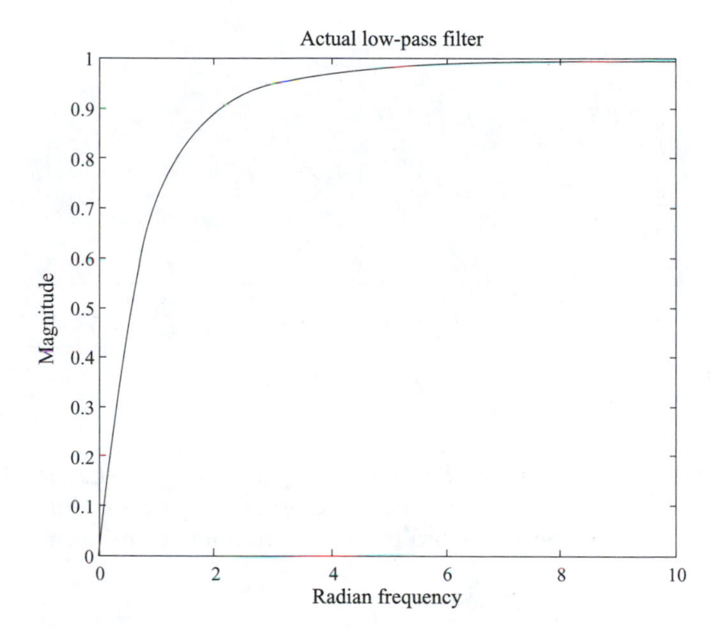

FIGURE 8.7
Actual high-pass filter.

where w_0 in this case is the frequency at which the filter starts passing frequencies; it is the smallest frequency the filter is designed to pass. It will pass all higher frequencies at least theoretically.

Graphically, the ideal high-pass filter will have the form as shown in Figure 8.6.

Again, it is not realistic to create the sharp edge at w_0. In real life, the filter may look like Figure 8.7.

If we consider the transfer function

$$\frac{Y(jw)}{X(jw)} = T(jw) = \frac{jw}{jw+1}$$

we can calculate w_0 as in the following.

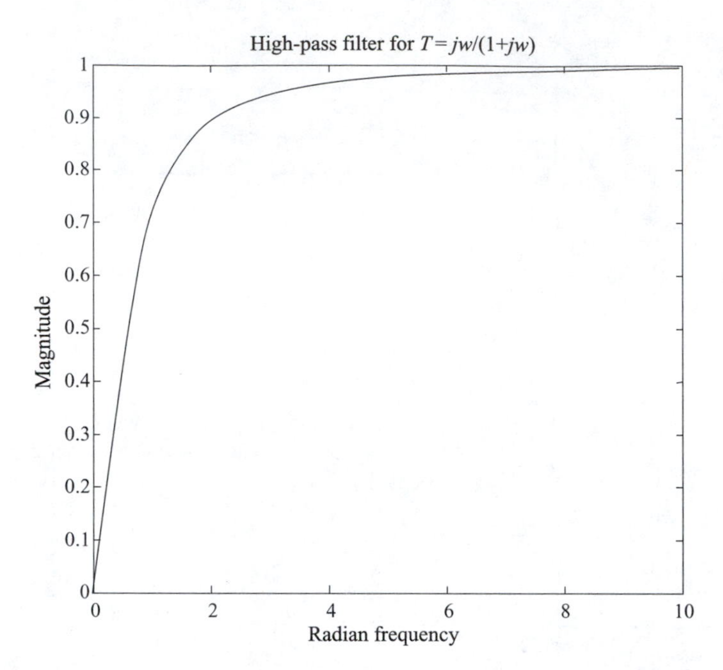

FIGURE 8.8
High-pass filter example.

Using the plot of $|T(jw)|$ vs. w as shown in Figure 8.8, we see that the maximum amplitude occurs at a very high frequency (∞) and it is unity.

We now use Equation 8.4 to find the cut-off frequency and write

$$\frac{\max(|T(jw)|)}{\sqrt{2}} = |T(jw)|_{w_0}$$

$$\frac{1}{\sqrt{2}} = |T(jw)|_{w_0} = \frac{w_0}{\sqrt{1 + w_0^2}}$$

Squaring both sides gives

$$\frac{1}{2} = \frac{w_0^2}{1 + w_0^2}$$

Solving the above equation, while considering the positive side of the w-axis, gives $w_0 = 1$.

Using calculus, we can find the derivative of $|T(jw)|$ with respect to w and get

$$\frac{d}{dw}\left(\frac{w}{\sqrt{1 + w^2}}\right) = \left\{\frac{2}{2(1 + w^2)^{1/2}}\right\} / \{1 + w^2\}$$

Setting the above equation equals to zero results (the denominator cannot be zero) in

$$\left\{\frac{2}{2(1+w^2)^{1/2}}\right\} = 0$$

This last equation results in $w = \infty$.

At $w = \infty$, $|T(j\infty)| = 1$, which is the maximum magnitude of $|T(jw)|$. Following the procedure above, we can solve for w_0 to get the same result, $w_0 = 1$.

8.4.1 An Example of a High-Pass System

1. Using passive circuit elements

   Consider the following circuit in the Laplace transform domain as shown in Figure 8.9.

2. Using active circuit elements

   The circuit in Figure 8.10 is given. Try to derive the transfer function for Figure 8.10 and prove that it is a high-pass filter.

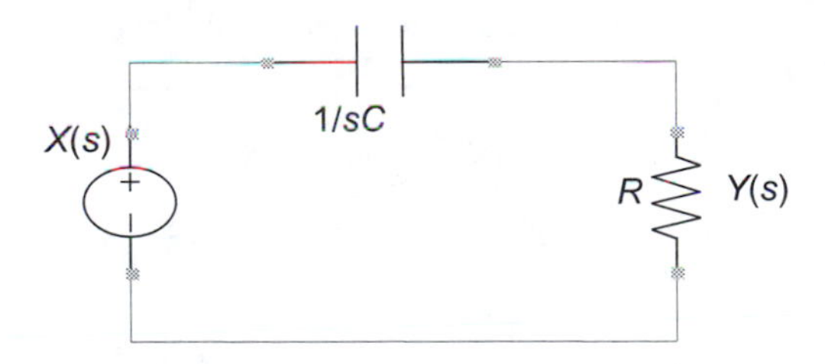

FIGURE 8.9
Passive high-pass filter.

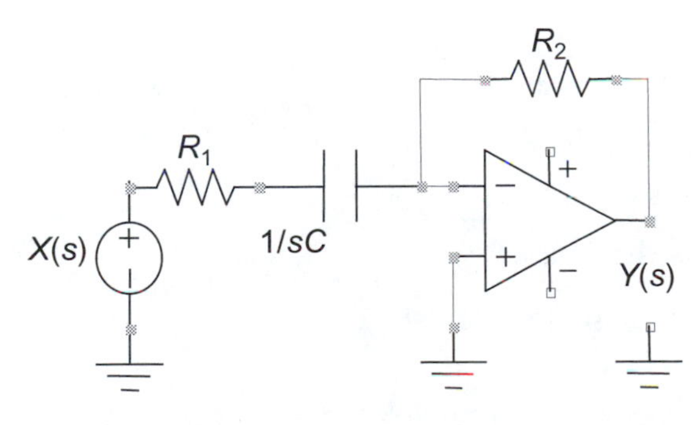

FIGURE 8.10
Active high-pass filter.

Qualitatively, at high frequency, the capacitor in Figure 8.9 acts as a short circuit, and therefore

$$Y(s) = X(s)$$

At low frequency, the capacitor is an open circuit, and

$$Y(s) = 0$$

For the circuit in Figure 8.9, quantitatively, the transfer function can easily be derived using voltage divider techniques as

$$T(s) = \frac{Y(s)}{X(s)} = \frac{R}{R + 1/Cs} = \frac{s}{s + 1/RC} \tag{8.9}$$

With $s = jw$ the transfer function becomes

$$T(jw) = \frac{Y(jw)}{X(jw)} = \frac{R}{R + 1/Cjw} = \frac{jw}{jw + 1/RC}$$

with $w_0 = 1/RC$ (show that for yourself). In this case we can rewrite the transfer function as

$$T(jw) = \frac{Y(jw)}{X(jw)} = \frac{R}{R + 1/Cjw} = \frac{jw/w_0}{1 + jw/w_0}$$

This filter has the following characteristics:

1. $|T(j0)| = 0$
2. $|T(jw_0)| = \frac{1}{\sqrt{2}} \approx .707$
3. $|T(j\infty)| = 1$

8.4.2 What Is the Phase?

For the first order high-pass system above

$$\langle T(jw) = 90° - \tan^{-1}(wRC)$$

For any value of w, the phase is always positive. Thus the output leads the input; $y(t)$ leads $x(t)$.

8.4.3 What Is the DC Gain?

Putting the transfer function describing the filter in a form where the coefficient of the s variable is unity we get

$$T(s) = \frac{Y(s)}{X(s)} = \frac{s}{s + 1/RC}$$

Letting s approach zero, we see that the DC gain (at zero frequency) is zero.

Example 8.1

Design a low-pass filter that passes the range of frequencies from zero to 50k rad/sec that are present in the signal $x(t)$. Notice that $x(t)$ may contain frequencies above the 50k rad/sec range desired. The filter should introduce an additional gain of 10.

Solution

From the design requirements it is seen that the cut-off frequency of the low-pass filter desired is $w_0 = 50{,}000$ rad/sec. We can use the RC circuit discussed before for the low-pass filter with the transfer function parameters set as in the following.

We will use the unity gain transfer function

$$T(s) = \frac{Y(s)}{X(s)} = \frac{1/RC}{s + 1/RC}$$

where $w_0 = 1/RC = 50{,}000$.

Choosing a value for R or C first and then finding the other will finish the design.

If we let $R = 10^8$ ohms, then $C = (1/5)(10)^{-12}$ farads.

In this case the transfer function will be

$$T(s) = \frac{Y(s)}{X(s)} = \frac{(5.10)^4}{s + (5.10)^4}$$

Since we are interested in additional gain of 10, we can multiply the transfer function of the filter by 10, and the final design is

$$T(s) = \frac{Y(s)}{X(s)} = 10\,\frac{(5.10)^4}{s + (5.10)^4}$$

Example 8.2

Design a high-pass filter that passes the range of frequencies from 50k rad/sec to infinity (ideally) that are present in the signal $x(t)$. Notice that $x(t)$ may contain frequencies below the 50k rad/sec range desired. The filter should introduce an additional gain of 10.

Solution

From the design requirements it is seen that the cut-off frequency of the high-pass filter desired is $w_0 = 50{,}000\ \text{rad/sec}$. We can use the RC circuit discussed previously for the high-pass filter with the transfer function parameters set as in the following.

We will use the unity gain transfer function

$$T(s) = \frac{Y(s)}{X(s)} = \frac{s}{s + 1/RC}$$

where $w_0 = 1/RC = 50{,}000$.

Choosing a value for R or C first and then find the other will finish the design.

If we let $R = 10^8$ ohms, then $C = (1/5)(10)^{-12}$ farads.

In this case the transfer function will be

$$T(s) = \frac{Y(s)}{X(s)} = \frac{s}{s + (5.10)^4}$$

Since we are interested in additional gain of 10, we can multiply the transfer function of the filter by 10, and the final design is

$$T(s) = \frac{Y(s)}{X(s)} = 10\frac{s}{s + (5.10)^4}$$

8.5 The Ideal All-Pass Filter

The all-pass filter has the following characteristics:

$$|T(jw)| = \left|\frac{Y(jw)}{X(jw)}\right| = A \text{ for all } w \tag{8.10}$$

This filter passes all frequencies presented in any input signal $x(t)$. Graphically it is represented as shown in Figure 8.11.

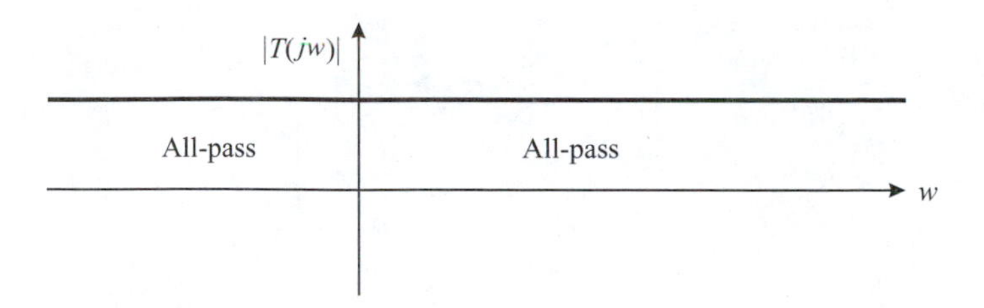

FIGURE 8.11
Ideal all-pass filter.

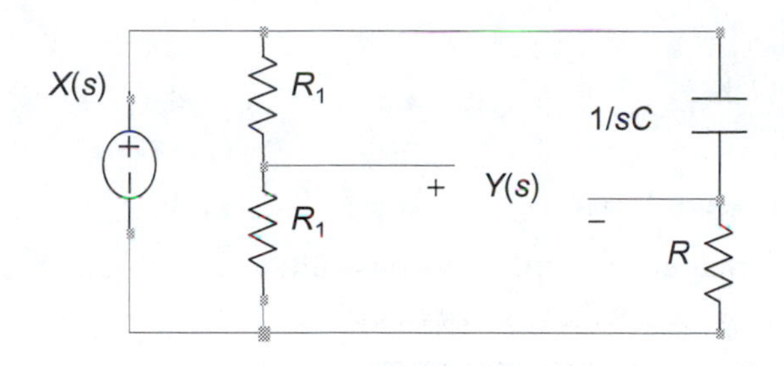

FIGURE 8.12
Passive all-pass filter.

8.5.1 An Example of an All-Pass Filter

1. Using passive circuit elements

   The circuit in Figure 8.12 is considered with its transfer function presented below.

2. Using active circuit elements

   Derive the transfer function for the circuit in Figure 8.13 and show that it represents an all-pass filter.

Using circuit analysis, we can find the transfer function for Figure 8.12. The derivation is an exercise.

$$T(s) = \frac{1}{2} - \frac{s}{s + 1/(RC)} \tag{8.11}$$

The magnitude of the transfer function is

$$|T(jw)| = \frac{1}{2}\left|\frac{1 - sRC}{1 + sRC}\right| = \frac{1}{2} \qquad \text{for all } w$$

Notice that this all-pass filter reduces the magnitude of the input signal $x(t)$ by $1/2$, but still passes all frequencies in the signal. The magnitude of the signal $x(t)$ can be adjusted by passing the output of the all-pass filter through a system of gain two if we want to preserve the magnitude of the signal $x(t)$.

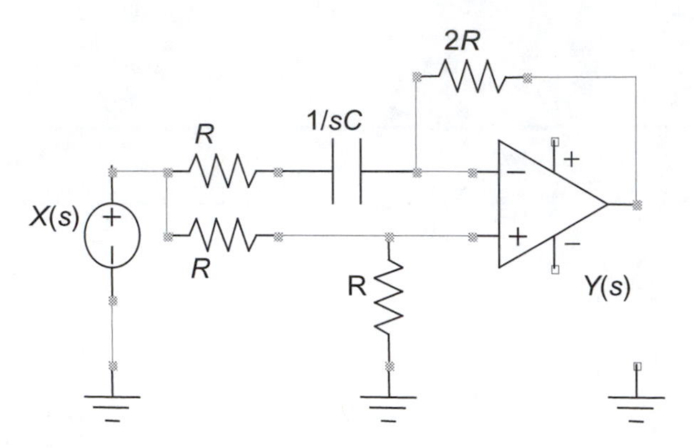

FIGURE 8.13
Active all-pass filter.

8.5.2 What Is the Phase?

The transfer function given in the all-pass filter example above can be written as

$$T(s) = -\left[\frac{1}{2}\right]\frac{s - 1/RC}{s + 1/RC}$$

$$\langle T(jw) = -180 - 2\tan^{-1}(wRC)$$

We can see that the phase angle is always negative, and therefore, the output lags the input $x(t)$ for all w.

8.5.3 What Is the DC Gain?

The DC gain is the value of $T(jw)$ at $w = 0$. In our all-pass example, the DC gain is $\frac{1}{2}$.

Example 8.3

Design an all-pass filter that will reduce the gain of the input signal $x(t)$ by a factor of $\frac{1}{4}$ while passing all the frequencies present in the input signal.

Solution

We can use one all-pass filter similar to the one we discussed earlier to reduce the magnitude of $x(t)$ by $\frac{1}{2}$ and then pass the output of the same filter through another all-pass filter of the same characteristics to reduce the magnitude of the signal $x(t)$ by another $\frac{1}{2}$. Yet, after passing $x(t)$ through the two

FIGURE 8.14
System for Example 8.3.

cascaded filters, the magnitude will be reduced by $\frac{1}{4}$ and all the frequencies in $x(t)$ will pass. Notice that any values for R and C will give a gain of $\frac{1}{2}$. For the first all-pass filter we can choose $R = 10^8$ ohms and $C = (1/5)(10)^{-12}$ farads. Therefore, the first all-pass filter will have the following transfer function

$$T_1(s) = -\left[\frac{1}{2}\right]\frac{s-(5.10)^4}{s+(5.10)^4}$$

For the second all-pass filter we can have the same transfer function or we can choose different values for R and C since different values for R and C will not have any effect on the gain of the filter.

$$T_2(s) = -\left[\frac{1}{2}\right]\frac{s-(5.10)^4}{s+(5.10)^4}$$

The final transfer function that will reduce the magnitude of $x(t)$ by $\frac{1}{4}$ and pass the entire frequency range present in $x(t)$ is

$$T(s) = T_1(s)T_2(s) = \left\{-\left[\frac{1}{2}\right]\frac{s-(5.10)^4}{s+(5.10)^4}\right\}\left\{-\left[\frac{1}{2}\right]\frac{s-(5.10)^4}{s+(5.10)^4}\right\}$$

Graphically, the filtering can be expressed as shown in Figure 8.14.

8.6 The Ideal Band-Pass Filter

The ideal band-pass filter has the following characteristics:

$$|T(jw)| = \left|\frac{Y(jw)}{X(jw)}\right| = \begin{cases} A & (w_1 < w < w_2) \\ 0 & \text{otherwise} \end{cases} \tag{8.12}$$

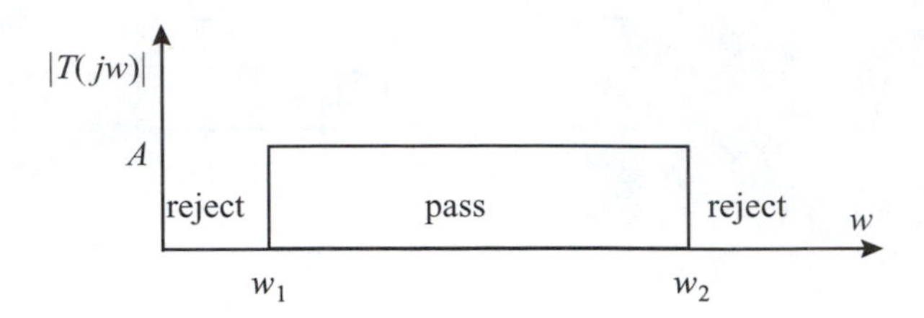

FIGURE 8.15
Ideal band-pass filter.

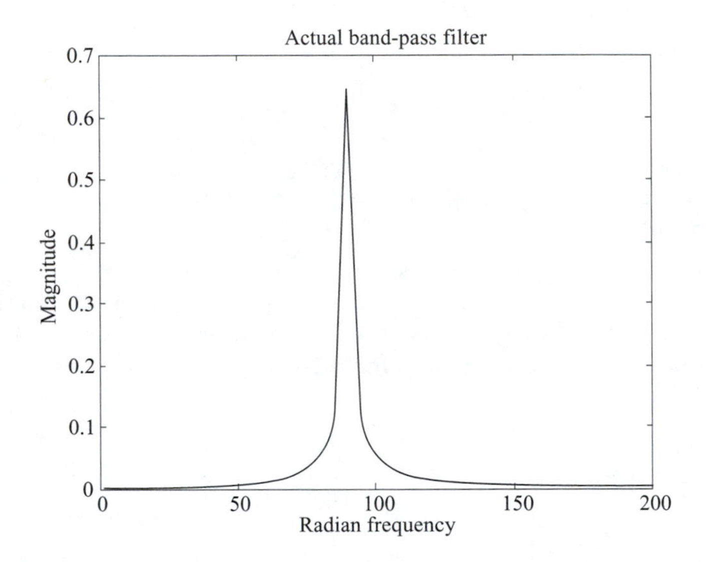

FIGURE 8.16
Actual band-pass filter.

where $[w_1, w_2]$ is the range of the frequencies the band-pass filter is allowed to pass. This filter is depicted graphically in Figure 8.15.

Again, the sharp changes at w_1 and w_2 are not realistic. A realistic band-pass filter can be constructed as having the magnitude graph shown in Figure 8.16.

8.6.1 An Example of a Band-Pass Filter

1. Using passive circuit elements

   An example of a band-pass filter is shown in Figure 8.17.
2. Using active circuit elements

   Consider the circuit in Figure 8.18.

Try to derive the transfer function for the circuit in Figure 8.18 and see if it represents a band-pass filter. For the circuit in Figure 8.17, qualitatively, at

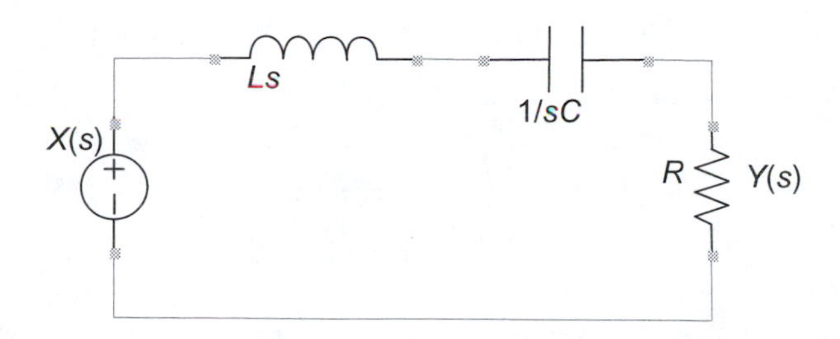

FIGURE 8.17
Passive band-pass filter.

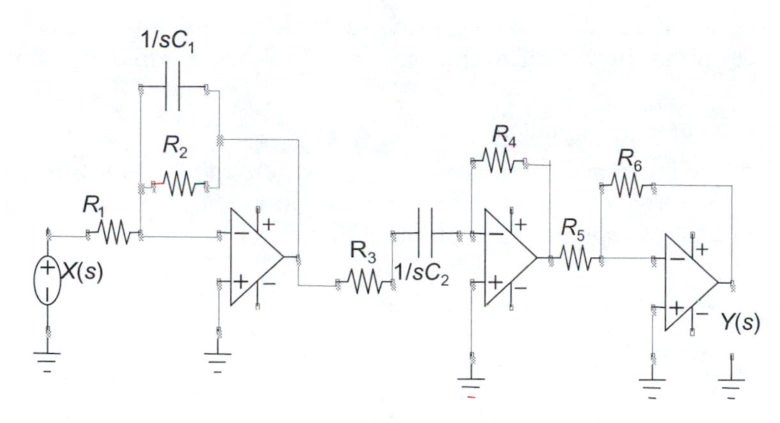

FIGURE 8.18
Active band-pass filter.

high frequency, the capacitor acts as a short circuit and the inductor acts as an open circuit, and therefore

$$Y(s) = 0$$

At low frequency, the capacitor is an open circuit and the inductor is a short circuit, and

$$Y(s) = 0$$

What happens when we operate in between these two extremes?

Quantitatively, the transfer function relating $X(s)$ to $Y(s)$ can be derived using circuit analysis techniques to obtain

$$T(s) = \frac{(R/L)s}{s^2 + (R/L)s + 1/(LC)} \tag{8.13}$$

With $s = jw$, the magnitude of the transfer function can be calculated as

$$|T(jw)| = \frac{w\frac{R}{L}}{\sqrt{(1/LC - w^2)^2 + (w\frac{R}{L})^2}}$$

In the case of a band-pass filter, we will look at the following parameters:

1. The cut-off frequencies w_1 and w_2

   These can be calculated using the same procedure that we used to determine the cut-off frequencies for the low-pass and the high-pass filters.
2. The center frequency w_c

   The center frequency is the frequency where the transfer function $T(s)$ is purely real and at which the magnitude of $T(s)$ is maximum. It is also the geometric mean of w_1 and w_2 or

$$w_c = \sqrt{w_1 w_2}$$

3. The bandwidth, β

   The bandwidth is the width of the frequency band in the Band-pass filter.
4. The quality factor, Q

   The quality factor is defined as the ratio of the center frequency to the bandwidth

$$Q = w_c/\beta$$

   It is a measure of the bandwidth of the pass-band.

The center frequency can be calculated by making $T(s)$ purely real. We can do that in the circuit at hand by setting

$$jw_cL + \frac{1}{jw_cC} = 0 \tag{8.14}$$

Solving the above equation leads to

$$w_c = \sqrt{\frac{1}{LC}} \tag{8.15}$$

We next calculate the cut-off frequencies w_1 and w_2. $T(jw)$ has its maximum at w_c. Therefore, the maximum magnitude of $T(jw)$ is $|T(jw_c)|$.

$$|T(jw_c)| = \frac{\sqrt{\frac{1}{LC}}\frac{R}{L}}{\sqrt{(1/LC - 1/LC)^2 + \left(\sqrt{\frac{1}{LC}}\frac{R}{L}\right)^2}} = 1$$

To find w_1 and w_2 we divide $|T(jw_c)|$ by the square root of 2 and set it equal to $|T(jw)|$ as

$$|T(jw)| = \frac{w\frac{R}{L}}{\sqrt{(1/LC - w^2)^2 + \left(w\frac{R}{L}\right)^2}} = \frac{1}{\sqrt{2}}$$

With some manipulations we get the following quadratic equation in w

$$w^2L \pm wR - 1/C = 0$$

By solving the above equation, we get the two cut-off frequencies

$$\begin{aligned} w_1 &= -\frac{R}{2L} + \sqrt{R/(2L)^2 + (1/LC)} \\ w_2 &= \frac{R}{2L} + \sqrt{R/(2L)^2 + (1/LC)} \end{aligned} \tag{8.16}$$

For the bandwidth, β, we have

$$\beta = w_2 - w_1 = R/L \tag{8.17}$$

and for the quality factor

$$Q = w_c/\beta = \sqrt{\frac{L}{CR^2}} \tag{8.18}$$

8.6.2 What Is the Phase?

We can see that the phase angle at low frequencies is close to 90° and at high frequencies it is close to −90°. At low frequencies the output leads the input, and at high frequencies the output lags the input.

8.6.3 What Is the DC Gain?

The magnitude of $T(s)$ at low frequencies is close to zero. This value is the DC gain of the filter.

Example 8.4

Design a band-pass filter that has cut-off frequencies at 2kHz and 20kHz.

Use the RLC circuit that was considered previously with appropriate values for R, L, and C.

Solution

From the equation

$$w_c = \sqrt{\frac{1}{LC}}$$

we see that

$$L = \frac{1}{Cw^2c}$$

Notice that

$$w_c = 2\pi f_c$$

and

$$f_c = \sqrt{f_1 f_2} = 6.3 \times 10^3 \text{ Hz}$$

$$(w_c)^2 = (2\pi f_c)^2 = 16\pi^2 \times 10^7$$

Let C = 1 × 10^{-6} farads. Then

$$L = \frac{1}{Cw_c^2} = (10^6)/(16\pi^2 \times 10^7) = 10/16\pi^2 = 63.3 \text{ mH}$$

The quality factor is

$$Q = \frac{f_c}{f_2 - f_1} = 0.35$$

For R we use the equation

$$R = \sqrt{L/(CQ^2)} = 718.84 \text{ ohms}$$

Having values for R, L, and C, the design is complete.

8.7 The Ideal Band-Reject Filter

The ideal band-reject filter has the following characteristics:

$$|T(jw)| = \left|\frac{Y(jw)}{X(jw)}\right| = \begin{cases} 0 & w_1 < w < w_2 \\ A & \text{otherwise} \end{cases} \tag{8.19}$$

where $[w_1, w_2]$ is the range of the frequencies the band-reject filter will reject. Graphically, this filter is depicted in the picture shown in Figure 8.19.

Again, as we mentioned before, the sharp edges are not realistic and we can approximate sudden changes as in the graph given in Figure 8.20.

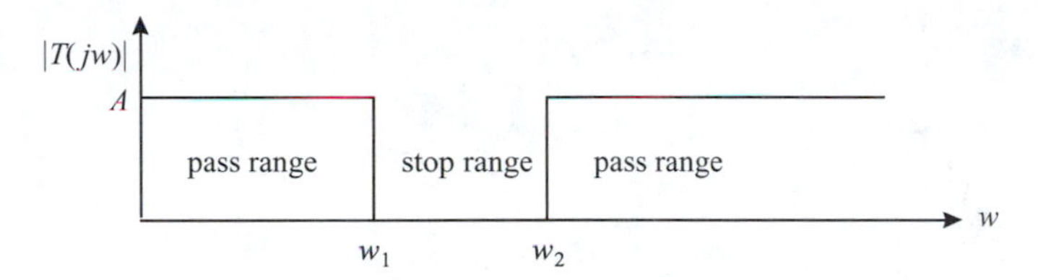

FIGURE 8.19
Ideal band-reject filter.

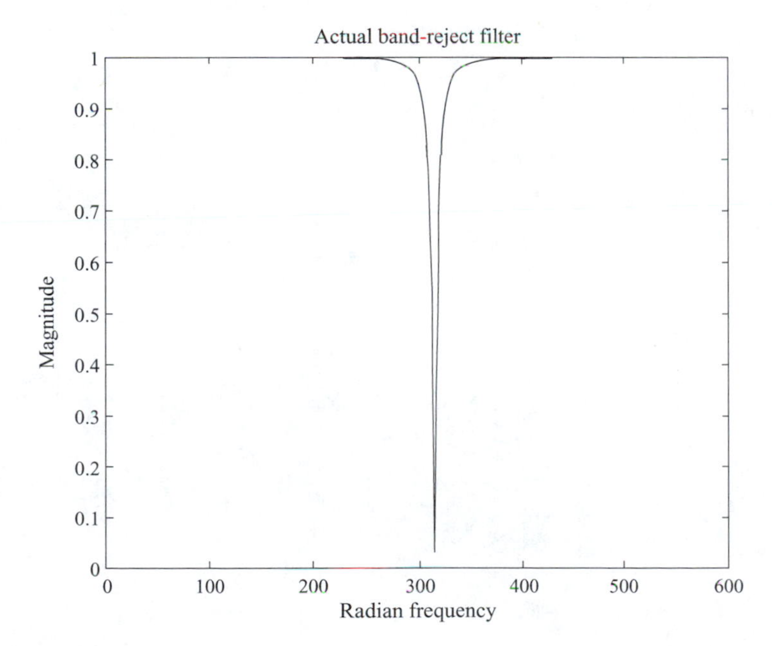

FIGURE 8.20
Actual band-reject filter.

8.7.1 An Example of a Band-Reject Filter

1. Using passive circuit elements

   Consider the circuit representing a band-reject filter as shown in Figure 8.21.

2. Using active circuit elements

   Consider the circuit as shown in Figure 8.22. This circuit is the same circuit we used for the band-pass filter, but the output, $y(t)$, is taken as the voltage across the inductor and the capacitor in series.

Qualitatively, at high frequency, the capacitor acts as a short circuit and the inductor acts as an open circuit, and therefore

$$Y(s) = X(s)$$

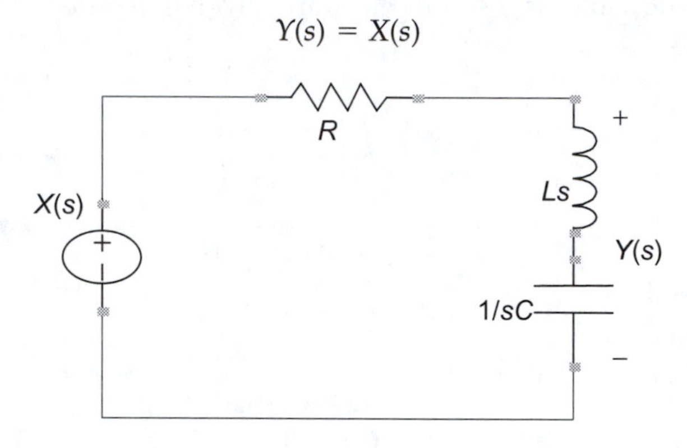

FIGURE 8.21
Passive band-reject filter.

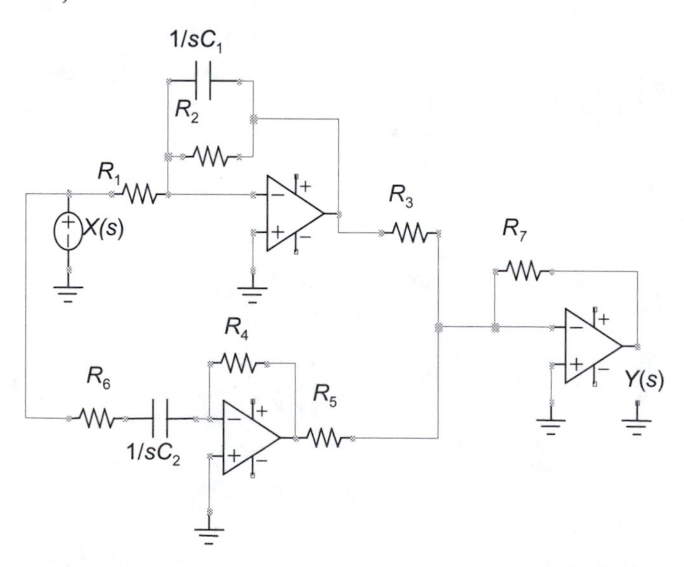

FIGURE 8.22
Active band-reject filter.

At low frequency, the capacitor is an open circuit and the inductor is a short circuit, and

$$Y(s) = X(s)$$

What happens when we operate between these two extremes?

Quantitatively, the transfer function relating $X(s)$ to $Y(s)$ can be derived using circuit analysis techniques to obtain

$$T(s) = \frac{s^2 + \frac{1}{LC}}{s^2 + (R/L)s + 1/LC} \tag{8.20}$$

Notice here that the poles of $T(s)$ for the band-reject are the same as the poles for the band-pass. The difference is in the location of the zeros in the two filters.

With $s = jw$, the magnitude of the transfer function can be calculated as

$$|T(jw)| = \frac{\frac{1}{LC} - w^2}{\sqrt{(1/LC - w^2)^2 + \left(w\frac{R}{L}\right)^2}} \tag{8.21}$$

In the case of a band-pass filter, we looked at some parameters that determine the shape of the magnitude and the phase plots.

1. The cut-off frequencies w_1 and w_2

   These can be calculated using the same procedure that we used to determine the cut-off frequencies for the band-pass filter.

2. The center frequency w_c

   The center frequency is the frequency where the transfer function $T(s)$ is purely real. It is also the frequency at which the magnitude of $T(s)$ is minimum (in the stop band). It is also the geometric mean of w_1 and w_2 and given as

$$w_c = \sqrt{w_1 w_2}$$

3. The bandwidth, β

   The bandwidth is the width of the reject band in the band-reject filter.

4. The quality factor, Q
   The quality factor is defined as the ratio of the center frequency to the bandwidth

$$Q = wc/\beta$$

   It is a measure of the bandwidth of the reject band.

The center frequency can be calculated by making $T(s)$ purely real. We can do that in the circuit at hand by setting

$$jw_cL + \frac{1}{jw_cC} = 0 \tag{8.22}$$

Solving the above equation leads to

$$w_c = \sqrt{\frac{1}{LC}} \tag{8.23}$$

Notice here that $|T(jw)|$ at w_c is minimum and is ideally 0.

We next calculate the cut-off frequencies w_1 and w_2. $T(jw)$ has its minimum at w_c.

$$|T(jw)| = \frac{\frac{1}{LC} - w^2}{\sqrt{(1/LC - w^2)^2 + \left(w\frac{R}{L}\right)^2}} = \frac{1}{\sqrt{2}}$$

With some manipulations we get the following quadratic equation in w

$$w^2L \pm wR - 1/C = 0$$

By solving the above equation we get the two cut-off frequencies

$$\begin{aligned} w_1 &= -\frac{R}{2L} + \sqrt{(R/2L)^2 + (1/LC)} \\ w_2 &= \frac{R}{2L} + \sqrt{(R/2L)^2 + (1/LC)} \end{aligned} \tag{8.24}$$

For the bandwidth, β, we have

$$\beta = w_2 - w_1 = R/L \tag{8.25}$$

and for the quality factor

$$Q = w_c/\beta = \sqrt{\frac{L}{CR^2}} \tag{8.26}$$

8.7.2 What Is the Phase?

We can see that the phase angle at low frequencies is close to 0° and at high frequencies it is close to 0° again. Around the reject band it is bounded by −90° and 90°.

8.7.3 What Is the DC Gain?

The magnitude of $T(s)$ at low frequencies is close to one, and the DC gain is therefore unity.

Example 8.5

Design a band-reject filter that has a center frequency of 500 Hz and a bandwidth of 200 Hz.

Solution

We can use the RLC circuit that was considered above with appropriate values for R, L, and C.

We can calculate the quality factor from the equation

$$Q = \frac{f_c}{f_2 - f_1} = 2.5$$

and the inductance L from

$$L = \frac{1}{Cw_c^2}$$

With $C = 1 \times 10^{-6}$ farads we have

$$L = \frac{1}{Cw_c^2} = \frac{1}{(2\pi(500))^2(10^{-6})} = 101 \text{ mH}$$

For the resistance R we use the equation

$$R = \beta L = 2\pi(200)(0.101) = 127 \text{ ohms.}$$

Having values for R, L, and C, the design is complete.

8.8 Some Insights: Filters with High Gain vs. Filters with Low Gain and the Relation between the Time Constant and the Cut-Off Frequency for First Order Circuits and the Series RLC Circuit

If we are interested in designing filters with a gain higher than unity we need to use circuits with active elements. Circuits with all passive elements, in most cases, give us a maximum gain of unity. If we are interested in using passive circuits with gains less than unity we can place additional loads across the output terminals.

If we seek circuits with gains higher than unity we need to use circuits with active elements such as operational amplifiers.

The time constant, τ, for first order filters is the parameter that characterizes the shape of the transient response. For RL and RC filters, the time constant is L/R and RC, respectively. We noticed that for RL or RC filters the cut-off frequency is $w_0 = 1/\tau$.

For an RLC series circuit, the overdamped, underdamped, and the critically damped transients are determined by the neper frequency α, and the resonant frequency w_c. The neper frequency is $R/2L$ rad/sec and the resonant frequency is $\sqrt{\frac{1}{LC}}$ rad/sec. α and w_c are time domain characteristics. The bandwidth, $\beta = 2\alpha$, is a frequency characteristic. Recall that the quality factor $Q = w_c/\beta$. The time when the response becomes underdamped is when $Q = \frac{1}{2}$. As Q increases in value the peak at w_c becomes sharper and the bandwidth smaller. This indicates an underdamped response. If Q is low and β is wide, it is an indication of overdamped transient response.

8.9 End-of-Chapter Examples

EOCE 8.1

Consider the circuit in Figure 8.23.

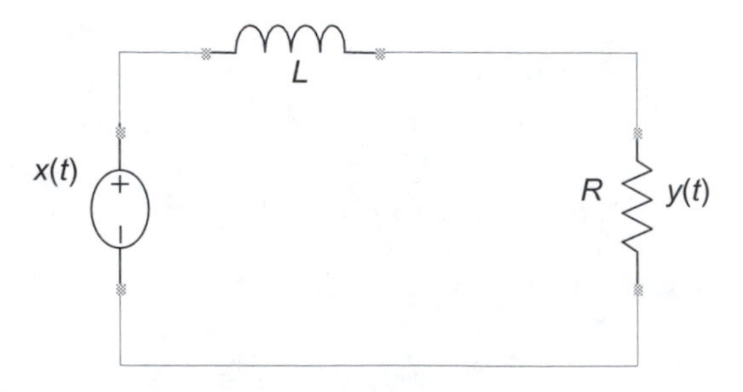

FIGURE 8.23
Circuit for EOCE 8.1.

Find values for R and L so that at the output $y(t)$ we will observe frequencies below 20π rad/sec.

Solution

We will start by investigating what kind of filter is given in Figure 8.23. We will put the circuit in Laplace domain and find the transfer function relating the input $x(t)$ to the output $y(t)$ as shown in Figure 8.24.

The transfer function is obtained by the voltage divider method as

$$Y(jw) = \frac{X(jw)(R)}{R + jwL}$$

or

$$T(jw) = \frac{R}{R + jwL} = \frac{(R/L)}{R/L + jw}$$

We can see that this circuit is a low-pass filter with a maximum gain of unity, and a cut-off frequency of R/L. Prove that.

It is desired that we pass frequencies below 20π rad/sec, which means that the desired cut-off frequency is 20π rad/sec.

Let $L = 1000$ mH, then $R = 20\pi$ ohms.

Now we can use MATLAB to plot the frequency response, $|T(jw)|$ vs. w, to check the quality of the design. We write the following MATLAB script.

```
R = 20*pi;
L = 1;
w = 0 : 0.1 : 100*pi;
T = R./(R + j*w*L);
Mag_T = abs(T);
```

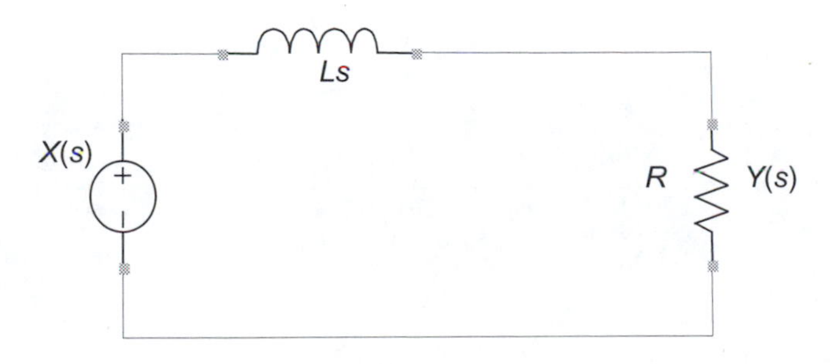

FIGURE 8.24
Circuit for EOCE 8.1.

```
plot(w,Mag_T);
xlabel('w in rad');
ylabel('Magnitude of the transfer function');
% press the mouse 3 times to get values on the graph
[x,y] = ginput (3)
```

The output is

```
x =
47.7885
147.4038
400.4808
y =
0.8012
0.3890
0.1524
```

The plot is shown in Figure 8.25.
We can see that after 10 Hz the magnitude drops drastically.

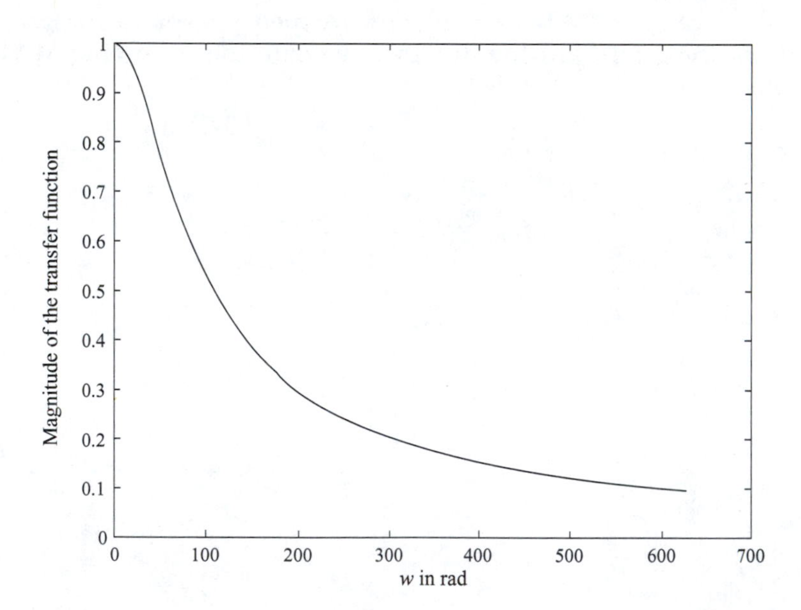

FIGURE 8.25
Plot for EOCE 8.1.

EOCE 8.2

The maximum gain that can be achieved with almost all passive circuits is unity. Suppose we reconsider EOCE 8.1 and, in addition, we need to attenuate the input signal by 50%. Can we modify the existing circuit as described in EOCE 8.1 to accomplish this?

Solution

Let us add another resistor in parallel with R across the output $y(t)$. The circuit is shown in Figure 8.26, where R_L is the added resistive load. The transfer function in this case is

$$Y(jw) = \frac{X(jw)(R_{eq})}{R_{eq} + jwL}$$

$$T(jw) = \frac{R_{eq}}{R_{eq} + jwL} = \frac{(R_{eq}/L)}{R_{eq}/L + jw}$$

where

$$R_{eq} = \frac{RR_L}{R + R_L}$$

It is clear that adding a load resistor will not change the maximum magnitude of the transfer function; it is still unity and the attempt fails.

Let us instead add another resistor in series with the input source, $x(t)$, as shown in Figure 8.27.

The transfer function in this case is

$$T(jw) = \frac{R/L}{jw + R_s/L + R/L}$$

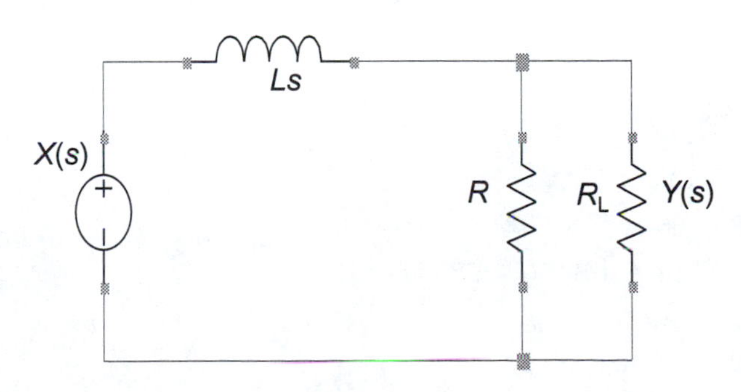

FIGURE 8.26
Circuit for EOCE 8.2.

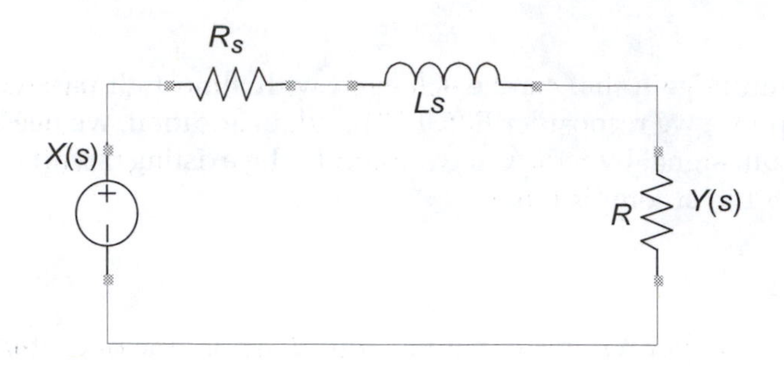

FIGURE 8.27
Circuit for EOCE 8.2.

with cut-off frequency at

$$\frac{R_s + R}{L}$$

and maximum gain of

$$\frac{R}{R_s + R}$$

We desire a magnitude of 0.5. This can be accomplished by setting $R = R_s$. The desired cut-off frequency is 20π rad/sec.

$$\frac{R_s + R}{L} = 20\pi = \frac{2R}{L}$$

With $R = 20\pi$, we select a new L of 2 H.

The two circuits in EOCE 8.1 and EOCE 8.2 are similar but have different gains. Both have the same cut-off frequency. This can be seen using the following graph generated by the MATLAB script.

```
R = 20*pi;
L2 = 2;
L1 = 1;
Rs  = 20*pi;
w = 0 : 0.1 : 100*pi;
T1 = R./(R + j*w*L1);
T2 = R./(R + j*w*L2 + Rs);
Mag_T1 = abs(T1);
Mag_T2 = abs(T2);
plot(w,Mag_T1,w,Mag_T2);
xlabel('w in rad');
ylabel('Magnitude of the transfer functions');
```

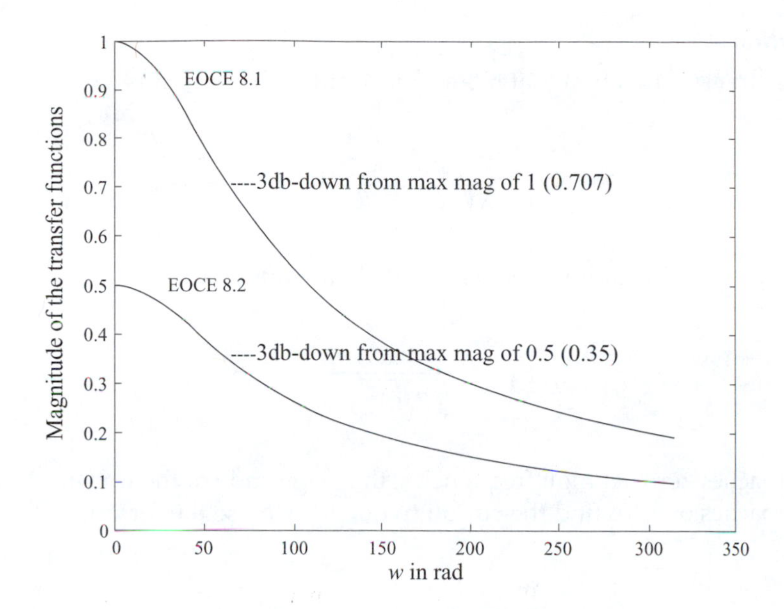

FIGURE 8.28
Plots for EOCE 8.2.

```
gtext('EOCE 8.1');
gtext('----3db-down from max mag of 1 (0.707)');
gtext('EOCE 8.2');
gtext('----3db-down from max mag of 0.5 (0.35)');
```

The output is shown in Figure 8.28.

EOCE 8.3

Consider the circuit shown in Figure 8.29, with $y(t)$ as the output and $x(t)$ as the input. Show that the transfer function that relates the input to the output represents a high-pass filter. Select values for R and L so the filter will attenuate frequencies below 10 Hz.

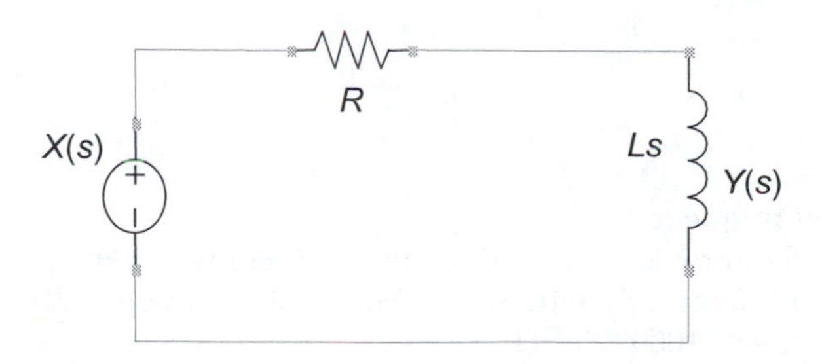

FIGURE 8.29
Circuit for EOCE 8.3.

Solution

The transfer function that relates the input to the output is

$$T(jw) = \frac{Y(jw)}{X(jw)} = \frac{Ls}{R + Ls} = \frac{s}{R/L + s}$$

We can see that at low frequencies the magnitude of $T(jw)$

$$\frac{w}{\sqrt{(R/L)^2 + w^2}}$$

approaches zero. At high frequencies, the magnitude of the transfer function approaches one. To find the cut-off frequency we use the formula

$$\frac{\max(|T(jw)|)}{\sqrt{2}} = |T(jw)|_{w_0}$$

or

$$\frac{1}{\sqrt{2}} = |T(jw)|_{w_0} = \frac{w_0}{\sqrt{(R/L)^2 + w_0^2}}$$

Squaring both sides gives

$$\frac{1}{2} = \frac{w_0^2}{(R/L)^2 + w_0^2}$$

where

$$w_0 = \frac{R}{L}$$

is the cut-off frequency.

We want the filter to attenuate frequencies below 10 Hz, so we use w_0 as the cut-off frequency. By letting $R = 1\text{K}$, we will have $20\pi = 1000/L$. This gives L a value of $1000/20\pi$ H.

Let us plot the magnitude of $T(jw)$ vs. w using the MATLAB script

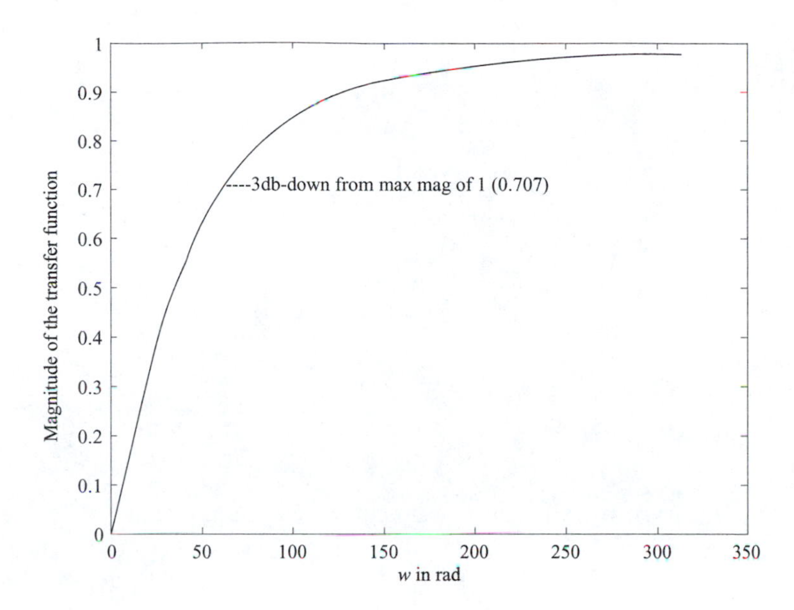

FIGURE 8.30
Plot for EOCE 8.3.

```
R = 1000;
L = 1000/(20*pi);
w = 0 : 0.1 : 100*pi;
T1 = w./(R/L + j*w);
Mag_T1 = abs(T1);
plot(w,Mag_T1);
xlabel('w in rad');
ylabel('Magnitude of the transfer function');
gtext('----3db-down from max mag of 1 (0.707)');
```

The plot is shown in Figure 8.30.

EOCE 8.4

Consider the series RLC circuit in Figure 8.31, with $y(t)$ as the output and $x(t)$ as the input. We have mentioned in this chapter that this filter is a pass-band. We will select values for R, L, and C so that the output will receive only the range of frequencies between 10 and 20 Hz.

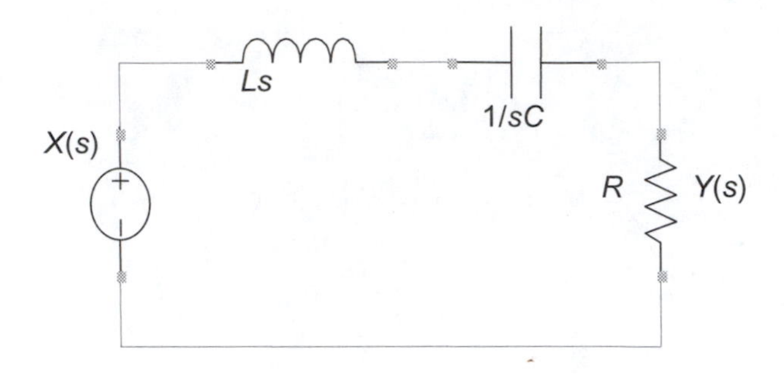

FIGURE 8.31
Circuit for EOCE 8.4.

Solution

We start the design by calculating the center frequency as $w_c = 2\pi f_c$ and $f_c = \sqrt{f_1 f_2} = 14.14$ Hz. $(w_c)^2 = (2\pi f_c)^2 = 8186.2$. Let $C = 1 \times 10^{-6}$ farads, then

$$L = \frac{1}{Cw_c^2} = 122.156 \text{ H}$$

and the quality factor is

$$Q = \frac{f_c}{f_2 - f_1} = 1.414$$

For R we use the equation

$$R = \sqrt{L/(CQ^2)} = 7816.4 \text{ ohms}$$

Having values for R, L, and C, the design is complete.

Notice that the center frequency is independent of R. Let us vary R and give it values of 10, 100, 1000, and the designed value of 7816.4 ohms. Let us fix L and C as in the design. We will write the MATLAB script to demonstrate the effect of varying R and observing the bandwidth and the magnitude of each transfer function that corresponds to each R.

```
clf
R = [10  100  1000  7816.4]
L = 122.156;
C = 0.000001;
```

```
w = 0 : 1 : 200;
for i = 1 : 4
T1 = ((j*w) * (R(i)/L))./(-w.^2 + j*w*(R(i)/L) + 1/(L*C));
Mag_T1 = abs(T1);
plot(w,Mag_T1);
hold on
end
title('second order RLC series filter')
xlabel('w in rad');
ylabel('Magnitude of the transfer functions');
axis([50 150 0 1.25])
gtext('R = 10');gtext('R = 100');
gtext('R = 1000');gtext('R = 7816.4');
```

The plots are shown in Figure 8.32.

We can see that the center frequency is the same for each R used, but for R = 10 ohms, the maximum magnitude is very small. This magnitude increases as the value of R increases.

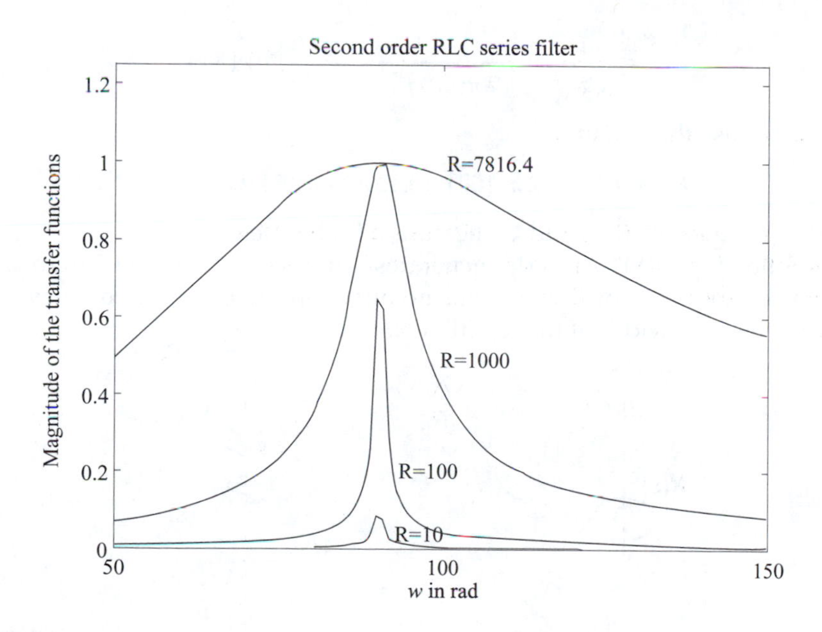

FIGURE 8.32
Plots for EOCE 8.4.

EOCE 8.5

Design a band-reject filter that has a center frequency of 50 Hz and a bandwidth of 100 Hz.

Solution

We can use the RLC circuit that was considered earlier and which is given again as Figure 8.33.

The transfer function relating the input to the output is

$$T(s) = \frac{s^2 + \frac{1}{LC}}{s^2 + (R/L)s + 1/LC}$$

The quality factor is

$$Q = \frac{f_c}{f_2 - f_1} = 0.5$$

and the inductance is

$$L = \frac{1}{Cw_c^2}$$

With $C = 1 \times 10^{-6}$ farads, L is

$$L = \frac{1}{Cw_c^2} = \frac{1}{(2\pi(50))^2(10^{-6})} = 10.13\,\text{H}$$

For R we use the equation

$$R = \beta L = 2\pi(100)(10.13) = 6364.9 \text{ ohms}$$

Having values for R, L, and C, the design is complete.

The following MATLAB script generates multiple plots for the design and also varies the value of R which has no effect on the central frequency but controls the bandwidth of the rejection band.

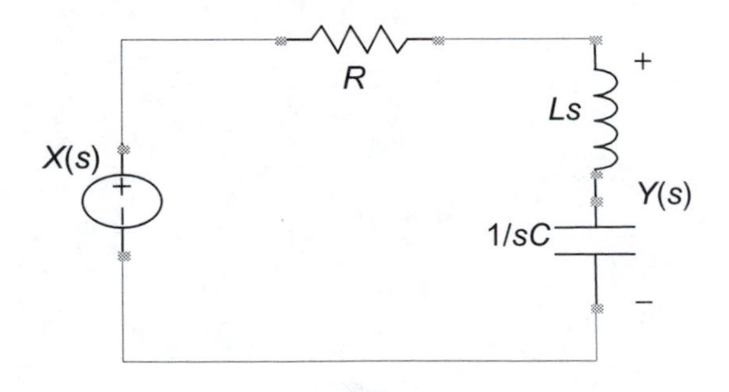

FIGURE 8.33
Circuit for EOCE 8.5.

```
clf
R = [10 100 1000 6364.9]
L = 10.1321;
C = 0.000001;;
w = 0 : 1 : 200*pi;
for i = 1 : 4
T1 = (-w.^2 + 1/(L*C))./(-w.^2 + j*w*(R(i)/L) + 1/(L*C));
Mag_T1 = abs(T1);
plot(w,Mag_T1);
hold on
end
title('second order RLC series filter')
xlabel('w in rad');
ylabel('Magnitude of the transfer functions');
axis([50 500 0 1.1])
gtext('R = 10');gtext('R = 100');
gtext('R = 1000');gtext('R = 7816.4');
```

The plots are in Figure 8.34.

Pay close attention to the way the resistance R is shaping the magnitude of the transfer function seen in Figure 8.34. The smaller the R, the narrower the

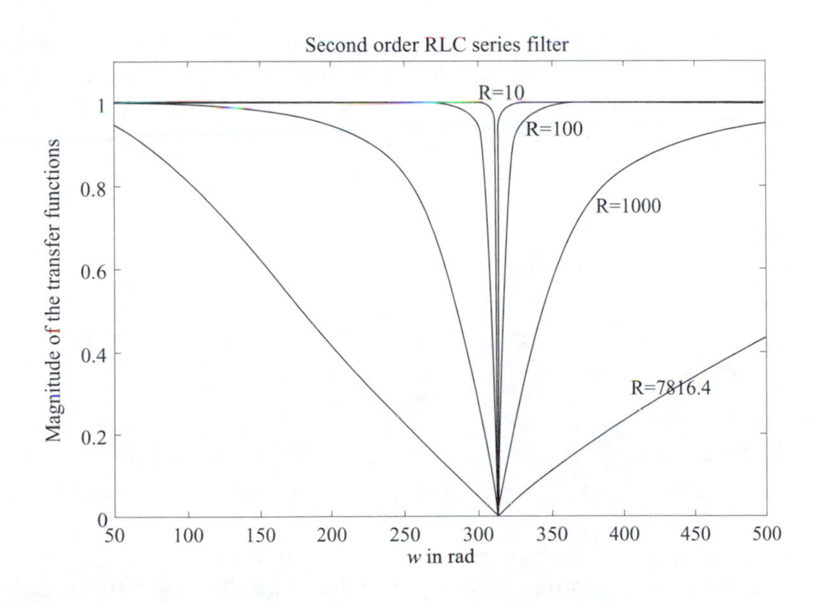

FIGURE 8.34
Plots for EOCE 8.5.

rejection, or the more selective the filter becomes in terms of rejecting a specific band of frequencies.

EOCE 8.6

Consider the following filters.

$$H_1(s) = \frac{1}{s+1}$$

$$H_2(s) = \frac{s}{s+1}$$

$$H_3(s) = \frac{1}{s^2+s+1}$$

$$H_4(s) = \frac{s}{s^2+s+1}$$

$$H_5(s) = \frac{s^2}{s^2+s+1}$$

$$H_6(s) = \frac{s+1}{s^2+s+1}$$

Use MATLAB to study the phase and magnitude characteristics of these filters. These filters are given as transfer functions relating inputs to outputs. The MATLAB function we use here is called "bode." "Bode" produces the semi-log magnitude plots and the phase plots in degrees, both vs. the radian frequency w.

Solution

For transfer function H_1 we have the MATLAB script

```
clf
num = [1];
den = [1  1];
bode(num,den);
title('The filter representing H1(s)')
```

The magnitude and the phase plots are shown in Figure 8.35.

The plot indicates that this is a low-pass filter since the magnitude starts to attenuate at high frequencies.

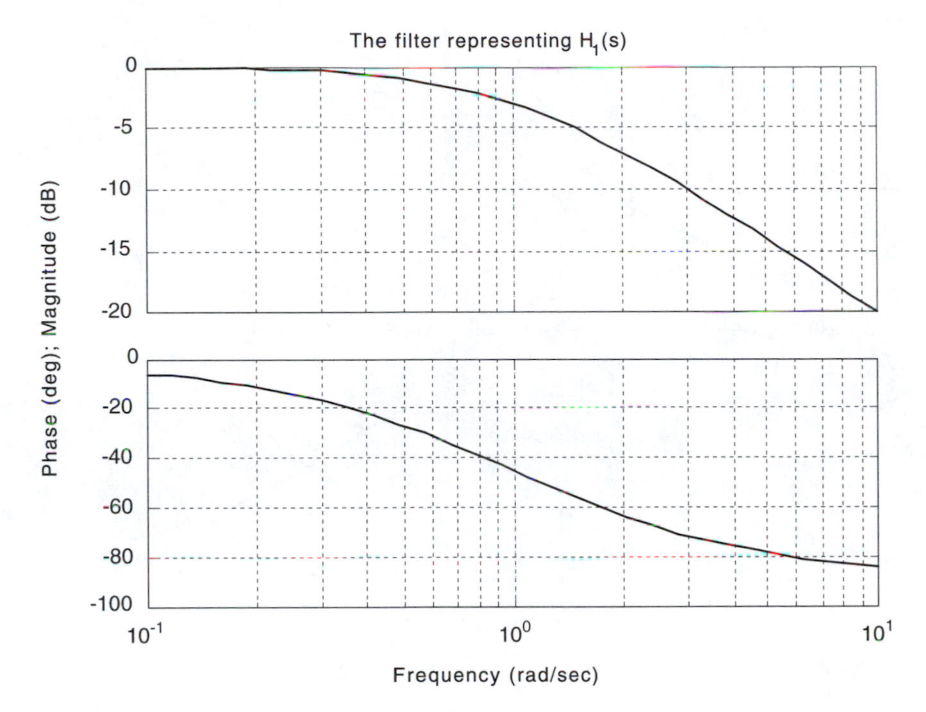

FIGURE 8.35
Plots for EOCE 8.6.

For transfer function H_2 we use the script

```
clf
num = [1  1];
den = [1  1];
bode(num,den);
title('The  filter  representing  H2(s)')
```

The magnitude and the phase plots are shown in Figure 8.36.

This is clearly a high-pass filter since it emphasizes the magnitude at higher frequency bands. For the filter representing H_3, the plot for the magnitude and the phase are shown in Figure 8.37.

This is also a low-pass filter since it attenuates the magnitude at higher frequencies.

For the filter represented by H_4 the magnitude and the phase plots are shown in Figure 8.38.

This is a band-pass filter since the magnitude is emphasized at the midrange frequencies.

For H_5 the magnitude and the phase plots are shown in Figure 8.39.

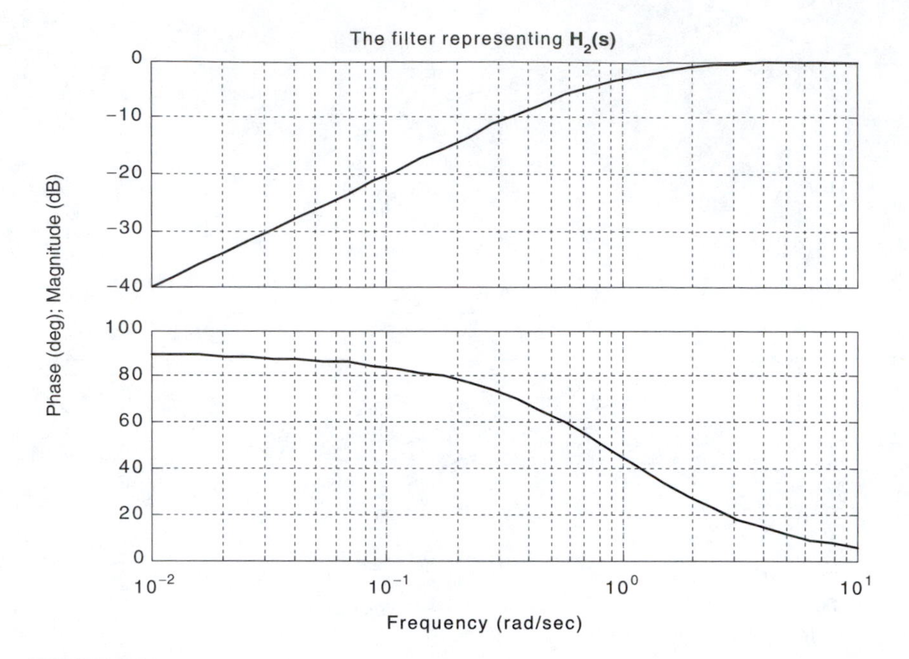

FIGURE 8.36
Plots for EOCE 8.6.

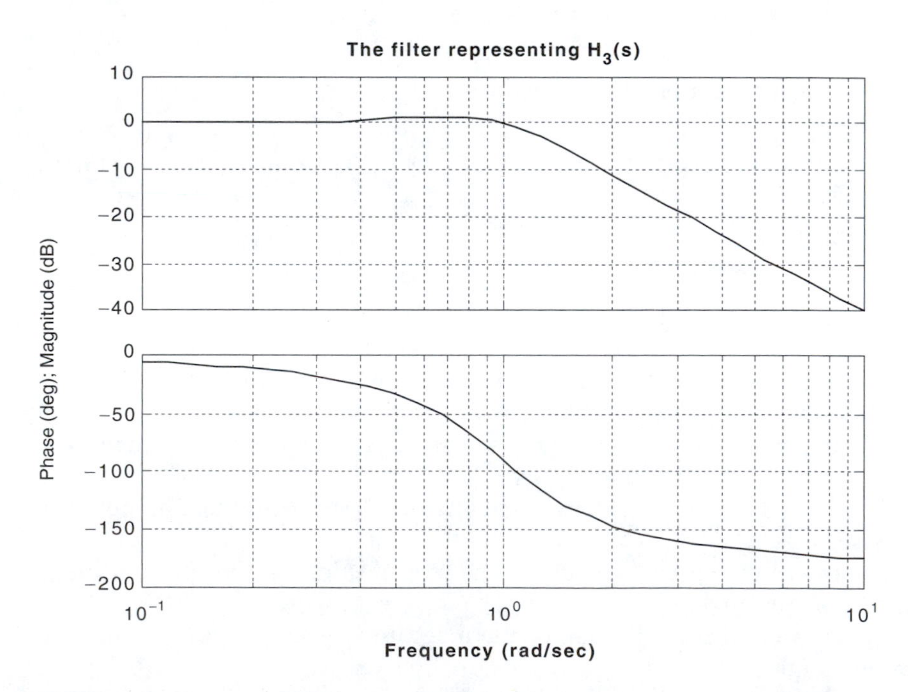

FIGURE 8.37
Plots for EOCE 8.6.

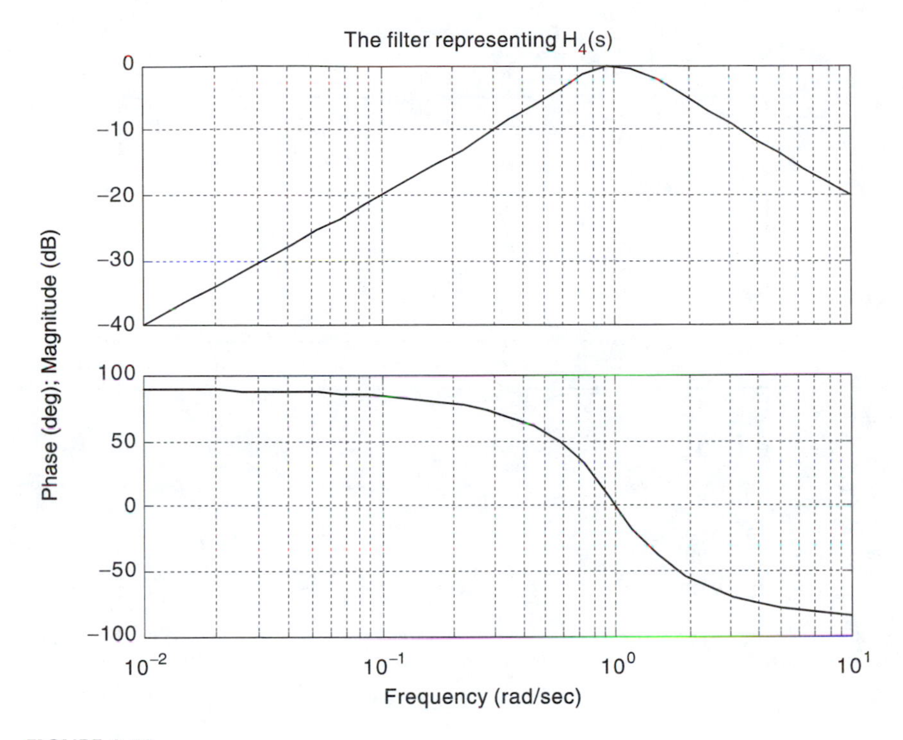

FIGURE 8.38
Plots for EOCE 8.6.

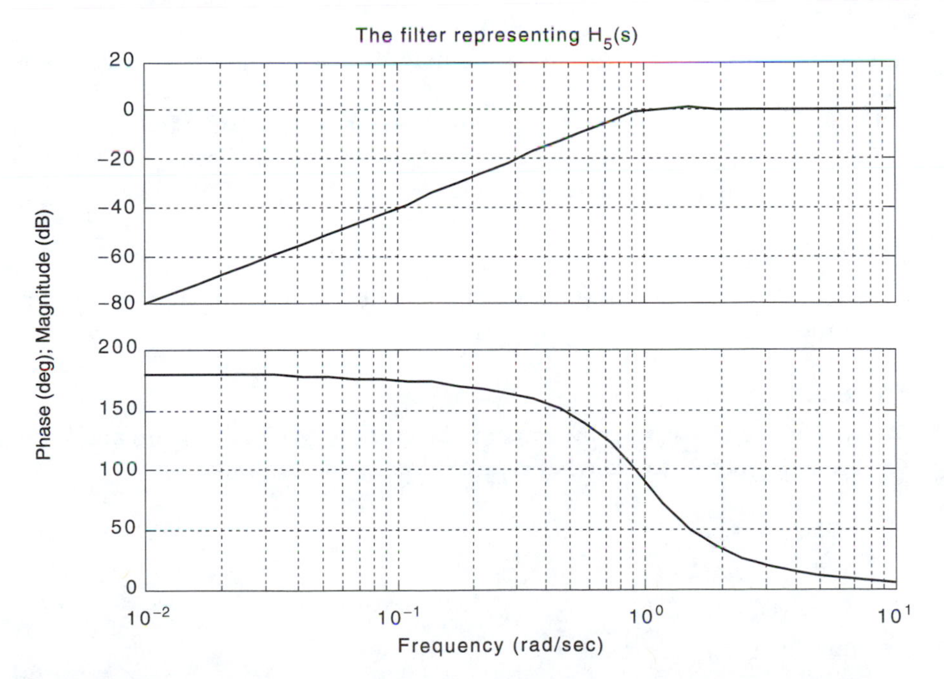

FIGURE 8.39
Plots for EOCE 8.6.

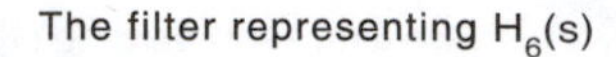

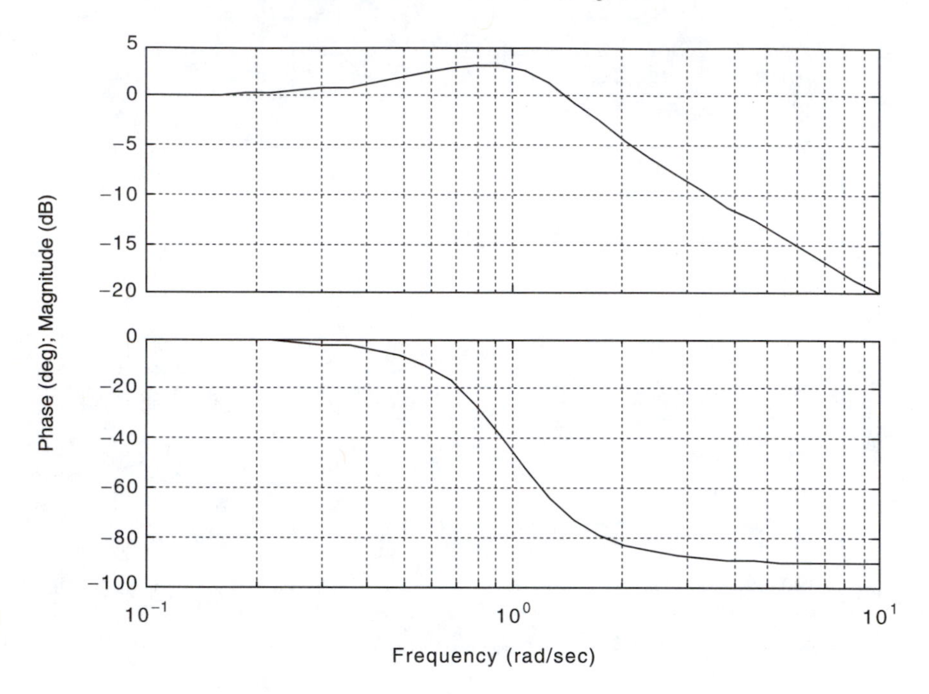

FIGURE 8.40
Plots for EOCE 8.6.

This is a high-pass filter since the magnitude at higher frequencies is emphasized.

Finally for H_6 the magnitude and the phase plots are shown in Figure 8.40.

This final filter is low-pass since its magnitude is emphasized at low frequencies and attenuated at high frequencies.

EOCE 8.7

Consider EOCE 8.6 again. What will happen if you pass $x(t)$ through system 1 and then system 2?

Solution

We will have H_1H_2 as the new system. We will use MATLAB again and write the following script to see what type of filter is this combination.

```
clf
num1 =  [1  1];
den1 = [1  1  1];
num2 = [1  0];
den2 = [1  1];
```

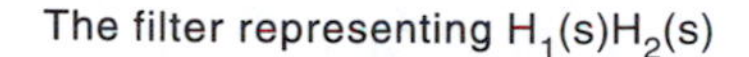

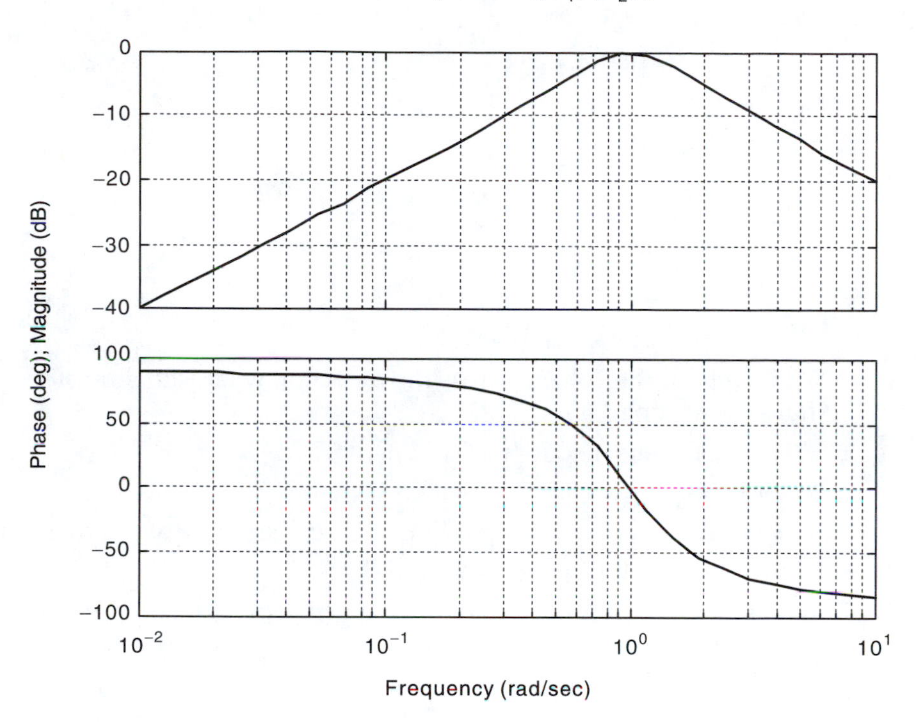

FIGURE 8.41
Plots for EOCE 8.7.

```
num = conv(num1, num2); % multiply num1 with num2
den = conv(den1, den2);
bode(num, den);
title('The filter representing H1(s)H2(s)')
```

The plots are in Figure 8.41.
This is a band-pass filter as it emphasizes frequencies in the mid-range.
Try the other combinations.

8.10 End-of-Chapter Problems

EOCP 8.1

Consider the system in Figure 8.42.

1. Taking $y(t)$ as the output voltage across the resistor, what is the transfer function relating $Y(s)$ to $X(s)$?

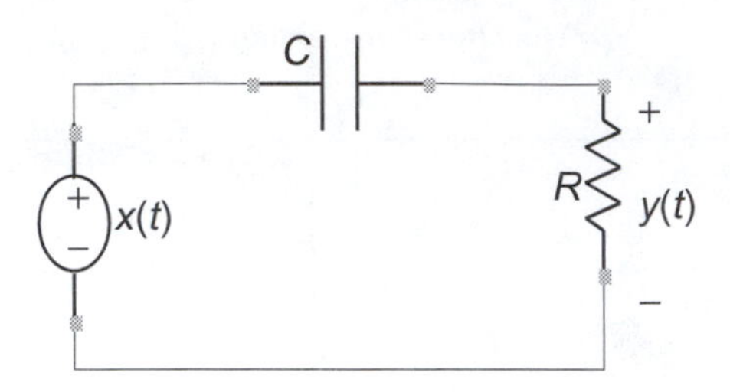

FIGURE 8.42
Circuit for EOCP 8.1.

2. Taking $y(t)$ as the voltage across the capacitor, what is the transfer function relating $Y(s)$ to $X(s)$?
3. What are the eigenvalues in 1 and 2?
4. What type of filters are represented in 1 and 2?
5. What are the effects of the zeros in 1 and 2 on the nature of the filter?
6. What is the cut-off frequency in 1 and 2?
7. If $R = 1k$ and $C = 1$ micro farad, what is an approximation to the output $y(t)$ in the filter represented in 1 if the input $x(t) = 10\sin(10t) - \sin(1000t) + 12$?
8. If $R = 1\text{k}$ and $C = 1$ micro farad, what is an approximation to the output $y(t)$ in the filter represented in 2 if the input $x(t) = 10\sin(10t) - \sin(1000t) + 12$?

EOCP 8.2

Consider the circuit in Figure 8.43.

1. Taking $y(t)$ as the voltage across the capacitor, what is the transfer function relating $Y(s)$ to $X(s)$?
2. What type of filter is represented in Figure 8.43?
3. What is the cut-off frequency if it exists?

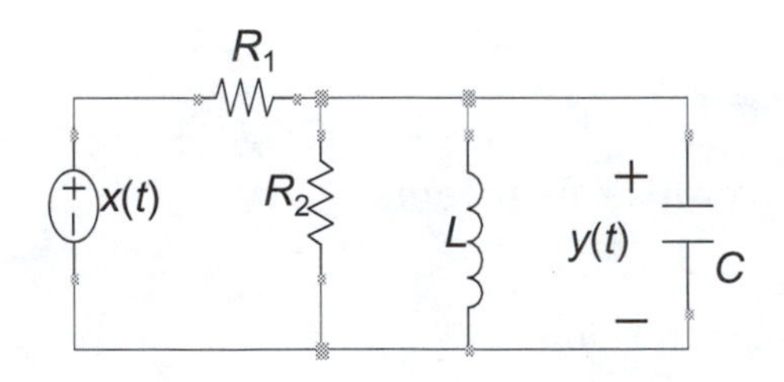

FIGURE 8.43
Circuit for EOCP 8.2.

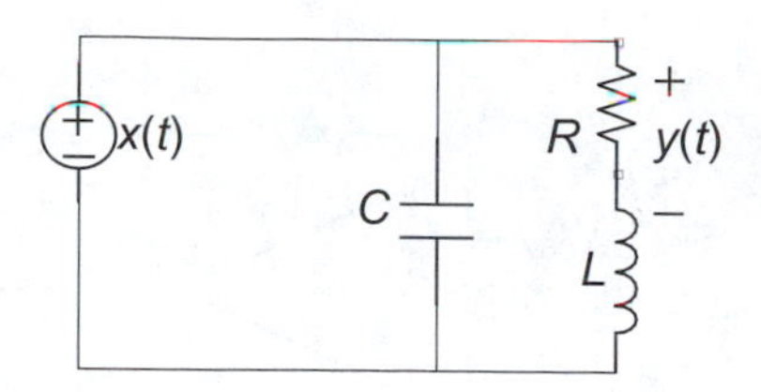

FIGURE 8.44
Circuit for EOCP 8.3.

EOCP 8.3

Consider the system in Figure 8.44.

1. Taking $y(t)$ as the voltage across the resistor, what is the transfer function relating $Y(s)$ to $X(s)$?
2. What type of filter is represented in Figure 8.44?
3. What is the cut-off frequency if it exists?

EOCP 8.4

Consider the system in Figure 8.45.

1. Taking $y(t)$ as shown in Figure 8.45, what is the transfer function relating $Y(s)$ to $X(s)$?
2. What type of filter is represented in Figure 8.45?
3. What is the cut-off frequency if it exists?

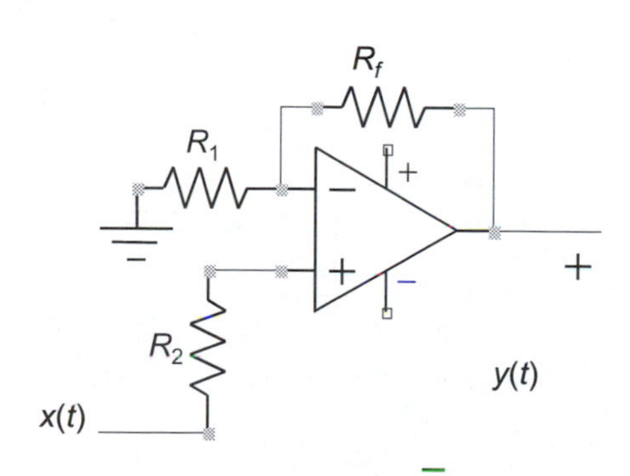

FIGURE 8.45
Circuit for EOCP 8.4.

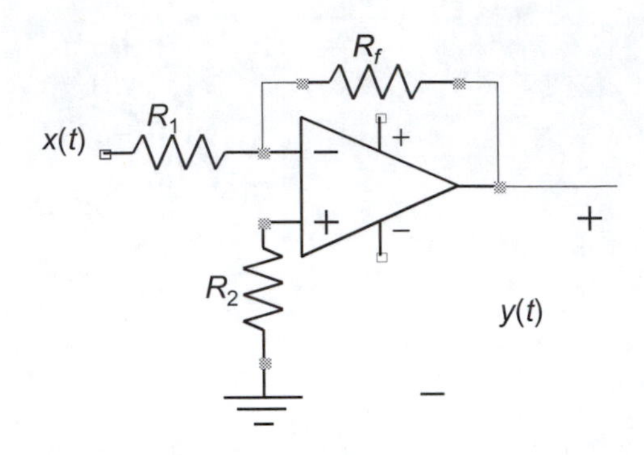

FIGURE 8.46
Circuit for EOCP 8.5.

EOCP 8.5

Consider the system in Figure 8.46.

1. Taking $y(t)$ as shown in Figure 8.46, what is the transfer function relating $Y(s)$ to $X(s)$?
2. What type of filter is represented in Figure 8.46?
3. What is the cut-off frequency if it exists?

EOCP 8.6

Consider the system in Figure 8.47.

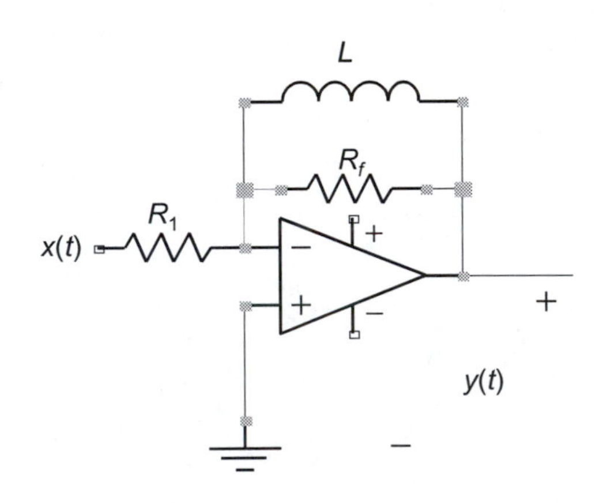

FIGURE 8.47
Circuit for EOCP 8.6.

1. Taking $y(t)$ as seen in Figure 8.47, what is the transfer function relating $Y(s)$ to $X(s)$?
2. What type of filter is represented in Figure 8.47?
3. What is the cut-off frequency if it exists?
4. What is the output $y(t)$ if $x(t) = 10\sin(10t) - \sin(1000t) + 12$?

EOCP 8.7

Consider the system in Figure 8.48.

1. Taking $y(t)$ as shown in Figure 8.48, with $L_2 = 0$, what is the transfer function relating $Y(s)$ to $X(s)$?
2. What type of filter is represented in Figure 8.48?
3. What is the cut-off frequency if it exists?

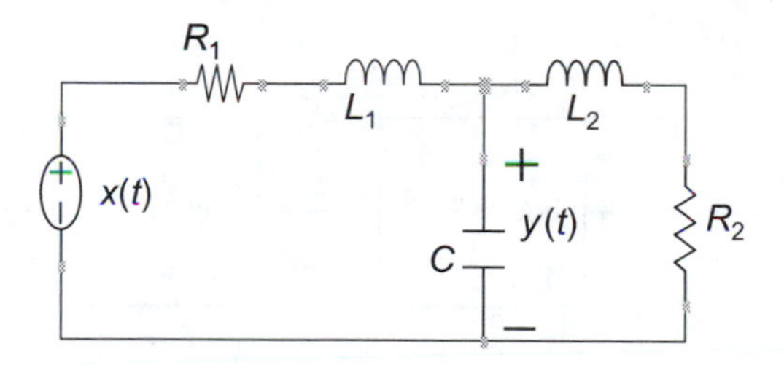

FIGURE 8.48
System for EOCP 8.7.

EOCP 8.8

Consider the system in Figure 8.49.

1. Taking $y(t)$ as shown in Figure 8.49, what is the transfer function relating $Y(s)$ to $X(s)$?
2. What type of filter is represented in Figure 8.49?
3. What is the cut-off frequency if it exists?

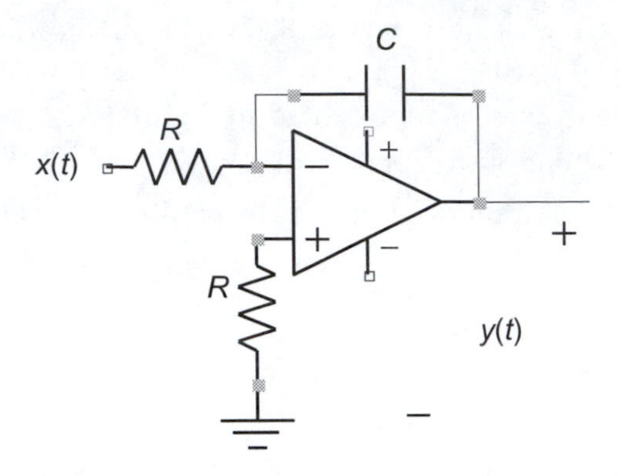

FIGURE 8.49
System for EOCP 8.8.

EOCP 8.9

Consider the system represented in the block diagram as shown in Figure 8.50.

1. Find the transfer function relating $Y(s)$ to $X(s)$ as shown in Figure 8.50.
2. What is the filter represented in Figure 8.50?

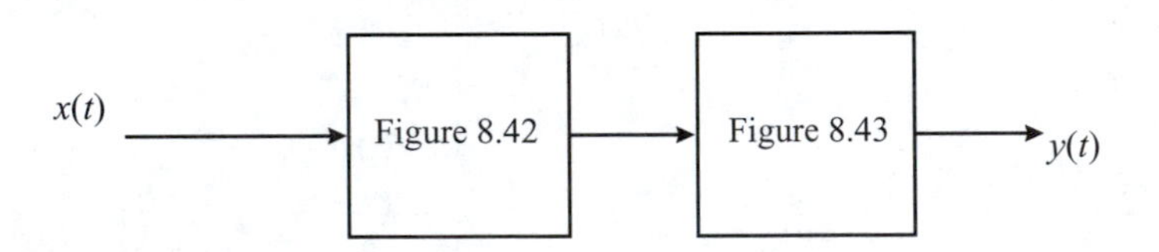

FIGURE 8.50
System for EOCP 8.9.

EOCP 8.10

Consider Figures 8.42 and 8.43 as connected in parallel with $x(t)$ and $y(t)$ as the input and the output, respectively.

1. Find the transfer function relating $Y(s)$ to $X(s)$.
2. What is the filter represented as the result of this arrangement?

EOCP 8.11

Consider the system as shown in Figure 8.51.

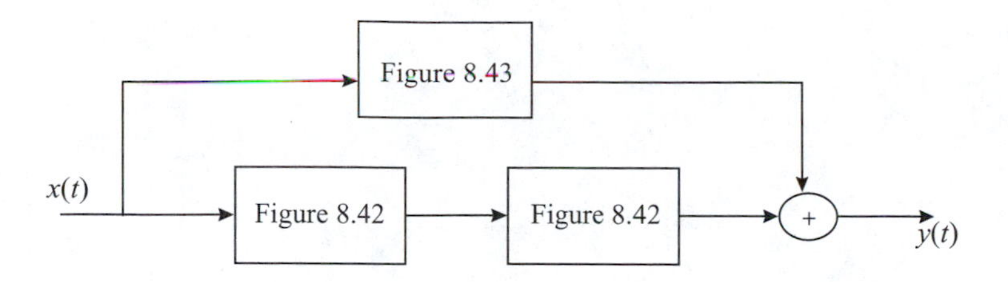

FIGURE 8.51
System for EOCP 8.11.

1. Find the transfer function relating $Y(s)$ to $X(s)$.
2. What kind of filter does it represent?

EOCP 8.12

Consider the following systems represented by the transfer functions relating the output to the input of some dynamic systems.

1. $\dfrac{1}{s^2 + 5s + 6}$

2. $\dfrac{s}{s^2 + 4s + 4}$

3. $\dfrac{s^2}{s^2 + 7s + 10}$

4. $\dfrac{s^3 + 1}{s^2 + 7s + 10}$

5. $\dfrac{s^3 + s^2 + 1}{s^2 + 7s + 10}$

a) Plot the magnitude of each system vs. frequency. You may use MATLAB to obtain the "bode" plots.
b) Remember that the cut-off frequency is found by locating the maximum magnitude on the "bode" plot and going 3db down the magnitude scale. Find the cut-off frequencies for the given systems.
c) Characterize the filters given in this problem.

EOCP 8.13

Consider the following systems.

1. $\dfrac{1}{7s + a}$

2. $\dfrac{1}{s^2 + as + b}$

3. $\dfrac{s^2}{s^2 + as + b}$

4. $\dfrac{s^2 + s + 1}{s^2 + as + b}$

a) For the first system, design to have a maximum DC gain of 10 and a cut-off frequency of 10k. Find the a value and the additional gain desired.

b) For the system in 2, it is desired that we have a DC gain of 5 and a cut-off frequency of 10k. We want the design to resemble a low-pass filter with positive a and b values.

c) For the system in 3, it is desired to obtain a maximum DC gain of 5 and a cut-off frequency of 10k. We want the design to resemble a high-pass filter with positive a and b values.

d) For the system in 4, and for negative values for a and b, it is desired that this filter rejects the band of frequencies between 0.9 rad/sec and 1.1 rad/sec. We also desire an absolute DC gain of 1. Find the values for a and b.

References

Close, M. and Frederick, K. *Modeling and Analysis of Dynamic Systems,* 2nd ed., New York: Wiley, 1995.

Cogdell, J.R. *Foundations of Electrical Engineering,* 2nd ed., Englewood Cliffs, NJ: Prentice-Hall, 1996.

Denbigh, P. *System Analysis and Signal Processing,* Reading, MA: Addison-Wesley, 1998.

Golubitsky, M. and Dellnitz, M. *Linear Algebra and Differential Equations Using MATLAB,* Stamford, CT: Brooks/Cole, 1999.

Harman, T.L., Dabney, J., and Richert, N. *Advanced Engineering Mathematics with MATLAB,* Stamford, CT: Brooks/Cole, 2000.

The MathWorks. *The Student Edition of MATLAB,* Englewood Cliffs, NJ: Prentice-Hall, 1997.

Nilson, W.J. and Riedel, S.A. *Electrical Circuits,* 6th ed., Englewood Cliffs, NJ: Prentice-Hall, 2000.

Phillips, C.L. and Parr, J.M. *Signals, Systems, and Transforms*, 2nd ed., Englewood Cliffs, NJ: Prentice-Hall, 1999.

Pratap, R. *Getting Started with MATLAB 5,* New York: Oxford University Press, 1999.

Valkenburg, M.E. *Analog Filter Design,* Philadelphia: W.B. Saunders, 1982.

Woods, R.L. and Lawrence, K.L. *Modeling and Simulation of Dynamic Systems,* Englewood Cliffs, NJ: Prentice-Hall, 1997.

Wylie, R.C. and Barrett, C.L. *Advanced Engineering Mathematics,* 6th ed., New York: McGraw-Hill, 1995.

Ziemer, R.E., Tranter, W.H., and Fannin, D.R. *Signals Systems Continuous and Discrete,* 4th ed., Englewood Cliffs, NJ: Prentice-Hall, 1998.

9

Linearization of Nonlinear Systems

CONTENTS

9.1 Introduction

Almost all physical systems are somewhat nonlinear. Springs typically become stiffer near the end of their range of motion. The fluid capacitance of a reservoir usually increases with stored volume. Friction phenomena tend to be nonlinear. The analytical solution techniques used in previous chapters in this book depend fundamentally on the assumption of linearity of the systems. The superposition techniques we used are not applicable in cases in which the systems under investigation are nonlinear.

It is known that most components found in physical systems have nonlinear characteristics. In practice, we can find some devices that have moderate nonlinear characteristics, or the nonlinear properties of these devices may occur if the devices are driven into certain nonlinear regions. If we model these devices by linear systems we may get accurate analytical results over a relatively wide range of operating conditions. But there are many situations where physical devices can possess strong nonlinear characteristics. In these cases modeling these strongly nonlinear devices by linear systems can be valid only for a limited range of operation.

Let us look at four ways to analyze nonlinear systems.

1. We can find an analytical solution to the nonlinear system under consideration.
2. We can use software application tools to compute numerical solutions for specific parameters in a nonlinear system.
3. We can solve the equations describing the nonlinear system numerically using computers.
4. We can replace the nonlinear system by a linear system that behaves nearly the same as the nonlinear system.

The last way stated above is called linearization, and it is this method that will be used in this chapter.

9.2 Linear and Nonlinear Differential Equations

An nth-order ordinary differential equation is said to be linear in y if it can be written in the form

$$a_n(t)\frac{d^n}{dt}y(t) + a_{n-1}(t)\frac{d^{n-1}}{dt}y(t) + \cdots + a_1(t)\frac{d}{dt}y(t) + a_0(t)y(t) = x(t) \tag{9.1}$$

where $a_0, a_1, \ldots, a_n$ are functions on some interval of t, and $a_n(t) \neq 0$ on that interval.

Therefore, an ordinary differential equation is linear if

1. The unknown function and its derivatives algebraically occur to the first degree only.
2. There are no products involving either the unknown function and its derivatives or two or more derivatives.
3. There are no transcendental functions involving the unknown function or any of its derivatives.

A differential equation (system) that is not linear is called a nonlinear differential equation (system). Solutions of nonlinear differential equations, like linear differential equations, may approach zero, become unbounded, or remain bounded as t gets large.

Example 9.1

Consider the following differential equations. Which are linear and which are not?

1. $\frac{d^2}{dt}y(t) - 3\frac{d}{dt}y(t) + 3y(t) = t^3$

2. $t\frac{d^3}{dt}y(t) + y(t)e^t + 5 = 0$
3. $\left(\frac{d}{dt}y(t)\right)^3 + 2y(t) = t$
4. $y(t)\frac{d}{dt}y(t) + 3t = 0$
5. $\frac{d^2}{dt}y(t) + 5y(t) = \cos(y(t))$

Solution

1. The first equation is linear in $y(t)$.
2. The second equation is linear in $y(t)$.
3. The third equation is nonlinear because the first derivative of the unknown function is to the third degree.
4. The fourth equation is nonlinear because the unknown function is multiplied by its derivative.
5. The fifth equation is nonlinear because it involves the cosine function of the unknown (transcendental function).

9.3 The Process of Linearization

We will start with a simple example to illustrate the process of approximating a nonlinear equation (system) at an operating point. Then we will give a structured process for linearizing systems.

Example 9.2

Consider the freely swinging (frictionless) pendulum of length l as shown in Figure 9.1. Linearize the system.

Solution

Using Newton's law of motion, we obtain the following nonlinear equation

$$\frac{d^2}{dt}\theta(t) + (g/l)\sin(\theta(t)) = 0 \tag{9.2}$$

For small θ, $\sin(\theta)$ approaches θ. In this case we can rewrite the equation as the second order linear differential equation

$$\frac{d^2}{dt}\theta(t) + (g/l)\theta(t) = 0 \tag{9.3}$$

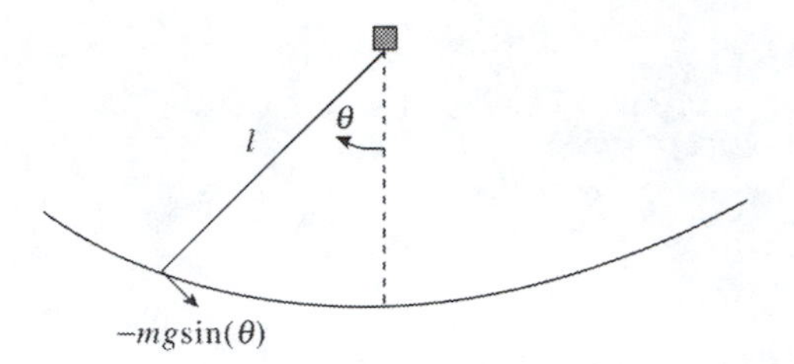

FIGURE 9.1
System for Example 9.2.

where g is the force of gravity and l is the length of the pendulum.

The general solution of the last equation is

$$\theta(t) = c_1 \cos\left(\frac{g}{l} t\right) + c_2 \sin\left(\frac{g}{l} t\right) \tag{9.4}$$

where c_1 and c_2 are to be determined.

Therefore, if θ is close to zero (the equilibrium point as shown in Figure 9.1), the linear equation solution we approximated will behave somewhat the same as the solution of the nonlinear equation at that equilibrium point θ.

9.3.1 Linearization of a Nonlinear System Given by a Differential Equation

To put a nonlinear system described as a differential equation in a linearized form we do the following:

1. Determine the operating point of the system described by the differential equation.
2. Write the linear variables as the sum of their nominal and incremental variables.
3. Use the first two terms in the Taylor series expansion to rewrite the nonlinear terms at the operating point.
4. Clean the differential equation describing the system.
5. Find the initial conditions for the new linear differential equation using initial conditions for the original nonlinear differential equation.

The operating point of the system will be the equilibrium point where all variables in the system are constant and equal to their nominal values. Nominal values for inputs are selected as their average values. The sum of the nominal and the incremental solutions for the output in a system is only an approximation to the solution of the nonlinear system.

Example 9.3

Consider the following nonlinear differential equation

$$\frac{d^2}{dt}y + \frac{d}{dt}y + y^2 = x(t)$$

with initial conditions

$$y(0) = \frac{d}{dt}y(0) = 0$$

where $x(t)$ in the input. The nonlinear element in the differential equation is obviously y^2.

Assume that the input is

$$x(t) = A\sin(t)$$

Linearize the system.

Solution

The nominal value for the input is its average value of zero, so $x_n = 0$.

To find the operating point of the system, we follow the procedure outlined above where we take all variables in the system as constants and write the nonlinear equation as

$$\frac{d^2}{dt}y_n + \frac{d^2}{dt}y_n + y_n^2 = x_n = 0$$

The nominal values are constants; therefore, all their derivatives are zero. Then we can write $y_n^2 = 0$, which gives y_n the value of zero.

Next, we write the linear terms in the nonlinear differential equation as a sum of their nominal and incremental values

$$\left(\frac{d^2}{dt}y_n + \frac{d^2}{dt}\Delta y\right) + \left(\frac{d}{dt}y_n + \frac{d}{dt}\Delta y\right) + y^2 = x_n + \Delta x$$

where Δ is a small increment in the value of the variable. Since the derivatives of the nominal values are all zero, we get

$$\frac{d^2}{dt}\Delta y + \frac{d}{dt}\Delta y + y^2 = \Delta x$$

Then we linearize the term y^2 at the nominal operating point using the Taylor series expansion formula

$$f(x) = f(x_n) + \frac{d}{dt}f(x)\bigg|_{x_n} \Delta x + \frac{d^2}{dt}f(x)\bigg|_{x_n} (\Delta x)^2 + \text{higher order terms} \tag{9.5}$$

to get

$$y^2 = (y_n)^2 + 2y_n\Delta y$$

Since $y_n = 0$, then $y^2 = 0$.

At this point we have the linearized incremental differential equation

$$\frac{d^2}{dt}\Delta y + \frac{d}{dt}\Delta y = \Delta x$$

Its initial conditions are calculated as in the following.

$$\begin{aligned} \Delta y(0) &= y(0) - y_n \\ \frac{d}{dt}\Delta y(0) &= \frac{d}{dt}y(0) - \frac{d}{dt}y_n \end{aligned}$$

Remembering that the nominal values are constants, the initial conditions become

$$\begin{aligned} \Delta y(0) &= 0 \\ \frac{d}{dt}\Delta y(0) &= 0 \end{aligned}$$

We now get the final form of the linearized system as

$$\frac{d^2}{dt}\Delta y + \frac{d}{dt}\Delta y = \Delta x$$

with $\Delta y(0) = 0$, $\frac{d}{dt}\Delta y(0) = 0$, and Δx is an incremental input. The solution

$$\Delta y + y_n$$

is an approximation to the solution, y, for the nonlinear differential equation given.

9.3.2 Linearization When $f(\underline{z})$ Is a Function of the State Vector Only

Let us look at the following nonlinear system described by the following nonlinear state-equation

$$\frac{d}{dt}\underline{z} = f(\underline{z})$$

where $f(\underline{z})$ is $n \times 1$ and a nonlinear function of the state vector.

We will start by finding the equilibrium points of

$$\frac{d}{dt}\underline{z} = f(\underline{z})$$

and describing their behavior at that point. The local behavior of the nonlinear system

$$\frac{d}{dt}\underline{z} = f(\underline{z})$$

near a hyperbolic equilibrium point $\underline{z}_0$ is determined by the behavior of the linear system

$$\frac{d}{dt}\underline{z} = A(\underline{z})$$

with the matrix

$$A = \frac{\partial}{\partial \underline{z}} f(\underline{z}_0)$$

The linear function

$$A\underline{z} = \frac{\partial}{\partial \underline{z}} f(\underline{z}_0)\underline{z}$$

is called the linear part of $f(\underline{z})$ at $\underline{z}_0$. $\underline{z}_0$ is called an equilibrium point or critical point of $\frac{d}{dt}\underline{z} = f(\underline{z})$ if $f(\underline{z}_0) = \underline{0}$.

The equilibrium point $\underline{z}_0$ is called the hyperbolic equilibrium point of

$$\frac{d}{dt}\underline{z} = f(\underline{z})$$

if none of the eigenvalues of the matrix

$$A = \frac{\partial}{\partial \underline{z}} f(\underline{z}_0)$$

has zero real part. The linear system

$$\frac{d}{dt}\underline{z} = A(\underline{z})$$

where

$$A = \frac{\partial}{\partial \underline{z}} f(\underline{z}_0)$$

is the linearization of

$$\frac{d}{dt}\underline{z} = f(\underline{z})$$

at $\underline{z}_0$. An equilibrium point $\underline{z}_0$ of

$$\frac{d}{dt}\underline{z} = f(\underline{z})$$

is called a sink if all of the eigenvalues of the matrix

$$A = \frac{\partial}{\partial \underline{z}} f(\underline{z}_0)$$

have negative real parts. The equilibrium point is called a source if the eigenvalues have positive real parts. If $\underline{z}_0$ is a hyperbolic equilibrium point and at least one eigenvalue of

$$A = \frac{\partial}{\partial \underline{z}} f(\underline{z}_0)$$

has a positive real part and at least one other has a negative real part, then $\underline{z}_0$ is called a saddle point.

Example 9.4
Consider the following nonlinear state-equation

$$\frac{d}{dt}\underline{z} = f(\underline{z})$$

where

$$f(\underset{\sim}{z}) = \begin{bmatrix} z_1^2 - z_2^2 - 1 \\ 2z_2 \end{bmatrix}$$

Describe the equilibrium points.

Solution

We can see that $f(\underset{\sim}{z}) = \underset{\sim}{0}$ at

$$\underset{\sim}{z} = \begin{bmatrix} 1 \\ 0 \end{bmatrix}$$

and

$$\underset{\sim}{z} = \begin{bmatrix} -1 \\ 0 \end{bmatrix}$$

Now we check these equilibrium points and look at

$$\frac{\partial}{\partial \underset{\sim}{z}} f(\underset{\sim}{z}) = \begin{bmatrix} \frac{\partial}{\partial z_1}(z_1^2 - z_2^2 - 1) & \frac{\partial}{\partial z_2}(z_1^2 - z_2^2 - 1) \\ \frac{\partial}{\partial z_1}(2z_2) & \frac{\partial}{\partial z_2}(2z_2) \end{bmatrix} = \begin{bmatrix} 2z_1 & -2z_2 \\ 0 & 2 \end{bmatrix}$$

At the point (−1,0)

$$\frac{\partial}{\partial \underset{\sim}{z}} f(-1,0) = \begin{bmatrix} -2 & 0 \\ 0 & 2 \end{bmatrix}$$

and at the point (1,0)

$$\frac{\partial}{\partial \underset{\sim}{z}} f(1,0) = \begin{bmatrix} 2 & 0 \\ 0 & 2 \end{bmatrix}$$

Therefore, we conclude, since the eigenvalues for

$$\begin{bmatrix} -2 & 0 \\ 0 & 2 \end{bmatrix}$$

are 2 and −2 and the eigenvalues for

$$\begin{bmatrix} 2 & 0 \\ 0 & 2 \end{bmatrix}$$

are 2 and 2, that (1,0) is a source and (−1,0) is a saddle point.

Example 9.5

Consider the following nonlinear state-equation

$$\frac{d}{dt}\underset{\sim}{z} = f(\underset{\sim}{z})$$

where

$$f(\underset{\sim}{z}) = \begin{bmatrix} z_1 - z_1 z_2 \\ z_2 - z_1^2 \end{bmatrix}$$

Describe the equilibrium points.

Solution

We can see that $f(\underset{\sim}{z}) = \underset{\sim}{0}$ at

$$\underset{\sim}{z} = \begin{bmatrix} 0 \\ 0 \end{bmatrix}, \quad \underset{\sim}{z} = \begin{bmatrix} -1 \\ 1 \end{bmatrix}, \quad \text{and} \quad \underset{\sim}{z} = \begin{bmatrix} 1 \\ 1 \end{bmatrix}.$$

We check the equilibrium points by first taking the partial derivative of $f(\underset{\sim}{z})$ with respect to $\underset{\sim}{z}$.

$$\frac{\partial}{\partial \underset{\sim}{z}} f(\underset{\sim}{z}) = \begin{bmatrix} \frac{\partial}{\partial z_1}(z_1 - z_1 z_2) & \frac{\partial}{\partial z_{21}}(z_1 - z_1 z_2) \\ \frac{\partial}{\partial z_1}(z_2 - z_1^2) & \frac{\partial}{\partial z_2}(z_2 - z_1^2) \end{bmatrix} = \begin{bmatrix} 1 - z_2 & -z_1 \\ -2z_1 & 1 \end{bmatrix}$$

At the equilibrium point (0, 0) we have

$$\frac{\partial}{\partial \underset{\sim}{z}} f(0,0) = \begin{bmatrix} 1 & 0 \\ 0 & 1 \end{bmatrix}.$$

The eigenvalues are at 1 and 1. Since all eigenvalues are positive, this point is a source.

At the equilibrium point (1,1) we have

$$\frac{\partial}{\partial \underline{z}} f(1,1) = \begin{bmatrix} 0 & -1 \\ -2 & 1 \end{bmatrix}$$

The eigenvalues are at -1 and 2, and this point is a saddle point.
At the equilibrium point $(-1,1)$ we have

$$\frac{\partial}{\partial \underline{z}} f(-1,1) = \begin{bmatrix} 0 & 1 \\ 2 & 1 \end{bmatrix}$$

The eigenvalues are at -1 and 2, and this point is a saddle point.

9.3.3 Linearization When $f(\underline{z})$ Is a Function of the State Vector and the Input $\underline{x}(t)$

Consider the nonlinear state and output equations.

$$\begin{aligned} \frac{d}{dt}\underline{z} &= f(\underline{z}, \underline{x}) \\ \underline{y} &= h(\underline{z}, \underline{x}) \end{aligned} \tag{9.6}$$

If we know of a nominal solution to the above equations for $\underline{z}_n(t)$ and $\underline{y}_n(t)$ with $\underline{x}_n(t)$ as input vector, and if we slightly perturb the vectors $\underline{z}(t)$, $\underline{y}(t)$, and $\underline{x}(t)$, then we can describe this perturbation mathematically as

$$\Delta \underline{z} = \underline{z} - \underline{z}_n \tag{9.7}$$

$$\Delta \underline{x} = \underline{x} - \underline{x}_n \tag{9.8}$$

$$\Delta \underline{y} = \underline{y} - \underline{y}_n \tag{9.9}$$

The nonlinear equations above can be written with the help of Taylor series expansion while neglecting second order terms and higher as

$$\begin{aligned} \frac{d}{dt}\underline{z} = \frac{d}{dt}\underline{z}_n + \frac{d}{dt}\Delta\underline{z} &= f(\underline{z}_n + \Delta\underline{z}, \underline{x}_n + \Delta\underline{x}) \\ &= f(\underline{z}_n, \underline{x}_n) + \left(\frac{\partial}{\partial \underline{z}} f\right)_n \Delta\underline{z} + \left(\frac{\partial}{\partial \underline{x}} f\right)_n \Delta\underline{x} \end{aligned} \tag{9.10}$$

and

$$\begin{aligned} \underline{y} &= \underline{y}_n + \Delta \underline{y} = h(\underline{z}_n + \Delta \underline{z}, \underline{x}_n + \Delta \underline{x}) \\ &= h(\underline{z}_n, \underline{x}_n) + \left(\frac{\partial}{\partial \underline{z}} h\right)_n \Delta \underline{z} + \left(\frac{\partial}{\partial \underline{x}} h\right)_n \Delta \underline{x} \end{aligned} \tag{9.11}$$

where $(\ )_n$ means the derivatives are evaluated at the nominal solutions.
Since

$$\frac{d}{dt}\underline{z}_n = f(\underline{z}_n, \underline{x}_n) \tag{9.12}$$

and

$$\underline{y}_n = h(\underline{z}_n, \underline{x}_n) \tag{9.13}$$

then we can write

$$\frac{d}{dt}\Delta \underline{z} = \left(\frac{\partial}{\partial \underline{z}} f\right)_n \Delta \underline{z} + \left(\frac{\partial}{\partial \underline{x}} f\right)_n \Delta \underline{x} \tag{9.14}$$

and

$$\Delta \underline{y} = \left(\frac{\partial}{\partial \underline{z}} h\right)_n \Delta \underline{z} + \left(\frac{\partial}{\partial \underline{x}} h\right)_n \Delta \underline{x} \tag{9.15}$$

where

$$\left(\frac{\partial}{\partial \underline{z}} f\right) = \begin{pmatrix} \frac{\partial}{\partial z_1} f_1 & \frac{\partial}{\partial z_2} f_1 & \cdots & \frac{\partial}{\partial z_n} f_1 \\ \frac{\partial}{\partial z_1} f_2 & \frac{\partial}{\partial z_2} f_2 & \cdots & \frac{\partial}{\partial z_n} f_2 \\ \vdots & \vdots & \cdots & \vdots \\ \frac{\partial}{\partial z_1} f_n & \frac{\partial}{\partial z_2} f_n & \cdots & \frac{\partial}{\partial z_n} f_n \end{pmatrix} \tag{9.16}$$

and

$$\left(\frac{\partial}{\partial \underline{x}} f\right) = \begin{pmatrix} \frac{\partial}{\partial x_1} f_1 & \frac{\partial}{\partial x_2} f_1 & \cdots & \frac{\partial}{\partial x_n} f_1 \\ \frac{\partial}{\partial x_1} f_2 & \frac{\partial}{\partial x_2} f_2 & \cdots & \frac{\partial}{\partial x_n} f_2 \\ \vdots & \vdots & \cdots & \vdots \\ \frac{\partial}{\partial x_1} f_n & \frac{\partial}{\partial x_2} f_n & \cdots & \frac{\partial}{\partial x_n} f_n \end{pmatrix} \tag{9.17}$$

Also

$$\left(\frac{\partial}{\partial \underset{\sim}{z}} h\right) = \begin{pmatrix} \frac{\partial}{\partial z_1} h_1 & \frac{\partial}{\partial z_2} h_1 & \cdots & \frac{\partial}{\partial z_n} h_1 \\ \frac{\partial}{\partial z_1} h_2 & \frac{\partial}{\partial z_2} h_2 & \cdots & \frac{\partial}{\partial z_n} h_2 \\ \vdots & \vdots & \cdots & \vdots \\ \frac{\partial}{\partial z_1} h_n & \frac{\partial}{\partial z_2} h_n & \cdots & \frac{\partial}{\partial z_n} h_n \end{pmatrix} \tag{9.18}$$

and

$$\left(\frac{\partial}{\partial \underset{\sim}{x}} h\right) = \begin{pmatrix} \frac{\partial}{\partial x_1} h_1 & \frac{\partial}{\partial x_2} h_1 & \cdots & \frac{\partial}{\partial x_n} h_1 \\ \frac{\partial}{\partial x_1} h_2 & \frac{\partial}{\partial x_2} h_2 & \cdots & \frac{\partial}{\partial x_n} h_2 \\ \vdots & \vdots & \cdots & \vdots \\ \frac{\partial}{\partial x_1} h_n & \frac{\partial}{\partial x_2} h_n & \cdots & \frac{\partial}{\partial x_n} h_n \end{pmatrix} \tag{9.19}$$

Example 9.6

The plane-polar two-body equations of motion for a satellite are

$$\frac{d^2}{dt}r - r\left(\frac{d}{dt}\theta\right)^2 + \mu/r^2 = u_r$$

$$r\frac{d^2}{dt}\theta + 2\frac{d}{dt}r\frac{d}{dt}\theta = u_\theta$$

where r is the radial distance, θ is the polar angle, (u_r, u_θ) is the control acceleration and $\mu = \mu_0 \text{ km}^3/\text{sec}^2$.

Develop the linearized perturbed equations of motion describing the deviation of the satellite from a reference orbit.

Solution

We will assume that the linearization is carried at the nominal values $r = r_n$, $\frac{d}{dt}r = \frac{d}{dt}r_n$, $\theta = \theta_n$, $\frac{d}{dt}\theta = \frac{d}{dt}\theta_n$ and $u_r = (u_r)_n = (u_\theta)_n = 0$.

We will also let

$$\Delta \underset{\sim}{z} = \left[\Delta r \ \frac{d}{dt}\Delta r \quad \Delta\theta \ \frac{d}{dt}\Delta\theta\right]^T$$

with

$$\frac{d}{dt}\underline{z} = f(\underline{z}, \underline{x})$$
$$\underline{y} = h(\underline{z}, \underline{x})$$

and by letting $z_1 = r$, $z_2 = \frac{d}{dt}r$, $z_3 = \theta$, and $z_4 = \frac{d}{dt}\theta$, we have the state equation as

$$f = \frac{d}{dt}\underline{z} = \begin{pmatrix} \frac{d}{dt}z_1 \\ \frac{d}{dt}z_2 \\ \frac{d}{dt}z_3 \\ \frac{d}{dt}z_4 \end{pmatrix} = \begin{pmatrix} z_2 \\ z_1 z_4^2 - \frac{\mu}{z_1^2} + u_r \\ z_4 \\ \frac{-2z_2 z_4}{z_1} + \frac{u_\theta}{z_1} \end{pmatrix}$$

At this point we can find $\left(\frac{\partial}{\partial \underline{z}} f\right)$ as

$$\left(\frac{\partial}{\partial \underline{z}} f\right) = \begin{pmatrix} \frac{\partial}{\partial z_1} f_1 & \frac{\partial}{\partial z_2} f_1 & \frac{\partial}{\partial z_3} f_1 & \frac{\partial}{\partial z_4} f_1 \\ \frac{\partial}{\partial z_1} f_2 & \frac{\partial}{\partial z_2} f_2 & \frac{\partial}{\partial z_3} f_2 & \frac{\partial}{\partial z_4} f_2 \\ \frac{\partial}{\partial z_1} f_3 & \frac{\partial}{\partial z_2} f_3 & \frac{\partial}{\partial z_3} f_3 & \frac{\partial}{\partial z_4} f_3 \\ \frac{\partial}{\partial z_1} f_4 & \frac{\partial}{\partial z_2} f_4 & \frac{\partial}{\partial z_3} f_4 & \frac{\partial}{\partial z_4} f_4 \end{pmatrix}$$

Evaluating the partial derivatives we will get

$$\left(\frac{\partial}{\partial \underline{z}} f\right) = \begin{pmatrix} 0 & 1 & 0 & 0 \\ z_4^2 + \frac{2\mu}{z_1^3} & 0 & 0 & 2z_1 z_4 \\ 0 & 0 & 0 & 1 \\ \frac{2z_2 z_4 - u_\theta}{z_1^2} & \frac{-2z_4}{z_1} & 0 & \frac{-2z_2}{z_1} \end{pmatrix}$$

For $\left(\frac{\partial}{\partial \underline{x}} f\right)$ we have

$$\left(\frac{\partial}{\partial \underline{x}} f\right) = \begin{pmatrix} \frac{\partial}{\partial u_r} f_1 & \frac{\partial}{\partial u_\theta} f_1 \\ \frac{\partial}{\partial u_r} f_2 & \frac{\partial}{\partial u_\theta} f_2 \\ \frac{\partial}{\partial u_r} f_3 & \frac{\partial}{\partial u_\theta} f_3 \\ \frac{\partial}{\partial u_r} f_4 & \frac{\partial}{\partial u_\theta} f_4 \end{pmatrix}$$

where by evaluating the partial derivatives again we get

$$\left(\frac{\partial}{\partial \underline{x}} f\right) = \begin{pmatrix} 0 & 0 \\ 1 & 0 \\ 0 & 0 \\ 0 & \frac{1}{z_1} \end{pmatrix}$$

Therefore, the linearized equations can finally be put in the following form

$$\frac{d}{dt}\Delta \underline{z} = \begin{pmatrix} 0 & 1 & 0 & 0 \\ z_4^2 + \frac{2\mu}{z_1^3} & 0 & 0 & 2z_1z_4 \\ 0 & 0 & 0 & 1 \\ \frac{2z_2z_4 - u_\theta}{z_1^2} & \frac{-2z_4}{z_1} & 0 & \frac{-2z_2}{z_1} \end{pmatrix}_{\substack{\text{evaluated at the}\\ \text{nominal values}\\ \text{given}}} \Delta \underline{z} + \begin{pmatrix} 0 & 0 \\ 1 & 0 \\ 0 & 0 \\ 0 & \frac{1}{z_1} \end{pmatrix}_{\substack{\text{evaluated at the}\\ \text{nominal values}\\ \text{given}}} \Delta \underline{x}$$

where $\underline{x} = [u_r \;\; u_\theta]^T$ is the input to the system.

The above linear state equation can be solved using the techniques presented in Chapter 6 and it will approximate the solution of the nonlinear equation at the nominal values given.

9.4 Some Insights: The Meaning of Linear and Nonlinear

It is important to study the real meaning of linear and nonlinear systems so that we can use the right models to represent the systems under consideration. The linear system has two properties: the property of superposition and the property of homogeneity. The property of superposition means that the total response of a system to many inputs is the sum of the individual outputs each is due to an individual input acting alone. The property of homogeneity means that if the input is multiplied by a scalar, we find the output due to that particular input and then multiply the measured output by that scalar.

We first examine any system under consideration. If any nonlinear components are present, we have to linearize the system before we establish the relation between the input and the output (the transfer function).

If we are given a system to analyze, we must first look for nonlinearities and then write the differential equation describing the system. We then linearize the system about the equilibrium point. In the case of the pendulum discussed in this chapter, the equilibrium point is the angle $\theta = 0$ rad, when the system is at rest. Then we linearize the nonlinear differential equation and take the Laplace transform to get the transfer function that describes the linearized system.

The linearization process depends on the magnitude of the input used, the initial conditions, and the choice of the operating point (the equilibrium point).

Consider the following nonlinear first order differential equation

$$\frac{d}{dt}y = f(y) = 1 - \sqrt{y}$$

The equilibrium point is the solution of $f(y) = 0$; $y = 1$ is such a point.

Let us linearize this equation at $y = 1$, using the Taylor series and taking the first two terms only to get

$$f(y) \approx f(y_n) - \frac{1}{2\sqrt{y_n}}\Delta y$$

With $y_n = 1$ and $\Delta y = y - y_n$ we have

$$f(y) \approx 0 - \frac{1}{2}(\Delta y) = -\frac{1}{2}(y - 1)$$

Now let us plot $f(y)$ and its approximation and notice their behavior. Using MATLAB we write the script

```
clf
y = 0:0.01:4;
fy = 1-sqrt(y);
deltafy = -1/2*y;
fyapprox = deltafy + 1/2;
plot(y, fy, y, fyapprox)
```

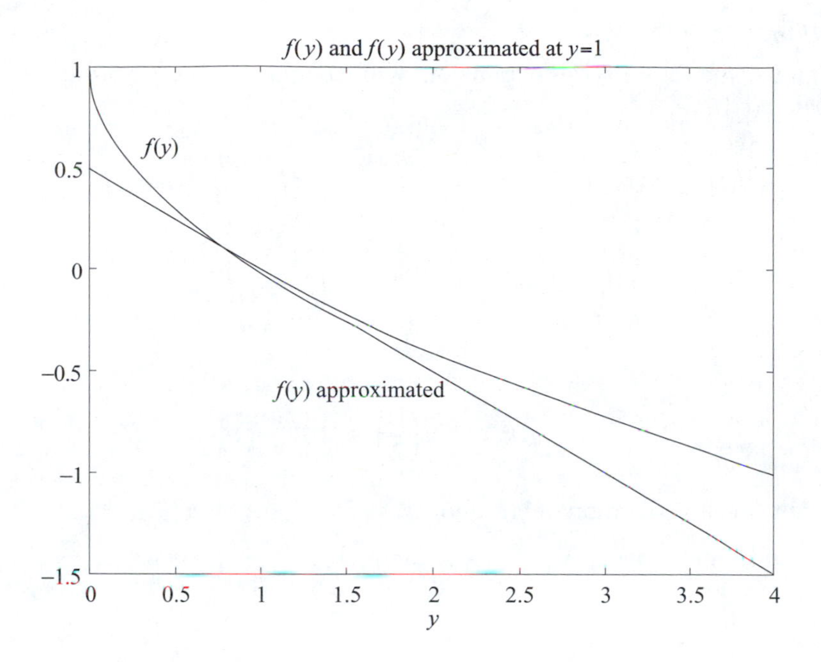

FIGURE 9.2
Approximating a function.

```
xlabel('y');
title('f(y) and f(y) approximated at y= 1');
gtext('f(y)')
gtext('f(y) approximated')
```

The plots are shown in Figure 9.2.

You can see in Figure 9.2 that $f(y)$ and its approximation at $y = 1$ are identical in the range from [0.5, 1.5]. This is the region of linear operation for this particular operating point, $y = 1$.

9.5 End-of-Chapter Examples

EOCE 9.1

Linearize the function

$$y(t) = 10\cos(t)$$

about the point $t = \pi/2$.

Solution

We can use the Taylor series expansion, while neglecting higher order terms, to write $y(t)$ as

$$y(t) = y(t_0) + \frac{d}{dt}y(t)\bigg|_{t_0}(t - t_0) = y(t_0) + \frac{d}{dt}y(t)\bigg|_{t_0}\Delta t$$

$$\frac{d}{dt}y(t)\bigg|_{t_0} = -10\sin(t)\bigg|_{t_0} = -10$$

and

$$y(t_0) = y\left(\frac{\pi}{2}\right) = 0$$

Finally, the linearization of $y(t)$ about t_0 is

$$y(t) = -10\Delta t$$

or

$$y(t) = -10\left(t - \frac{\pi}{2}\right)$$

EOCE 9.2

Consider the following nonlinear differential equation

$$\frac{d^2}{dt}y(t) + 2\frac{d}{dt}y(t) + \cos(y) = A\sin(t)$$

with initial conditions

$$y(0) = \frac{d}{dt}y(0) = 0$$

Linearize the above equation about its nominal point.

Solution

The nominal value for the input is its average value

$$(A\sin(t))_n = 0$$

To find the equilibrium point of the system we write the differential equation as

$$\frac{d^2}{dt}y_n(t) + 2\frac{d}{dt}y_n(t) + \cos(y_n) = 0$$

We know that the nominal values are constants, so their derivatives are zeros. Therefore, the equation above reduces to

$$\cos(y_n) = 0$$

A solution to this last algebraic equation is $y_n = \pi/2$.
Therefore, $y_n = \pi/2$ is a nominal point.
Using the relation

$$y = y_n + \Delta y$$

we can rewrite the nonlinear equation as

$$\left(\frac{d^2}{dt}\Delta y + \frac{d^2}{dt}y_n\right) + 2\left(\frac{d}{dt}\Delta y + \frac{d}{dt}y_n\right) + \cos(y) = x_n + \Delta x$$

The above equation reduces to

$$\frac{d^2}{dt}\Delta y + 2\frac{d}{dt}\Delta y + \cos(y) = \Delta x$$

Next we linearize $\cos(y)$ using the Taylor series to get

$$\cos(y) = \cos(y_n) - \sin(y_n)\Delta y = -\Delta y$$

Then the final linearized equation is

$$\frac{d^2}{dt}\Delta y + 2\frac{d}{dt}\Delta y - \Delta y = \Delta x$$

At this point we need initial conditions for $\Delta y(0)$ and $\frac{d}{dt}\Delta y(0)$.

$$\Delta y(0) = y(0) - y_n = -\pi/2$$
$$\frac{d}{dt}\Delta y(0) = \frac{d}{dt}y(0) - \frac{d}{dt}y_n = 0$$

This final linearized equation can be solved for Δy and then y can be found since

$$y = \Delta y + \frac{\pi}{2}$$

The linearized system is unstable, which can happen in some situations even if the nonlinear system is stable.

We can use MATLAB to check stability of the linearized system and type at the prompt

```
roots([1  2  -1]) % roots of the linearized system
```

with the result

```
ans =
-2.4142
0.4142
```

You can see that one of the eigenvalues of the linearized system is positive, which indicates instability.

EOCE 9.3

We will now reconsider Example 9.3 (discussed previously) to show that the linear system derived from the nonlinear system approximates the nonlinear system for inputs of small magnitudes and differs from it for inputs of large magnitudes.

Solution

The equations needed are given below.

For the nonlinear system we have

$$\frac{d^2}{dt}y + \frac{d}{dt}y + y^2 = x(t)$$

with initial conditions

$$y(0) = \frac{d}{dt}y(0) = 0$$

and

$$x(t) = A\sin(t).$$

For the linearized system we have

$$\frac{d^2}{dt}\Delta y + \frac{d}{dt}\Delta y = \Delta x$$

with the initial conditions

$$\Delta y(0) = 0$$
$$\frac{d}{dt}\Delta y(0) = 0$$

Since $\Delta x = x(t) + x_n$ and $x_n = 0$, we have

$$\Delta x = A \sin(t)$$

Let us consider the nonlinear system first and try to put it in state equation form.

We will let $z_1(t) = y(t)$ and $z_2(t) = \frac{d}{dt} y(t)$. Therefore, we get

$$\frac{d}{dt} z_1 = z_2$$
$$\frac{d}{dt} z_2 = -z_2 - z_1^2 + A \sin(t)$$

with initial conditions

$$z_1(0) = y(0) = 0$$

and

$$z_2(0) = \frac{d}{dt} y(0) = 0.$$

Let us define the nonlinear system using MATLAB as the function "eoce3non."

```
function zdot=eoce3non(t,z);% return the state derivatives
global A;
zdot=[z(2);-z(2)-z(1).^2 + A*sin(t)];
```

Next, we consider the linear system.

Let $z_1(t) = y(t)$ and $z_2(t) = \frac{d}{dt} y(t)$. Therefore, we have

$$\frac{d}{dt} z_1 = z_2$$
$$\frac{d}{dt} z_2 = -z_2 + A \sin(t)$$

with initial conditions

$$z_1(0) = \Delta y(0) = 0$$

and

$$z_2(0) = \frac{d}{dt}\Delta y(0) = 0.$$

Let us define the linearized system using MATLAB as the function "eoce3lin."

```
function zdot=eoce3lin(t,z);% return the state derivatives
global A;
zdot=[z(2);-z(2)+A*sin(t)];
```

Now let us define another script that we will call the MATLAB function "ode23" and use the functions we just defined, eoce3lin and eoce3non, to plot the solution in both cases.

```
global A;
A = 0.2;
clf % clears the plots
tspan = [0 20]; % the time span for the simulation
z0 = [0 0]; % initial condition vector
[t,z] = ode23('eoce3non',tspan, z0);
plot(t,z(:,1),'g*');
hold on
[t,z] = ode23('eoce3lin',tspan, z0);
plot(t,z(:,1),'b+');
title('solution of the systems A = 0.2');
gtext('linearized system')
gtext('nonlinear system')
xlabel('Time (sec)');
```

The plots are given for different values of A in Figures 9.3, 9.4, 9.5, and 9.6.

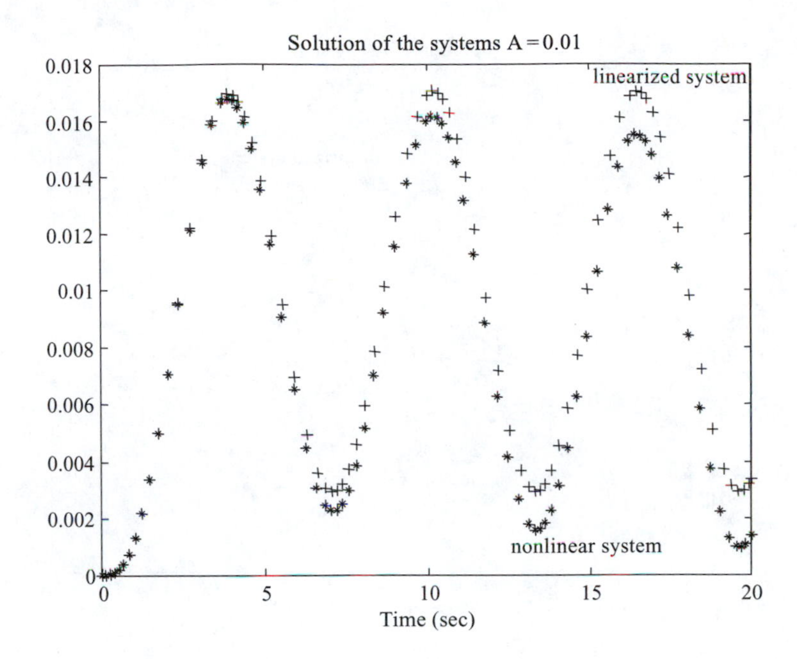

FIGURE 9.3
Plots for EOCE 9.3.

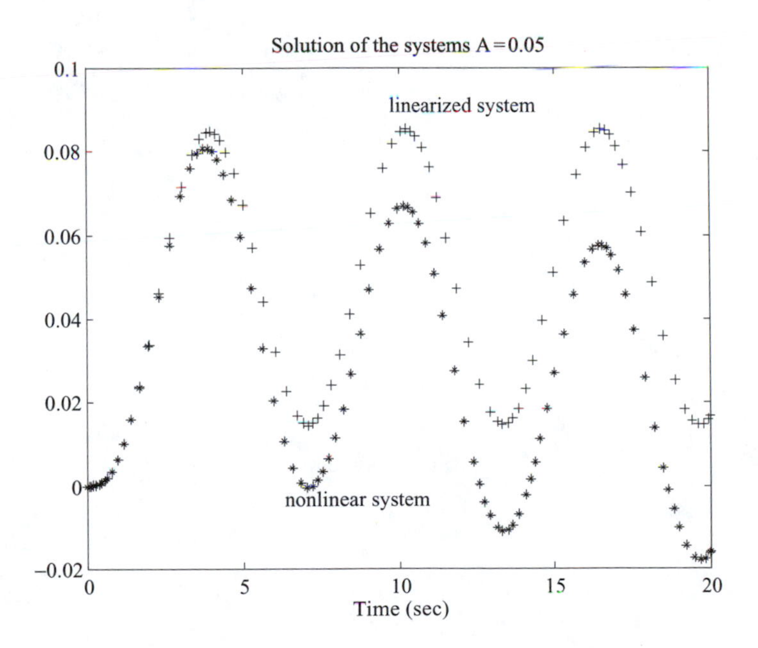

FIGURE 9.4
Plots for EOCE 9.3.

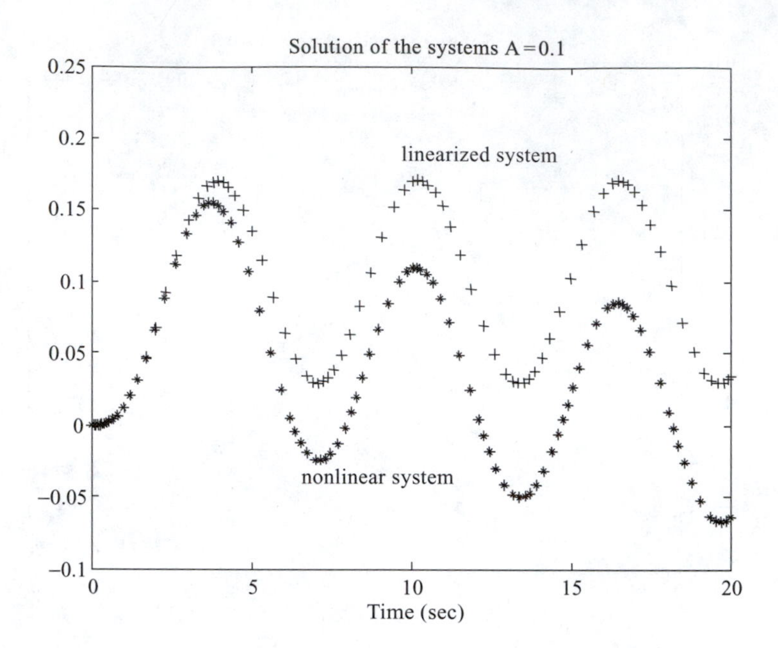

FIGURE 9.5
Plots for EOCE 9.3.

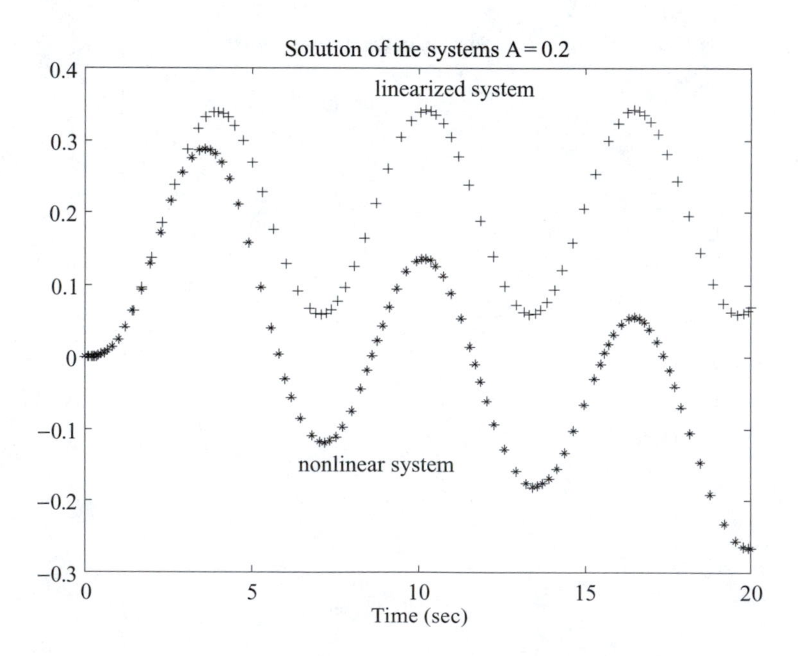

FIGURE 9.6
Plots for EOCE 9.3.

As you can see in the plots, the linearized system approximates the nonlinear system well for small variations in the input. As we said before, the variation in the input magnitude can impact the approximation.

EOCE 9.4

Consider the pendulum with some viscous damping as shown in Figure 9.7.

Using Newton's law of motion, we obtain the following nonlinear equation

$$\frac{d^2}{dt}\theta(t) + (\beta/m)\frac{d}{dt}\theta + (g/l)\sin(\theta) = 0$$

Let $\theta(0) = 0.2$ and $\frac{d}{dt}\theta(0) = 0$.

Plot θ vs. time for this nonlinear system. Linearize the system about its equilibrium position.

Solution

For the nonlinear and the linearized system we can use the MATLAB function "ode23" to plot θ vs. time. We first need to write the system in state-space.

Let $z_1(t) = \theta$ and $z_2(t) = \frac{d}{dt}\theta$. Therefore,

$$\frac{d}{dt}z_1 = z_2$$

$$\frac{d}{dt}z_1 = -\frac{\beta}{m}z_2 - \frac{g}{l}\sin(z_1)$$

with

$$z_1(0) = \theta(0) = 0.2 \quad \text{and} \quad z_2(0) = \frac{d}{dt}\theta(0) = 0 \ \text{t}$$

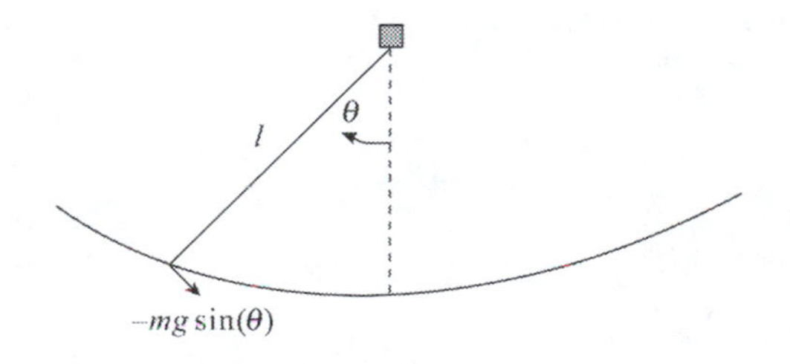

FIGURE 9.7
System for EOCE 9.4.

With $l = 0.6$, $g = 9.81$, $\beta = 0.02$, and $m = 0.2$, we have

$$\frac{d}{dt}z_1 = z_2$$

$$\frac{d}{dt}z_2 = -0.1z_2 - 16.35\sin(z_1)$$

Let us now linearize the system about its operating point.

Since the input force is zero, its nominal value is its average value, which is also zero.

The operating point is found by rewriting the nonlinear differential equation in terms of the nominal constant values of its variables. In this case the differential equation is reduced to

$$(g/l)\sin(\theta) = 0$$

This gives $\theta = 0$ radians as the vertical equilibrium position. So $\theta_n = 0$.

We now find the initial conditions for the incremental variables. We know that

$$\theta = \Delta\theta + \theta_n$$

so

$$\theta(0) = \Delta\theta(0) + 0 = 0.2$$

and

$$\frac{d}{dt}\theta(0) = \frac{d}{dt}\Delta\theta(0) + \frac{d}{dt}\theta_n = 0$$

Therefore, $\Delta\theta(0) = 0.2$ radians and $\frac{d}{dt}\Delta\theta(0) = 0$ radians.

Let us now linearize the nonlinear element (θ) using the Taylor series and write

$$\sin(\theta) = \sin(\theta_n) + \cos(\theta_n)\Delta\theta = \sin(0) + \Delta\theta = \Delta\theta$$

The final linearized differential equation is

$$\frac{d^2}{dt}\Delta\theta + (\beta/m)\frac{d}{dt}\Delta\theta + (g/l)\Delta\theta = 0$$

with the initial conditions

$$\Delta\theta(0) = 0.2$$

$$\frac{d}{dt}\Delta\theta(0) = 0$$

Now we write the state-space equations for this linearized differential equation.

Let $z_1(t) = \Delta\theta$ and $z_2(t) = \frac{d}{dt}\Delta\theta$. Therefore,

$$\frac{d}{dt}z_1 = z_2$$

$$\frac{d}{dt}z_2 = -\frac{\beta}{m}z_2 - \frac{g}{l}z_1$$

With $l = 0.6$, $g = 9.81$, $\beta = 0.02$, and $m = 0.2$, we have

$$\frac{d}{dt}z_1 = z_2$$

$$\frac{d}{dt}z_2 = -0.1z_2 - 16.35z_1$$

Now let us define the two systems using the two MATLAB functions, "eoce4non" and "eoce4lin," as

```
function zdot = eoce4non(t,z);
zdot = [z(2);-0.1*z(2)-16.3*sin(z(1))];
```

and for the linearized system

```
function zdot = eoce4lin(t,z);
zdot = [z(2);-0.1*z(2)-16.3*z(1)];
```

Now we write the script that will use eoce4non and eoce4lin as well as ode23.

```
clf
tspan = [0 20]; % the time span for the simulation
z0 = [0.2 0]; % initial condition vector
[t,z] = ode23('eoce4non', tspan, z0);
plot(t,z(:,1),'b*'); % determine swinging range
hold on
[t,z] = ode23('eoce4l in', tspan, z0);
plot(t,z(:,1));
```

```
title('The linearized and the nonlinear systems...
with z0 = [0.2  0]');
gtext('linearized model');
gtext('nonlinear model')
xlabel('Time (sec)');
ylabel('theta in radians')
```

The plots are shown in Figure 9.8.

It is clear that both systems are stable. With initial condition for θ close to 0, the two solutions agree as shown in Figure 9.8.

If you change the initial conditions from [.2 0] to [.8 0] (changing the range of swinging), we will obtain the plot in Figure 9.9 which clearly says that the two systems are not the same. The plots in Figures 9.10 and 9.11 are obtained when the initial conditions are changed to $\left[\frac{\pi}{3}, 0\right]$ and $\left[\frac{\pi}{2}, 0\right]$.

You can see in Figure 9.11 that the two solutions become out-of-phase as time elapses. It is clearly a bad approximation for a large initial angle $\theta(t)$.

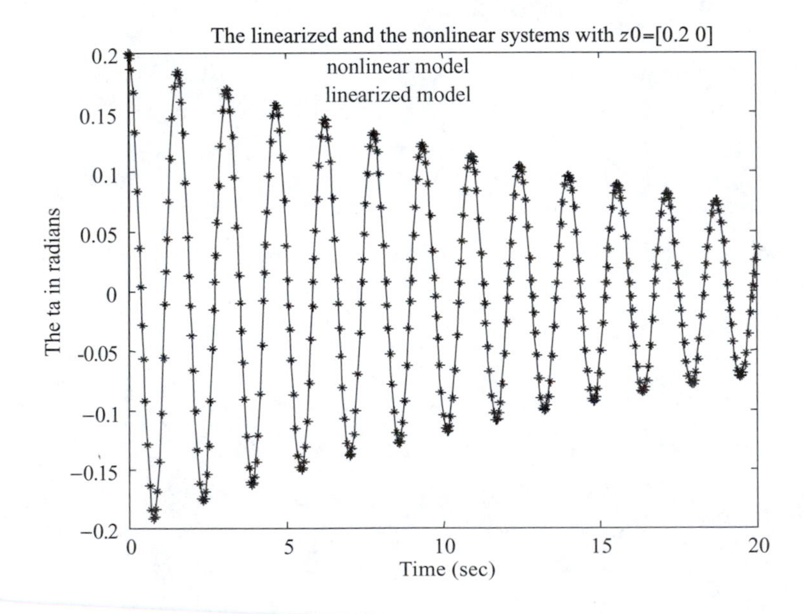

FIGURE 9.8
Plots for EOCE 9.4.

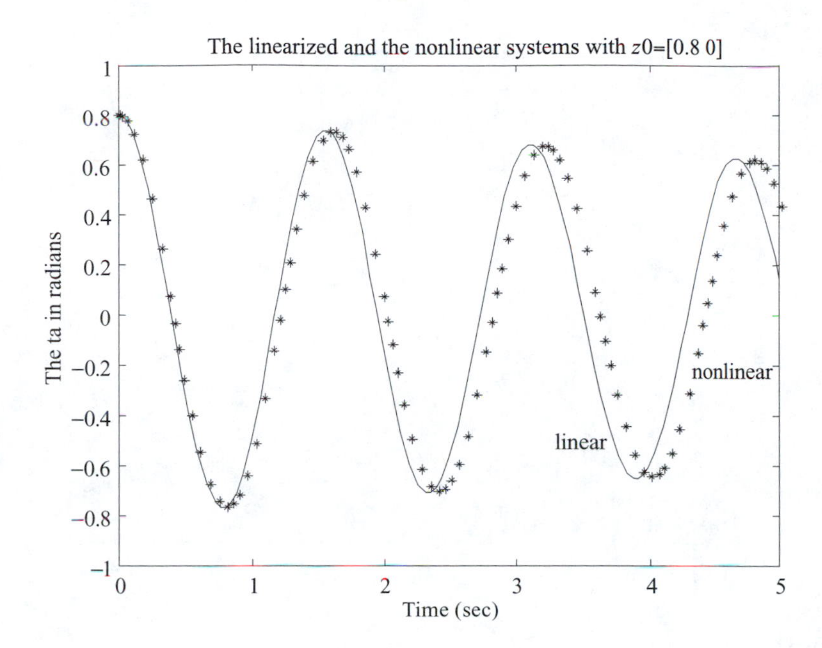

FIGURE 9.9
Plots for EOCE 9.4.

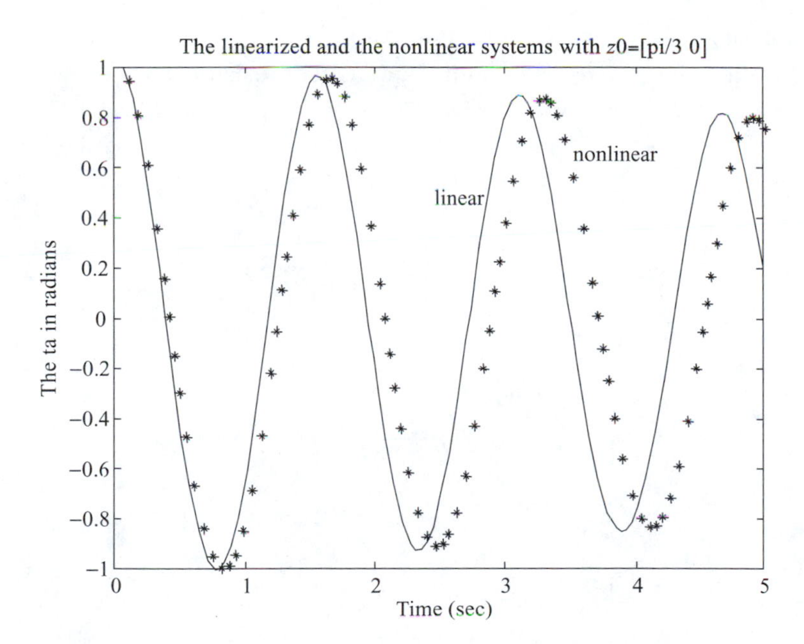

FIGURE 9.10
Plots for EOCE 9.4.

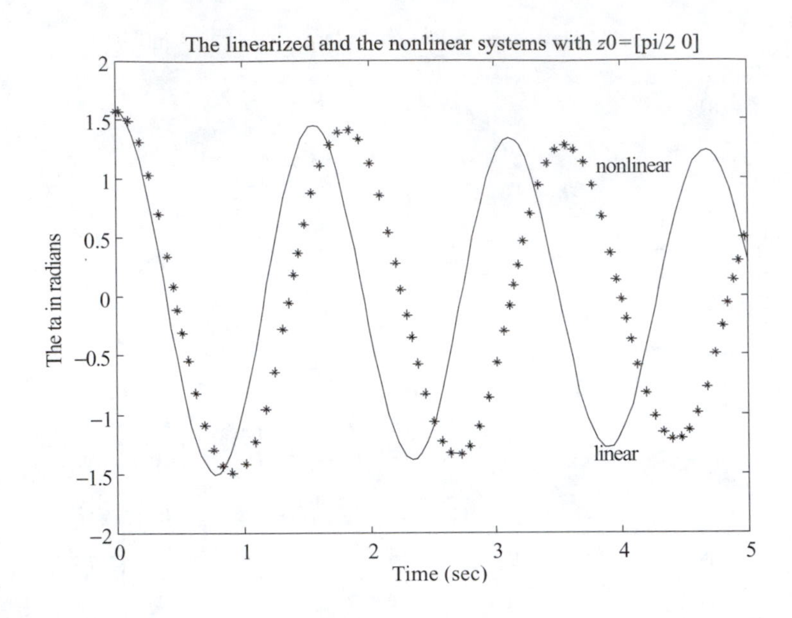

FIGURE 9.11
Plots for EOCE 9.4.

EOCE 9.5

Consider the first order circuit in Figure 9.12.

$y(t)$ is the voltage across the nonlinear element resistor. The current, going downward, in the nonlinear resistor is

$$I(t) = y^2(t)$$

With

$$x(t) = 10 + A\cos(t)$$

find a linear model for the circuit.

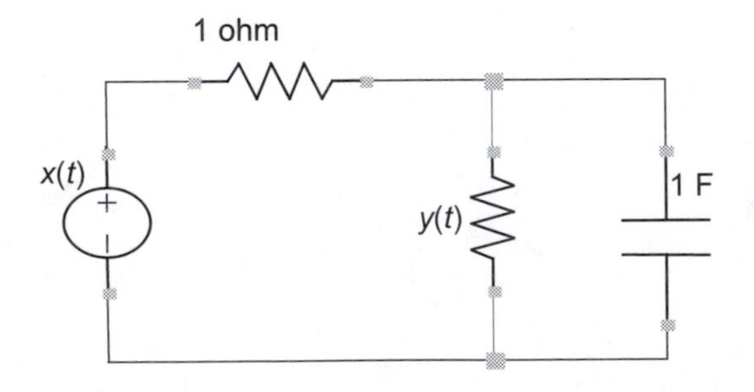

FIGURE 9.12
Circuit for EOCE 9.5.

Solution

We first need a differential equation that represents the system. Using nodal analysis we can write the following

$$I + y - x + \frac{d}{dt}y = 0$$

or

$$y^2 + y - x + \frac{d}{dt}y = 0$$

Rewriting the above differential equation we get

$$\frac{d}{dt}y + y^2 + y = x$$

The nominal value for x is $x_n = 10$ volts which is the average value for $x(t)$.
The operating point for the circuit is determined by rewriting the nonlinear differential equation in terms of its nominal values to get

$$y^2 + y = 10$$

with two equilibrium points $y_{n1} = -3.7016$ and $y_{n2} = 2.7016$.
For $y_n = 2.7016$, $I_n = (y_n)^2 = 7.2986$. Using the Taylor series we have

$$y^2 = y_n^2 + 2y_n\Delta y = 7.2986 + 5.4032\Delta y$$

Rewriting the linear parts in the nonlinear differential equation as the sum of their nominal and incremental values we get

$$\frac{d}{dt}(y_n + \Delta y) + 7.2986 + 5.4032\Delta y + (2.7016 + \Delta y) = 10 + A\cos t$$

Simplifying the above expression we get

$$\frac{d}{dt}\Delta y + 6.4032\Delta y = A\cos(t)$$

This is the incremental linear differential equation. The solution to this incremental differential equation plus the nominal value for y will approximate the solution of the nonlinear differential equation that represents the circuit given.

We also know that

$$I_n + \Delta I = y_n^2 + 5.4032\Delta y$$

But

$$I_n = y_n^2$$

Therefore,

$$\Delta I = 5.4032\Delta y$$

This implies that the approximation of the nonlinear resistor is 1/5.4032 ohms. The time constant for the linearized circuit is 1/6.4032 seconds. The linearized circuit is given in Figure 9.13.

For $y_n = -3.7016$, $I_n = (y_n)^2 = 13.7018$. Using the Taylor series we have

$$y^2 = y_n^2 + 2y_n\Delta y = 13.7018 - 7.403\Delta y$$

Rewriting the linear parts in the nonlinear differential equation as the sum of their nominal and incremental values we get

$$\frac{d}{dt}(y_n + \Delta y) + 13.7018 - 7.403\Delta y + (-3.7016 + \Delta y) = 10 + A\cos(t)$$

Simplifying the above expression we get

$$\frac{d}{dt}\Delta y - 6.4032\Delta y = A\cos(t)$$

This is the incremental linear differential equation. The solution to this incremental differential equation plus the nominal value for y should approximate

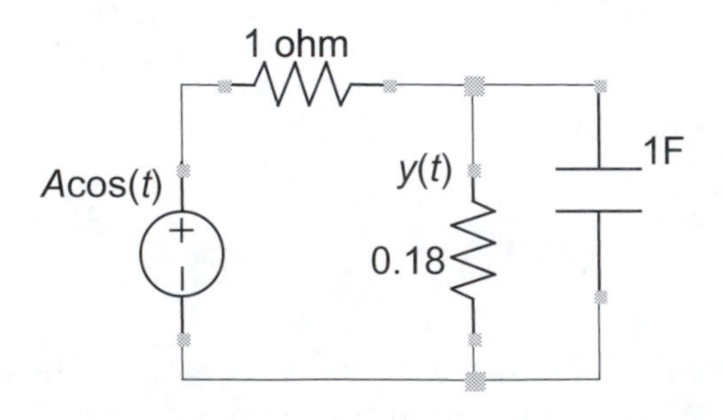

FIGURE 9.13
Circuit for EOCE 9.5.

the solution of the nonlinear differential equation that represents the circuit given. But this is not the case since the linearized model is unstable. Therefore, care must be taken when we try to linearize nonlinear models. This also shows that care must be exercised when we choose the nominal values for the variables in the system we investigate.

EOCE 9.6

Consider the chemical system shown in Figure 9.14, where z_1 and z_2 are the heights of the liquid in the two tanks. The solution is coming into tank 1 and leaving tank 2. Assume the system is operating and we stop letting the liquid in at $t = 0$ seconds. At this time assume $z_1(0) = 2.5$ meters and $z_2(0) = 1.5$ meter.

Assume that we have the following state-space equations that describe the system.

$$\frac{d}{dt}z_1(t) = 1 - \sqrt{z_1 - z_2}$$

$$\frac{d}{dt}z_2(t) = \sqrt{z_1 - z_2} - \sqrt{z_2}$$

Develop a solution to this nonlinear problem and then linearize at the equilibrium point.

Solution

Let us define the nonlinear system using MATLAB as

```
function zdot=eoce6non(t,z); % return the state derivatives

zdot = [1-sqrt(z(1)-z(2)); sqrt(z(1)-z(2))-sqrt(z(2))];
```

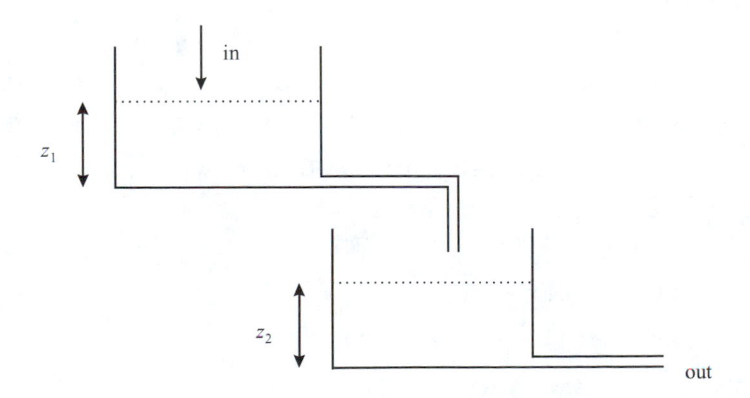

FIGURE 9.14
System for EOCE 9.6.

The nominal points are found by setting the right side of the nonlinear state-space system to zero to get $z_{1n} = 2$ and $z_{2n} = 1$. Using the Taylor series we can linearize the system about this nominal point as

$$\begin{bmatrix} \frac{d}{dt}\Delta z_1 \\ \frac{d}{dt}\Delta z_2 \end{bmatrix} = \begin{bmatrix} \frac{-1}{2\sqrt{z_1 - z_2}} & \frac{1}{2\sqrt{z_1 - z_2}} \\ \frac{1}{2\sqrt{z_1 - z_2}} & \frac{-1}{2\sqrt{z_1 - z_2}} - \frac{1}{2\sqrt{z_2}} \end{bmatrix}_{\text{nominal}} \begin{bmatrix} \Delta z_1 \\ \Delta z_2 \end{bmatrix}$$

The initial conditions for the incremental system are

$$\begin{aligned} \Delta z_1(0) &= z_1(0) - z_{1n} = 2.5 - 2 = 0.5 \\ \Delta z_2(0) &= z_2(0) - z_{2n} = 1.5 - 1 = 0.5 \end{aligned}$$

We now develop a function that describes the linearized system as in the following.

```
function zdot=eoce6lin(t, z); % return the state derivatives
zdot = [-0.5*z(1)+0.5*z(2);0.5*z(1) - 1*z(2)];
```

To produce the simulations we write the following script.

```
clf
z1n = 2;
tspan = [0 50]; % the time span for the simulation
z0 = [2.5 1.5]; % initial condition vector
[t,z] = ode23('eoce6non',tspan, z0);
plot(t, z(:,1),'g*');% height in the input tank
hold on
z0 = [0.5 0.5];
[t,z] = ode23('eoce6lin',tspan, z0);
plot(t,z(:,1)+z1n,'b+');% z1(t)=deltaz1(t)+z1n
title('The systems in EOCE 9.6 with initial con-
ditions...
[2.5 1.5]');
gtext('linearized system')
gtext('nonlinear system')
xlabel('Time (sec)');
```

The outputs are shown in Figures 9.15, 9.16, and 9.17.

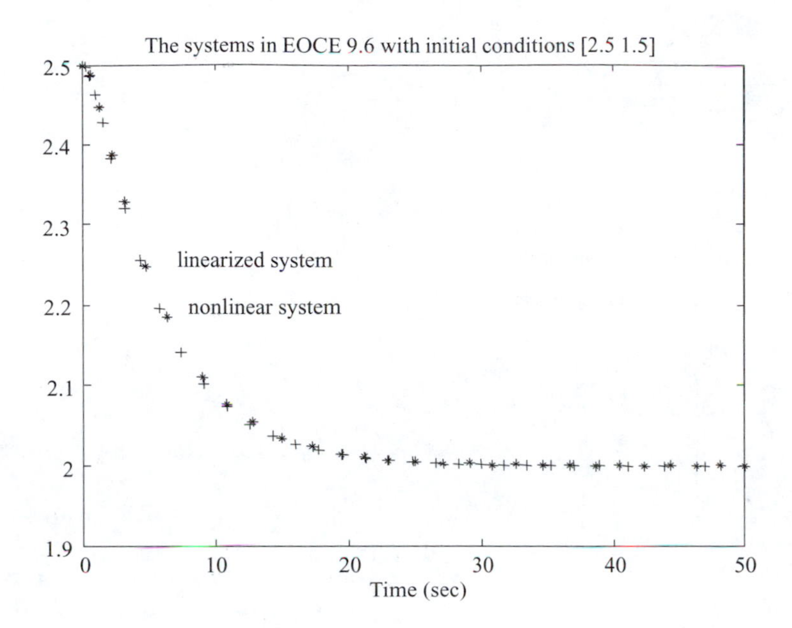

FIGURE 9.15
Plots for EOCE 9.6.

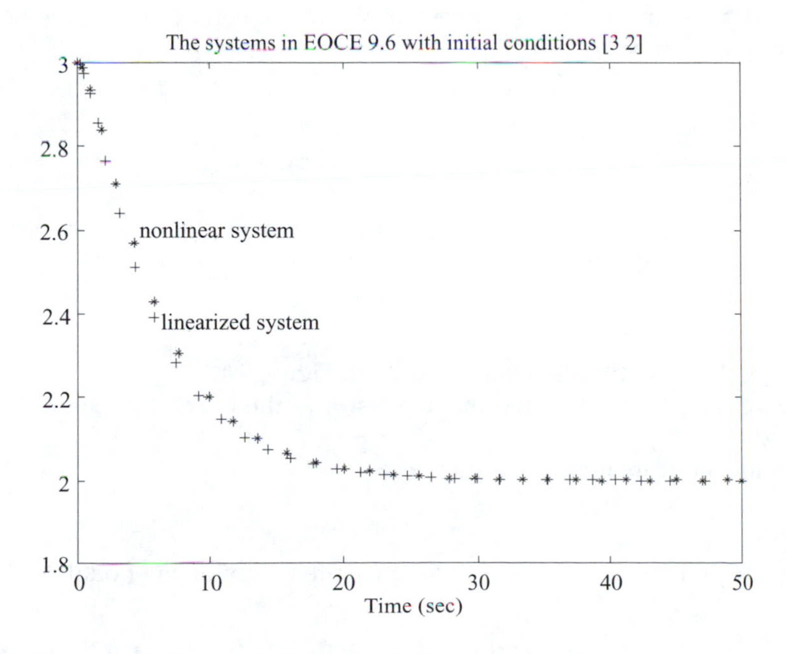

FIGURE 9.16
Plots for EOCE 9.6.

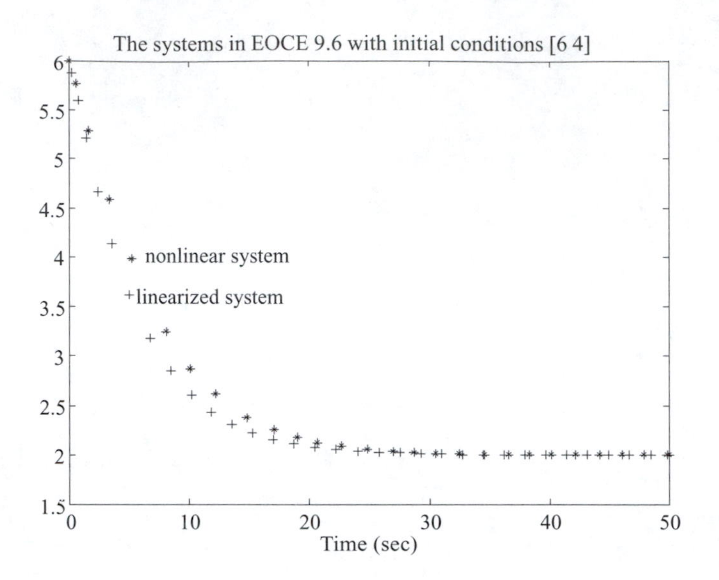

FIGURE 9.17
Plots for EOCE 9.6.

These plots reaffirm the points we discussed previously. We simply said that the approximation must be carried with the right initial conditions, right inputs, and accurate modeling. Remember also that when we change the initial conditions for the nonlinear system the initial conditions for the linearized system change as well.

$$\Delta z_1(0) = z_1(0) - z_{1n}$$
$$\Delta z_2(0) = z_2(0) - z_{2n}$$

You can also see that both the linearized and the nonlinear responses try to get to the nominal point $z_{1n} = 2$ meters.

EOCE 9.7

Consider the second order circuit shown in Figure 9.18.

The voltage across the nonlinear resistor is the current in the inductor squared.

Develop a linear circuit for Figure 9.18.

Solution

Using circuit analysis techniques we can write the following nodal equation

$$I_L = \frac{d}{dt}V_c + V_c$$

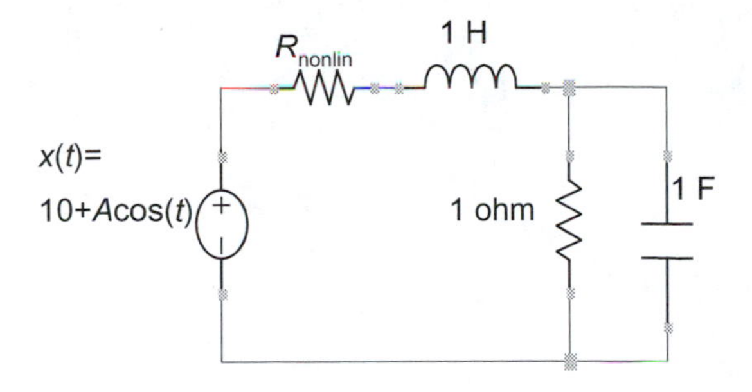

FIGURE 9.18
Circuit for EOCE 9.7.

where V_c is the voltage in the capacitor and I_L is the current in the inductor.

Summing the voltages in the outer loop will result in

$$x(t) = (I_L)^2 + \frac{d}{dt}I_L + V_c$$

Considering the voltage in the capacitor as a state and the current in the inductor as another, we arrive at the following state equations

$$\frac{d}{dt}V_c = -V_c + I_L$$

$$\frac{d}{dt}I_L = x - V_c + I_L^2$$

The nominal value for the input is $x_n = 10$ volts. Assuming that $x(t) = 0$ for $t < 0$, let us find the initial conditions for the state equations of the nonlinear system.

In the state equations, with derivatives set to zero, we can write

$$V_c(0) = I_L(0)$$

$$V_c(0) = -(I_L(0))^2$$

which implies

$$V_c(0)[1 + V_c(0)] = 0$$

Either $V_c(0) = 0$ or $V_c(0) = -1$. Since the input to the circuit is zero for a long time, we will reject the second value assuming no initial energy stored in the storage elements. Therefore $V_c(0) = 0$ and $I_L(0) = 0$. Let $z_1 = V_c$ and $z_2 = I_L$.

In this case the nonlinear state equations are

$$\frac{d}{dt}z_1 = z_2 - z_1$$

$$\frac{d}{dt}z_2 = x - z_1 - z_2^2$$

with initial conditions

$$z_1(0) = 0$$

$$z_2(0) = 0$$

With the nominal value for x as $x_n = 10$ volts, let us find the nominal values for the other variables. In the nonlinear state equations where derivatives are set to zero, with $x_n = 10$, the nonlinear states can be written as

$$z_{2n} = z_{1n}$$

$$(z_{2n})^2 + z_{2n} - 10 = 0$$

The solution of the above quadratic equation is

$$z_{1n} = 2.7016 = z_{2n}$$

and

$$z_{1n} = -3.7016 = z_{2n}$$

We still have the nonlinear term z_2^2 present in the nonlinear set of state equations. We can use the Taylor series to write

$$z_2^2 = z_{2n}^2 + 2z_{2n}\Delta z_2$$

Now, going back to the nonlinear set and substituting for the linear states the sum of their nominal and incremental values, and for the nonlinear term its Taylor series expansion at the nominal point, we can write

$$\frac{d}{dt}(z_{1n} + \Delta z_1) = z_{2n} + \Delta z_2 - z_{1n} - \Delta z_1$$

$$\frac{d}{dt}(z_{2n} + \Delta z_2) = x_n + \Delta x - z_{1n} - \Delta z_1 - z_{2n}^2 - 2z_{2n}\Delta z_2$$

By simplifying the above equations we get

$$\frac{d}{dt}\Delta z_1 = \Delta z_2 - \Delta z_1$$

$$\frac{d}{dt}\Delta z_2 = \Delta x - \Delta z_1 - 2z_{2n}\Delta z_2$$

with $\Delta x = A\cos(t)$. $z_1(0) = z_{1n} + \Delta z_1(0)$ and $z_2(0) = z_{2n} + \Delta z_2(0)$. Therefore, the linearized set of state equations can be put in the final form as

$$\frac{d}{dt}\Delta z_1 = \Delta z_2 - \Delta z_1$$

$$\frac{d}{dt}\Delta z_2 = -\Delta z_1 - 2z_{2n}\Delta z_2 + A\cos(t)$$

with the initial conditions

$$\Delta z_1(0) = z_{1n}$$

$$\Delta z_2(0) = z_{2n}$$

We can now use MATLAB to implement these equations to see how and if the linearized set approximates the nonlinear set for the two operating points: 2.7016 and −3.7016.

For the nonlinear system we define the following function in MATLAB.

```
function zdot=eoce7non(t,z); % return the state derivatives
global A;
zdot = [z(2) - z(1);10 + A*cos (t)-z(1)-z(2).^2];
```

and for the linearized system we write the function

```
function zdot=eoce7lin(t,z); % return the state derivatives
global A;
zdot = [z(2) - z(1); -z(1) - 2*2.7016*z(2) + A*cos (t)];
```

Finally we write the following MATLAB script that will call the above two functions with the help of the "ode23" solver function.

```
clf
global A;
A = 1;
tspan = [0 20]; % the time span for the simulation
z0n = [0 0];
[t,z] = ode23('eoce7non',tspan, z0n);
plot(t,z(:,1),'b*'); % the voltage in the capacitor
hold on
plot(t,z(:,2),'b+'); %The current in the inductor
z01 = [-2.7017 -2.7016];
[t,z] = ode23('eoce7lin',tspan, z01);
```

```
plot(t,z(:,1)+2.7016);% for linearized
plot(t,z(:,2)+2.7016);% for linearized
title('The linearized and the nonlinear systems for
z1n = z2n = 2.7016 and A = 1');
gtext('*nonlinear,Vc(t)');
gtext('+nonlinear, IL(t)')
xlabel('Time (sec)');
```

The plots are shown in Figures 9.19, 9.20, and 9.21.

You can see from these figures that as A gets bigger the two solutions deviate from each other. Try values for $A = 11$ and higher.

For the nominal value of -3.7016, we can repeat the simulations to get the graph in Figure 9.22 which indicates instability for the linearized system for even a small value for A.

This indicates again that we should be very careful when we select our nominal points in case we have more than one or even if we have only one.

To complete this problem we need to find an approximate value for the nonlinear resistor.

As seen in the circuit

$$V_{nonlinR} = (I_L)^2 \approx (I_{Ln})^2 + 2I_{Ln}(\Delta I_L)$$

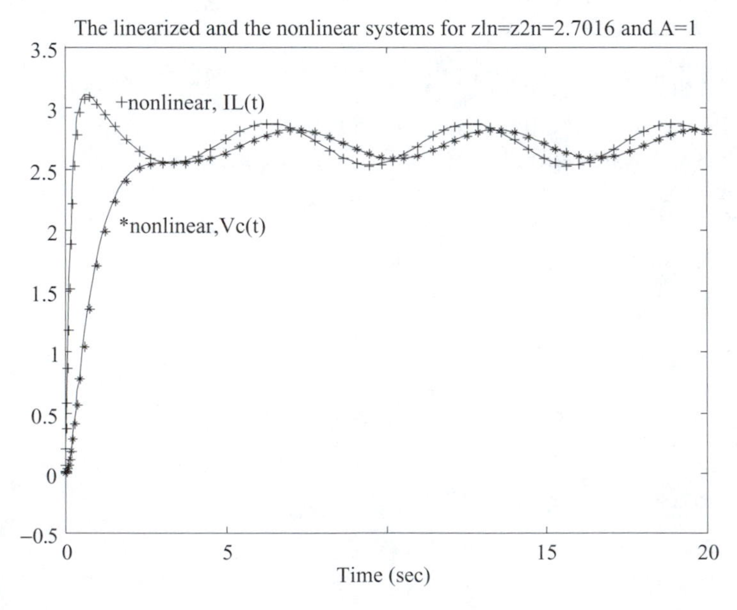

FIGURE 9.19
Plots for EOCE 9.7.

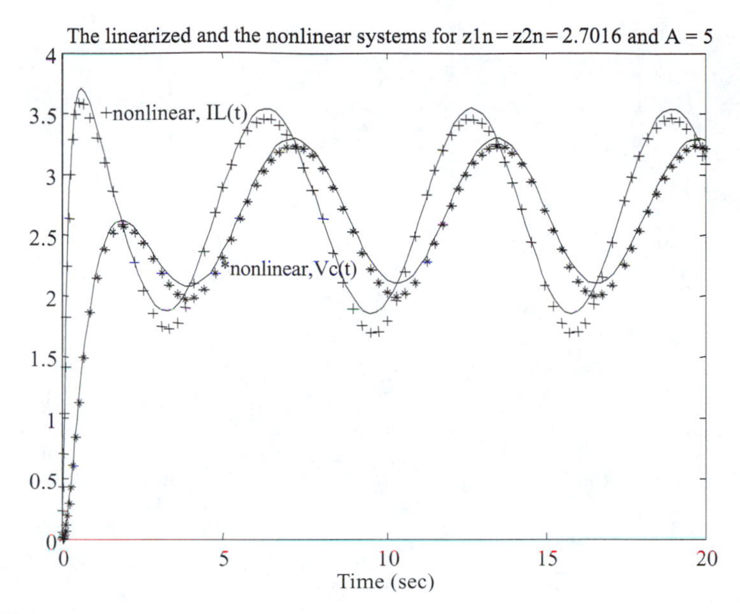

FIGURE 9.20
Plots for EOCE 9.7.

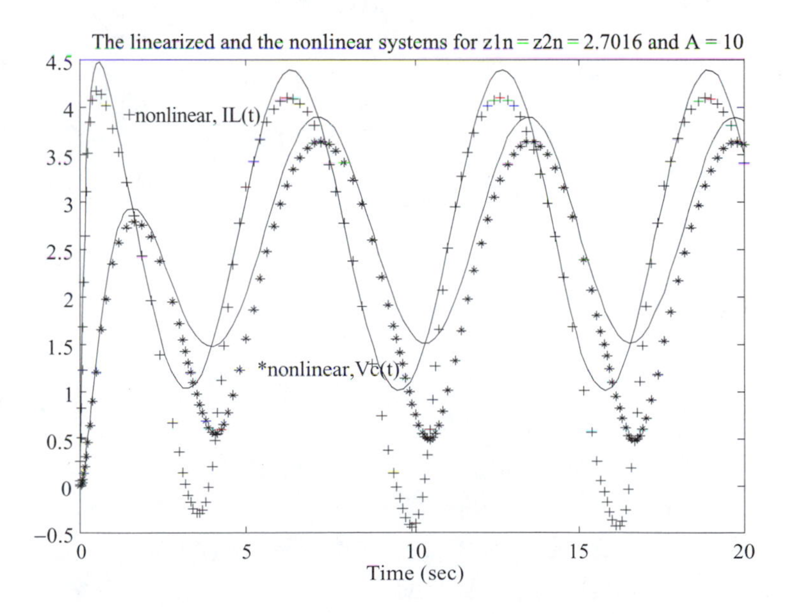

FIGURE 9.21
Plots for EOCE 9.7.

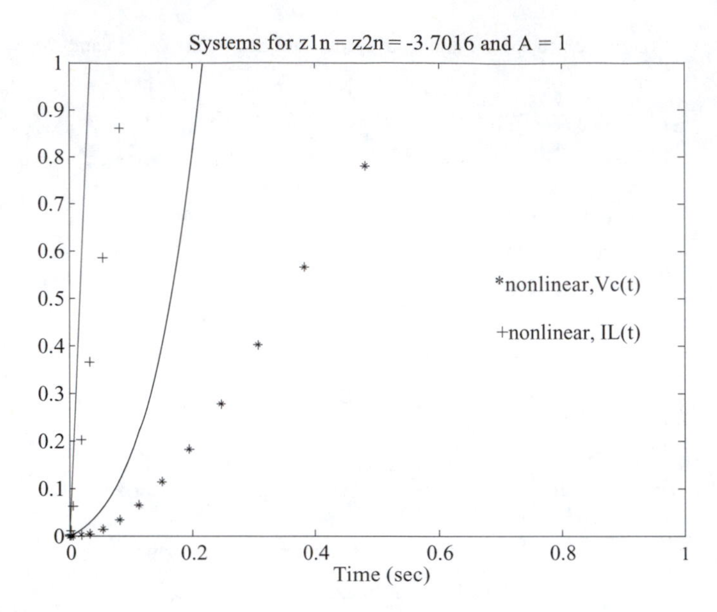

FIGURE 9.22
Plots for EOCE 9.7.

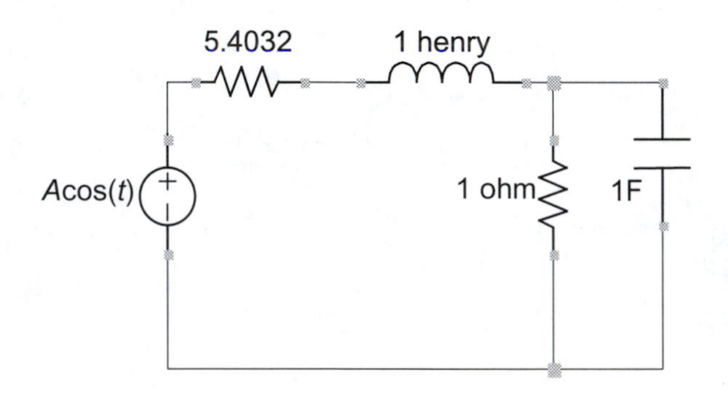

FIGURE 9.23
Circuit for EOCE 9.7.

or

$$V_{nonlinR} - (I_{Ln})^2 \approx 2I_{Ln}(\Delta I_L)$$

$$\Delta V_{nonlinR} = V_{nonlinR} - (I_L)^2 \approx 2I_{Ln}(\Delta I_L)$$

$$\Delta V_{nonlinR} \approx 2I_{Ln}(\Delta I_L) = 2(2.7016)\Delta I_L$$

This means, using Ohm's law, that R_{nonlin} is approximately 5.4032 ohms. The linearized circuit is then shown in Figure 9.23.

9.6 End-of-Chapter Problems

EOCP 9.1

Consider the following system.

$$y(t) = tx(t)$$

Let $x(t) = |t|$. Linearize the system for the operating point $t_n = 3$.

EOCP 9.2

Consider the following systems.

1. $y(t) = A\sin(t),\ t_n = 3\pi/2$
2. $y(t) = .4t^3,\ t_n = 3/2$
3. $y(t) = \sin^2(t) + 1,\ t_n = 3\pi/5$

Find a linearized model for the systems given.

EOCP 9.3

Consider the following system represented by the differential equation

$$\frac{d^2}{dt}y(t) + \frac{d}{dt}y(t) + y^2(t) = x(t) + 12$$

Assume a nominal value of zero for $x(t)$.

1. Linearize the system assuming zero initial conditions.
2. Is the linearized system stable?

EOCP 9.4

Consider the system in Figure 9.24.

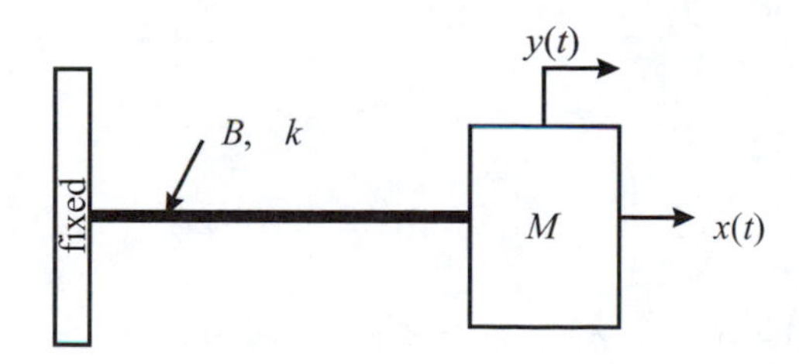

FIGURE 9.24
System for EOCP 9.4.

The rod to the left of the mass M can be modeled by a nonlinear translational spring that will produce a force equal to $3y^2(t)$ opposite to $M\frac{d^2}{dt}y(t)$, in parallel with a translational damper of constant B. The system can be represented as

$$M\frac{d^2}{dt}y(t) + B\frac{d}{dt}y(t) + 3y^2(t) = x(t)$$

Assume a nominal value of zero for $x(t)$.

1. Linearize the system assuming zero initial conditions.
2. Is the linearized system stable?

EOCP 9.5

Consider the following system in state-space form.

$$\frac{d}{dt}z_1(t) = z_2(t)$$

$$\frac{d}{dt}z_2(t) = -2z_1(t) - 3z_2(t) + \cos(t)$$

Assume $z_1(0) = 0$ and $z_2(0) = 0$.

1. Linearize the system given knowing that $\cos(t)$ is the input to the system.
2. Is the linearized system stable?

EOCP 9.6

Consider the following system.

$$\frac{d}{dt}y(t) + .2y^3(t) = 10 + 2\sin(t)$$

Assume

$$x(t) = 10 + 2\sin(t)$$

1. Find all the operating points for the system.
2. Find the linearized system corresponding to each operating point.
3. Are the linearized systems stable?

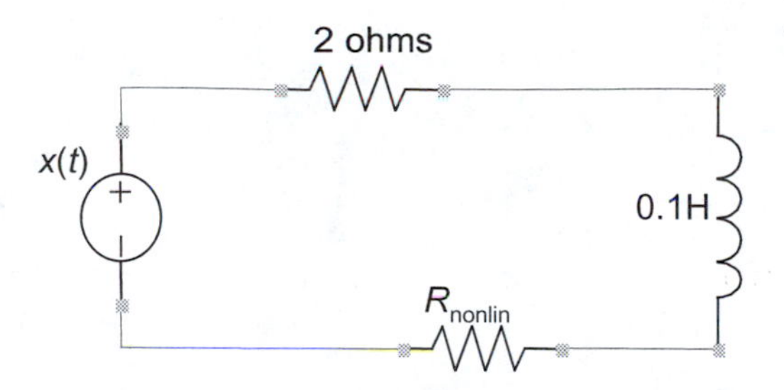

FIGURE 9.25
System for EOCP 9.7.

EOCP 9.7

Consider the system in Figure 9.25.

The voltage across the nonlinear resistor is related to the current through it as

$$V(t) = I^3(t)$$

where $I(t)$ is flowing into the negative terminal of the input $x(t)$. Assume that

$$x(t) = 11 + 3\sin(t)$$

and the initial conditions are all zeros.

1. Write the nonlinear differential equation in $I(t)$ for the system.
2. Linearize the system around the operating point.

EOCP 9.8

Consider the system in Figure 9.26.

The voltage across the nonlinear resistor is related to the current through it as

$$V(t) = e^{I(t)}$$

where $I(t)$ is flowing to the right into the inductor. Assume that

$$x(t) = 3\sin(t)$$

and the initial conditions are all zeros.

1. Write the nonlinear differential equation in $I(t)$ for the system.
2. Linearize the system around the operating point.

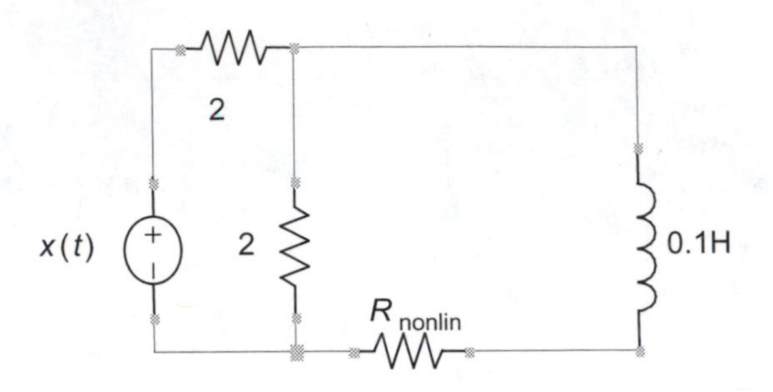

FIGURE 9.26
System for EOCP 9.8.

EOCP 9.9

Consider the circuit in Figure 9.27.

The voltage across the nonlinear resistor is related to the current through it as

$$V(t) = 2I^2(t)$$

where I(t) is flowing to the right into the inductor. Assume that

$$x(t) = 3\cos(t) + 5$$

and the initial conditions are all zeros.

1. Write the nonlinear differential equation in $I(t)$ for the system.
2. Linearize the system around the operating point.

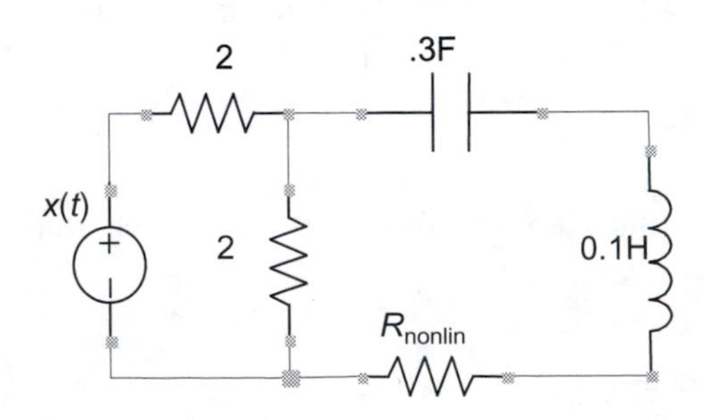

FIGURE 9.27
System for EOCP 9.9.

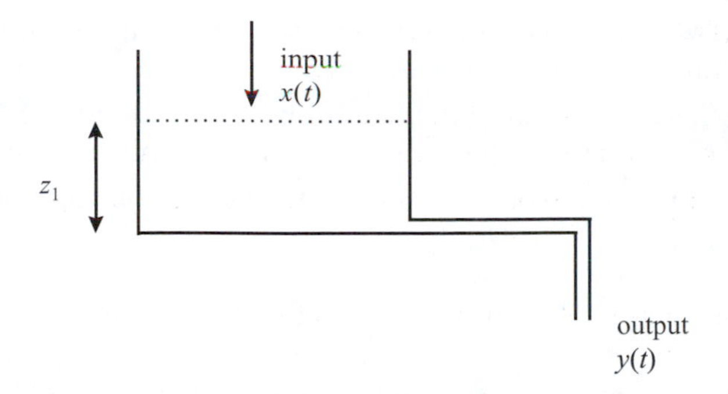

FIGURE 9.28
System for EOCP 9.10.

EOCP 9.10

Consider the chemical system in Figure 9.28.

$x(t)$ is the input solution into the tank and $y(t)$ is the output solution from the tank.

Let

$$y(t) = 3\sqrt{z_1(t)}$$

Let C be the capacitance of the tank. Then the system can be described by the differential equation

$$C\frac{d}{dt}z_1 + 3\sqrt{z_1(t)} = x(t)$$

1. Linearize the system presented by the nonlinear differential equation.
2. Is the linearized system stable?

EOCP 9.11

Consider an RL circuit with an input voltage $x(t)$ and an output current through the inductor and the resistor as $y(t)$. Assume that the resistance in the circuit varies as the sum of a constant 100 ohms plus the square of the current $y(t)$ as

$$R = 100 + y^2(t)$$

With an inductance of 10 henry and an input voltage, $x(t)$, of 100 volts, the differential equation describing the system is

$$\frac{d}{dt}I(t) = 10 - 10I(t) - .1I^2(t)$$

1. Use MATLAB to plot $I(t)$ vs. time for the nonlinear differential equation describing the circuit.
2. If the resistor is linear with the nonlinear term dropped, $R = 100$, use MATLAB to plot the current $I(t)$ vs. time on the same graph.
3. Comment on the plots with respect to the speed of the response.

EOCP 9.12

A pendulum swings freely with an initial angle $\theta(0) = \theta_0$ and an initial velocity $\frac{d}{dt}\theta(0) = 0.$

The equation describing the motion of the pendulum is

$$\frac{d^2}{dt}\theta(t) + g/l\sin(\theta) = 0$$

where l is the length of the pendulum and g is the gravitational acceleration.

1. Use MATLAB to plot the angular displacement for initial angular displacements of $\pi/8$ and $\pi/4$ radians.
2. If θ is small we can say that $(\theta) = \theta$. Plot, using MATLAB, the angular displacement for initial angular displacements of $\pi/8$ and $\pi/4$ radians.
3. What can you say about the two plots in 1 and 2?

EOCP 9.13

Consider the first order circuit shown in Figure 9.29.

If $x(t)$ is a sinusoidal input with small amplitude and the current in the non-linear resistor is related to the voltage as

$$I_R(t) = 3e^{.5V_R}$$

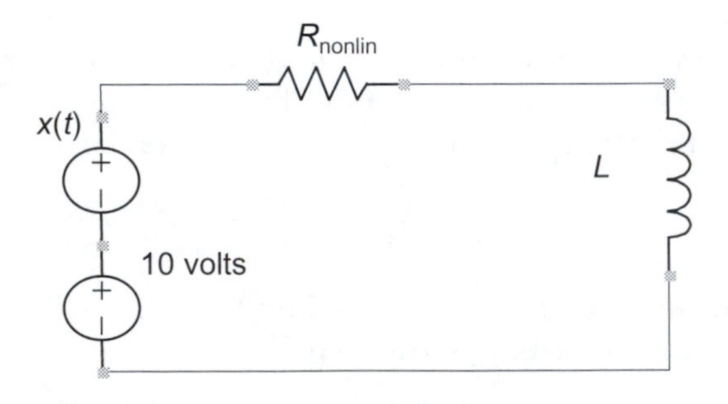

FIGURE 9.29
System for EOCP 9.13.

1. Find the transfer function relating the voltage across the inductor to the input signal $x(t)$.
2. Is the system stable?

EOCP 9.14

Consider the following system described by the differential equation

$$\frac{d^2}{dt}y(t) + 10\frac{d}{dt}y(t) + 3y = \cos(y)$$

Here the input is a function of the output. Linearize the system near $y = 1$ rad.

EOCP 9.15

Consider the third order system

$$\frac{d^3}{dt}y(t) + 3\frac{d^2}{dt}y(t) + 2\frac{d}{dt}y(t) + y(t) = 0$$

1. Linearize the system around $y(t) = 0$.
2. Put the linearized system in state-space form.
3. Plot $y(t)$ vs. time for the nonlinear system.
4. Plot the output $y(t)$ vs. time for the linear system for a small value near the operating point $y = 0$

References

Close, M. and Frederick, K. *Modeling and Analysis of Dynamic Systems*, 2nd ed., New York: Wiley, 1995.

Golubitsky, M. and Dellnitz, M. *Linear Algebra and Differential Equations Using MATLAB*, Stamford, CT: Brooks/Cole, 1999.

Harman, T.L., Dabney, J., and Richert, N. *Advanced Engineering Mathematics with MATLAB*, Stamford, CT: Brooks/Cole, 2000.

The MathWorks. *The Student Edition of MATLAB*, Englewood Cliffs, NJ: Prentice-Hall, 1997.

Pratap, R. *Getting Started with MATLAB 5*, New York: Oxford University Press, 1999.

Woods, R.L. and Lawrence, K.L. *Modeling and Simulation of Dynamic Systems*, Englewood Cliffs, NJ: Prentice-Hall, 1997.

Wylie, R.C. and Barrett, C.L. *Advanced Engineering Mathematics*, 6th ed., New York: McGraw-Hill, 1995.

Index

E

F

G

H

I

N

O

P

R

S

T

U

V

W

Y

Z

RECEIVED
JUL 2 4 2001